CHIESA
E SOCIETÀ

9

Direttore di Collana
Prof.ssa Emilia Palladino

JACQUINEAU AZETSOP S.J. - PAOLO CONVERSI
(eds.)

Foundations of Integral Ecology

G&BPress
Pontificia Università Gregoriana
Pontificio Istituto Biblico

Cover e impaginazione: Yattagraf srls - Tivoli (RM)

Gregorian & Biblical Press
Piazza della Pilotta 4, 00187 - Roma
www.gbpress.org - info@biblicum.com

ISBN 978-88-7839-**479**-7

to Jacquineau

LAUDATO SI'
OBSERVATORY

Foundations of Integral Ecology is the first book of the *Laudato Si' Observatory*. It was established in 2018 at the Faculty of Social Sciences of the Pontifical Gregorian University thanks to the commitment of the late Jacquineau Azetsop, the then Dean. It is an inter-Faculty body since it involves also the active participation of the Faculties of Theology and Philosophy of the University.

The *Laudato Si' Observatory* was instituted as a response to the encyclical of Pope Francis *Laudato Si'* (LS), published on Pentecost Sunday, 24 May 2015. This speaks of the urgent need to take action in response to the "socio-ecological crisis" that would constitute a world-wide "change of course" (LS, §61, §163) leading to the promotion of an "integral ecology" (LS, §10). The *Laudato Si' Observatory* seeks to answer to this call in an academic and interdisciplinary way, noting the claim of the encyclical that promoting integral ecology represents a great cultural, spiritual and educational challenge that will involve a long path of renewal (LS, §202).

In this perspective, the first book of the *Laudato Si' Observatory* focuses on the "foundations" of the notion of integral ecology, as seen from the perspective of the areas in which the contributing authors specialize, especially: social science, philosophy and theology. Thanks is extended to all these authors, who participated in a variety of online meetings as the project for this book was developed. A particular acknowledgment goes to Prof. Peter Lah, SJ, Pro-Dean of the Faculty of Social Sciences of the Pontifical Gregorian University, for his support to this endeavor, as well as to Prof. Gerard Whelan, SJ, and Dr. Ciara Murphy for helping in the work of translating the chapters of this book which were first composed in Italian and needed to be translated.

CONTENTS

INTRODUCTION

Paolo Conversi

In *Laudato Si'* Pope Francis states "Humanity still has the ability to work together in building our common home"[1], and "there is reason to hope that humanity at the dawn of the twenty-first century will be remembered for having generously shouldered its grave responsibilities"[2]. This book is an effort to respond to "the urgent challenge to protect our common home [which] includes a concern to bring the whole human family together to seek a sustainable and integral development, for we know that things can change"[3].

Contributors to this book explore how, "we have the freedom needed to limit and direct technology; we can put it at the service of another type of progress, one which is healthier, more human, more social, more integral"[4]. These reflections acquire even more weight in the face of the Covid-19 pandemic that is putting a strain on the current society. They are also written as Russia has invaded Ukraine, reminding of us how difficult it is to genuinely cooperate in the building of a common home, instead of opting for division and violence.

Highlighting numerous and complex challenges to be faced, the pandemic itself brings us to

> seize this time of trial as a *time of choosing* [...] a time to choose what matters and what passes away, a time to separate what is necessary from what is not [...] We have realized that we are on the same boat, all of us fragile and disoriented, but at the same time important and needed, all of us called to row together, each of us in need of comforting the other [...] We too have realized

[1] Pope Francis, Encyclical Letter *Laudato Si' on Care for Our Common Home*, 24 May 2015, n. 13.
[2] Pope Francis, *Laudato Si'*, n. 165.
[3] Pope Francis, *Laudato Si'*, n.13.
[4] Pope Francis, *Laudato Si'*, n. 112.

DOI 10.32060/C.S.9.19-31

> that we cannot go on thinking of ourselves, but only together can we do this […] We were caught off guard by an unexpected, turbulent storm [… which] exposes our vulnerability and uncovers those false and superfluous certainties around which we have constructed our daily schedules, our projects, our habits and priorities. It shows us how we have allowed to become dull and feeble the very things that nourish, sustain and strengthen our lives and our communities. The tempest lays bare all our prepackaged ideas and forgetfulness of what nourishes our people's souls; all those attempts that anesthetize us with ways of thinking and acting that supposedly 'save' us, but instead prove incapable of putting us in touch with our roots and keeping alive the memory of those who have gone before us. We deprive ourselves of the antibodies we need to confront adversity. In this storm, the façade of those stereotypes with which we camouflaged our egos, always worrying about our image, has fallen away, uncovering once more that (blessed) common belonging, of which we cannot be deprived: our belonging as brothers and sisters[5].

It is time of choosing and the "ecological crisis", at the root of which is the moral crisis of our society, represents a privileged moment of stimulus to conversion and concrete decisions; it also opens the door to an interdisciplinary, local and international dialogue[6], in the awareness that 'everything is interrelated', 'everything is intimately connected'[7].

Seriously addressing the causes of this ethical and socio-environmental crisis is a "great cultural, spiritual and educational challenge which stands before us, and it will demand that we set out on the long path of renewal"[8], highlighting the need for a real 'change of course'[9]. However, it must be acknowledged that we need to develop the necessary culture to face this crisis and there is also a need to build leadership[10].

With this work we intend to respond to this call of Pope Francis through the deepening of one of the main intuitions of *Laudato Si'*: *integral ecology*. It is a complex and multidimensional concept, focused on a long-term perspective; it goes beyond environmental dimension, it opens the door to a trans-disciplinary dialogue and it offers ample space for research and development[11].

---

[5] Pope Francis, *Extraordinary moment of prayer*, 27 March 2020.

[6] Pope Francis, *Laudato Si'*, nn. 216ss.

[7] Pope Francis, *Laudato Si'*, nn. 91, 137 and 238.

[8] Pope Francis, *Laudato Si'*, n. 202.

[9] Pope Francis, *Laudato Si'*, nn. 61, 163.

[10] Pope Francis, *Laudato Si'*, n. 53.

[11] Interdicasterial Working Group of the Holy See on Integral Ecology, *Journeying towards Care*

A careful and correct analysis of the concept of integral ecology follows in the footsteps indicated by *Veritatis gaudium*. By recognizing a "true epochal shift, marked by a wide-ranging 'anthropological' and 'environmental crisis'"[12], it asks the academic community to network and to open transdisciplinary paths, exercised with wisdom and creativity, underlying "new ways of relating to God, to others and to the world around us, and inspiring essential values"[13].

The aim of the present work is to bring the different sciences into a dialogue in the light of *Laudato Si'*. This dialogue proceeds according to three perspectives: socio-scientific, philosophical and theological, to better understand the "prism" of integral ecology. Here, it is helpful to return to an image often employed by Pope Francis: that of a "many-faceted polyhedron whose different sides form a variegated unity, in which 'the whole is greater than the part'"[14]. Each perspective is investigated through several essays, keeping alive the attention to facilitate within each contribution an interdisciplinary dialogue.

The author of each essay was asked to give an answer, in the context of her/his own analysis, to the question: "what are the foundations of integral ecology?", trying to identify operational paths for the future. From all the essays it is clear that Pope Francis' call for a paradigm shift is without doubt a radical change in relationships at the global level. Just as the crisis is multidimensional, so is the call for responsibility.

As underlined by *Jacquineau Azetsop*, integral ecology decentralizes man from himself and opens up a vast relational field that closes the doors to the epistemic fragmentation born of modernity. In this perspective, he analyzes the eco-social approach and the relational anthropology that shapes integral ecology, and draws the implications of this analytical endeavor for global health: even though integral ecology has an ethical overtone, he asserts, the framework that sustains it can be used as a tool for analysing global health issues and promoting multilevel intervention with a dynamic interplay between social institutions and human agency. What is needed is a new language and

for Our Common Home. Five Years after Laudato Si'. Rome: Libreria Editrice Vaticana, 31 May 2020, pp. 8-11.

[12] Pope Francis, Apostolic Constitution *Veritatis gaudium on Ecclesiastical Universities and Faculties*, Proemio, 27 December 2017, n. 3.

[13] Pope Francis, *Veritatis gaudium*, n. 4.

[14] Pope Francis, Encyclical Letter *Fratelli tutti on Fraternity and Social Friendship*, 3 October 2020, n. 215, quoting his Apostolic Exhortation *Evangelii gaudium on the Proclamation of the Gospel in Today's World*, 24 November 2013, n. 237.

a new vision which acknowledge the fundamental interconnectedness of all, and may serve as the basis for addressing public and global health challenges, well aware that individualism cannot be considered as a framework for the world economics and global health.

A new vision is also what *Paolo Conversi* calls for in his paper, where he embarks upon an in-depth study of the economic foundations of integral ecology. A broader vision of human existence which prompted Pope Francis to argue vehemently for "a humanism capable of bringing together the different fields of knowledge, including economics, in the service of a more integral and integrating vision"[15]. On the basis of six pillars of the economic "foundations" of integral ecology, he calls to reconsider economic structures and to convert our economic system in a direction more respectful of creation and integral human development of present and future generations, by adopting a new development paradigm. He argues in favour of a courageous cultural revolution that brings about a new economic synthesis towards a new Christian humanism enlightened by integral ecology. What is needed is a "conversion" of the economic system, as the prevailing one, devoid of ethical concerns, does not lead to a more just social order but leads to a "throwaway culture", which is deeply uneconomic and unsustainable.

Rethinking economy also requires to rethink work: our society needs not only work, but also an exploration of its anthropological meaning, decoupling the notion of work from financial remuneration. It is important to acknowledge the centrality of work within the paradigm of integral ecology: work appears to be closely connected to the construction of that common good from which an integral ecology is inseparable. By presenting the "Future of Work – Labour after *Laudato Si'*" Project, *Paolo Foglizzo* and *Pierre Martinot-Lagarde* explore the challenges confronting today's world of work. Work is a form of relationship and the task of "thinking" about work also relates to theology and spirituality; this is the lesson learned, for example, from the experience of monasticism. Work includes care and care implies work, which is a factor of dignifies humanization. The articulation of the principle of the dignity of the human person (which is at the core of Catholic Social Teaching - CST) with the more secular notion of decent work (which is a powerful political and advocacy tool), shows how fruitful dialogue can be among CST and other philosophical and political approaches as a push for social change.

[15] Pope Francis, *Laudato Si'*, n. 141.

Social change is asked by a comprehensive integral and integrating vision of environmental problems. It requires an in-depth investigation of interactions between social systems, as well as between individual and collective behaviours. It recalls the importance of the role of law in a globalised society in constant movement. Law, with its own language and its tools - aimed at sanctioning, promoting, controlling -, assumes a decisive weight in the building and spread of a juridical culture of care for the environment. Assuming the law's point of view of or among the States, *Francesco Gambino* discusses the socio-juridical foundations of integral ecology. Law cannot be only punitive but rather possesses a promotional function; law is not made barren by jargon and uniform measures but it is open to the complexity of local problems. It is necessary to know how to intercept interacting systems, cultural differences, spaces of communication. In this perspective, he argues that the approach taken by a jurist must be interdisciplinary and ending up by broadening his/her horizons, calling into question aspects related not only with the history of law, but also with sociology and anthropology studies. It is also a cultural effort through which the language of the law can contribute.

The weakness of individualism is even stronger when we have to answer to the question to manage natural global commons to ensure fairness among persons or State users while at the same time protecting and nurturing this common resource. The essay of *Paul Tang Abomo* focuses on underlining how the *Laudato Si'*s response to the "tragedy of the commons" is more effective than neo-realist and neo-liberal institutionalist approaches, especially when the scope of this tragedy is extended to global entities such as the Earth, oceans, the atmosphere. In responding to global commons, the solutions based on property rights, public management or self-enforcing institutions give rise to some limits connected with cost enforcement, group interests or free-riders problems. The integral ecology approach calls for change of attitude, from a materialistic and anthropocentric relation based on dominion to a new relation with creation based on stewardship and "ecological conversion". Integral ecology calls for this shift and stresses the interconnectedness of all of creation. This calls for new responsibilities.

Responsibility is also a key word in the *Ciara Murphy*'s paper on the contribution of faith-based movement in the building of an integral ecology approach at local level, specifically in Ireland. In ecosystems, different organisms occupy different niches or functions. This too is true for social movements. Foundational to this is dialogue and collaboration, which develop networks, creating webs of understanding. It is a dreadful mistake to imagine the current ecological crisis can be solved with individual action alone. She points out that a reflection on "foundational principles" for integral ecology that do not emerge from a practice of dialogue and real-world engagement in the complexities of the ecological movement may prove limited and require careful interpretation. In this perspective, the field of eco-theology has the potential to articulate 'why' – ethically, philosophically, and theologically – there is the need to do everything in our power to protect and conserve our natural environment.

Moving to a more philosophical perspective, *Gaetano Piccolo* lays his analysis on those metaphysical foundations which are unavoidable for a healthy relationship between humanity and the creation. His investigation goes through a greater understanding of the throwaway culture and of the deviation resulting from the 'technocratic paradigm'. These are two issues also involved in the debate on the answer to the Covid-19 pandemic. Human being is placed before a limit which characterises his existence. The world is not at our disposal; rather, it is the space which enables us to give meaning to existence. It appears urgent to educate ourselves for a new relationship with things and with others. It represents a precious and irreplaceable occasion to ask oneself and to find the meaning of existence, becoming aware that, as Lévinas reminded us, responsibility for the other is incumbent on me.

Remaining in the field of relationship, it is important to remember that life is a function of both the organism and its environment. The boundary between them is porous. *Louis Caruana*'s paper focuses on the debate concerning the very idea of environment and the way life is intelligible in terms of relations. He underlines how the emphasis on the importance of the philosophical category of relation obliges us to consider the essence of humanity not in terms of specific attributes of the human individual, each individual taken as an isolated unit, but in terms of the network of relations that individuals have with the environment in which they live and flourish, in other words, with the environment in a broad sense. Separating the organism from its environment will lead to distorted understanding. The integral ecological perspective requires the adoption of a "causal interdependence" approach, where the environment is not immune to the degradation we see in human society, and, in the opposite sense, human society is not immune to environmental degradation. We need to consider also personal or moral degradation, which refers to egoism, indifference to the suffering of others, lack of foresight and responsibility regarding future generations. Human society, culture, and politics are not immune to personal degradation, and the individual person is not immune to social, cultural, and political degradation.

Ecology and anthropology are, indeed, built together: "There can be no ecology without an adequate anthropology"[16]. This is one of the main points from which *Lorella Congiunti* builds her essays, trying to draw some peculiar principles of integral ecology. Anthropology is internal to the ecological discourse itself, within an interdisciplinary, never reductionist, methodology, illuminated by Faith. She argues that we need a new anthropology and that the search for this anthropology passes through the identification of the human roots of the ecological crisis. An adequate anthropology highlights the relationships of the human person with other people, with the natural environment, with God, taking care of work correctly understood[17]. In these terms, integral ecology requires a complex, synthetic, reconciling method that is able to grasp all the dimensions of reality and arrive at the essence of the human person, addressing a complex method in which each specialization constitutes one face of the polyhedron. It is a method of understanding and action that is based on a metaphysical, ontological vision, which is committed to understanding reality.

The search for an anthropology adequate for integral ecology is also the focus of the work of *Jakub Gorczyca*. The deterioration of the natural environment originates from human roots, which is to be found in a misguided anthropocentrism that leads to a misguided lifestyle characterised by individualism, moral relativism, consumerism, technocratic pragmatism, and the "throwaway culture which affects the excluded just as it quickly reduces things to rubbish"[18]. Consequently, the remedy for this deterioration has to be found in the area of human relations and actions lived authentically and, on the theoretical level, in an anthropology that is able to think of the human being in a way that is radically different from that which has led to the present crisis. He tries to answer the question about what kind of philosophical anthropology can be suitable for integral ecology? The right response is based on the attentive building and development of a culture of encounter, of dialogue, of responsibility, of respect, of neighbouring, of gift and of gratitude. The development of such a culture,

---

[16] Pope Francis, *Laudato Si'*, n. 118.
[17] Pope Francis, *Laudato Si'*, n. 125.
[18] Pope Francis, *Laudato Si'*, n. 22.

according to the demands of integral ecology, cannot be done without the education of conscience, which represents that secret place where the voice of the truth echoes and calls the subject to continual conversion.

The question of developing an adequate anthropology to ground a notion of integral ecology is also a focus of the work of *Gerard Whelan*. He explores two aspects of this issue: the philosophical foundations of Pope Francis as he proposes this notion of integral ecology; and how these foundations might be deepened with reference to the thought of Robert Doran, a disciple of the better-known Bernard Lonergan. After tracing two of the major philosophical influences on Pope Francis, romantic philosophy and a realist metaphysics, he suggests that greater coherence and explanatory depth can be brought to these foundations by employing an anthropology based on "intellectual conversion", and "psychic conversion".

Pope Francis's address in *Laudato Si'* is intended to be a dialogue with all[19]. In this direction, *Adrien Lentiampa* underlines the value of welcoming Africans to enter actively into this dialogue to enrich the discussion on the protection of our common home. Starting from an African approach to rationality, he hatches a new understanding of our "being-in-the-world", and of our way of approaching and treating the ecological question. Such a renewal approach to ecology is possible if one points to the cultural values which have always animated Africa and been the source of its life; values which have led Africans to inhabit the world in a way that respects and protects the environment. Instead of the culture of waste – and then of death – which has a hold over the mind-set of our contemporaries, this African rationality offers us a culture of life. It is a culture which promotes life, not only at the individual level, but life in its different aspects which also includes the world around us. An African culture at the service of integral ecology, teaching us a conception of happiness which is not summed up in the accumulation of the goods of nature but aspires to a balanced life where there come into play personal accomplishment as well as communitarian achievement in a clean environment.

Beyond dialogue between cultures, a further way of dialogue is through different schools of thought. In his essay, *Ferenc Patsch* studies the delicate question whether or not it is possible to find connections between the Heideggerian philosophy and the Magisterial teaching of Pope Francis, taking into consideration in particular his concept of "integral ecology". He tries to high-

---

[19] Pope Francis, *Laudato Si'*, n. 3.

light some aspects of a possible "Heideggerian influence" in the Pope Francis' thoughts, finding overlaps, for with the Pope's critique of the technocratic paradigm and a general disposition of stressing an integrated approach to life and to processes of people building. However, he reaches the conclusion that, despite of all the remarkable similarities, there are not enough elements to label the magisterium of Pope Francis as "Heideggerian", but some parallelisms are most probably more than just *meta-literal* ones.

Care for our "common home" and for "universal fraternal love" are not only questions of technical-legal adjustments, but also of culture, paradigms of thought. In search of the foundations of integral ecology paradigm, *Elisabetta Casadei* investigates what mental structure lies behind it and what type of intelligence and sensitivity it reflects. In her paper she studies modes of cognitive and emotional perception, of relations, of thought and action which can belong to both males and females, but in different measures and expressed in different ways and to different degrees. It is a matter of responsible leadership, capable of decoding cultural processes, developing modes of communication, combining technological and social innovation, sharing of knowledge and focusing on complex and long-term objectives.

Shifting to a more theological perspective, *Prem Xalxo* argues that an authentic and integral human development entails ethical and theological considerations, which also become the foundations of integral ecology. In order to be able to face the ethical and theological challenges of the contemporary world, a thorough reading of the book of creation facilitates doing theology and understanding the meaning of *fides quaerens intellectum*. It challenges us to overcome ecological sin and to respond to the call to ecological conversion and to promote the reconciliation of all creation "in Christ and for Christ". An ethical-theological approach to integral ecology is an earnest invitation to take responsibility and to strengthen an ecological conscience to reinforce the value of life, dignity and human existence in communion with God the Creator, with one another, and with all creation, stemming from their deep interconnectedness.

This deep interconnectedness makes it necessary for integral ecology to be based on a serious dialogue between sciences – natural sciences, social sciences, ecclesiastical and religious sciences – which conforms to a method that respects the just autonomy and competence of each academic discipline. To support this assumption, *René Micallef* studies the way *Laudato Si'* engages other sciences in critical dialogue. Even if this Encyclical is a remarkable point of arrival in a long process, and a great model for "integral ecology", he underlines that it is also a point of departure for a project that is still taking shape for a more historically-conscious and cross-disciplinary methodology. This project should be aimed at addressing human persons in their current historical condition, promoting an ethics that humbly faces a particular theme, constructing an argument on the basis of perfectible and fallible empirical knowledge, remaining close to the experience of the faithful and the "people of good will" who are ready to allow their hearts and minds to be moved to ecological conversion. The enduring success of "integral ecology", he argues, lies in the progressive rediscovery and deepening of the appropriate, semi-inductive and interdisciplinary methodology that *Laudato Si'* amply showcases.

This approach is helpful also to guide the discernment of the Church of the 21st century in carrying out its task of proclaiming the crucified and risen Christ. In this process, *Giacomo Costa* argues that two key words are central in the magisterium of Pope Francis: synodality and integral ecology. They are technical terms belonging to semantic and disciplinary universes that are normally considered distant and mutually unrelated; this runs the risk of overshadowing a rich and complementary understanding of them, and especially of disconnecting us from the experience of faith in which they are rooted. He intends to show how they are an articulation of the same basic vision: the notions of "journeying together" and "caring for the common home" complement and illuminate each other; for both, the joy of the Gospel is the main point of reference; both refer to the need for a conversion, both ecological and synodal, that deeply touches the Church, its hierarchical structures, and its approach to fulfilling its *diakonia* in service to the world.

Laudato Si' dedicates a number of passages to the theme of ecological conversion. *Bruno Bignami* underlines that Christian spirituality allows us to gain a mysticism capable of animating and motivating the commitment of our times. There is need for a 'way of thinking, of feeling, and of living", even before ideas and innovative proposals. Without this dimension which warms the heart, conversion appears in terms of inconclusive moralism. Spirituality nourishes a passion for care of the world: it moves interiorly, it encourages, it gives meaning to personal and community commitment. From it, different levels of the one ecological conversion spring up. The first is the interior one, but it is not enough. The involvement of the community is necessary because "social problems must be addressed by community networks and not simply by the sum of individual good deeds [...]. The ecological conversion needed to bring about lasting change is also a community conversion"[20]. This involves many aspects that are well studied in his essay.

Laudato Si' also reminds us that the doctrines of creation and redemption form a unified whole: nature and nurture, creating and healing, human and nonhuman worlds are intimately interconnected; humanity and its 'common home' with all its creatures can be conceived as a complex and multilayered network. There are three fundamental relationships which, according to the biblical creation narratives, are constitutive of human life in this network: with God, with our neighbor, and with nature; all three are affected by sin. As inseparable from the divine act of creation, *Ligita Ryliškytė* argues, salvation involves the restoration and healing of all three relationships. By drawing on biblical scholarship and Lonergan's theories of history and generalized emergent probability, she explores how integral ecology is to be founded on re-thinking redemption as the restoration of the dynamic world order: this approach allows to reclaim the biblical notion of redemption as the new creation of all things in Christ. This is also a call to an integral ecology conversion, which "clearly respects its human and social dimensions [and is] capable of taking into account every aspect of the global crisis"[21].

Another way to deepen the concept of integral ecology is through an inter-religious dialogue. This is the approach taken by *Bryan Lobo*. Hinduism, in its immense potentiality to be eco-friendly, can be linked to the fundamental vision of Pope Francis *Laudato Si'* Encyclical, that invites the readers to care for nature (Creation). In his essay he presents some aspects of Hinduism having the proposals of *Laudato Si'* as a backdrop. Since Pope Francis insists on entering "into dialogue with all people about our common home"[22], dialoguing with Hindus through their scriptures, religious beliefs and practices, and their environmental activism, would be a further step towards united efforts for ecological health since, according to the Pope, "other religions [...] have expressed deep concern and offered valuable reflections on issues which all of us find disturbing"[23], and Hinduism has a lot to offer in the ongoing reflection and activity for ecological concerns.

With this last contribution, as well as that of Adrien Lentiampa, one recalls how much the development of integral ecology can benefit by an in-depth

---

[20] Pope Francis, *Laudato Si'*, n. 219.

[21] Pope Francis, *Laudato Si'*, n. 137.

[22] Pope Francis, *Laudato Si'*, n. 3.

[23] Pope Francis, *Laudato Si'*, n. 7.

dialogue with different cultures, for a renewed discourse on "ecology", to be understood as a *study on our common home*, on the basis of its etymological meaning (in ancient Greek the word "οικος" means home, and the word "λογος" means study, reflection, knowledge). Favoring this inter-cultural dialogue is very beneficial for a better understanding of integral ecology and its foundations. Adrien Lentiampa remembers that, as underlined by Marie Vivane Tsangu Makumba, life comes to us first from our ancestors, sources of fertility, before resulting from the conjugal union of our parents who were invited in their turn to strengthen it by transmitting it to their children[24]. Where the dominion of the subject, denounced by Lévinas, sees the implement, enjoyment, and nourishment as the only possible relations to establish with the world[25], and whereas, for the technocratic culture, flourishing consists in possession and consumption, the African culture will tell us that "one can only flourish when being in a close and harmonious relationship with others and with the universe"[26].

The search of ways for strengthening this harmonious relationship, as well as for favouring an individual and community conversion, have been among the main issues underlined by these contributions, aimed at deepening the foundations of integral ecology. From these reflections it is clear that this prophetic concept needs a new language and new vision. In this perspective, integral ecology calls for a *new humanism* based on a "culture of care" which permeates all of society[27]. Care for the environment and care for our neighbors, near or far in space and time; but also care for themselves and for the Creator. A new *humanism* capable of bringing together the different fields of knowledge, in the service of a more integral and integrating vision.

It is interesting here to recall some of the main innovations developed during the period of *Humanism* in the XV century. In that period we have witnessed: 1) the spreading of a cultural renewal based on the centrality of man; 2) the rediscovering of the classics and of the use of "reason" which makes the human person a more responsible and consciousness protagonist of the world around

---

[24] Marie Viviane Tsangu Makumba, *Pour une introduction à l'africanologie*. Fribourg: Ed. Universitaires de Fribourg, 1995, pp. 135-136.

[25] Emmanuel Lévinas, *Totality and Infinity. An essay on exteriority*. Pittsburgh: Duquesne University Press, 2000, pp. 131-133.

[26] Bénézet Bujo, *Quelle Église pour un christianisme authentiquement africain? Universalité dans la diversité*. Fibourg: Schwabe Verlag, 2020, p. 54.

[27] Pope Francis, *Laudato Si'*, n. 231.

her/him; 3) the promotion of a new integral education and the flourishing of arts adopting an interdisciplinary approach; 4) the revaluation of nature through enlarged human capabilities to transform it, while respecting it; 5) the development of work as a way to fulfill the human person; 6) and the broadening of a new communication, through the invention of printing, facilitating the circulation of ideas.

Historia magistra vitae (history is a teacher of life), Cicero reminds us in *De oratore* (II, 9, 36). It is astounding to see how many points in common there are between that period, which gave rise to the Renaissance, and our age: 1) acting responsibly around the centrality of the human person[28]; 2) rediscovering the contributions of different cultures, starting from their main classical thinkers, for a culture of care able to mould the concept of integral ecology[29]; 3) deepening the benefits of the complex exercise of inter-disciplinary and trans-disciplinary[30]; 4) fostering the flourishing and the sharing of ingenuity and creativity aimed at strengthening the alliance between humans and nature[31]; 5) recovering the sense and meaning of economics and of human work[32]; 6) developing new and very effective means of communication, increasing interconnectedness through in particular the accelerate digitalization of the world and the spread of Internet[33].

These issues are also at the basis of an integral ecology conversion, which invite us to face with courage and perseverance the stimulating challenge to choose and to embark on a new Renaissance for humanity, based in the implementation of integral ecology.

[28] See for instance the essays in this book of Caruana, Congiunti, Gorczyca, Whelan and Xalxo.
[29] See for instance the essays in this book of Lentiampa, Lobo and Patsch.
[30] See for instance the essays in this book of Azetsop, Gambino and Murphy.
[31] See for instance the essays in this book of Piccolo, Costa, Bignami, Tang Abomo and Ryliškytė.
[32] See for instance the essays in this book of Conversi and Foglizzo.
[33] See for instance the essays in this book of Micallef and Casadei.

ECOSOCIAL IMAGINATION, NEOLIBERALISM AND COVID-19 PANDEMIC

Towards a New Social Ecology of Global Health

Jacquineau Azetsop, SJ

> Neoliberal capitalism has been exposed for its inhumanity and criminality, and COVID-19 has shown that there can be no health policy without solidarity, industrial policy and state capacity... We must come out of this crisis with a better society[1].

1. Introduction

Never in the recent history of humankind, have the limits of industrialized countries emerged with such a sharpness. The covid-19 crisis and the inadequate response of both health systems and economies have shown that the neoliberal perception of society as a business model has failed. Noam Chomsky argues that "The pandemic is a vivid illustration of the great damage that the neoliberal era has done to the large majority of the population. We should, however, have no illusions about belief in markets or *homo economicus*. The core principle of neoliberalism is to shift decisions from governments, which are to some extent subject to public influence, to private tyrannies that are completely unaccountable to the public – and according to neoliberal doctrine, must be devoted solely to self-enrichment"[2].

[1] Alfredo Saad-Filho, "From Covid-19 to the end of Neoliberalism", *Critical Sociology*, 2020, 46/4-5, p. 482.

[2] Lena Matsiori and Noam Chomsky: *We are paying for the logic of capitalism*, 2020. https://www.ekathimerini.com/opinion/interviews/254221/noam-chomsky-we-are-paying-for-the-logic-of-capitalism/

DOI 10.32060/C.S.9.33-52

Indeed, several internationally renowned intellectuals rose to emphasize that the covid-19 pandemic may have prompted the process that may lead to the decay of neoliberalism[3]. After four decades of neoliberalism, the pandemic has struck states exhausted by market dictatorship. Because of the search for short-term profitability, neoliberalism has encouraged deindustrialization through the globalization of production and built fragile financial structures. This disintegration of the global economy has rendered the wealthier neoliberal economies unable to produce enough face masks and personal protective equipment for their healthcare workers. These shortcomings were not only caused by the inability to produce due to changing technologies or China's trade policy, but also by deliberately chosen economic policies[4]. The erosion of public services and social protection systems, caused by decades of neoliberal attacks, has made societies more vulnerable to such a virus, as the means necessary to demonstrate solidarity and care are not functioning, are underfunded or simply do not exist[5].

The concept of integral ecology is sustained by an eco-social framework that assumes the fundamental relationality or interconnectedness inherent to the cosmos harmony and human existence. Just as all spheres of society, all aspects of human existence are interconnected. This holistic approach to human existence is articulated through the notion of "integral ecology" which radically challenges the one-sided postmodern anthropological and epistemological approach to human existence and economics that fuels the crisis of contemporary rationality that Pope Francis talks about in *Laudato Si'*.

My task is threefold: firstly, after having demonstrated that the ecology crisis is a both an anthropological and epistemological crisis, I intend to outline the main features of the eco-social imagination that sustains the "integral ecology" approach. Secondly, I intend to draw the pragmatic and epistemological implications of the eco-social approach to global health in opposition to neoliberalism today. I will end up concluding that the eco-social perspective calls for a radical transformation of the relations between global health and economy, since neoliberalism has sought to turn public health into a commodity

[3] Joseph Stiglitz, *The End of Neoliberalism and the Rebirth of History*, 2019. Available at https://www.socialeurope.eu/the-end-of-neoliberalism-and-the-rebirth-of-history.

[4] Alfredo Saad-Filho, "From Covid-19 to the end of Neoliberalism", *Critical Sociology*, 46/4-5, 2020, pp. 477-485.

[5] Pedro Salgado, *COVID as the (Second) Death of Neoliberalism?* 2020. https://www.irgac.org/2020/09/14/covid-as-the-second-death-of-neoliberalism/

for profit. The neoliberal perspective persistently prioritizes competition and consumerism over free access and social justice, leaving citizens to the whims of private healthcare providers and medical services all over the world. That is why people struggled to cope as demand surged during the Covid-19 pandemic. Furthermore, as a philosophical trend, neoliberalism is sustained by an individualistic market-based anthropology and a reductionist epistemology which can help respond to global challenges that require public provision of service, institutional solidarity and compassion. Instead, the epistemological and anthropological shift that the *Laudato Si'* eco-social approach provides us with appropriate theoretical and anthropological groundings for studying global health challenges because it brings to the forefront the connections that exists between health and other aspects of human existence.

Thirdly, the human ecology, therefore, is not merely another single and isolated scientific discipline, but more importantly a holistic field of intellectual inquiry through which one studies the fundamental unity of the many dimensions of what it means to be a human being that lives unavoidably in a larger ecology. This broader vision of human existence prompted Pope Francis to argue vehemently for "a humanism capable of bringing together the different fields of knowledge, including economics, in the service of a more integral and integrating vision"[6]. This paper is far from being an analysis of *Laudato Si'*. Instead, I intend to analyze the eco-social approach and the relational anthropology that shapes integral ecology, and draws the implications of this analytical endeavor for global health. I contend that even though integral ecology has an ethical overtone, the framework that sustains it can be used as a tool for analysis of global health issues and promote multilevel intervention with a dynamic interplay between social institutions and human agency. Human responsibility to the environment is not just an ethical one, but a cosmic one, where the very integrity of creation and humanity within it are at stake.

2. The ecologic crisis and Covid-19 pandemic

The fundamental cause of the ecological crisis is the subjectivism of modernity, its excessiveness and its idols: money, progress, profit and power. Modern culture has been built on the distortion of transcendence, anthropological reductionism and the objectification of nature. With the absence of

[6] Pope Francis, Encyclical Letter *Laudato Si' on Care for Our Common Home*, 24 May 2015, n. 141.

the verticality that gives meaning and unites everyone in one human family, human beings have become the epicenter of universe and the measurement of everything with the human reason exercising power over nature and impressing its brands on history. With the advent of secularization which later became secularism, the historical dimension will be radicalized, generating a greater social and symbolic fragmentation that leads to the compartmentalization of the human knowledge and existence. On a practical level, technology seems to have become the only heir of the Enlightenment reason, whose promises it takes on, unlimitedly strengthening a logic of domination; on the other hand, on the level of thought, the will to power generates a radical nihilism, raising human freedom beyond any order external to the subject, to the extreme point where the death of God and the death of man go hand in hand. Compulsive consumerism and cynical disenchantment become the opposite ingredients of a new cultural cocktail[7]. Modern rationality is dangerous for the future of human society. Some aspects of this crisis are shocking. How can we tolerate the damage of pollution and its effects on the quality of life, world hunger, the tyranny of excess consumption, new forms of slavery structurally maintained in order to sustain a high standard in some parts of the world, wars financed by blood diamonds or oil and gas acquired at the price of human blood, etc.? The human being's capacity for consumption and for destruction is frightening.

The world is facing a break with the order of nature which is the profound sign of modernity and postmodernity. Technology has become a homogeneous, one-dimensional[8], ideological and totalitarian ideology. In the Anthropocene era, human beings are better able to manipulate nature and life. A new condition, which attributes an unprecedented responsibility to the human being[9]. With the modern focus on the human subject, Normative and axiological prescription have been challenged. The lack of objective norms, principles and truths opens the door to all types of derivatives including unlimited exploitation of earthly resources, pollution, exploitation of the poor and wars. The human being becomes the beginning and the end of power, particularly

[7] Luigi Alici, "Natura e persona: lo 'sguardo diverso' di Papa Francesco", in Giuseppe Notarstefano, *Abiterai la Terra: Commento all'enciclica Laudato si'*. Roma: Fondazione Apostolicam Actuositatem, 2015, p. 52.

[8] Pope Francis, *Laudato Si'*, n. 106.

[9] Chiara Giaccardi e Maura Magatti, "Educarsi alla cura: coltivare, custodire, cantare", in Leonardo Boff et al., *Curare la madre terra: Commento all'enciclica Laudato Sì di Papa Francesco*. Bologna: Memi, 2015, p. 44.

through the myth of fascinating technology and unlimited progress. *Laudato Si'* does not condemn progress[10] nor does it condemn growth but insists on a fair distribution of resources as a way of questioning individualist ideologies and epistemologies that shape contemporary understanding of justice[11].

3. Neoliberalism and Covid-19

During this health crisis, the world, especially most Europeans and North Americans, has rediscovered the importance of public services and the role of the state. To manage the crisis, the world went back to a sort of state's capitalism. Even highly neoliberal economies could not avoid this apparent regression. The economic consequences of this global health crisis are rather serious. The world is now facing an economic crisis. After what the world experienced so far, one can easily assert that the neoliberal solutions of cutting public services to pay for the debts are simply unacceptable. The world cannot undermine the public services that have been used to save life during the Covid-19 pandemic. The Covid-19 crisis, however, has fully exposed why neoliberal economies are unfit to deal with a pandemic such as this. The current health crisis has illustrated how destructive neoliberalism has been for health services. How are we to explain why European Union countries now have fewer hospital beds, more exhausted and underpaid healthcare workers? Isn't it mostly due to EU governments and institutions acceptance of the neoliberal dogma of putting the interests of a wealthy minority before the interests of the general public?

3.1 Profiting from healthcare

As Covid-19 virus spreads globally, it must surely count as a stinging indictment of the varied forms of social austerity and healthcare economizing characteristic of the diffusion of neoliberalism over the last four decades[12]. As pandemic goes on in Europe and North America, decades of privatization and government cuts to health systems have led to severe under investment, lack of preparedness, woeful coordination and fatal shortages of medicines and testing equipment. This is a system that has not only eroded the democratic

[10] Pope Francis, *Laudato Si'*, nn. 132-136.

[11] Pope Francis, *Laudato Si'*, n. 193.

[12] Vicente Navarro, "Neoliberalism as a class ideology; or, the Political causes of the growth of inequalities", in *International Journal of Health Services* Vol. 37, No. 1, 2007, p. 47.

ideals of equality and popular sovereignty, but has also created a political and economic context in which the looming pandemic puts a severe strain on medical workers and hospitals that lack ventilators and other essential equipment to treat patients and limit the number of deaths caused by the virus. There is no doubt that neoliberalists have sought to turn public health into a commodity from which investors can earned money[13]. By persistently prioritizing competition and consumerism over free access, leaving citizens to the whims of private healthcare providers and medical services, governments around the world are now struggling to cope as demand surges during a pandemic that many had warned about, and governments had chosen to ignore.

Public provision of universal health care is the best way to fight global epidemics and protect public health. Health cannot be commodified, for it is not a good like any other good. Health ought to remain in public sector or in non-profit sectors so that individuals would not depend upon the market. To achieve such a goal, governments need to promote universal access to quality healthcare and provide basic dimensions of well-being to all citizens that will be matched by substantial increases in government funding and investment. The current crisis has revealed how the entire world has failed to grasp how destructive the neoliberal policies and austerity measures have been, and why current social systems are unfit to cope with public crises. This situation reminds us of the effects of Structural Adjustment Programs (SAPs) in Africa, Latin America and some parts of Asia in the 80s and 90s. SAPs resulted in reduced government health spending and the emergence of more actors of the global civil society doing what governments used to do. One result of this was the fragmentation health systems at local levels due to undermined local control of health programs. This has also contributed to social inequality, underfunded public health systems and the exodus of public healthcare workers to the private sectors. This, in turn, fuelled a brain drain of medical personnel and undermined the fight against HIV and AIDS, as well as public health more generally, in developing countries.

After forty years of neoliberalism, the cuts of public funding have deeply affected medical research and the development of vaccines and medicines. This disposition has pushed governments away from this field and allowed

[13] José Gusmão, *Covid-19 exposes the destructive legacy of neoliberalism: 6 ways neoliberalism made this pandemic worse*, 2020. https://www.guengl.eu/covid-19-exposes-the-destructive-legacy-of-neoliberalism/

in the European and North-American big pharmaceutical companies. These companies have been given the exclusive rights and lucrative licenses. They dictate the price to desperate governments around the world as being seen in the case of the Covid-19 vaccine provision. The importance of public research and development activities are important to ensure population access to medicine and treatment as well as to ensure national security. Leaving this spot to private investors is definitely dangerous to both public safety and national security.

3.2. No state intervention and competition over solidarity

The cutting of public funds for services that guarantee the social well-being has had important consequences for public health. These services are of vital importance for limiting the enormously negative impact of the epidemic on people's quality of life[14]. The most striking aspect of the pandemic in Europe and North America has been the lack of medical equipment and facilities to treat the sick. During the early weeks of the crisis, the lack of help from EU states to one another was particularly striking. The solidarity that had long been trumpeted by the EU Commission was as good as worthless as first, Italy and then Spain were overwhelmed. Sadly, neoliberal competition has long replaced solidarity and coordination. In times of crisis, EU member states were unable to coordinate or put the necessary measures in place to support one another, not just in health but also in economic and social recovery.

To address the threat of neoliberalism to our fragile planet and promote the well-being of future generations, the fundamental change of the world's production, consumption and lifestyles is needed. The culture of individual interest alone is detrimental to carrying out such a change. The culture of interdependence is the foundation of the humanization project. The culture of relationships, complementarity, synergies contributes to go beyond technology, to overcome the ambiguities of a future based solely on technological innovations[15].

---

[14] Navarro Vincente, *The financial and economic crises and their impact on health and social well-being*. New York: Routledge, 2014, p. 272.

[15] Luigi Fusco Girard, "Ambiente e governo del territorio: il ruolo della città", in Giuseppe Notarstefano, *Abiterai la Terra: Commento all'enciclica Laudato si'*. Roma: Fondazione Apostolicam Actuositatem, 2015, p. 112.

3.3. The rise of emerging diseases

It is evident that "Climate change may not seem to have an obvious link to the Covid-19 outbreak, but when one considers the part played by neoliberalism in the destruction of natural habitat, and the loss in biodiversity, the increase in the risks of zoonotic diseases is directly related to the destruction of our planet"[16]. Indeed, Covid-19 is just the latest in a long line of emerging zoonotic diseases or pathogens caused by viruses, bacteria, fungi or parasites that spread from animals to humans. About 70% of emerging human pathogens can be traced back to animals[17]. The first observation that needs to be made is that expansion of neoliberalism (with its commercialization of medicine alongside other dimensions) has contributed to the fact that since the 1980s, the world has witnessed a few epidemics. Well-known viruses such as Ebola, Middle East respiratory syndrome (MERS), severe acute respiratory syndrome (SARS), the West Nile virus, dengue fever and, of course, bird flu are all zoonotic examples which have killed thousands. The application of policies such as the deregulation of globalized capital and labor, alongside policies of social austerity, is one of the factors that has most contributed to the expansion of such diseases on both sides of the North Atlantic[18]. Shaped by neoliberal principles, the increasing demand for food and natural resources has not only prompted deforestation and the destruction of our common house, but has also pushed people to areas where they are in increased contact with animals, birds, bats, wild species of grain, etc. thus increasing the risk for zoonotic infections.

Unsurprisingly, human beings are now in closer contact with animals than ever through intensive farming, the trading of livestock and their widespread transportation, as well as cohabiting in close proximity with other species in market places. Likewise, the unregulated harvesting of wild species has also contributed to the spread. All of these are driven by the neoliberal dogma of free trade and competition, and we are paying the price for it.

---

[16] José Gusmão, 2020.

[17] José Gusmão, 2020.

[18] Navarro Vincente, *The financial and economic crises and their impact on health and social well-being*. New York: Routledge, 2014, p. 272.

3.4. Assault on democracy and social justice

It has been a central pedagogical principle of neoliberalism that individual responsibility is the only way to address social problems, and consequently, there is no need to address broader systemic issues, hold power accountable or embrace matters of collective responsibility. As a politics of containment, neoliberalism privatizes and individualizes social problems, i.e., hand washing and wearing a mask as a way of containing the pandemic[19]. By depoliticizing public health issues, neoliberalism undermines any real democratic attempt to address the social roots of health crises. I am talking about process that seeks to eliminate health threats not only based on personal agency and political equality, but more importantly through a collective solidarity so as to promote the right to health and provide people with the basic dimension of well-being. No doubt that, democracy appears to be a challenge to market individualism.

Additionally, neoliberalism's emphasis on commercial values rather than social and democratic values, its virulent ideology of extreme competitiveness, irrational selfishness, and its impatience with matters of ethics, justice, and truth has undermined critical thought and the power of informed judgment. The social contract has been all but eliminated while notions of the public good, social responsibility, social obligations and democratic forms of solidarity are under attack. It is then understandable why neoliberal capitalism can be understood as being the underlying pandemic feeding the current global shortage of hospitals, medical supplies, beds and robust social welfare provisions in countries. Under such circumstances, the social sphere and its interconnections become an object of either financial exploitation or utter disdain, or both[20]. What is lost in this depoliticizing discourse of neoliberalism and made clear in the current pandemic is that our lives are indeed interconnected for better or worse. The financial crisis of 2008 made visible the plague of neoliberalism that has for over 40 years ravaged the public good and imposed misery and suffering upon the poor and others considered excess, waste or dangerous. With its merging of brutal austerity policies, financialization of the economy, the concentration of power in few hands, and the language of racial and social cleansing, neoliberalism has morphed into a form of fascist politics in most liberal countries[21].

---

[19] Henry A. Giroux, *The Covid-19 Pandemic is exposing the plague of neoliberalism*, 2020. https://truthout.org/articles/the-covid-19-pandemic-is-exposing-the-plague-of-neoliberalism/

[20] Henry A. Giroux, *The Covid-19 Pandemic*, 2020.

[21] Charles Wright Mills, *The sociological imagination*. New York: Oxford University Press, 1959, p. 9.

4. The eco-social imagination, a relational perspective

Against the atomistic anthropology that sustains neoliberal thinking, *Laudato Si'* offers a thicker approach to personhood and a broader framework to understand and address global challenges including public health crises.

4.1. An Ecosocial approach to the ecological crisis

> The sociological imagination enables us to grasp history and biography and the relations between the two within society… No social study that does not come back to the problems of biography, of history and of their intersections within a society has competed its intellectual journey… For that imagination is the capacity to shift from one perspective to another – from the political to the psychological; from examination of a single family to comparative assessment of the national budgets of the world; from the theological school to the military establishment; from consideration of an oil industry to studies of contemporary poetry… It is the capacity to range from the most impersonal and remote transformations to the most intimate features of the human self – and to see the relations between the two[22].

When Wright Mills first coined the term "sociological imagination", he meant to point to "the awareness of the relationship between personal experience and the wider society"; Peter Berger refers to it as "seeing the general in the particular". Mills identified "troubles" (personal challenges) and "issues" (larger social challenges), also known as biography, and history, respectively. Mills asserts the difference between personal troubles and public issues; he contends that personal troubles are private matters which are backed by people's willful behavior within their immediate relations. Further, Mills maintains that public issues (social issues) go beyond one's private sphere and are entrenched in the organizations and institutions of society, and of history[23]. The sociological imagination is making the connection between personal challenges and larger social issues so as to point out how personal trajectories and biographies are shaped by the collective and global trajectory at the different level. Mills' sociological imagination allows individuals to see the relationships between events in their personal lives (biography) and events in their society (history). In other words, this mindset provides the ability for individuals to realize the relationship between their personal experiences and the larger society in which they

[22] Charles Wright Mills, *The sociological imagination*, 1959, pp. 3–7.
[23] Charles Wright Mills, *The sociological imagination*, 1959, p. 9.

live their lives. With his sociological imagination approach, Mills encourages people to think about their lives not just on an individual level, but also considering societal, biological, and historical context. Societal context tells us about our culture—when we consider it, we think about how our desires, actions, and thoughts are shaped by our community and how that community is changing. Biological context tells us about how "human nature" impacts our desires and needs. And lastly, historical context considers our place in time; how have events of the past led up to where we are currently?

By affirming that everything is related[24], Pope Francis sees the crisis of human society within a larger crisis, the ecological crisis. With such a vision, individual trajectories are connected to social destiny, a larger ecology shaped by social structures and relationships. In this sense, individuals' lives are shaped, in one way or another, by their immediate surroundings, respective families, friends and relatives, your country and the world as a whole. This fundamental relationality of the human existence is sustained by an eco-social understanding of the human condition in a world where competing interests, under the dominance of triumphalist capitalism (neoliberalism) and thin understandings of the human person, place human life and the future of the planet in harm's way. The interconnectivity of all the aspects of human existence explicitly paves the way for "a strategy for real change" that requires "rethinking processes in their entirety, for it is not enough to include a few superficial ecological considerations while failing to question the logic which underlies present-day culture"[25]. The solution to the crisis is what is called "integral ecological", an approach that provides a larger vision of what it means to be a society; a vision which is grounded in an eco-social framework. The "integral ecology" refers to a wide range of issues regarding the "relationship existing between nature and the society which lives in it"[26]. The Meaning of the concept of society is as broad as the one of integral ecology. It refers not only to human communities living within political boundaries, but also to the human community as a whole and as the infrastructure of relations between the existing political units on the planet[27]. The recourse to the ecological crisis then

---

[24] Pope Francis, *Laudato Si'*, n. 138.

[25] Pope Francis, *Laudato Si'*, n. 197.

[26] Pope Francis, *Laudato Si'*, n. 139.

[27] Pasquale Ferrara, "Sustainable International Relations. Pope Francis' Encyclical *Laudato Si'* and the Planetary Implications of Integral Ecology", in *Religions* 2019, *10*(8), 466, https://doi.org/10.3390/rel10080466.

offers an opportunity to make a strong claim on the declining quality of human life, the dominance of the neoliberal perspective on the world's economy, the ever-growing global inequalities, the loss of the sense of communality and the technological hegemony that marginalizes social concerns. The notion of integral ecology is both a rhetorical and theoretical tool aiming at providing a "vision capable of taking into account every aspects of the global crisis"[28].

Although the concept of integral ecology is comparable to the notion of "integral development" which focuses on the development of every man and of all men[29], the first one seems to be more innovative due to its rootedness in the fundamental relationality that exists between all the aspects of human existence. In other words, the originality of this concept comes from that fact that it accounts for the interactions between nature and society. Thus, there is only one crisis, which is both environmental and social. Pope Francis' call for a paradigm shift is without doubt a radical change in relationships at the global level. Just as the crisis is multidimensional, so is the call for responsibility. Such a change may have significant implications for environmental issues on the wider spectrum of security, development, economic and ethical issues in our contemporary global society. This wider approach seems to echo the strong claim of the Brundtland Report, a document that forcefully outlines the interrelated factors causing environmental degradation:

> The environment does not exist as a sphere separate from human actions, ambitions, and needs and attempts to defend it in isolation from human concerns have given the very word "environment" a connotation of naivety in some political circles. The word 'development' has also been narrowed by some into a very limited focus, along the lines of 'what poor nations should do to become richer', and thus again is automatically dismissed by many in the international arena as being a concern of specialists, of those involved in questions of "development assistance[30].

---

[28] Pope Francis, *Laudato Si'*, n. 137.

[29] Saint Paul VI, *Encyclical Letter Populorum progressio on the Developmen of Peoples*, 26 March 1967, n. 14.

[30] Brundtland Commission, *Report of the World Commission on Environment and Development: Our Common Future*, 1987, p. 13.

4.2. Integral ecology, the rationality of being-together

Purely technical solutions do not give satisfaction, because they only concern with the symptoms of the crisis without touching its root causes. The ecological crisis, in fact, is not only a physical or material problem, but an anthropological one with moral, cultural and spiritual implications. The correct answer is therefore an "integral ecology", because "everything is connected"[31]. Faced with the technocratic paradigm, the fragmentation of knowledge (for example: the economy conceived as an autonomous science disconnected from society), the internal fragmentation of man, a certain search for economic efficiency that excludes the poor and the weak, the paradigm of inclusion and unity stands as the key from which humanity needs to heal the ecosystem. Integral ecology decentralizes man from himself, questions the outrageous violence of capitalism and opens up a vast relational field that closes the doors to the epistemic fragmentation born of modernity. This interpretation is based on the common nature of the human being. We speak of sociability naturally shared by all human beings. In fact, if everything is connected, then we ourselves, as human beings, are also united as brothers and sisters in the same pilgrimage[32]. The call to unity is based on the "human root of the ecological crisis and, more precisely, starting from the practical relativism of modern man who, with his exasperated individualism, considers everything else, all God's creatures, functional to himself"[33]. Therefore, it calls for unity, of the return to the connatural relationality of the human being, it is a call to a certain human authenticity that is expressed both in politics, in the economy or in the use of technology.

To show how he responds to a complex crisis with a complex response, the Pope uses a systemic approach. Systemic thinking, as indicated by the themes of the encyclical, offers an interconnection, a general framework, a holistic perspective, considering all of creation. Such a perspective can help overcome the nature-society dualism originating from the Cartesian and mechanistic understanding of the human subject and the social world. Against the dichotomist and individualistic western approach, the Pope's eco-social imagination may intermingle well with existing age-old non-western approaches to personhood and society to offer concrete solutions to the crisis in a way which relevant to these local cultures.

---

[31] Pope Francis, *Laudato Si'*, n. 138.

[32] Pope Francis, *Laudato Si'*, n. 92.

[33] Flaminia Giovanelli, "Presentazione", in Giuseppe Notarstefano, *Abiterai la Terra: Commento all'enciclica Laudato si'*. Roma: Fondazione Apostolicam Actuositatem, 2015, p. 8.

The concept of "personhood", which is posited in most health interventions is the Liberal ideal of an unencumbered, rational, self-interested, and self-sufficient subject whose values are of a purely individualistic sort. Persons are perceived as separate from one another; their interests are private and must be respected and accommodated as far as possible. The dominant epidemiological paradigm, the one that considers the individual as the central unit of study, even in low-income countries, preconizes so-called methodological and ethical individualism with an understanding of the human person that has its roots in the Liberal moral and political tradition of the Enlightenment. Within this philosophical tradition—represented by Locke, Hume, Mill and others—the ethical justification of actions or practices strongly depends on the putatively "free" decisions of individuals, i.e. an action or practice that can only be ethically justified when undertaken without any coercive influence and entered into by free and informed agreement. Freedom from coercion and understanding of informed consent appear to be the core determinants of choice and individual responsibility.

The idea of human beings conceived as free and rational entities is alien to most non-western cultures in which the moral vision of the human person is essentially relational. Individuals are never totally independent or purely rational, separate and self-interested. Confucian[34], Jewish[35], African[36], Brazilian[37] and Filipino[38] moral visions of the human person and society are essentially relational. Moral progress is understood within the roles and obligations that an individual person has to carry out in the society of which s/he is a member. It is, therefore, obvious that these cultures have developed a non-contractual view of society where social intercourses are regulated by the virtues of mutual care and benevolence and not by claims of individual rights. In Filipino society, for example, society, family, and church assume a moral and religious significance no longer found in the West. In these non-western societies, the family is considered to be the social unit of greatest value[39].

---

[34] Donald Tsai, "Medicine and Society: Personhood and Autonomy in Multicultural Health Care", in *American Medical association Journal of Ethi*cs 10, 2008, pp. 171-176.

[35] Yohanna Barth-Rogers and Alan Jotkowitz "Executive Autonomy, Multiculturalism and Traditional Medical Ethics", in *Am J Bioethics* 2009, p. 9, 39-40.

[36] Godfrey Tangwa, "Bioethics: An African Perspective", in *Bioethics* 10, 1996, pp. 183-200.

[37] Marcio F. Dos Anjos, "Bioethics in the Developing World: A Liberation Theology Perspective", in *J Med Philos* 21/6, 1996, pp. 681-684.

[38] Angeles T. Alora and Josephine M. Lumitao, *Beyond Western Bioethics. Voices from the Developing World.* Washington: Georgetown University Press, 2001.

[39] Ibid.

Hence, the primary focus of moral responsibility is the family and not the individual. Among African thinkers, John Mbiti[40], Bénézet Bujo[41] and Laurenti Magesa[42] stress relationality as a distinctive feature of our common humanity. Recently Magobe Ramose has framed it as follows: "to be human is to affirm one's humanity by recognizing the humanity of others"[43]. Without falling into another extreme, however, namely that of overemphasizing the communal spirit against freedom, we argue that paying attention to these moral traditions may shape social science and epidemiological research differently.

There are irrefutable evidentiary proofs that historical circumstances and structural factors shape the environment within which freedom is exercised. Very often, the relationships between factors that favor infection are synergetic, and their sources remain embedded in major economic and social institutions, public policies, and infrastructural arrangements in given circumstances. They are linked to historical relationships of power distribution and property ownership, systemic racism or tribalism, neighborhood segregation, absolute poverty, gender inequality, and the feminization of poverty. Freedom and behavior can only partly explain the clustering of diseases within socioeconomically challenged neighborhoods or population groups if other factors are not taken into consideration. Considering the multiplicity of constraints on the human being's abilities to responsibly control her or his behavior, we humbly need to accept individuals' limitations and propose some venues for collective agency and structural solutions.

4.3. The need for an anthropological recovery

There will be no new relationship with nature without a new human being. There is no ecology without an adequate anthropology[44]. Therefore, integral ecology is supported by an anthropology of the relationship. To resolve the ecological crisis there must be an anthropological awakening, because the

---

[40] John Mbiti, *African Religions and Philosophy*. Nairobi: East African educational Publishers, 1969.

[41] Bénézet Bujo, *Foundations of an African Ethics: Beyond the Universal Claims of Western Morality*. New York: Herder & Herder Book, 2000.

[42] Laurenti Magesa, *African Religion: The Moral Traditions of Abundant Life*. Nairobi: Paulines Publications Africa, 1998.

[43] Mogobe B. Ramose, "Globalization and Ubuntu", in: Coetzee PH and Roux APJ, eds. *The African Philosophy Reader*. London: Routledge, 2003, pp. 622-642.

[44] Pope Francis, *Laudato Si'*, n. 118.

individualism that determines exchanges and relationships between subjects national and global communities to the laws of the market. All this marks the future of the peoples and has clearly shown its limits. So, *Laudato Si'* is imbued with this relational anthropology. Terminology expressing relationality is used more than ten times by Pope Francis in *Laudato Si'*. With a renewed anthropology, we can adequately understand that, on the one hand "to think" means to think with… and also to think for ... for the sake of others, and on the other hand, "to be" means "to be with", because it is quite possible to seek our own good without endangering the welfare of others, without destroying "the common house" we all share.

Neoliberal values of individualism and freedom provide the ideological bases for reducing society's responsibility in health promotion by emphasizing responsibility for one's choice and the willpower to control the circumstances that shape one's life[45]. Both freedom of choice and willpower are closely related to the values of internal control and self-discipline emphasized by Max Weber in his approach to the Protestant work ethic[46].

The economic claim for individual freedom is market individualism which is defined as confidence in the free market as a regulatory mechanism and as a belief in the importance of allowing individuals to make choices within the market and accepting responsibility for consequences that arise from them. Market individualism has been so dominant that the call to individual responsibility has overshadowed the power of collective agency. People, especially those at the bottom of society, often need government support to move out of poverty.

The market constrains State-intervention in the areas of individual and social pathology and welfare, thereby favoring the growth of unjustifiable inequalities. These inequalities have emerged especially in rich countries such as the United States of America during this time of Covid-19 pandemic. Insufficient action by governments in addressing socioeconomic inequalities in health is simply unjust. Since the government's role is to establish a "perimeter of equity", the litmus test for the morality of any society is the institutional form by which it cares for and nurtures its most vulnerable members. The primary focus of a policy intervention should be on the behavior of the policy-maker

---

[45] Christian S. Crandall and Rebecca Martinez, "Culture, Ideology, and Anti-fat Attitudes", in *PSPB* 22, 1996, pp.1165-76.

[46] Christian S. Crandall and Colby Cohen, "The personality of the stigmatizer. Cultural World View, conventionalism, and Self-Esteem", in *J Res Pers*. 28/4, 1994, pp. 461-80.

and the attendant social institutions, not on the vulnerable groups themselves. Emphasis is placed on health and on the level of vulnerability of those who are victims of socioeconomic exclusion because, historically, the major determinants of health are not personal choices but risks that are external to individuals. Since the community is the context in which people live and decide their fate, efforts to reduce exposure to health hazards should primarily focus on the rules, policies, and norms that define the social environment.

The behavior change solution is often supported through the medium of the market philosophy. One of the key principles in the idea that the markets are the most efficient arbiters of economy is that individuals will, if unencumbered by regulation, seek to advance their own economic interest. Under this system, it is not surprising that positing individuals as the "vehicles" of "perfect economic rationality" is tantamount to making them the focus for all aspects of behavioral responsibility. However, the history of medicine shows that the decline in infectious diseases and the increase in life expectancy did not result from the discovery of penicillin. The decline in infectious diseases predates scientific medicine and may more correctly be attributed to improved sanitation, clean water supplies, better nutrition, and less crowded living conditions. Even today, antibiotics and pesticides may certainly provide relief but they do not alter the fundamental conditions that serve as a bedrock for the rise of major epidemics.

The fragmentation of practical reason in modernity is sustained by methodological and epistemological individualism which, in turn, opens the doors to socio-economic and political and ethical individualism. I believe that with an anthropological perspective that recognizes and respects the complexity of reality, our comprehension of personhood should open up to a more holistic vision of the world. We are speaking here of a holism reshaped on the anvil of an anthropological vision that harmoniously brings together the individual and society. With such a way of thinking, that is, essentially relational thinking, it could help overcome the lack of transcendence and the dictatorship of immediacy in which practical reason is mired today. It is truly an invitation to break the chains of epistemological and metaphysical superficiality and socioeconomic indifference to which our technocratic culture has confined us to think long-term in order to create a more just world.

Laudato Si' is not fundamentally an encyclical on climate change or global warming. Rather, it is a text that addresses the destruction of our fragile world. It identifies causes of this as including the development of modern technology linked to unlimited freedom and consumerism. It also stresses that the victims of this behavior include the human poor. In fact, the subtitle of the encyclical reflects what the pope would like to share with the world: the need to take care of our common home. So, the key question that raises such a perspective is to know how to get there, in a world dominated by essentially individualistic, nationalist thoughts and attitudes. To face the challenges of development and sustainability, a reductionist system of thought that is still focused on the good of a minority will not serve as an intellectual resource to create a more just world. The world is an ecosystem, a set of connected realities. Therefore, one cannot act on one part without willingly or unwillingly affecting the others[47].

4.4. Ecological conversion as a precondition for structural intervention

The dominant cultures in our contemporary society are powerful vehicles which spread market individualism. This cultural influence extends to spirituality. An example of this is the so-called "prosperity Gospel". Here, eternal reward is associated with earthly successes with the current economic system. Little value is placed on a sense of community, a care for the more vulnerable, and the need to build a life-promoting sociopolitical environment. The world probably lacks the needed categories and vision to articulate and develop a collective dimension of a process of conversion. An approach that does not only focus on individual change for the sake of one's salvation but more deeply on attitudes that call upon a dialectical and multi-level posture between individual heart and social institutions at all levels. The multidimensionality of ecological conversion seems to be one of the high points of *Laudato Si'* because it brings about a deeper spiritual and providential challenge to enrich the way in which we look at ourselves and the world[48]. We look at ourselves as part of a large ecology in which our individual actions determine the fate of our common house and the state of our common house determines the welfare of individuals. This fundamental interconnectedness inevitably shapes our relation with God and determines our faithfulness to the creator of all things.

Similarly, our understanding of morality can no longer focus on the individual but ought to be rooted the lager matrix that shapes the human behavior and

---

[47] Stefano Zamagni, "Un'ecologia integrale: Civilizzare l'economia e custodire il creato", in Giuseppe Notarstefano, *Abiterai la Terra: Commento all'enciclica Laudato si'*. Roma: Fondazione Apostolicam Actuositatem, 2015, p. 27.

[48] Giacomo Costa, "Azione, conversione, contemplazione: un'enciclica da mettere in pratica", in Leonardo Boff et al. *Curare la madre terra: Commento all'enciclica Laudato si' di Papa Francesco*. Bologna: Memi, 2015, p. 58.

welfare. Grounded in a thicker anthropology and a complex epistemology, the ecological rationality calls for an understanding of sin, grace, transgression, love, examination of conscious, prayer, law, devotion and piety which is essentially (to remain faithful to our dense nature) relational. Societal challenges such as pandemics and epidemics would never have been successfully addressed with categories and theories that does not account for the complex nature of their etiology. How are we to put an end to emergent diseases and epidemics without addressing the neoliberal model of society that places finance at the center of the global society while human beings simply become means of economic success. At the same time, the magnitude of the crisis offers new possibilities in which people can begin to rethink what kind of society, world, and future we want to inhabit. What we do not want to do is to go back to a system that equates democracy and capitalism. We must move beyond merely modifying the system, because the current crisis has deeper political and economic roots and demands a complete restructuring of society. Not only does the world needs adequate language and tools to understand social phenomena and pathologies, but the way we inhabit the world ought to be shaped by ways of life, individual values and public norms that help us face the elephant in the room. Spirituality, as a way of life and a way of encountering God, would help address the anthropological fragmentation and the epistemological relativism that shape our times if it enables us to develop a new language to name how we ought to live and then new ways of engaging our common house.

The eco-social imagination that sustains the integral development approach cannot be reduced to a mere normative question regarding ecological responsibility nor can it be confined to technological solutions. Furthermore, it cannot be confined to environmental issues. With regards to global health, this approach forces us to search for new analytical tools which are radical, trans-disciplinary and trans-paradigmatic. First and foremost, the centrality of the questions related to the eco-social approach is trans-sectorial and transnational. It is even more apparent in this time of Covid-19 that communicable disease has no frontier. In addition, the socio-global inequities that often favor the spread of diseases do not only affect individuals but entire countries. They challenge public health researchers to consider an ethics of international relations and to take public health diplomacy seriously[49].

[49] Vicente Navarro, "Neoliberalism as a class ideology; or, the Political causes of the growth of inequalities", in *International Journal of Health Services* Vol. 37, No. 1, 2007, pp. 47-62.

5. Conclusion

> Neoliberalism kills. We need to find a better way. The idea of a communitarian economics in which—locally, nationally, and globally—people have a real say in what kind of social institutions they have and how these are run is one way to address the planet's health problems. There are others. The crucial issue, however, is to accept as Navarro has argued that health must be political and that fundamental to any genuine progress in addressing poverty, inequality, and ill health at a global level is to recognize that, first, neoliberalism is at the root of these problems and, second, some alternative must be found. That is the debate on which public health must embark. Whatever the solution, what is needed is a new political economy of public health[50].

The current viral pandemic cannot be discussed outside of the economic and social crisis the world is in. What is needed is a new language to comprehend the current pandemic crisis. This language must provide a sustained critique of neoliberal individualism with its divisive and exclusivist rhetoric that weakens social cohesion all around the world. The new language that needs to be developed ought to be sustained with a vision of society and the human person that help create a new policy environment within which the underlying causes of poverty, class domination, environmental destruction, Global health crises and a resurgent racism within and outside countries are addressed. A radical vision is needed, a perspective that will aim at creating a new political and economic social order within countries as well as a new world order. The Covid-19 pandemic challenges the world to talk about society and global challenges in a different way, with a language which is different from neoliberal principles that has led the world astray. In order to avoid another pandemic such as covid-19 or regional epidemics such as MERS, Ebola and SARS, the world has no other solution than to acknowledge that individualism, as I have mentioned throughout this reflection, cannot be considered as a framework for the world economics and global health. A radical approach to individualism cannot help build a just world, instead the Pope Francis' eco-social perspective, which acknowledge the fundamental interconnectedness of all, may serve as the basis for addressing public and global health challenges.

---

[50] Gavin Mooney, "Neoliberalism is bad for our health", in *International Journal of Health Services* 42/3, 2012, pp. 383-401; Pope Francis, *Laudato Si'*, n. 51.

THE ECONOMIC FOUNDATIONS OF INTEGRAL ECOLOGY

Paolo Conversi

1. Introduction

The Encyclical of Pope Francis *Laudato Si' on care for our common home* is a heartfelt invitation to reconsider the foundations of the market economy model that is in vogue today[1]. This contribution aims to analyze what may be described as the economic "pillars" of integral ecology, one of the central and most innovative concepts of *Laudato Si'*[2].

One definition of economics describes it as, "the study of how societies use scarce resources to produce valuable goods and services and distribute them among different individuals"[3]. It aims to identify the best allocation of resources between alternative uses. It is based on two key concepts: the first is the concept of "scarcity" of resources, which are limited compared to the "limitlessness" of human desires and human needs to be satisfied through the use of these same resources; the second is the concept of "efficiency", that is making the best use of limited resources to satisfy the aforementioned needs and desires. The main purpose of economic science is, therefore, to improve the conditions of people in their daily life and to solve the problems

---

[1] Stefano Zamagni, "Un'ecologia integrale: civilizzare l'economia e custodire il mercato", in *Abiterai la terra. Commento all'enciclica Laudato si'*, ed. AAVV. Rome: AVE, 2005, p. 25.

[2] Interdicasterial Working Group of the Holy See on Integral Ecology, *Journeying towards Care for Our Common Home. Five Years after Laudato Si'*. Rome: Libreria Editrice Vaticana, 31 May 2020, pp. 5-18 and 216-20.

[3] Paul A. Samuelson, William D. Nordhaus, Carlo A. Bollino, and Paolo Polinori, *Economia*. Milan: McGraw-Hill Education, 21st Edition, 2019, p. 1.

DOI 10.32060/C.S.9.53-92

of the economic organization in the face of scarce and limited resources to respond to human needs and desires[4].

However, this goal is to be achieved through the interaction of complex systems in the process aimed at guaranteeing a real improvement in the development of individuals and the communities in which they live. Economics itself has often found it difficult to interact with these complex systems[5]. Pope Francis criticizes the way "economic growth, for its part, tends to produce predictable reactions and a certain standardization with the aim of simplifying procedures and reducing costs"[6]. He adds:

---

[4] It is appropriate to mention here *Maslow's pyramid of needs*, according to which needs can be divided into five categories: 1) *physiological needs*: food, water, warmth, rest ... those needs related to individual physical survival and the first to be satisfied due to the instinct for self-preservation; 2) *safety needs*: security, protection, predictability, freedom from fear ... they must guarantee a certain sense of tranquility; 3) *belongingness needs*: relationships and friendships, to love and to be loved, to be part of a group, to cooperate, to participate ... this category represents everyone's aspiration to be an element of the community; 4) *esteem needs*: to be respected, approved, recognized ... the individual wants to feel competent and productive; 5) *self-actualization needs*: achieving one's full potential, self-fulfillment ... a desire to become everything one is capable of becoming through his/her own mental and physical faculties. See, Abraham H. Maslow, *Motivation and Personality*. New York: J. & J. Harper, 1954. Later, Maslow expanded this first five-stage model to include *cognitive needs*, *aesthetic needs* and *transcendence needs*. See, Abraham H. Maslow, *Religions, values, and peak experiences*. New York: Penguin, 1970.

[5] Some scholars define complex (adaptive) systems as those "composed of multiple individual elements that interact with each other yet whose aggregate properties or behaviour is not predictable from the elements themselves" (Uri Wilensky and William Rand, *An Introduction to Agent-Based Modeling: Modeling Natural, Social, and Engineered Complex Systems with Netlogo*. Cambridge, MA: MIT Pres, 2015, p. 6). "The interactions of these elements [...] often lead to results not directly predictable from the intentions or actions of any single agent [...] The social sciences, especially economics, have not always looked at the world through the lens of complexity, often preferring top-down, equilibrium-based models rather than the bottom-up, agent-based models used in complexity research [...]. However, even Earth science models that include environmental dynamics in a complex way often represent the socio-economic (human) world as a simple process of macroeconomic optimization. As a result, many important features of complexity – such as interactions and feedback among human and ecological systems, economic and social networks, and even human agency – are left out (Jonathan Donges, Ricarda Winkelmann, Wolfgang Lucht, Sarah E. Cornell, James G. Dyke, Johan Rockström, Jobst Heitzig and Hans Joachim Schellnhuber, "Closing the Loop: Reconnecting Human Dynamics to Earth System Science" in *The Anthropocene Review* 4(2), 2017, pp. 151-7). [...] Human society is linked through many networks not just trade and information but also politics and infrastructure. Human behavior – shaped by norms and values – causes changes in the functioning of the Earth system, which in turn has feedback effects on human norms, values and behaviours. If we study the natural world and human world separately, ignoring the loops both within and between them, we risk missing emergent phenomena such as critical tipping points". UNDP, *Human Development Report 2020: The Next Frontier: Human Development and Athropocene*. http://hdr.undp.org/sites/default/files/hdr2020.pdf, pp. 54-5.

[6] Pope Francis, Encyclical Letter *Laudato Si' on Care for Our Common Home*, 24 May 2015, n. 141.

> This suggests the need for an "*economic ecology*" capable of appealing to a broader vision of reality. [...] We urgently need a *humanism* capable of bringing together the different fields of knowledge, including economics, in the service of a more integral and integrating vision. [...] There is an interrelation between ecosystems and between the various spheres of social interaction, demonstrating yet again that 'the whole is greater than the part'[7].

This statement by Pope Francis, which will be further developed later in this work, already offers many points for reflection that help to elucidate the economic foundations of integral ecology. The goal of this work is to contribute to a better understanding of the "prism" of integral ecology, starting from a clearer definition of one of its principle dimensions, or lenses, the economic one.

Among the themes addressed in the Encyclical indicated by Pope Francis in paragraph 16, there are several that could be considered as "pillars" of the economic "foundations" of integral ecology: the critique of new paradigms and forms of power derived from technology; the call to seek other ways of understanding the economy and progress; the conviction that everything in the world is connected; the intimate relationship between the poor and the fragility of the planet; the problem of a "throwaway culture"; and the proposal of a new lifestyle[8]. To these aspects we could add a sixth "pillar", clearly present in the Encyclical, namely the question of work.

Embarking upon an in-depth study of the economic foundations of integral ecology, it seems appropriate, indeed necessary, to analyses these pillars that permeate *Laudato Si'*.

Before beginning such a reflection, it could be useful to go through a brief historical overview of the international community's efforts in promoting and guaranteeing the main objective of economics: a real improvement in the development of people and their communities.

2. An Historical Overview

"Human advance is conditioned by our conception of progress"[9]. Starting from this assumption, and considering the results obtained thus far, it can be

[7] Pope Francis, *Laudato Si'*, n. 141, quoting his Apostolic Exhortation *Evangelii gaudium on the Proclamation of the Gospel in Today's World*, 24 November 2013, n. 237.

[8] Pope Francis, *Laudato Si'*, n. 16.

[9] UNDP, *Human Development Report 1996. Economic Growth and Human Development*. New York: Oxford University Press, 1996, p. III.

argued that the conception of human progress adopted after World War II was almost inadequate: too often it had been associated exclusively with that of economic growth.

Driven by the broad expectations arising from the decolonization process during the 1950s and 1960s, the United Nations designated the 1960s-70s as the United Nations' first decade of development: States and people should have intensified "their efforts to mobilize and to sustain support for the measures required on the part of both developed and developing countries to accelerate progress towards self-sustaining growth of the economy of the individual nations and their social advancement"[10]. Thirty years later, in March 1995, many Heads of State and Government met in Copenhagen for the *World Summit for Social Development* in order to identify the measures to be taken to win the fight against poverty, unemployment and social exclusion[11], three problems that, still far from being eradicated, underlined the limits of the approaches used to try to defeat them.

At the end of the last century, there was much awareness that the world had become more polarized, and the gaps between the poorest and richest people and countries have continued to widen[12]. This polarization was not only happening between nations, but also within nations themselves. Moving from the economic to the more "human" sphere, data at the end of the 20th century showed that the Human Development Index (HDI)[13] had declined in 21 countries from 1990 to 1999[14]. The situation of "human degradation" was

---

[10] United Nations General Assembly, A/RES/1707(XVI). *United Nations Development Decade: a programme for international economic co-operation*, 19 December 1961, paragraph 1.

[11] United Nations, A/CONF.166/L.3. *Programme of Action of the World Summit for Social Development*. Copenhagen, 1995.

[12] The share of income of the richest 20% of the world population had grown between 1960 and 1997 from 70% to 86% of world income, while that of the poorest 20% had decreased from 2.3% to just 1.0%; in the last four decades of the twentieth century, the gap between the share of the richest and the poorest has more than doubled, going from the ratio of 30:1 in 1960 to that in 1997 of 74:1. UNDP, *Human Development Report 1999. Globalization with a Human Face*. New York: Oxford University Press, 1999, pp. 2-3 and 36-8.

[13] The Human Development Index (HDI) has been rather successful in serving as an alternative measure of development, supplementing GDP (Gross Domestic Product). It measures a country's average performance in the three basic dimensions of human development: longevity, education and income per capita. It is a synthetic index that contains three variables: life expectancy at birth, school results (adult literacy rate and combined enrolment ratio at primary, secondary and tertiary levels), and adjusted per capita income in PPP$ (purchasing power parity in US dollars).

[14] UNDP, *Human Development Report 1999*, pp. 134-7.

also noted in the so-called "first world" countries: the Human Poverty Index of these countries (HPI-2) showed that human poverty and exclusion are hidden phenomena within the "coldness of statistics", since 1 out of 8 people in industrialized countries experienced at least one aspect of human poverty[15].

The beginning of the current century witnessed the adoption in 2000 of the *Millennium Development Goals*[16], which were succeeded in 2015 by the *Sustainable Development Goals*[17]. Even before the Covid-19 pandemic impact, in the 2020 *Sustainable Development Goals Report*, as well as in the *Special Editions of the United Nations Secretary General - Progress towards SDG Report*, we can see many concerns about slow or poor progress towards achieving these goals[18]. These concerns are even more serious in light of the heavy environmental impacts caused by human activities[19], as well as of the negative implications that the Covid-19 pandemic has on the realization of the SDGs themselves[20].

Although it should not be forgotten that some key indicators of human development have improved[21], there are still strong warnings that highlight the

---

[15] UNDP, *Human Development Report 1999*, pp. 149-150. The Human Poverty Index (HPI) measures the deprivation of basic human development in the three dimensions of the HDI. For greater theoretical validity, a distinction is made between developing countries (HPI-1) and industrialized countries (HPI-2). The variables used in the first case are: the percentage of people not expected to survive to age 40, the adult illiteracy rate, the percentage of people without access to safe water and to health services, as well as the percentage of underweight children under five. HPI-2 considers instead the percentage of people not expected to survive to age 60, the adult functional illiteracy rate, the percentage of people living below the income poverty line (50% of median personal disposable income), and the long-term unemployment rate (twelve months or more).

[16] United Nations General Assembly, A/RES/55/2: *Millennium Declaration*, 8 September, 2000.

[17] United Nations General Assembly, A/RES/70/1: *Transforming our world: the 2030 Agenda for Sustainable Development*, 21 October 2015.

[18] United Nations, *Sustainable Development Goals Report 2020*; United Nations Secretary General, E/2019/68. *Special Edition: Progress towards the Sustainable Development Goals*, 2019.

[19] Intergovernmental Panel on Climate Change (IPCC), *Special report on the impacts of global warming of 1.5 °C above pre-industrial levels and related global greenhouse gas emission pathways, in the context of strengthening the global response to the threat of climate change, sustainable development, and efforts to eradicate poverty*, 2018.

[20] United Nations Department of Economic and Social Affairs (UNDESA), *Sustainable Development Outlook 2020*.

[21] In the last thirty years of the twentieth century, in just over a generation, life expectancy has increased by 17 years, the infant mortality rate has halved, the undernourishment rate has decreased by 1/3, the enrollment of primary and secondary schools have increased by 150%, the share of rural households without access to water has dropped from 60% to 25%. At the beginning of the 21st century, there has been further improvements, where the proportion of people living in situations of extreme poverty has halved from 2010 to 2019, from 14.3% to 7.1% of the world population; maternal mortality decreased by 38% from 2000 to 2017; the global under-5 mortality rate fell from

difficulty of promoting "effective" development at the human, social and environmental level in large sections of the population. "Wealth has increased, but together with inequality, with the result that new forms of poverty are emerging"[22]. These data clearly show that the paradoxical situation still remains in which *abundance coexists with scarcity*. We live, in fact, in a world of "opulence" never seen before:

> Never has humanity had such power over itself, [but] nothing ensures that it will be used wisely, particularly when we consider how it is currently being used. [...] There is a tendency to believe that every increase in power means 'an increase of progress itself', an advance in 'security, usefulness, welfare and vigour; ...an assimilation of new values into the stream of culture', as if reality, goodness and truth automatically flow from technological and economic power as such. The fact is that 'contemporary man has not been trained to use power well' because our immense technological development has not been accompanied by a development in human responsibility, values and conscience[23].

These reflections by Guardini, quoted by Pope Francis, are still very relevant.

Humanity is witnessing the "amalgamation" of some issues that risk weakening the "foundations" of our society. Such issues refer to phenomena of deprivation, degradation and oppression in all countries. They concern, among other things, the persistence of extreme poverty and the inability to meet basic needs, the continuous occurrence of famines, the difficulty of strengthening resilience and adaptation to increasingly frequent and intense natural disasters, the violation of elementary and fundamental political freedoms, as well as the continuous threats to the environment and to the sustainability of economic and social life[24]. These data go hand in hand with different types

---

76 deaths per 1,000 live births in 2000 to 42 in 2015, and to 39 in 2018; the proportion of the global population using safely managed drinking water services increased from 61% in 2000 to 71% in 2017. UNDP, *Human Development Report 1995. Gender and Human Development*. New York: Oxford University Press, 1995. UNDP, *Human Development Report 1997. Human Development to Eradicate Poverty*. New York: Oxford University Press, 1997; UNDP, *Human Development Report 2002: Deepening Democracy in a Fragmented World*. New York: Oxford University Press, *2002*; United Nations, *SDGs Report 2020*.

[22] Pope Francis, Encyclical Letter *Fratelli tutti on Fraternity and Social Friendship*, 3 October 2020, n. 21.

[23] Pope Francis, *Laudato Si'*, 2015, nn. 104-5, quoting Romano Guardini, *Das Ende der Neuzeit*, 9th ed. Würzburg, 1965; English: *The End of the Modern World*, Wilmington, 1998, pp. 87-8.

[24] Consider, for example, the worrying increase in food insecurity which went from 22.4% in

of existing inequalities and disparities[25], which from one side "restrict participation in decision-making, limit the potential for innovation and increase vulnerability"; from the other side, they draw attention not only to the overall size of the pie, but also to the relative size of its slices, and to the oven[26]. Even

2014 to 25.9% in 2019, reaching almost 1/4 of humanity; or to numerous other data such as: 2.2 billion people around the world in 2017 still lacked safely managed drinking water and 4.2 billion people worldwide, 55% of the world population, still lacked safely managed sanitation. The proportion of urban population living in slums rose to 24% in 2018, or over 1 billion people. The global material footprint grew from 73.2 billion metric tons in 2010 to 85.9 billion metric tons in 2017, a 17.4% increase (the material footprint is the amount of primary materials required to meet basic needs for food, clothing, water, shelter, infrastructure and other aspects of life; it is an indicator of the pressure put on the environment to support economic growth and to satisfy the material needs of people). From 2010 to 2019, e-waste generated globally grew from 5.3 to 7.3 kilograms per capita annually, meanwhile the environmentally sound recycling of e-waste increased at a much slower pace, from 0.8 to 1.3 kilograms per capita annually. The percentage of food lost after harvesting and during transport, storage and processing stands at 13.8% globally. Furthermore, young workers are exposed to poverty more systematically than adults, a result of inadequate earnings and deficits in job quality. In 2019, 12.8% of workers between the ages of 15 and 24 lived in poverty, compared with 6.3% of workers over the age of 24. The atmospheric concentrations of the greenhouse gases carbon dioxide (CO_2), methane (CH_4), and nitrous oxide (N_2O) have all increased since 1750 due, in large part, to human activity. In 2011, the concentrations of these greenhouse gases exceeded the pre-industrial levels by about 40%, 150%, and 20%, respectively. In 2018 the global anthropogenic fossil CO_2 emissions increased by 1.9% compared to 2017. Approximately half the live coral cover on coral reefs has been lost since the 1870s, with accelerating losses in recent decades. FAO, IFAD, WFP, UNICEF and WHO, *The State of Food Security and Nutrition in the World 2020: Transforming Food Systems for Affordable Healthy Diet*, FAO 2020, p. 50; UNESCO, FAO, UNDP, UN-Habitat, UNIDO, UNU, WHO, WMO, Word Bank Group, *The United Nations World Water Development 2020, Water and Climate Change,* UNESCO 2020; WHO and UNICEF, *Progress on household drinking water, sanitation and hygiene 2000–2017. Special focus on inequalities*, New York, 2019; UNEP, *Global Environmental Outlook*, 1997-2017; UNDP, *Human Development Reports*, 1990-2020; United Nations, *Sustainable Development Goals Report 2020*; Intergovernmental Panel on Climate Change (IPCC), Working Group I Contribution to the IPCC Fifth Assessment Report, *Climate Change 2013: The Physical Science Basis*, Summary for Policymakers; European Commission, JRC Science for Policy Report: *Fossil CO_2 and GHG Emissions of All World Countries*, 2019; IPBES, *The global assessment report on Biodiversity and Ecosystem Services, Summary for Policymakers*, 2019.

25 "The gain in life expectancy at age 70 from 1995 to 2015 in very high human development countries was more than twice that in low human development countries [...and] the proportion of the adult population with tertiary education is growing more than six times faster in very high human development countries than in low human development countries, and fixed broadband subscriptions are growing 15 times faster". UNDP, *Human Development Report 2019. Beyond income, beyond averages, beyond today: Inequalities in human development in the 21st century*, 2019, pp. 7 and 10.

26 UNDP, *"Human Development Report 2020"*, pp. 5 and 6.

economics is called to pay attention to these issues, since, as mentioned, it is aimed at improving the conditions of people in their daily lives, as well as at solving the problems of economic organization in the face of scarce and limited resources to respond to human's needs and desires.

It appears evident that the greatest risk of these phenomena is not only a persistent "pocket" of poverty, but also the social exclusion of a part of the world population, engendering a sort of social Darwinism, where the "strongest" are those who prevail and a climate of intense competition is created between the participants, who would be "forced" to fight for individual survival at the expense of the most vulnerable; the logic of individual survival would thus seem to prevail over the logic of co-existence, leading to the erosion of social contracts. In this way, a part of the population would lose the sense of belonging and of social commitment, giving rise to disarray and dissociation from a society that has placed it on the sidelines and for which it would become a potential threat. Within this paradigm, some people would be considered a liability, being of no use, since they are a burden or cost to society; similarly to environmental waste, they represent "social waste", which "pollute". Such a dynamic is exacerbated by the accelerated consumption of nature which can create interpersonal conflicts and tensions in the relationship between humanity and nature itself. This "loss of nature" is especially detrimental to the poorest and weakest.

In addition to the serious moral implications of this "outcome", it is also paradoxical from the economics point of view, in that the "economic waste" of human resources, caused by an economically inefficient attitude, tends to reduce the human being to a *homo œconomicus competitor*, who does not care about inequalities or the impact of his/her actions to the earth's fine equilibrium.

> Poverty, decadence and suffering in one part of the earth are a silent breeding ground for problems that will end up affecting the entire planet. If we are troubled by the extinction of certain species, we should be all the more troubled that in some parts of our world individuals or peoples are prevented from developing their potential and beauty by poverty or other structural limitations. In the end, this will impoverish us all[27].

[27] Pope Francis, *Fratelli tutti*, n. 137.

3. The critique of new paradigms and forms of power derived from technology. First "pillar": Reconciling short and long-term perspectives

These data and reflections highlight an inability of the economic system to promote a real and fair sharing of the improvements gained since the second post-war period.

One of the reasons of this inability can be found in the difficulty of this system to understand the "interrelation between ecosystems and between the various spheres of social interaction" in front of which *Laudato Si'* recalls, as mentioned, the urgent "need [of] a *humanism* capable of bringing together the different fields of knowledge, including economics, in the service of a more integral and integrating vision"[28].

For too long the concept of development has undergone a *prominent economic characterization*; for too long we have focused on the overall size of the pie and underestimated the relative size of its slices and the oven: the concept of progress placed at the basis of development was that of an economic, unlimited, unilinear, homogeneous progress, without taking into account the strong differences between the various national contexts, and without considering that the ideological paradigm of unlimited growth inevitably leads to "violence of limitlessness" and the frustration of acquisition, competition and the sense of time[29]. Furthermore, this paradigm inevitably clashes with the finite nature of the world, which "is linked not only to the limits of the world's physical and natural resources, but, more important, to the limits determined by human and social interdependence, complexity, and cohesion that interact with physical and natural limits"[30]. In the words of *Laudato Si'*:

> It is based on the lie that there is an infinite supply of the earth's goods, and this leads to the planet being squeezed dry beyond every limit. It is the false notion that 'an infinite quantity of energy and resources are available, that it is possible to renew them quickly, and that the negative effects of the exploitation of the natural order can be easily absorbed'[31].

---

[28] Pope Francis, *Laudato Si'*, n. 141.

[29] Giorgio Ruffolo, *Lo sviluppo dei limiti*. Bari: Laterza, 1994.

[30] The Group of Lisbon, *Limits to Competition*. Cambridge, MA: The MIT Press, 1994, p. 8.

[31] Pope Francis, *Laudato* Si', 106, quoting the Pontifical Council for Justice and Peace, *Compendium of the Social Doctrine of the Church*. Rome: Libreria Editrice Vaticana, 2004, n. 462.

Starting from this economic-centered perspective, political decisions in many countries have often been based on the identity of economic growth equal to well-being[32]. Too often the means (economic growth) have been confused with the aim (integral human development)[33].

The quantitative aspect of growth has frequently prevailed over the qualitative one, thus undervaluing the fact that economic growth tout court can be distorted and even counterproductive not only from an environmental and social point of view, but also, paradoxically, from an economic one, especially if it is assessed by adopting a long-term perspective. As has been shown by empirical evidence, an excessive focus on economic growth and therefore on the Gross Domestic Product (GDP) can lead to very worrying phenomena such as: job-less growth, where the economy grows without expanding the possibilities of new jobs; ruthless growth, when the fruits of growth benefit the rich classes; voiceless growth, highlighting an absurd trade-off between economic growth and extension of participation; rootless growth, where it causes the loss of people's cultural identity; and also futureless growth, to the detriment of future generations, including the young people of today[34].

This pre-eminent economic characterization leads to the dominance of the efficient and short-term logics of an economic system that tends to be "deaf" to the "voices" of other systems. It leads to the prevalence of its ruling law: the return on investment in terms of profit, which must be as immediate as possible. This law, if applied in a fully autonomous and isolated manner, leads the economy to favor any technological development solely on the basis of

---

[32] The identity "economic growth" = "well-being" was combined during the 1960s with the theories based on the so-called *trickle-down process*, according to which economic growth led by some privileged social classes would be followed by well-being for all through a dripping effect; the well-being would be gradually and automatically "filtered" even towards those who at first could not enjoy its benefits. Among the ideas to support these theories was also Kuznets' curve, which posits that in the process of growth / development, the distribution of its benefits first should have worsened and then improved at a later time (Simon Kuznets, "Economic Growth and Income Inequality", in *The American Economic Review*, Vol. 45, No. 1, March 1955, pp. 1-28). Despite numerous criticisms addressed to this approach and the various alternatives proposed, among which the strategy of basic needs of the 1970s should be mentioned (Paul Streeten et al. *First Things First: Meeting Basic Needs in the Developing Countries*. New York: Oxford University Press, 1981), these theories had a strong influence also in the following years.

[33] Aristotle's words come to mind, according to which "the life of profit is a life of constraint and wealth is not clearly the desired good: in fact it is only a useful thing and a means in view of something else". Aristotle, *Etica Nicomachea*. Milan: BUR, 1994, Book 1, Section 3.

[34] UNDP, *Human Development Report 1996*, pp. 56-65.

profit, without paying attention to any negative consequences for humans and the environment. In the words of *Laudato Si'*:

> Some circles maintain that current economics and technology will solve all environmental problems, and argue, in popular and non-technical terms, that the problems of global hunger and poverty will be resolved simply by market growth. They are less concerned with certain economic theories which today scarcely anybody dares defend, than with their actual operation in the functioning of the economy. They may not affirm such theories with words, but nonetheless support them with their deeds by showing no interest in more balanced levels of production, a better distribution of wealth, concern for the environment and the rights of future generations. Their behaviour shows that for them maximizing profits is enough. Yet by itself the market cannot guarantee integral human development and social inclusion[35].

The Pope further states:

> The principle of the maximization of profits, frequently isolated from other considerations, reflects a misunderstanding of the very concept of the economy. As long as production is increased, little concern is given to whether it is at the cost of future resources or the health of the environment; as long as the clearing of a forest increases production, no one calculates the losses entailed in the desertification of the land, the harm done to biodiversity or the increased pollution. In a word, businesses profit by calculating and paying only a fraction of the costs involved. Yet only when 'the economic and social costs of using up shared environmental resources are recognized with transparency and fully borne by those who incur them, not by other peoples or future generations' can those actions be considered ethical[36].

Focus on short-term profit is moving us in the wrong direction, amplifying the problems of inequality and of environmental degradation. We can therefore identify a first pillar of the economic foundations of integral ecology, which can be seen as a first response to the technocratic paradigm: the need to adopt a "systemic" vision capable of reconciling short-term needs with those that have a longer-range vision: the need to thwart the distorted "alliance between the economy and technology [that] ends up sidelining anything

---

[35] Pope Francis, *Laudato Si'*, n. 109.

[36] Pope Francis, *Laudato Si'*, n. 195, quoting Benedict XVI, *Encyclical letter Caritas in veritate on Integral Human Development in Charity and Truth*, 29 June 2009, n. 50.

unrelated to its immediate interests"[37], and risks to lead to counterproductive results even for the economic sphere in the long term.

From the point of view of political economy, this requires, first of all, that profit should take into consideration a longer-range vision. One way to achieve it is to avoid market failures due to environmental externalities, the cost of which should not be borne by the community. It means to internalize the environmental costs incurred for the production of the goods and services for which that same profit is obtained. This can allow for the final price to be an "effective economic value", by integrating and incorporating the negative environmental externalities that production activity is generating; the ecological constraint cannot be neglected in the models of economic growth, which risk to be blinded by an immediate result affected by the serious defect of short-term myopia. Off course, this is not a simple exercise. The carbon tax debate is a clear example of the difficulties of environmental cost's internalization in the market logic[38].

Internalizing all environmental costs incurred in the production process requires a careful assessment of its environmental and even social impacts; it calls for the "real economic value" of the cost of a product, by integrating in this "cost" also the "value" of environmental goods and services' "scarcity" the final products have affected. This economic value should indeed reflect the real "sacrifice" of resources that the production of such a good required the producing firm and the community to endure. This evaluation demands a careful interaction of economic actors with political and social ones.

Only in these terms we can understand how to respond effectively to the laws of scarcity and efficiency to which economics is called, aimed at improving the conditions of people and solving the problems of economic organization in the face of scarce and limited resources to respond to human needs and desires:

> Caring for ecosystems demands far-sightedness, since no one looking for quick and easy profit is truly interested in their preservation. But the cost of the damage caused by such selfish lack of concern is much greater than the economic benefits to be obtained. Where certain species are destroyed or seriously harmed, the values involved are incalculable. We can be silent witnesses to

---

[37] Pope Francis, *Laudato Si'*, n. 54.

[38] World Bank is working on it through its "Carbon Pricing Dashbord: https://carbonpricingdashboard.worldbak.org. See also Thomas Sterner, "Impuestos medioambientales", in *Eco Integración de América Latina. Ideas Inspirads por la Enciclica Laudato si'*, ed. BID, I&C, n. 41, Año 21, Marzo 2017, pp. 44-9.

terrible injustices if we think that we can obtain significant benefits by making the rest of humanity, present and future, pay the extremely high costs of environmental deterioration[39].

4. The call to seek other ways of understanding the economy and progress. Second "pillar": Acting responsibly to "cultivate and care for our common home"

However, the internalization of environmental externalities is necessary, but not by itself sufficient, especially in order to overcome the aforementioned identity:

economic growth = well-being.

Being concerned about the consequences of the adopted development model, the international community has been increasingly urged to bring out the limits of this economic model, which over time have become evident not only from the social and environmental point of view, but also from the economic one. This growing awareness of the limits of the predominant economic approach has gradually led to the creation of a new vision of development, which has emerged in the international community.

"Times of crisis are also times of radical questions in which new theoretical models or new paradigms flourish"[40]. Over time, development science has given ever more attention to the ethical-political issues of development[41], and many scholars have made increasingly clear that within the study of economics the analysis of the positive consequences of ethics on economics is taking on a strong connotation, based on the fact that the richness of ethical considerations that could be relevant to the economy is much greater than has been traditionally recognized: the restrictions imposed by welfarism and consequentialism, as well as the need for a narrow conception of rational decisions, have made impossible many significant considerations in economic evaluation[42].

---

[39] Pope Francis, *Laudato Si'*, n. 36.

[40] Louis Baeck, "Cinquant'anni di cultura dello sviluppo: teorie e pratiche in questione", in *Abitare la società globale: per una globalizzazione sostenibile*, ed. Roberto Papini, Antonio Pavan, and Stefano Zamagni. Rome: Edizioni Scientifiche Italiane, 1997, p. 94.

[41] See, among others, Denis Goulet, *Development Ethics: a Guide to Theory and Practice*. New York: Apex Press, 1995.

[42] Amartya Sen. *Etica ed economia*. Bari: Laterza, 1987, p. 89. The importance of taking into account the recognition of the moral foundation of economic activity is echoed in the words of Aristotle, according to whom ethics is the "science of acting", since it directs knowledge not to pure contem-

It became increasingly clearer that a "human" development needs to go "far beyond income and growth to cover the full flourishing of all human capabilities. It [has to emphasize] the importance of putting people – their needs, their aspirations, their choices – at the centre of the development effort. [In this perspective, h]uman development can be expressed as a process of enlarging people's choices"[43], a process that leads to the expansion of the possibilities for all individuals to express their potential and their creativity. It has been even more underlined that the driving force of economic growth must be the person, while the theories of human capital have received much attention from the economic world, placing an important emphasis on the role of education. On these basis, emphasis placed on GDP as the main indicator of development has been gradually integrated through other more qualitative aspects, such as: life expectancy, literacy, food security, capabilities[44].

Gradually a new concept of development emerges which is based on the assumption that at the center of the development process itself is the notion of "quality of life". If the object of development is the quality of life, the subject of development is not the individual tout court, but it becomes the person understood as the center of a series of social relations with other people: the person in the community and the community of the people are therefore the subjects of development[45].

---

plation, but to praxis and action (Aristotle, *Etica Nicomachea*, 1994). In this regard, the contribution of the economist Amartya Sen should be highlighted. He works out a critique of economic rationality intended as the maximization of personal interest in the face of the plurality of reasons guiding the choices of individuals. He considers utilitarian theory as a limited theory, since it takes into account only the results achieved and not the freedom to achieve them. In the latter case, we should adopt an assumption that is based on the differences of human beings and their capabilities to transform primary goods into the freedom to pursue their objectives. A social commitment to the individual freedom must be based on the goal of increasing the capabilities that different people possess. The choice between different social assets is influenced by their aptitude to promote human capabilities. Full consideration of individual freedom must go beyond the capacities related to private life and must pay attention to other aims of the person, such as certain social goals not directly connected with the life of the individual. A society implies interaction and our lives depend on each other: no man is an island in itself; the idea of mutual dependence can only lead to that of mutual responsibility. Amartya Sen, *La libertà individuale come impegno sociale.* Bari: Laterza, 1997, pp. 25 and 52.

[43] UNDP, *Human Development Report 1996*, p. 49.

[44] There are numerous synthetic index on development. In addition to HDI and HPI already mentioned, we can recall some which have been inspired by *Laudato Si'*, as the Fordham Francis Index developed by Fordham University in 2016 (https://www.fordham.edu/info/26662/fordham_francis_index), or the "Ecología Integral Humanista Index", developed by Universidad Católica de Costa Rica in 2017 (https://www.observatoriolaudatosi.cr/%C3%ADndice.html).

[45] Giovanni Sarpellon, "Sviluppo, disuguaglianza, esclusione sociale", in *Abitare la società glo-*

Hence, the importance of a new orientation for the economy arises, recalling the need, as indicated by *Laudato Si'* in the aforementioned paragraph 141, of an *economic ecology*, capable of inducing us to consider reality in a broader way. Indeed, there is a profound etymological proximity between the terms "ecology" and "economics", even if they are concepts and disciplines that since the Industrial Revolution of the 18th century have become dissociated from each other. Since then, on the one hand, there has been a prevalence of an economic perspective unable to adapt itself in the needed interaction with the various ecosystems. But, on the other hand, even earth sciences, climatology, astronomy ... , the so-called "hard sciences", have shown difficulty in relating with other sciences and, in particular, with the human and social sciences. Only recently, the potential impact of phenomena such as climate change on human societies has required greater interaction among hard sciences, social sciences, ethics and anthropology, making the interdisciplinary approach an essential component of any kind of analysis.

Ecology and economics have indeed the same etymological root: "ecology" comes from ancient Greek words "οικος" (house, home) and "λογος": study / reflection / knowledge of the house; "economy" from "οικος" and "νομος": rules / management of the house. In both cases, there is a common attention to the "home", the "common home". We cannot manage an asset properly if we do not know it adequately. From this perspective, economics becomes something that goes far beyond the reductionist and myopic approach of maximizing profit at any cost. In this direction, it seems indispensable, in elaborating an adequate development model, to acquire an integral and multidisciplinary vision of this concept, which must be observed not only through the lens of economic science, but also through a broader and more forward-looking interdisciplinary vision. It becomes necessary to go beyond the dominant "limited" paradigm based on economic growth, on faith in science and on the law of consumerism.

Even if "each age tends to have only a meagre awareness of its own limitations"[46], the new development paradigm must be based on the awareness that growth needs to develop a *sense of limitation*. Within the concept of limitation is inherent the concept of value, where the limit should be considered not as a

bale: per una globalizzazione sostenibile, ed. Roberto Papini, Antonio Pavan, and Stefano Zamagni. Rome: Edizioni Scientifiche Italiane, 1997, p. 111.

[46] Pope Francis, *Laudato Si'*, n. 105.

barrier but as a "threshold for", as a launching pad. It is from the emergence of limits that the actual possibility of a paradigm shift arises. In such a context, the limits should be seen not as a "dam", but as a "embankment" that directs and guides the fluctuating river of globalization[47]. From this perspective, "the establishment of a legal framework which can set clear boundaries and ensure the protection of ecosystems has become indispensable; otherwise, the new power structures based on the techno-economic paradigm may overwhelm not only our politics but also freedom and justice"[48].

In order to strengthen this new regulatory system, it is therefore essential to become aware of the ecological and social limits and to base it on them. In the first case, what must prevail is the *awareness of responsibility*[49]: if the range of action of our power expands, the range of action of our responsibility must also expand. As far as the social limit is concerned, it is the *awareness of solidarity* that must be spread on the basis of that interdependence at an international level that makes the "never in my backyard" motto loses its logical sense, since there is no longer an outsider on which to unload the own waste.

Awareness of responsibility and solidarity calls for new models of progress to arise and for a change in models of global development, entailing a "responsible reflection on 'the meaning of the economy and its goals with an eye to correcting its malfunctions and misapplications'"[50]. This renewed reflection on the meaning of economics cannot follow the way of an annihilation of the technological power required by the most radical "ecologists", who show an attitude of aesthetic despotism. Nor should it pursue the path of anesthetization of conscience claimed by the "efficiencyists", who instead require an anesthetic authoritarianism. It calls for a third way, the *development of consciousness*[51].

A new development model based on the "enhancement of limits" and on the "development of consciousness" should be focused on leveraging the mandate of cultivating and caring for our common home (tilling and keeping - cf. *Gen* 2,15). Two deeply interrelated attitudes, which represent a response to the need

[47] Giulio Querini and Paolo Conversi: "Globalizzazione e sviluppo umano sostenibile: la valorizzazione dei limiti ambientali", in *Sviluppo economico, scambi internazionali, integrazione regionale e ambiente*, ed. Ferruccio Marzano. Rome: eUroma, 2002, p. 283.

[48] Pope Francis, *Laudato Si'*, n. 53.

[49] Hans Jonas, *Il principio responsabilità: un'etica per la civiltà tecnologica.* Turin: Einaudi, 1979, 175-9, 282-7.

[50] Pope Francis, *Laudato Si'*, n. 194, quoting Benedict XVI, *Message for the 2010 World Day of Peace: If You Want to Cultivate Peace, Protect Creation*, n. 9.

[51] Giorgio Ruffolo, *Lo sviluppo dei limiti*, 1994.

to recover the sense of the economy and its purposes. Here we find a second pillar of the economic foundations of integral ecology: recovering the relationship of reciprocity between human beings and nature on the basis of the conscientious actions of "cultivating and caring for our common home".

From the point of view of political economy, it means making the most of scarce resources, without remaining passive with respect to natural rhythms, but at the same time, caring and preserving them, not exploiting them: "each community can take from the bounty of the earth whatever it needs for subsistence, but it also has the duty to protect the earth and to ensure its fruitfulness for coming generations"[52].

Economy must, indeed, "care for" our common home in the face of its enemies: the external desert, but also the internal ones provoked by a "tyrannical" anthropocentrism[53]. Finding means to combat inequalities and environmental degradation should be an important part of the way economics can help to solve the problems of the economic organization in the face of scarce and limited resources to respond to human needs and desires. On the other hand, to achieve this, the economy would have to advance in cultivating our common home, making it flourish in all its forms.

5. The conviction that everything in the world is connected. Third "pillar": Strengthening a cooperative alliance between humans and nature

Cultivating and caring for our common home requires not only rapprochement, but also a strong interaction between economics and other humanistic and scientific disciplines. From this perspective, it may be useful to consider the evolution of the relationship between humans and the natural environment, which has changed over time.

For a long time, this relationship has been characterized by a conditioning of nature towards the human person. In the aftermath of the Industrial Revolution, in the middle of the eighteenth century, humanity gradually begins to free itself from the conditioning imposed by nature on human existence and activity. From a conditioned human being, he initiates to transform himself into a dominating human being, setting his relationship with nature on exploitation, even at the limits of exhaustion; quantitative damage threatens to change into

[52] Pope Francis, *Laudato Si'*, n. 67.

[53] Pope Francis, *Laudato Si'*, n. 68.

qualitative damage through irreversible destruction of the ecosystem equilibrium, reaching also the so-called "planetary boundaries"[54]. This domination of nature takes place due to the development of an increasingly specialized technique; however, the main effect of this "new orientation" lies in the fact that it is the human being himself who remains subjugated by the new demands from the technological system. Where the "machine" is defined as a material system equipped with tools that takes the place of the human being as such (not just his single organs) in order to perform functions and operations for which he lacks the strength, the ability or even the memory and the intelligence. As such, the "machine" itself becomes an object in its own "right", acquires a certain autonomy and independence, as much as to impose its laws on subsequent inventions and to condition its own inventor. We then arrive at the paradoxical situation in which it is the created object itself that dominates the human person, creates ethical value and deprives him/her of a totally independent ability to choose and evaluate. Technology, and therefore the economy, which is its guide indicating the directions to take, acquire primacy over the other social and physical sciences[55]. The technical environment, the "technosphere", created by the human being, transforms the person and requires an adaptation to it not unlike that of the animal to its environment.

There are important consequences to take into consideration, which also have a strong impact on economic science itself. Faced with this tendency to self-regulation, the technological system also enters into interconnection with the rest of the polytechnic universe and with other systems. The existing interferences between the technical-economic system and the social, demographic, political or military systems are quite evident, entailing a continuous feedback between the different systems. A particular relationship is instead established between the technological system and the global ecosystem: the former tends to engulf the natural environment, without being stopped by external feedback from the latter. Just think of the phenomena of urbanization or work automation, which tend to cancel the relationship between human beings and nature.

---

[54] Johan Rockström, Will Steffen, et al, "Planetary Boundaries: Exploring the Safe Operating Space for Humanity", in *Ecology and Society* 14 (2), 2009", pp. 1-32. Based on scientific research, the Authors argue that since the Industrial Revolution human activity has gradually become the main driver of global environmental change: once human activity passes certain thresholds, the planetary boundaries, there is a risk of irreversible and abrupt environmental change.

[55] Peter Henrici, "Essere umano e natura nell'era tecnologica", in *Ecologia umana e valori etico-religiosi*, ed. Pedro Beltrão. Rome: Editrice PUG, 1985, pp. 82 and 86.

However, some limitations emerge in this connection: the technological system will always have to draw from the natural ecosystem, from which it will never be independent; in this sense the ability of the first to listen carefully to the signals of the second becomes essential for the very survival and sustainability of the "technosphere". This appears even more evident if we consider that the latter, by definition, is much simpler than the natural ecosystem, making it much more unstable and precarious than the ecosphere; it is sufficient that one of its components malfunctions to undermine the whole system, while the natural ecosystem, with its indefinite complexity, has almost infinite compensatory mechanisms and therefore reveals an admirable and almost unshakable stability[56]. The technological system therefore lacks these compensatory mechanisms and those instances of feedback from the outside, capable of correcting the dysfunctions to which it could encounter, thus denoting a dangerous *deafness*; it appears even more worrying if we emphasize once again the strong influence and conditioning capacity that the technological ecosystem has on the human being.

Based on this relationship of different dominance between "technosphere", human beings and the "ecosphere", it is appropriate to reflect on Georgescu-Roegen's economic interpretation of the law of entropy[57]. This law states that in a closed system the quality of matter-energy tends to deteriorate due to its use within the economic process. According to the laws of thermodynamics, the basic energy cannot be consumed or destroyed, but no type of work or action can be undertaken without causing a degradation of energy from more to less useful. The matter-energy which constitutes every natural resource used in the economic process is indeed qualitatively different from that which forms the waste: in the first case the resources are organized according to ordered schemes, so they will indicate the presence of a low entropy; after the transformation process, there will be a greater disorder that will lead to an increase in the degree of entropy.

It seems evident that the law of entropy has a profound relevance for economic theory, when the entropic degradation becomes the foundation of any scarcity. Once used in production, natural resources undergo an irreversible process of degradation: they become high entropy, i.e. waste. This theory therefore questions the "technosphere" itself, which is based on an economic

[56] Peter Henrici, "Essere umano e natura nell'era tecnologica", p. 89.
[57] Nicholas Georgescu-Roegen, *Energia e miti economici*. Turin: Borlinghieri, 1982.

process that gives rise to a linear flow defined as "entropic": it begins with the withdrawal of resources and, passing through producers and consumers, ends with the generation and introduction of waste into nature. The economic implications of this entropic flow are numerous, as it will be seen later in this work. Taking as a basis the law of entropy, it could be said that the miracle of modern industrial growth was founded on an incredibly fast withdrawal of natural capital, and therefore of low entropy resources, injecting high entropy into the environment, undermining the conditions of survival and reproduction of other living species and also, ultimately, of the human person.

It becomes clear that the post-industrial revolution period is characterized by enormous technological power, accompanied, however, by extreme human misery, both from an environmental and a social point of view. This "coexistence" is indicative of the failure of a "distorted" anthropocentric conception that led to the adoption of a short-sighted, "deaf" and self-destructive production system. It is due to the lack of respect for the earth's ecological cycle, which sees the last link in the chain in the organisms that convert waste into organic matter.

In this direction, following Commoner's intuition, we can draw a fundamental lesson from nature: nothing can survive on the planet if it does not become a cooperative part of a larger and more global whole[58]; human beings, on the other hand, have broken this "circle of life". They did so because they were driven not by biological necessities, but by a social organization designed to "conquer nature".

This is where a light of optimism shines. Taking into consideration the very nature of this socio-environmental crisis, we discover that it assumes a fundamentally moral character[59]. This crisis is not the product of human biological capabilities, which could hardly be changed in time. Rather it derives from the social actions of the human being, which, as such, are subject to much more rapid changes, especially in an era like ours, in which the opportunities offered by the sciences of communication, telematics and robotics are virtually infinite. This awareness would therefore bring out a third relationship between human beings and nature, characterized no longer by a conditioning of one towards the other, but by the responsibility of the first to the second, based on a "*cooperative alliance*".

---

[58] Barry Commoner, *Il cerchio da chiudere*. Milan: Garzanti, 1971, p. 355.

[59] Saint John Paul II, *Message for the 2010 World Day of Peace: Peace with God the Creator, Peace with All of Creation*, n. 6.

In this context, responsibility is formed by placing limitless development of education, training and intelligence at the foundation of the limits to growth, based on this cooperative alliance between human beings and nature. In present times, characterized by a rapid development of means of communication, there are plenty of possibilities for the flourishing and sharing of ingenuity and creativity aimed at strengthening this alliance. Although the dangers of *infodemia* and of the so-called *infoglut* are always lurking[60], it is important to underline that this alliance represents an immense source of enrichment with multiplier and catalytic effects on the way of working on economic well-being to be decoupled from the environmental degradation.

This means also working in strengthening the balance between humans and nature, between short term and long term, between speed and stability, between cultivating and caring for our common home.

Here we encounter a third pillar for the economic foundations of integral ecology: to place entropic degradation as the foundation of every scarcity and to leverage the responsibility of the human being towards the environment and our common home, based on an unlimited development of education and interdisciplinary and interdependent knowledge.

From the point of view of political economy, it calls for "decoupling the production of goods and services from unsustainable, wasteful or uncaring treatment of humans, nature and animals (do better), and decoupling the satisfaction of human needs from the imperative to deliver ever more economic output (do well)"[61].

> A new type of thinking is essential if mankind is to survive and move to higher levels. Education is the way societies have built to transmit the knowledge of the past, but also to prepare the young generation for the challenges of the future. Adults of tomorrow are today in schools, and they are the ones who will leave in the whole 21st century, bearing its evil but possibly being agents of change[62].

[60] *Infodemia* is the dissemination of an excessive amount of information from different sources, sometimes not scrutinized with accuracy, which makes it difficult to orient oneself on a given topic. *Infoglut* is about the great potential offered by current communications systems which can transform information into perverse instruments of human confusion. Paolo Conversi, "Il paradigma dell'ecologia integrale per fronteggiare le sfide del XXI secolo e la necessità di un approccio educativo" in *Educatio catholica: La cura della casa comune / Care for our common home*, Anno VI – 3/2020, Congregazione per l'Educazione Cattolica, p. 64.

[61] Enrst Ulrich von Weizsäcker and Anders Wijkmn, *Come On! Capitalism, Short-termism, Population and the Destruction of the Planet – A Report to the Club of Rome*. New York: Springer, 2018, p. 57.

[62] Pierre Léna, "Educating for Sustainable Development & Climate Change: A Challenge for Sci-

However, this process of change is and will be long and continual, since it will have to clash with all those radicalizations that are at the basis of the current predominant paradigm. Nevertheless, it must be noted that a paradigm shift from traditional environmental education to multidisciplinary integral ecology education can already be observed in education science contexts[63]. A paradigm shift, which is based on an integral ecology multi-perspective approach: an interdisciplinary perspective (the connection between different fields of science), a global perspective (the interaction between "technosphere" and "ecosphere", as well as within different regions and cultures), an operative perspective (effectiveness and impact analysis of implemented actions) and a future perspective (adoption of a long-term vision). It is a paradigm shift which involves the formal educational structures as well as the informal educational arrangements.

6. The intimate relationship between the poor and the fragility of the planet. Fourth "pillar": Implementing "eco-efficiency"

We are indeed faced with a problem that is difficult to solve. Once again, there is the need to bring economics closer to ethics, to complement a short term view with a longer range one, to overcome the defects of myopia and deafness. Once again, the need emerges to implement an open, constructive and participatory dialogue between those who belong to the economic world and the political-social actors. The main purpose of this dialogue is to identify the principles, activities and attitudes that can be at the basis of not only a "sustainable economy", but also a "human economy", truly attentive to its impact on the most fragile and vulnerable components of our society in the short term but also in the long term: the environment and the poorest.

Environmental degradation and human and ethical degradation are intimately connected[64]. Even if affluence is the biggest polluter (greenhouse gas emission, resource consumption and the use of land in a way that destroys soil quality are often companions of affluence), it is well known that the poor

---

ence Academics", in *Science and Sustainability, Impacts of Scientific Knowledge on Human Society and Its Environment*, ed. Werber Arber, Joachim von Braun and Marcelo Sánchez Sorondo. Rome: Libreria Editrice Vaticana, 2020, p. 69.

[63] Interdicasterial Working Group of the Holy See on Integral Ecology, *Journeying towards* ..., pp. 45-84.

[64] Pope Francis, *Laudato Si'*, n. 56.

are contextually the makers, but also the main victims of environmental degradation[65]. The fight against poverty and the promotion of environmentally sound integral human development are two sides of the same coin, given that the mechanisms that destroy the environment are the same ones that generate poverty, inequity and exclusion.

Following this direction, it is appropriate to consider the relationship between the environment and economy at the microeconomic level. At first glance, it is clear that activities aimed at improving and protecting the environment could have a profound negative impact on the economy: any use of pollution controlling mechanisms, or any "cleaner" waste disposal method, would have the direct consequence of increasing costs with a negative impact on profits. On the other hand, it should be considered that there could also be some substantial benefits, such as, for example, an increase in jobs in those industries that produce goods and services compatible with the protection of the environment. The final outcome, however, would be rather differentiated and would not be likely to lead to that necessary reciprocation of dialogue. This was in fact the path followed by the first environmental movements, which in this way made the environmental problem perceived as a threat to the business world. The immediate result was to highlight a conflict between two radically opposed conceptions.

However, such approach did not take into account what is the common root of economic and environmental problems, and has been characterized by the difficulty for businesses to perceive the internalization of environmental goals not as a threat or sunk cost, but as an investment for improving efficiency, as well as their consistency with the new conditions of demand and the market. Business can indeed play a fundamental role in the achievement of environmentally sound integral human development by promoting a "human and sustainable economy" which can satisfy both its own interests and those of the planet. The keystone of this change in attitude towards the environment lies in the concept of *competitive advantages*. Business can achieve competitive advantages vis-à-vis its competitors in different ways, resulting in benefits not only for the company itself, but also for the environment and the poor, who would benefit from better environmental sustainability, as well as a greater access to the economic system.

---

[65] See, among others, Andryan Setyadharma, et al, *The trade-off between Poverty and Environmental Degradation: Evidence from Indonesia*, 2020, IOP Conference Series: Earth and Environmental Science 448-012065.

First of all, a company's competitive advantage can refer to its ability to anticipate the future direction of global market demand. When the trend of this demand seems to be moving more and more in the direction of greater environmental awareness on the part not only of consumers, but also of the legislative actor, the companies that will be the first to satisfy these requests will be able to acquire an important market niche destined to expand in the future. The costs incurred to make their activities increasingly "sustainable" will be offset by the competitive advantages they will obtain in the future. Ecology has now abandoned its "niche" and is becoming an important demand factor. Several elements have begun to satisfy this new sensitivity such as, for example, the organic products offered in the food sector, the energy-saving products made by household appliance manufacturers, the "soft tourism" approach adopted by the tourism sector, and so on. This goes even without thinking of the increasing attention to the ESG (Environment, Social and Governance criteria) and to the "environmental diligence" in the evaluation of companies by the investment and financial markets.

Another way by which a business could achieve competitive advantages refers not so much to the demand side as to the supply side, taking advantage of the possibility of decreasing costs through greater "eco-efficiency"; the latter can come both from saving energy resources and from the dematerialization process. In both cases, there would be the important benefit of reducing the dependence on the purchase of energy and raw materials and reducing costs and thus improving profits. This benefit would be obtained not only at the microeconomic level, but also at the macroeconomic level for those countries that are poor in these resources and therefore would be less subject to the need to import them, with obvious advantages on the national current account balance.

With regard to energy efficiency, we are witnessing a rapid growth rate of those technologies that can achieve improvement in this area: several studies show many examples of a fivefold increase in resource productivity, even within the four most energy-intensive sectors of construction, industry, transport and agriculture[66]. We can also see the fast spread of electric vehicles in private and public transport, or the rising access to renewable energy, be it wind, water, geothermal, solar. The energy industry is indeed currently undergoing massive

---

[66] Enrst Ulrich von Weizsäcker et alt., *Come On!*, pp. 144-9. See also Enrst Ulrich von Weizsäcker et alt, *Factor Five: Transforming the Global Economy by 80% Improvements in Resource Productivity*. London: Earthscan, 2009.

changes and global renewables have been increasing dramatically[67]. Switching from non-renewable energy to renewable energy would not only have the fundamental advantage of shifting the economic process from a linear direction to a more "circular" one, but would favor the promotion of an economy compatible with climate stability, thanks to the gradual decrease in the use of fossil fuels, one of the main greenhouse gas emitters. Even if new technologies and newly discovered reserves of coal, gas and oil have extended the dominance of fossil fuels, what we need in order to accelerate the transition from a fossil fuel-driven economy to one being powered by renewable energy is stronger efforts in R&D. Push factors as the Paris Agreement on climate change, entered into force at the end of 2016[68], are fundamental for advancing R&D in this field and promoting the decrease in the cost of renewable energies.

These developments will also benefit small entrepreneurs, who would have the ability to generate electricity independently, without being tied to large companies which are capable of investing big amounts of capital for oil or coal-fired power plants. Solar or wind energy does not require significant financial resources. In this way it would emerge the possibility of creating local economic activities also in developing countries, which would not be so dependent on the import of energy or its purchase from foreign companies, with consequent benefits both for the environment and for local businesses. Here again, international instruments which promote technology development and transfers, such as the Paris Agreement[69], can be very fruitful.

Moreover, it must be considered that the use of non-renewable fossil fuels will entail a cost of extraction which is likely to increase over time, even if partially offset by technological advances. In turn, the use of renewable energies will never involve this risk of "scarcity" and "exhaustion". As such, solar or wind energy, which is abundant in almost all regions of the world,

---

[67] The biggest user of energy, China, is becoming the world renewable energy powerhouse and has recently announced that it will reach in 2060 climate neutrality. On 12 December 2020, at the High Level Virtual Climate Ambition Summit, many other countries has announced their commitment to reach carbon-neutrality in 2050. Australia, one of the countries most linked to the export of coal, has recently decided to build the largest solar power plant in the world, seeking to install panels capable of producing 10 gigawatts.

[68] UNFCCC, *Paris Agreement to the United Nations Framework Convention on Climate Change*, 13 December 2015. See also Paolo Conversi, "Laudato si' e la Cop21", in *Laudato si'. Linee di lettura interdisciplinari per la cura della casa comune*, ed. Humberto Miguel Yañez SJ. Rome: Gregorian & Biblical Press, 2017, pp. 270-2.

[69] UNFCCC, *Paris Agreement*, Art. 10.

can support the economy without requiring an increase in capital investment to cope with a growing scarcity of these resources, as they come from sources independent of the rate of consumption.

Furthermore, taking into consideration the dematerialization process, it should be noted that it is connected with the concept of intensity of use[70]. It is a strategy that aims to reduce the "material content" of the goods produced without diminishing their usefulness. Through this process the input productivity increases, since we can see the decreasing of their factor intensity (the amount of input used per unit of product). At the basis of the decline in the intensity of use, there are three factors: the changes in the final demand structure, the improvements in the efficiency of the use of materials, and the replacement with alternative materials and technologies. Even dematerialization, through technological progress and in the context of economic growth, can be identified as an essential requirement of a sustainable economy. Being "environmentally correct", in addition to having become synonymous with being "politically correct", is also becoming more and more synonymous with being "economically correct": the *polluter pays principle*, so feared by companies, is gradually turning into the *prevention pays principle*[71].

However, it must be emphasized that trying to achieve better "eco-efficiency" is not just a technology problem, it also requires profound changes in the assumptions and objectives that direct the activities of companies[72]. In this context, a fundamental incentive to this internalization of the environmental component by companies can and must come from government policy, which has the task of giving clear signals, with real political will, to follow the path that leads to a human and sustainable economy[73].

---

[70] The intensity of use is the ratio between the consumption of a material in physical terms and the gross domestic product deflated at constant prices. Riccardo Galli, "Innovazione tecnologica e smaterializzazione: situazione attuale e tendenze", in Legambiente, *Ecotrend: innovazione tecnologica, smaterializzazione, competitività: quali scenari per l'ecosviluppo*, 1995, p. 1.

[71] Enrico Sassoon and Cristina Rapisarda Sassoon, *Management dell'ambiente: la nuova priorità strategica degli anni novanta*. Rome: Il Sole 24 Ore Libri, 1993.

[72] Allan Miller, "Corporations as Agents for a Sustainable Society", in *Building Sustainable Societies: a Blueprint for a Post-Industrial World*, ed. Dennis C. Pirages. New York: M. E. Sharpe 1996, p. 247.

[73] The importance of the government's role in encouraging the business world to adopt sustainable and responsible management of its activities is strongly indicated in Agenda 21: "Governments should encourage the establishment and operations of sustainability managed enterprises. The mix would include regulatory measures, economic incentives and streamlining of administrative procedures to assure maximum efficiency", United Nations Conference on Environment and Development (UNCED), *Agenda 21,* 1992, para 30.19.

> Politics must not be subject to the economy, nor should the economy be subject to the dictates of an efficiency-driven paradigm of technocracy [...]. What is needed is a politics which is far-sighted and capable of a new, integral and interdisciplinary approach to handling the different aspects of the crisis. In other words, a healthy politics capable of reforming and coordinating institutions, promoting best practices and overcoming undue pressure and bureaucratic inertia. We cannot expect economics to do this, nor can we allow economics to take over the real power of the state[74].

Regulatory environmental norms can create significant incentives for companies to adopt "sustainable" behavior. The possibilities available to public decision-makers are many. Among them, quantitative and flexible market instruments aimed at internalizing environmental externalities certainly play a primary role[75]. In this context, it is also necessary to emphasize the important function of civil society activities that have played a strong role of influence for public decision makers at international, national or local level.

The signals coming from public policies must therefore be aimed at stimulating a modification of production technologies in such a way as to eliminate or strongly reduce their polluting capacities without diminishing or hindering their ability to produce the necessary goods and services.

Here is a fourth pillar of the economic foundations of integral ecology: promoting an eco-efficiency that allows the new "technosphere" to be: 1) solid from a production point of view with a strong and positive impact on the long-term economic performance of the production system; and 2) compatible with

---

[74] Pope Francis, *Fratelli tutti*, n. 177.

[75] Quantitative market instruments include environmental standards. Since they tend to stifle innovation, they can be useful when there are few major pollutants and when the technologies to control pollution or resources are uniform. In this way they would allow the application of impartial standards. However, they become necessary when it is essential to avoid exceeding a critical threshold and where there is a difficulty in the perception of environmental damage by individuals. Due to differences between States, these standards should be appropriate and specific to each country. Flexible market instruments, on the other hand, has as its objective not effectiveness, but efficiency; they refer to subsidies and corrective environmental taxes. The latter are criticized for being regressive, although they can be made fiscally neutral; they can give rise to what is called the "double dividend", since, in addition to generating revenue for the State, they help protect the environment. We are, here, in the presence of the so-called "Pigou approach"; this category also includes the principle according to which anyone who damages the environment must pay compensation, the *polluter pays principle*. William Baumol and Oates Wallace, *The Theory of Environmental Policy*. Cambridge: Cambridge University Press, 1988. See also, on the experience in Latin America on emissions control measures: Marcelo Caffera, "El control de la contaminación", in *Eco Integración de América Latina. Ideas Inspirads por la Enciclica Laudato si'*, ed. BID, I&C, n. 41, Año 21, Marzo 2017, pp. 68-79.

the ecosphere, to the benefit of the environment and the poorest, who would not only have a healthier and more accessible natural environment, but also greater economic opportunities.

7. The throwaway culture and the proposal of a new lifestyle. Fifth "pillar": Promoting a circular economy

As already mentioned, a sustainable economy requires the transition from "linear" production to "circular" production, following the example of the laws of nature.

The functioning of the natural ecosystem is indeed exemplary:

> Plants synthesize nutrients which feed herbivores; these in turn become food for carnivores, which produce significant quantities of organic waste which give rise to new generations of plants. But our industrial system, at the end of its cycle of production and consumption, has not developed the capacity to absorb and reuse waste and by-products. We have not yet managed to adopt a circular model of production capable of preserving resources for present and future generations, while limiting as much as possible the use of non-renewable resources, moderating their consumption, maximizing their efficient use, reusing and recycling them. A serious consideration of this issue would be one way of counteracting the throwaway culture which affects the entire planet, but it must be said that only limited progress has been made in this regard[76].

In the prevailing industrial ecosystem, at the end of the production cycle, there is not the ability to absorb the waste produced, which is "discharged" into the environment, creating serious pollution problems and sometimes irreversible damage to the natural ecosystem itself. These problems impact not only the environmental sphere, but also the social and economic ones, through the "slow-down" of human resources productivity, the emergence of symptoms of natural resources scarcity and increasingly significant problem of waste disposal.

Human beings have always produced "pollution" which, in situations of ecological balance, was "absorbed" by natural mechanisms and processes. Currently, however, the problem of pollution has assumed pathological dimensions, due, in particular, to the transformation of large cities, the consolidation of the industrialization process and the spread of the consumer society. There is no vast natural sink to absorb such an amount of waste: the greenhouse gases accumulation in the atmosphere is a clear example.

---

[76] Pope Francis, *Laudato Si'*, n. 22.

The pollution generated by humans can be produced through the production, disposal or abandonment of waste on the ground, through the emissions of gases into the atmosphere, or through the introduction of polluting material into surface, groundwater and waterways. In all these cases, in order to undertake the path that leads to a circular production system, we need to identify the various possibilities for reducing pollutants and waste. Various paths can be indicated: a change in household consumption through information and education activities or through tax incentives, a correction in the price system, adequate treatment of polluting products released into the environment, a change in production technologies.

Among the various possibilities aimed at facing the waste problem, two different types of initiatives emerge. The first focuses on the choice of the various methods of waste disposal and the most appropriate related techniques. Within this category, we can see three tendencies: the burial of waste through the use of landfills; the discharge of waste into water or the atmosphere or its abandonment on the ground; incineration, which has high investment and management costs, and in some cases can be responsible for the phenomenon of acid rain. In all three cases, a strong environmental and social impact is evident and can be exacerbated depending on the location of these disposal techniques and their structural characteristics.

In the second order of initiatives, the main aim is to reduce the size of the problem through the attempt to transform waste into "secondary" materials. It is a question of "enhancing" the waste resources that still hold a certain amount of "matter" and "energy". Waste could be recovered and become "useful" and suitable for satisfying human needs. In this perspective, there are some ways that can be put into practice: *saving*, aimed at reducing the amount of waste produced through better management, maintenance, production and consumption techniques; *reuse*, by using again the item, whether for its original purpose or to fulfil a different function; *recovery*, when the same waste can be used as an input in other production and consumption processes; *recycling*, when the production or consumption process is organized in such a way that the waste produced at one stage of the process is used as an input at a later stage.

In the past, waste mostly ended up in landfills and was then accumulated in the soil. The real revolution in waste management dates back to the last three decades. In many countries, the waste management system has become quite complex. The best waste is, of course, that which is not produced. Therefore, the "top" waste management strategy is prevention, i.e. reducing the "weight" of the waste produced and also its dangerousness. After prevention, there is preparation for the reuse of that product. Thirdly, we have recycling and recovery of materials collected by separate means, thanks to which the waste becomes a product again, but in a different form from the original one. After it, we find energy recovery, which is the possibility of obtaining energy or heat from waste through conversion process such as waste to energy plants. The last level is the disposal of the waste from which nothing useful can be obtained in controlled landfills or through low-efficiency incineration. Analyzing these processes, economic advantages that can come from correct waste management immediately appear, by considering the cost savings deriving from both a lower purchase rate of input and from a lower outlay to cover costs of waste disposal. Furthermore, the benefits for the natural environment are evident: recycling and reuse of materials save a lot of energy and limit pollution.

The search for production systems and goods designed to allow waste saving, reusing, recycling and the recovery of materials is an extremely valuable activity for the business world. For some time now, there have been companies that deal with waste treatment in order to obtain "secondary materials"[77], or to develop the new "ecodesign" sector aimed at implementing the "end of waste" perspective. Given that demand for basic materials is expected to increase rapidly, a revolution in the way we use basic materials and their substitutes is urgently needed.

It advocates a radical shift in mentality as well as production and consumption systems[78]. Waste re-evaluation for production is conditioned both by technologies and market structures. Often, in the past, companies have not found it convenient to invest resources in research activities for "circular" technologies: they had no return as pollution did not represent a cost for them, since it could be discharged to the external environment. In this way, the guiding role of the public decision makers is significant, as it could be in stimulating the expansion of public or private markets for secondary materials[79].

---

[77] John Young and Aaron Sachs, *The Next Efficiency Revolution: creating a sustainable materials economy*, Worldwatch Paper, n° 121, September 1994.

[78] Enrst Ulrich von Weizsäcker et alt., *Come On!*, pp. 140-4.

[79] Cynthia Pollok, *"Mining Urban Wastes: the potential for recycling"*, Worldwatch Paper, n° 76, April 1987. See also European Commission, *A New Circular Economy Action Plan for a Cleaner and More Competitive Europe*, COM (2020) 98 Final, 11 March 2020.

An interesting example is the commitment of the European Union to ban by 2021 some common single-use plastic items, such as cutlery and straw and its announcement of a collection-rate target for plastic bottles of 90% by 2029. Other countries, like China, Indonesia and Vietnam have closed their doors to importing plastic waste.

Undoubtedly, disrupting our current linear models of both production and consumption is a major task. To achieve this end, a strong intertwining between business, consumer and government is needed. They have to work together to develop resource strategies to avoid both resource constraints and increasingly serious problems of waste and pollution. The policies that the government can adopt in favor of a more correct behavior towards waste "goods" are many. First of all, cultural information and education policies are essential, the effects of which can be verified in the long term. Moreover, fiscal, administrative and economic instruments can be put in place[80]. It should also be remembered that one of the limits that hinder companies from undertaking an effective "circular" path lies in the fact that the main objective of each company, the maximization of its profit, is not achieved only through the minimization of costs, but also through "image and marketing costs" of the product which often means an elegant and more "polluting packaging"[81]. Once again, the importance of in-depth, interdisciplinary and "contextualized" environmental education re-emerges.

The educational aspect becomes even more important in the face of the fact that producing means combining resources, used on the basis of scientific and technical knowledge, to reach the levels required by demand, increasingly induced through refined and personalized sales strategies. In the actual economy mechanisms, the importance of the consumers' social task as subjects is recognized: they are increasingly called to become aware of their role, since their requests are, through direct and indirect information flows, received by entrepreneurs in order to decide its production capacity[82].

However, some dangers that could arise from excessive public intrusiveness in the waste sector should not be overlooked. On the one hand, it could stimulate the establishment of illegal forms of waste disposal with consequences

---

[80] Emilio Gerelli, *Le politiche ambientali di nuova generazione: il caso dei rifiuti solidi*, Camera di commercio, industria, artigianato e agricoltura di Pavia, 1994.

[81] John Young, *"Discarding the Throwaway Society"*, *Worldwatch Paper*, n° 101, January 1991.

[82] Giuseppe Notarstefano, "Il paradigma della custodia", in *Abiterai la terra. Commento all'enciclica Laudato si'*, ed. AAVV, Rome: AVE, 2005, pp. 116-7.

diametrically opposite to those desired[83]. On the other hand, too rigid forms of environmental regulation could give rise to international protectionist barriers, leading to the phenomenon of "eco-protectionism" much discussed above all by developing countries. In this context, also the different packaging methods and the various "eco-label" schemes[84] can cause non-tariff barriers, despite having obvious advantages in providing adequate information to consumers and encouraging them to be more environmentally conscientious.

It is clear how uneconomical the "throwaway culture" is. As mentioned at the beginning of this work, economics is the science that studies the best ways to use scarce resources for the production of goods and services that meet human needs and desires. It goes without saying that the production of "waste" indicates an inefficient use of the resources used. The current linear production model, based on the approach "take-make-waste"[85], with its consequences in terms of pollution, resource depletion, degradation of ecosystems and climate change, is in collision with nature and is not sustainable. It fails to give a careful evaluation of the entire "lifecycle" of a product and especially its "end of life". Moving to a circular production process benefits present generations as well as future ones. This transition can bring important social and economic benefits, turning challenges into opportunities, decoupling economic growth from resource usage and recoupling economic growth with societal progress[86].

Here is a fifth pillar of the economic foundations of integral ecology: the promotion of a circular economy, which also represents an important response to counter the "problems […] closely linked to a throwaway culture which affects the excluded just as it quickly reduces things to rubbish"[87].

The "throwaway culture" involves also the serious problem of food loss and waste, which often occurs due to lack of appropriate infrastructure, regula-

---

[83] Legambiente, *Dalle ecomafie all'ecosviluppo: l'impegno di Legambiente contro la criminalità organizzata*, 1996.

[84] The eco-label is a market tool that tends to reduce the environmental impact of products through a certification process which verifies the compliance of the product with the ecological requirements set. These requirements may vary from compliance with a single criterion such as recyclability to others based on the evaluation of the entire life cycle of the product.

[85] William McDonough and Michael Braungart, *Cradle to Cradle: Remaking the Way We Make Things*. Emmaus: Rodale Press, 2002.

[86] Peter Lacy, Jessica Long, and Wesley Spindler, *The Circular Economy Handbook. Realizing the Circular Advantage*. Palgrave Macmillan, 2020, pp. 19-30.

[87] Pope Francis, *Laudato Si'*, n. 22.

tions, profit-seeking, and negligence, as well as improper post-harvest handling and unfavorable climate conditions. It represents a waste of valuable resources used in production such as land, water, energy, agricultural chemicals and labour, which puts significant strains on the environment, economy and society[88]. Tackling food loss and waste is an important strategy to ensure food security and to combat climate change, and requires attention along with all aspects of wasteful production, transportation, packaging and energy usage along food supply chain. These are issues that a circular economy is trying to address systemically[89].

Several times, this work has made reference to the Industrial Revolution as the starting point of a distorted relationship between human beings and the environment, between economics and ecology. "Although the post-industrial period may well be remembered as one of the most irresponsible in history, nonetheless there is reason to hope that humanity at the dawn of the twenty-first century will be remembered for having generously shouldered its grave responsibilities"[90]. What many call the Fourth Industrial Revolution or sustainability revolution[91], based also in "disruptive technologies"[92] and founded on advances in artificial intelligence, robotics, biology and the energy transition can make an important contribution in making this wish of Pope Francis a reality.

8. The work dilemma and a just transition as the sixth pillar

In this perspective, in order to promote a real "human" and sustainable economy, it is important that the Fourth Industrial Revolution and the "energy transition" are implemented in parallel with the so-called "just transition", which requires particular attention to triggering a positive impact on the world of work. This can be identified as the sixth pillar of the economic foundations of integral ecology. In the "throwaway culture" "what is thrown away is not

[88] Food waste is responsible of 8% of the greenhouse gas emissions; if food waste were compared to a country, it would be the third most polluting, after China and the USA. UNDP, *Human Development Report 2020*, p. 86.

[89] Joachim von Braun, Marcelo Sánchez Sorondo and Roy Steiner, *Reduction of Food Loss and Waste*. Rome: Libreria Editrice Vaticana, 2020, pp. 12-9.

[90] Pope Francis, *Laudato Si'*, n. 165.

[91] Lacy, Long, and Spindler, *The Circular Economy Handbook*, pp. 43-52.

[92] Disruptive technologies are innovation that replaces and destroys the existing technology. Enrst Ulrich von Weizsäcker and Wijkmn, *Come On!*, pp. 44-8.

only food and dispensable objects, but often human beings themselves. Discarding others can take a variety of forms, such as an obsession with reducing labour costs whereby we lose the meaning of the "value" of the right and duty to work"[93].

A circular economy transition cannot ignore the importance of addressing the work dilemma:

> Any approach to an integral ecology, which by definition does not exclude human beings, needs to take account of the value of labour [...] If we reflect on the proper relationship between human beings and the world around us, we see the need for a correct understanding of work; if we talk about the relationship between human beings and things, the question arises as to the meaning and purpose of all human activity [...] Underlying every form of work is a concept of the relationship which we can and must have with what is other than ourselves[94].

This takes ever more significance in the face of rapid developments in robotics and various forms of automation. The latter are carrying and will carry even more in the future important advances and benefits for humanity, but they are also leading and will lead to the worrying phenomenon of technology-driven unemployment, to which we must respond with creativity and ingenuity.

Work is a necessity, it is part of the meaning of life, a way of maturation, human development and personal fulfillment. The value of work must therefore go beyond the mere economic dimension. We must not limit ourselves to considering that human resources are an important factor for productivity, quality and innovation capacity and that they represent just a component of the overall costs which are determined by wages, by the regulations governing the labor market and by social security schemes[95]. Work contains also deep intrinsic values, through which the essence of the human being is expressed. In a certain sense, "we are what we produce". "Through work man *not only transforms nature,* adapting it to his own needs, but he also *achieves fulfilment* as a human being and indeed, in a

[93] His Eminence Card. Pietro Parolin, Video-Message to the International Conference of the *Foundation Centesimus Annus ProPontefice* "The Milestones of an Integral Ecology for a Human Economy", 23 October 2020.

[94] Pope Francis, *Laudato Si'*, nn. 124 and 125.

[95] Carlo Azeglio Ciampi, *Sfida alla disoccupazione*. Bari: Laterza, 1996, p. 36.

sense, becomes 'more a human being'"[96]. From a perspective of integral ecology and integral human development, it is necessary to consider that human capital is the factor on which not only our wealth depends, but also and above all that of future generations. "To stop investing in people, in order to gain greater short-term financial gain, is bad business for society"[97]. In this perspective, it is essential to promote an economy that favors productive diversification and entrepreneurial creativity. "Business is a noble vocation, directed to producing wealth and improving our world. It can be a fruitful source of prosperity for the areas in which it operates, especially if it sees the creation of jobs as an essential part of its service to the common good"[98].

The concept of unemployment does not exist in ecosystems. In dealing with unemployment, it is always necessary to remember that, since in a well-structured economic system the participation of everyone in the production process increases the wealth of all through exchange and cooperation for the common good, involuntary unemployment is a violation to the fundamental rights of participation in social life, but also a manifestation of the economic inefficiency of our system[99]. Furthermore, it must be borne in mind that growing unemployment tends to accentuate political pressures in favor of forms of protection, weakening the ability of individual economies to seize the new opportunities that arise and to make the necessary adjustments, with strong negative effects on competitiveness. In this sense, it can be said that the fight against unemployment is not just a social issue, it is an economic imperative[100].

In general, on the aggregate level, the progressive increase in unemployment observed in recent years depends on many factors: the weakening of the development process, with strong repercussions on the demand for labor; the increase in the supply of work, attributable to demographic, economic and cultural factors, the restructuring process in the industry, which often originates from technological changes. Taking this last phenomenon into con-

---

[96] Saint John Paul II, *Encyclical letter Laborem Exercens on Human Work*, 14 September 1981, n. 9.

[97] Pope Francis, *Laudato Si'*, n. 128.

[98] Pope Francis, *Laudato Si'*, n. 129.

[99] Antonio Fazio, *Razionalità economica e solidarietà*. Bari: Laterza, 1996, p. 11. Involuntary unemployment is indeed an inability to make available to the community the work resource accessible to each adult; it can be seen as the deprivation of initial endowments for some members of society, p. 27.

[100] Carlo Azeglio Ciampi, *Sfida alla disoccupazione*, p. 91.

sideration, it must be remembered that there are various incentives that can promote the implementation of technological innovations. Among these, the two most relevant are the expansion of demand in the market, in which each company operates, and the increase in the unit cost of labor, driven by the relationship between wages and productivity. The variation of this cost must be compared, on the one hand, with the variation in the price of finished products and therefore its relative increase can involve a more rational use of workers for the same production, and, on the other hand, with the variation in the prices of machines. Also here, its relative increase can cause a lower use of labor per unit produced and a greater use of machines[101].

When these technological changes lead to labor savings in the production of goods already known, or allow the production of new goods that save labor in other branches, the so-called technological unemployment appears, which involves a phenomenon of technological displacement[102]: one of the concerns relating with disruptive digital innovation is associated with the elimination of jobs.

A "human and sustainable economy" should pay attention to this negative impact of technological displacement and deploy every effort to guarantee a fair and efficient use and evaluation of labor. Not only through new technologies aimed at the production of radically new goods and services, which would give rise to new jobs, as indeed has been observed for the "waste" industry; but also through the strengthening of fields where it is possible to see new development of jobs such as the ones related to education and human and social care. The massive transition to sustainability should take into account the importance to guarantee the just transition, to avoid jobs displacement and to evaluate the role of human capital not only from a strict economic point of view, but also from that of an "economic ecology".

---

[101] Luigi Frey, *La disoccupazione in Italia: il punto di vista degli economisti*. Milan: Franco Angeli, 1992; Paolo Sylos Labini, *Progresso tecnico e sviluppo ciclico*, Quaderni di economia del lavoro, n. 36, Milan: Franco Angeli, 1993.

[102] In this sense we can speak of the "paradox of paradise", where we consider that we are in a place where technology has advanced in such a way as to make it possible to produce material goods, practically at no cost; the obstacle that would be created is that in such a situation no one should be paid, thus going against the principles of the economy, since this paradise would cost nothing and there would be no money to distribute with work; the result would be that, paradoxically, our productive paradise could be seen as a social hell. Orio Giarini and Patrick Liedtke, *"The Employment Dilemma and the Future of work"*, Report to the Club of Rome, 1996, p. 90.

9. Towards a new Christian humanism enlightened by integral ecology

Looking for the economic foundations of integral ecology helps us to understand that the efficient use of scarce resources to produce valuable goods and services aimed at improving the conditions of people and their communities calls us to seek a courageous cultural revolution that brings about a new synthesis[103].

We need a "conversion" of the economic system, as the prevailing one, devoid of ethical concerns, does not lead to a more just social order but leads to a "throwaway culture", which is deeply uneconomic. This "conversion" recalls the need to promote an "*economic ecology*" capable of appealing to a broader vision of reality, a vision of "integral ecology" in which also the economic science plays a significant role with its interaction within complex systems. An economic science where the following six pillars of the economic "foundations" of integral ecology are recovered and well implemented:

1. Adopting a "systemic" vision capable of reconciling short-term needs with those that have a longer-range vision. It requires, first of all, a full *internalization of the environmental externalities* as a response to market failures that impede to the price to be a real indicator of scarcity.
2. Recovering the relationship of reciprocity between human beings and nature founded on the basis of the *responsible actions of cultivating and caring for our common home*. It requires to take into consideration the enhancement of limits as well as the development of consciousness based on the awareness of responsibility and solidarity, in order to regain the real sense and the meaning of economy and its purposes.
3. Placing entropic degradation as the foundation of every scarcity and focusing on an unlimited development of education, interdisciplinary and interdependent knowledge. It requires to *overcome a myopic, "deaf" and self-destructive production system*, to make flourishing and sharing ingenuity and creativity aimed at strengthening the alliance between humans and nature, to decouple economic from environmental degradation.
4. Leveraging *competitive advantages and eco-efficiency* that allow the new "technosphere" to be: 1) solid from a production point of view with a strong and positive impact on the long-term economic performance

---

[103] Pope Francis, *Laudato Si'*, nn. 113 and 114.

of the production system, and 2) compatible with the ecosphere, to the benefit of the environment and the poorest, who would not only have a healthier and more accessible natural environment, but also greater economic opportunities.

5. Promoting a *circular economy* as an effective way to counter a linear economy based on the uneconomical "throwaway culture".
6. Achieving a *just transition* as a central part of human and sustainable economy. It requires to counterbalance the technological unemployment and the economic inefficiency of an unfair human resource engagement.

These economic foundations of integral ecology call us to go much deeper in reconsidering economic structures and to convert our economic system in a direction more respectful of creation and integral human development of present and future generations, by adopting a new development paradigm[104], capable of "strengthening" the alliance between human beings and the environment. A virtuous economy, which revolves around the common good and pays particular attention to ethics and respect for the environment, is the main weapon to recover from the current crisis. We are facing a challenge of civilization and a change of perspective, which must place human dignity at the center of all our actions[105].

> Many things have to change course, but it is we human beings above all who need to change. We lack an awareness of our common origin, of our mutual belonging, and of a future to be shared with everyone. This basic awareness would enable the development of new convictions, attitudes and forms of life. A great cultural, spiritual and educational challenge stands before us, and it will demand that we set out on the long path of renewal[106].

---

[104] On 4 October 2021, many religious leaders gathered in the Vatican to adopt a *Joint Appeal* for the Twenty-sixth session of the United Nations Framework Convention on Climate Change (COP-26), where they underlined the "need to *change the narrative of development* and to adopt a new kind of economics: one that places human dignity at its center and that is inclusive; one that is ecologically friendly, caring for the environment, and not exploiting it; one based not on endless growth and proliferating desires, but on supporting life; one that promotes the virtue of sufficiency and condemns the wickedness of excess; one that is not only technologically driven, but is moral and ethical". Faith and Science: Toward COP26, *Joint Appeal*, 4 October 2021.

[105] Paolo Conversi, "The Church, Integral Ecology and Climate: the Challenge of a Change of Course", in *W Ṭrosce o Wspólny Dom Chrześcijanin Na Drogach Ekologii*, ed. Witold Kania. Katowice: Główny Instytut Górnictwa, 2020, pp. 25-7.

[106] Pope Francis, *Laudato Si'*, n. 202.

S. Paul VI's words seems to resound: we need a "new humanism which will enable our contemporaries [...] to find themselves"[107], and which favors the dissemination and implementation of a "culture of care" which permeates all of society[108]. Care for the environment and care for our neighbors, near or far in space and time. We need a new *humanism* capable of bringing together the different fields of knowledge, including economics, in the service of a more integral and integrating vision. It is interesting here to recall some of the main innovations developed during the period of *Humanism* in the XV century[109]. In that period we have witnessed: 1) the spreading of a cultural renewal based on the centrality of man; 2) the rediscovering of the classics and of the use of "reason" which makes the human person a more responsible and consciousness protagonist of the world around her/him; 3) the promotion of a new integral education and the flourishing of arts adopting an interdisciplinary approach; 4) the revaluation of nature through enlarged human capabilities to transform it, while respecting it; 5) the development of work as a way to fulfill the human person; 6) and the broadening of a new communication, through the invention of printing, facilitating the circulation of ideas.

Historia magistra vitae (history is a teacher of life), Cicero reminds us in *De oratore* (II, 9, 36). It is astounding to see how many points in common there are between that period, which gave rise to the Renaissance, and our age: 1) acting responsibly around the centrality of the human person, 2) rediscovering the contributions of different cultures, starting from their main classical thinkers, for a culture of care able to mould the concept of integral ecology; 3) deepening the benefits of the complex exercise of a trans-disciplinary approach[110]; 4) fostering the flourishing and the sharing of ingenuity and cre-

---

[107] Saint Paul VI, *Encyclical Letter Populorum progressio on the Developmen of Peoples*, 26 March 1967, n. 20.

[108] Pope Francis, *Laudato Si'*, n. 231.

[109] See, among others, Eugenio Garin, *La cultura del Rinascimento*. Bari: Laterza, 2010.

[110] *Multi-disciplinary* comes from the Ancient Latin *multus*, "very, many": it is intended to examine a problem by adding different perspectives and disciplines; in this process, a basic discipline may involve other disciplines to solve a problem; participants exchange knowledge and compare results, but fail to integrate them; the disciplines retain their distinctive character and the results remain rooted in the framework of the basic discipline. *Inter-disciplinary* comes from the Ancient Latin *inter*, "between, in the middle": it is aimed to integrate knowledge and methods of different disciplines using a synthesis of approaches; an interdisciplinary approach concerns collaborations between conflicting academic disciplines or research methods for new applications, new analyses or the creation of completely new disciplines. *Trans-disciplinary* comes from the Ancient Latin *trans*, "through, beyond": a new discipline is created that transcends the boundaries of the disciplinary

ativity aimed at strengthening the alliance between humans and nature; 5) recovering the sense and meaning of human work; 6) developing new and very effective means of communication, increasing interconnectedness through, in particular, the accelerate digitalization of the world and the spread of Internet.

These issues are also at the basis of the economic foundations of integral ecology, which invite us to face with courage and perseverance the stimulating challenge to embark on a new Renaissance for humanity.

---

perspective; a trans-disciplinary approach combines the interdisciplinary approach with a participatory approach; research paradigms involve also non-academic as participants in the process of achieving a common goal. *Integral ecology adopts essentially a trans-disciplinary approach.*

THE FUTURE OF WORK AFTER *LAUDATO SI'*

Paolo Foglizzo and Pierre Martinot-Lagarde, SJ*

The Encyclical *Laudato Si'*[1] (henceforth LS) dedicates a section consisting of nos. 124-129 to work, under the title "The need to protect employment"[2]. Even if it has not received the media attention of other parts of the text, the first lines are unequivocal in affirming the centrality of work within the paradigm of integral ecology. They start by stating: "Any approach to an integral ecology, which by definition does not exclude human beings, needs to take account of the value of labour, as Saint John Paul II wisely noted in his Encyclical *Laborem exercens*" (LS, 124). This section is therefore not a corollary, but a central element in the vision proposed by the Encyclical. This is no surprise, if we think of the constant attention that Pope Francis pays to the world of work and its problems, as we see in numerous speeches and documents[3]. For example, an encounter with the world of work is scheduled in most of his apostolic journeys. Among the most significant and authoritative positions

* The views and opinions expressed in this article are those of the authors and do not necessarily reflect the official policy or position of the International Labour Organisation.

[1] Pope Francis, Encyclical Letter *Laudato Si' on Care for Our Common Home*, 24 May 2015. References will appear in the text in round brackets. For all papal and Church documents, quotes and references point to the English text available in the Vatican website, www.vatican.va.

[2] When checked against the Spanish original "*Necesidad de preservar el trabajo*" and considering the actual content of the section, the English translation of this title appears inaccurate: "The need to protect labour" or "The need to protect work" probably would be better options.

[3] For a detailed analysis of Pope Francis's teachings about work and labour during the first five years of his pontificate, see Giacomo Costa and Paolo Foglizzo, eds., *Il lavoro è dignità. Le parole di Papa Francesco*. Rome: Ediesse, 2018. Several examples prove that the Pontiff's attention to the world of work has not diminished later.

DOI 10.32060/C.S.9.93-111

taken after the publication of the LS, we recall the *Message to Participants in the 108th Session of the International Labour Conference*[4], the Encyclical *Fratelli tutti*[5] (henceforth FT) and the *Video Message on the occasion of the 109th Session of the International Labour Conference*[6].

In FT, consideration of work and labour issues can be found throughout the whole text, with a significant concentration in chapter 3 (nn. 110, 116, 123, 127), which outlines the profile of an authentically inclusive and solidarity-based society, and in chapter 5 (nn. 162, 168-169), dedicated to outlining "A better kind of politics". In both cases, work appears to be closely connected to the construction of that common good from which, as we read in n. 156 of LS, "an integral ecology is inseparable".

1. An international project of Catholic-inspired organizations

This awareness of the centrality of work is the basis of "The Future of Work – Labour after *Laudato Si'*" Project[7], promoted by an international network of Catholic-inspired organizations, institutions, universities and research centres in view of the celebrations of the centenary of the foundation of the International Labour Organization (ILO) in 2019[8]. The Project articulates research, action and formation, aiming at three outcomes:

1. Outcome 1: to build a global ecclesial network involving other faith-based actors as well as social partners to share knowledge and experiences on the future of work.
2. Outcome 2: to conduct research and develop critical skills in order to contribute to the debate on the future of work, with LS as a source of inspiration.

---

[4] Pope Francis, *Message to Participants in the 108th Session of the International Labour Conference*, 10 June 2019.

[5] Pope Francis, Encyclical Letter *Fratelli tutti on Fraternity and Social Friendship*, 3 October 2020. References will appear in the text in round brackets.

[6] Pope Francis, *Video Message on the occasion of the 109th Session of the International Labour Conference*, 17 June 2021.

[7] For a detailed presentation of the initiative and the activities undertaken, see the Project's website https://futureofwork-labourafterlaudatosi.net.

[8] In the *Message to Participants in the 108th Session of the International Labour Conference* the Project is explicitly mentioned (see footnote 12) among the activities through which the Catholic Church takes part in the ILO's "attempt to redefine work in light of the new socio-economic and political realities, particularly those that are affecting the poor".

3. Outcome 3: to develop the capacity of the network partners to raise awareness, develop proposals and implement public advocacy on the issue of the future of work within the LS framework.

To reach these goals, during the 2017-2020 period, different kinds of activities were organized, including research investigations, knowledge-exchange seminars and conferences, capacity-building programs, advocacy and formation initiatives. Specifically, the research was organized along seven tracks, allowing partners to focus on a diversity of phenomena and choose the most appropriate methodology for each. The seven research tracks are:

1. Work, ecology and the environmental crisis, coordinated by CERAS (*Centre de recherche et d'action sociales*, France). It focuses on the issue of sustainability and how to combine the promotion of decent work with the much-needed just ecological transition.
2. Work, social justice and peace, coordinated by the *Universidad Iberoamericana de Puebla* (Mexico). It analyses the crucial role of employment and, in particular, the availability of decent work, in peace-building processes, especially in contexts marked by high rates of social violence such as Latin America.
3. Labour, demography and migration, coordinated by the ICMC (International Catholic Migration Commission). First, it investigates how employment (and lack of it) is one of the main drivers of migration. Second, it illustrates, with a diversity of testimonies, the often dramatic experiences of exploitation that migrants and refugees face in the world of work, especially in the informal economy.
4. Artificial intelligence, robotics and the future of decent work, the result of a partnership between the Lupina Foundation (Toronto, Canada), the Munk School of Global Affairs (University of Toronto, Canada) and the *Collegio Carlo Alberto* (University of Turin, Italy). It focuses on the impact of technological innovation processes (automation, artificial intelligence, robotics) on industries such as the automotive supply chain, the mining industry and personal services mediated by electronic platforms.
5. The future of enterprise and entrepreneurship after *Laudato Si'*, led by UNIAPAC (International Christian Union of Business Executives). It addresses the pragmatic unfolding of the many inspirations of LS, starting from the statement that entrepreneurial activity "is a noble vocation oriented to produce wealth and improve the world for all" (LS, n. 129).

6. Promotion of employment and social innovation in the context of *Laudato Si'*, coordinated by *Aggiornamenti Sociali* (Jesuit monthly journal based in Milan, Italy) and CeSPI (*Centro Studi di Politica Internazionale*, Rome, Italy). It aims at assessing if and how faith-based and value-based organizations successfully translate their inspiration into practice when they undertake initiatives to promote employment, with special attention to the integration of economic, social and environmental concerns.
7. Humanity at work, coordinated by *Observatoire de la Finance* (Geneva, Switzerland). It explores how global statistics capture and represent the enormous variety of situations in which work takes place, with particular attention to the many diverse forms of "invisible" work, from the informal and underground economy to "gratuitous" work (within the family, voluntary work, etc.).

Each team was fully in charge of its track, but they came together for interaction sessions and maintained constant dialogue, involving partners engaged in non-research activities. The fruit of this intense collective work was distilled into the report *Care is Work, Work is Care*[9], published in December 2020, whose content inspires these pages.

2. The challenges confronting today's world of work

In preparation for its centenary, the ILO had launched the "The Future of Work" initiative with the aim of highlighting the most important challenges for today's world of work, and of providing a framework within which to reformulate and relaunch its own mission. Such initiative attracted interest from other international organizations and actors in the world of work and in civil society, including numerous Catholic-inspired institutions and organizations. A document published by the COMECE Social Affairs Commission in 2018, under the title *Shaping the Future of Work*, explains the reasons of the Church's involvement:

> In the midst of industrialisation in the late 19th century, Pope Leo XIII shed light on the consequences of the new technologies and mass production on the human being. Today, the Church again feels committed to its mission

[9] *Care is Work, Work is Care*, Report of "The Future of Work – Labour after *Laudato Si'*" Project, ICMC, Geneva 2020, available at https://futureofwork-labourafterlaudatosi.net/2020/12/09/care-is-work-work-is-care-9-december-2020-global/ (consulted on 03/01/2022).

> to read the signs of the time – the new developments of digitalisation, artificial intelligence and ecological transition – and to call for the dignity of work for all[10].

In January 2019 the Commission set up by the ILO with the task of carrying out "The Future of Work" initiative published its final report, *Work for a Brighter Future*. It opens with these insightful words:

> New forces are transforming the world of work. The changes they entail call for decisive action. We face limitless opportunities to improve workers' quality of life, increase choice, close the gender gap, reverse the negative effects of global inequalities, and much more. However, none of this will happen automatically. If we do not act decisively we will move towards a world of wider inequalities and uncertainties[11].

The report spells out the major challenges confronting today's world of work:

- climate change and the consequent need to move swiftly towards a more sustainable economy, particularly in terms of energy (decarbonization, green economy, etc.);
- demographic imbalances between countries with an increasing number of young people and those with an aging population, with resulting pressures on the labour market in the former and on welfare systems in the latter and effects on migratory movements;
- the development of new technologies (artificial intelligence, automation and robotics), which will create new jobs while others will be lost in the transition and above all, will require workers to acquire new skills.

The analysis of "The Future of Work – Labour after *Laudato Si'*" Project coincides with these same points and adds a fourth one: the increase of inequality, insecurity and violence, a challenge that will be met only partially through the 2030 Agenda response. This latter seems fit to address imbalances which are – so to say – a side-effect of the way the global economy works, but it appears less apt to deal with asymmetries which are built into the economic core of the system and therefore can be described as structural (e.g. the core-periphery imbalances). Asymmetries related to wealth production and value extraction are a main cause of global instability, as they

---

[10] COMECE Social Affairs Commission, *Shaping the Future of Work*. Brussels: COMECE, 2018, p. 7.

[11] ILO – Global Commission on the Future of Work, *Work for a brighter future*, 2019, p. 10.

are at the roots of international conflicts (whether they break out in actual war[12] or not).

More recently, history itself added a fifth challenge: the COVID-19 pandemic. Its impact on the world of work at a global level has been devastating, as testified to by several international studies and reports, starting from the different editions of the *ILO Monitor: COVID-19 and the world of work*, and has proved more severe on women and on workers in the informal economy, thus exacerbating already existing inequalities. In his *Video Message on the occasion of the 109th Session of the International Labour Conference*, Pope Francis clearly spells out the dramatic consequences of the pandemic on specific segments of the world of work:

> We are called upon to prioritize our response to workers who find themselves on the margins of the labour market and who are still affected by the COVID-19 pandemic; low-skilled workers, day labourers, those who work in the informal sector, migrant and refugee workers, those who perform what are commonly referred to as "3Ds occupations": dangerous, dirty and degrading, and the list could go on. Many migrants and vulnerable workers, together with their families, usually remain excluded from access to national programmes for health promotion, disease prevention, treatment and care, as well as plans for financial protection and psychosocial services. This is one of the many cases of this philosophy of rejection that we have become accustomed to imposing in our societies. […] The lack of social protection measures in the face of the impact of COVID-19 has resulted in increased poverty, unemployment, underemployment, an increase of informal work, a delay in the inclusion of young people in the labour market, which is very serious, an increase in child labour, which is even more serious, vulnerability to human trafficking, food insecurity and increased exposure to infection among populations such as the sick and the elderly.

3. Rethinking work: the solid inspiration of *Laudato Si'*

No matter how powerful these forces transforming the world of work are, the future of the work is not something already written or predetermined. It will be what we, as humankind, want and are able to build, through concrete collective choices shaping society and the economy. This is why reflecting on

[12] The Russian invasion of Ukraine within the context of a growing tension between Russia and the Western world, is a good example to illustrate this. The same goes for the widespread and sometimes endemic instability of oil-rich countries and regions.

the meaning and purpose of work is of fundamental importance[13]; otherwise, any measure or reform we undertake could even backlash: unless we are clear on the ends, only a happy chance will enable us to choose the right means. In this sense, we can say that our society needs not only work, but also an exploration of its anthropological meaning. This is the lesson learned, for example, from the experience of monasticism:

> Saint Benedict of Norcia proposed that his monks live in community, combining prayer and spiritual reading with manual labour (*ora et labora*). Seeing manual labour as spiritually meaningful proved revolutionary. Personal growth and sanctification came to be sought in the interplay of recollection and work. This way of experiencing work makes us more protective and respectful of the environment; it imbues our relationship to the world with a healthy sobriety[14].

The task of "thinking" about work also relates to theology and spirituality. LS recalls: "Together with the awe-filled contemplation of creation which we find in Saint Francis of Assisi, the Christian spiritual tradition has also developed a rich and balanced understanding of the meaning of work, as, for example, in the life of Blessed Charles de Foucauld and his followers" (LS, n. 125).

Indeed, LS provides a new frame of reference within which to situate the many changes taking place and connect areas usually considered very separate. For instance, we can consider the following statement:

> If we reflect on the proper relationship between human beings and the world around us, we see the need for a correct understanding of work; if we talk about the relationship between human beings and things, the question arises as to the meaning and purpose of all human activity. This has to do not only with manual or agricultural labour but with any activity involving a modification of existing reality, from producing a social report to the design of a technological development. Underlying every form of work is a concept of the relationship which we can and must have with what is other than ourselves[15].

[13] A wide range of insights on the need to 'rethink labour', including not only the economic component but, more importantly, the social dimension, in order to answer the questions raised by integrating the centrality of human dignity and the common good within discussions on the future of work, can be found in de Alice La Rochefoucauld and Carlo M. Marenghi, eds., *Rethinking Labour. Ethical Reflections on the Future of Work*. Chambésy (CH): FCIV, 2018, available at www.fciv.org/downloads/WP10-Book.pdf (consulted on 03/01/2022). The volume is the result of a joint effort of The *Caritas in Veritate* Foundation and "The Future of Work – Labour after *Laudato Si'*" Project.

[14] Pope Francis, *Laudato Si'*, n. 126.

[15] Pope Francis, *Laudato Si'*, n. 125.

An innovative notion of work is expressed here, presenting it as "any activity involving a modification of existing reality", without any reference to an element often considered as the most important: financial compensation. This is a radical departure from common labour market discourse, which defines work as the contractual exchange between the worker's activity and its remuneration by the employer. In the *Message to Participants in the 108th Session of the International Labour Conference* and with a reference to LS, n. 128, Pope Francis emphasized the poverty of the latter approach: "Work is not just something that we do in exchange for something else. Work is first and foremost 'a necessity, part of the meaning of life on this earth, a path to growth, human development and personal fulfilment'". He reiterates this broader vision of work in FT:

> In a genuinely developed society, work is an essential dimension of social life, for it is not only a means of earning one's daily bread, but also of personal growth, the building of healthy relationships, self-expression and the exchange of gifts. Work gives us a sense of shared responsibility for the development of the world, and ultimately, for our life as a people[16].

In contrast, a reductionist vision of work eliminates our richest experiences, linked to direct action in a spirit of gratuitousness: via work, we express the meaning of existence and grow toward human flourishing. In his Encyclical *Caritas in veritate*, Benedict XVI had already warned against the risk of impoverishment implicit in the expulsion of the principle of gratuitousness from the economy, including the world of work:

> The great challenge before us, accentuated by the problems of development in this global era and made even more urgent by the economic and financial crisis, is to demonstrate, in thinking and behaviour, not only that traditional principles of social ethics like transparency, honesty and responsibility cannot be ignored or attenuated, but also that in commercial relationships the principle of gratuitousness and the logic of gift as an expression of fraternity can and must find their place within normal economic activity. This is a human demand at the present time, but it is also demanded by economic logic. It is a demand both of charity and of truth[17].

Imagining the future of work requires rethinking its relationship with remuneration, which cannot constitute its exclusive purpose, although this approach does

---

[16] Pope Francis, *Fratelli tutti*, n. 162.

[17] Benedict XVI, Encyclical letter *Caritas in veritate on Integral Human Development in Charity and Truth*, 29 June 2009, n. 36.

not imply pulling back from the fight for workers' rights. Decoupling the notion of work from financial remuneration cannot open the path to exploitation.

A second element of great relevance is the idea that work is a form of relationship "with what is other than ourselves". In other words, LS proposes a relational notion of work. Work involves many links: in the first instance, with people, but also with the environment and creation, in other words, the reality which is the object of transformation. This differs from the prevailing concept of work in our strongly individualistic culture. In the *Message to Participants in the 108th Session of the International Labour Conference* Pope Francis insists on this relational notion of work and its various dimensions:

> As well as being essential to the realization of the person, work is also fundamental to social development. My predecessor Saint John Paul II put it beautifully, when he explained that "work is work with others and work for others"[18]; and as its fruit, labour offers "occasions for exchange, relationship and encounter"[19]. [...] Yet our vocation to work is also inextricably connected with the way we interact with our environment and with nature. We are called to work, to "till and to keep" the garden of the world (cf. *Genesis* 2:15), that is, to cultivate the ground of the earth to serve our needs without failing to take care of and protect it. Work is a path to growth, but only when it is an integral growth that contributes to the entire ecosystem of life: to individuals, societies and the planet.

4. Work is care, care is work

As it is already manifest in the previous quote, this notion of work, which we may call "integral" as opposed to "reductionist", is deeply grounded in the tradition of Catholic Social Teaching: "Since work in its subjective aspect is always a personal action, an *actus personae*, it follows that the whole person, body and spirit, participates in it, whether it is manual or intellectual work"[20]. The concrete experience of people at work in all its diverse forms testify to the fact that work is filled with human meaning. Through work, we learn, we discover what we can do, we find satisfaction and take pride in "work well done". We may also meet failure and defeat. These are all human experiences

---

[18] Saint John Paul II, Encyclical letter *Centesimus annus*, 1 May 1991, n. 31.

[19] Pontifical Council for Justice and Peace, *Compendium of the Social Doctrine of the Church*, 2004, n. 273.

[20] Saint John Paul II, Encyclical letter *Laborem exercens on Human Work*, 14 September 1981, n. 24.

and through them, we grow in our humanity. Moreover, work is a fully human experience because it is the way for every human being to take part in the care for our common home and to contribute to the common good, i.e., to exercise the responsibility that makes us human.

But work is a human experience also because it inserts us into the concreteness of the world. Even when it does not involve manual activity, it means transforming reality, handling its materiality and coping with its limitations, for example, in terms of space and time. Furthermore, work puts us in contact with the world understood as the environment, compels us to face the issue of management of scarce and non-renewable resources or the impacts of our activities on nature. These are all fundamental dimensions of being human.

Finally, work is a human experience because it makes us meet other people: those with whom and those for whom we work, who more and more often are thousands of miles away. Contacts mediated by technology or the market remain human interactions. Furthermore, the world of work is the domain of shared dreams, hopes and ambitions as well as of conflict and struggle. It is through such interpersonal relationships that our personal identity matures and we grow in humanity. In this sense, work is one of the domains in life in which we can experience the deep truth of what Pope Benedict XVI says in *Caritas in veritate*, n. 53:

> As a spiritual being, the human creature is defined through interpersonal relations. The more authentically he or she lives these relations, the more his or her own personal identity matures. It is not by isolation that man establishes his worth, but by placing himself in relation with others and with God. Hence these relations take on fundamental importance.

If work is a factor of humanization because it allows us to build and cultivate relationships, then it is clear why it needs to include a dimension of care: without care, relationships die and become toxic for the people involved. Work is care, we can legitimately say, at least a work that is fully human, that is really *actus personae*.

The reverse sentence "care is work" is also true. As LS keeps insisting upon, care for our common home requires a radical transformation of the economic system and of the way in which production and consumption actually take place. No such transformation will be possible without work, i.e. without human "activity involving a modification of existing reality". While the world of work serves as an indispensable agent of transformation, it will necessarily undergo major changes throughout the process. Therefore, while it must become more and more capable of caring, it also needs to be taken care of. In particular, we need to avoid the burden of the transformation falling mainly on the shoulders of its weaker segments, namely marginal workers who may find it difficult to adapt their skills or acquire new ones, and risk being expelled from the labour market.

A first focus concerns sustainable jobs and workplaces, starting with a better assessment of the specific contribution of each worker's activity to the care of our common home. In particular, safety and security in the workplace, whether formal or informal, within the household or on the way to work, are to be achieved. A second focus is on the potentially positive contribution of businesses (from the smallest to the largest) to the care of the environment. Business should increase capacity to develop social and environment assessment of its practices and identify measures to address negative impacts. This should be built into innovative strategies. Business leaders have a special responsibility and should be supported in this mission.

The mutual relationship between care and work becomes even clearer when we consider work from a more fundamental, theological perspective, in the footsteps of *Laborem exercens*. From a Christian perspective, work is not only God's plan for women and men, but the option he made for himself. When the Son of God became a man, he spent most of his life working as a craftsman: "the eloquence of the life of Christ is unequivocal: he belongs to the 'working world', he has appreciation and respect for human work"[21]. The same holds true for Mary in the house of Nazareth and Joseph, to whom we still refer as the carpenter. This means that work has a specific place in the plan of salvation. For Jesus' disciples, work is a way to walk in the footsteps of their master: "Sweat and toil, which work necessarily involves in the present condition of the human race, present the Christian and everyone who is called to follow Christ with the possibility of sharing lovingly in the work that Christ came to do"[22].

Besides, "at the very beginning of Sacred Scripture, in the Book of *Genesis*, [...] the creation activity itself is presented in the form of 'work' done by God"[23]. In other words, he is first introduced as one who works. In several other passages, God is portrayed at work to keep his creation alive and protect it. For instance, in *Psalm* 104 we read: "You make the grass grow for the cattle and

[21] Saint John Paul II, *Laborem exercens*, n. 26.
[22] *Ibid.*, n. 27.
[23] *Ibid.*, n. 25.

plants for people's work to bring forth food from the earth [...]. Young lions roar for prey; they seek their food from God [...]. All of these look to you to give them food in due time. When you give it to them, they gather; when you open your hand, they are well filled" (*Psalm* 104:14.21.27-28). In this way, we learn that work is what God does when he takes care of creation. Since the very beginning he calls on human beings to join in and cooperate with his work of care for creation: "The Lord God then took the man and settled him in the garden of Eden, to cultivate and care for it" (*Genesis* 2:15). The working dimension of care goes hand in hand with the caring dimension of work!

In this light, we can discover a striking and a spiritually very enriching parallelism between *Laborem exercens* and LS. The former proposes work as a road to meet Jesus Christ, the Saviour and the Redeemer, and to follow him. In LS, care for our common home is the pathway to meet the same Jesus Christ as the Logos mysteriously filling the whole universe: "From the beginning of the world, but particularly through the incarnation, the mystery of Christ is at work in a hidden manner in the natural world as a whole, without thereby impinging on its autonomy" (LS, 99).

5. Human dignity at the roots of decent work

The transformative power of this notion of work appears when it is articulated with the concept of decent work, as the Project did in its position paper *A proposal to extend the Decent Work Agenda and address the current global crisis*[24]. This was published on 12 June 2019 and was addressed to the International Labour Conference taking place in Geneva for the approval of the Centenary Declaration on the Future of Work.

The notion of "decent work" is the cornerstone of the ILO's activities and was defined in its full sense in the *Declaration on social justice for a fair globalization*, adopted during the 2008 International Labour Conference:

> Promoting employment by creating a sustainable institutional and economic environment, [...] developing and enhancing measures of social protection, [...] promoting social dialogue and tripartism as the most appropriate methods (of translating economic development into social progress), respecting, promoting and realizing the fundamental principles and rights at work, which are of particular significance[25].

---

[24] The position paper is now available as Appendix D of the report *Care is Work, Work is Care.*
[25] ILO, *ILO Declaration on the social justice for a fair globalization*, 2008; retrieved from www.

Decent work has quickly been adopted as a kind of platform for action by actors of very different backgrounds such as trade unions, NGOs, grassroots and civil society movements and other international organizations. For example, it is referred to in the official formulation of the Sustainable Development Goal n. 8 – "Decent Work and Economic Growth" – adopted by the United Nations for the period 2015-2030. This had not been the case with Millennium Development Goals (2000-2015).

From the beginning, the Church has recognized the potential of this concept. On the occasion of the Jubilee of Workers (1 May 2000), to which ILO's then Director-General Juan Somavía took part, John Paul II launched an appeal for the creation of a coalition in favour of decent work, supporting the ILO strategy. Pope Benedict XVI took up the concept in *Caritas in veritate*, n. 63, where he explores the meaning of the dignity of work.

> What is meant by the word "decent" in regard to work? It means work that expresses the essential dignity of every man and woman in the context of their particular society: work that is freely chosen, effectively associating workers, both men and women, with the development of their community; work that enables the worker to be respected and free from any form of discrimination; work that makes it possible for families to meet their needs and provide schooling for their children, without the children themselves being forced into labour; work that permits the workers to organize themselves freely, and to make their voices heard; work that leaves enough room for rediscovering one's roots at a personal, familial and spiritual level; work that guarantees those who have retired a decent standard of living.

As we can start grasping by these words, interpreting the concept of decent work in the perspective of Catholic Social Teaching offers a contribution toward its enrichment in two directions. First, terms such as freedom, equality, dignity, used in the definition of decent work, can be understood in very different ways, especially in an increasingly plural world like ours, where anthropological visions are extremely diverse. For example, in the framework of a progressive and secularized individualism, their depth and scope are not the same as within the personalistic approach in which Catholic Social Teaching is grounded. According to the latter, human dignity refers fundamentally to the social nature of the human person and includes the prospect

---

ilo.org/wcmsp5/groups/public/---dgreports/---cabinet/documents/genericdocument/wcms_371208.pdf, consulted on 03/01/2022.

of contributing to the common good; therefore, it postulates the embedding of the person in the social fabric. John Paul II stated this clearly: "Even before the logic of the exchange of equivalents and forms of justice, which are its own, there is something that is due to man because he is man, by virtue of his eminent dignity. This something owed inseparably involves the possibility of surviving and making an active contribution to the common good of humanity"[26]. Secondly, the Decent Work Agenda progressively developed from a starting point, which is the assumption that work is the relationship between employees and employers, mediated by a legal contract. Its goal is to make the protection of formal workers' rights more enforceable through the improvement of legal regulations, given the unavoidable asymmetry in this type of relationship. In terms of social justice, what has thus been achieved is fundamental and this effort must not be weakened. Today however, it is urgent to give equal priority to other forms of work, often more precarious and less guaranteed but increasingly widespread, from the recently born gig economy to the more conventional informal one we have known for a long time.

The notion of work as a multidimensional relationship pushes for an extension of the Decent Work Agenda, because "work is only dignified when this manifold relationship is an effective expression of dignity"[27]. Respect for workers' rights in terms of pay, hours, security or freedom to unionize are not enough to define decent work, as most people are accustomed to thinking. Work cannot be deemed decent when it produces death or suffering for others; or when it causes environmental degradation or excessive consumption of non-renewable resources that jeopardize the opportunities for future generations to live in dignity. In particular, an extended or integral definition of decent work should simultaneously take into account four dimensions we already touched upon in the previous paragraphs:

- Work is a spiritual reality, through which women and men experience being creatures and are called to be co-creators. To be decent, work must promote workers' humanity, allowing them to express their capacities and their creativity, and to experience personal freedom even within contexts of limitations and contradictions.
- Work is a social reality, as it is cooperation with many others towards a common goal. Catholic Social Teaching never views workers simply

[26] Saint John Paul II, *Centesimus annus*, n. 34.

[27] "A proposal to extend the Decent Work Agenda and address the current global crisis", in *Care is Work, Work is Care*, p. 139.

as individuals, but places them within their core relationships, first with their families and then within the community to which they belong. This approach also considers the dignity of work within this relational framework and recognizes work as critical to the promotion of the well-being of the family and the development of society. When this dimension is disregarded, work cannot be defined decent.

- Work is an ecological reality: it affects nature and the environment; it can protect or destroy them. Decent work needs caring for our planet and the environment and using its resources responsibly.
- Work is an economic reality: it generates value and makes it available for the whole society, creating opportunities for development. Otherwise, it cannot be considered decent.

This articulation of the principle of the dignity of the human person with the more secular notion of decent work, which is also a powerful political and advocacy tool, shows how fruitful dialogue can be among Catholic Social Teaching and other philosophical and political approaches as a push for social change. Chapter 5 of LS strongly insists on the importance of this dialogue. On the one hand, decent work indicates a domain in which human dignity is concretely at stake, helping avoid its evaporating into a purely abstract notion. It is the kind of contribution to the development of Catholic Social Teaching described in the *Compendium of the Social Doctrine of the Church*:

> A significant contribution to the Church's social doctrine comes also from human sciences and the social sciences. In view of that particular part of the truth that it may reveal, no branch of knowledge is excluded. The Church recognizes and receives everything that contributes to the understanding of man in the ever broader, more fluid and more complex network of his social relationships. She is aware of the fact that a profound understanding of man does not come from theology alone, without the contributions of many branches of knowledge to which theology itself refers[28].

On the other hand, Catholic Social Teaching can work as a source of inspiration, pushing to expand the scope of the advocacy action and of the commitment to social justice:

---

[28] Pontifical Council for Justice and Peace, *Compendium of the Social Doctrine of the Church*, 2004, n. 78.

> The Church's social doctrine is a privileged instrument of dialogue between Christian communities and the civil and political community. It is an appropriate tool for promoting and cultivating attitudes of authentic and productive cooperation in ways adapted to the circumstances. The commitment of civil and political authorities, called to serve the personal and social vocation of mankind according to their own areas of competence and with the means available to them, can find in the social teaching of the Church an important support and a rich source of inspiration[29].

6. Work, care and dignity

In his *Video Message on the occasion of the 109th Session of the International Labour Conference*, Pope Francis takes up the care-work link (with a direct reference to the report *Care is Work, Work is Care*), with a strong emphasis on the dignity of workers and – not surprisingly – point out to the need of its implementation in daily life and economic practice:

> If work is a relationship, then it must include the dimension of care, because no relationship can survive without care. Here we are not just referring to the work of assistance: the pandemic reminds us of its fundamental importance, which perhaps we have overlooked. Care goes further; it must be a dimension of all work. Work that does not take care, that destroys Creation, that endangers the survival of future generations, does not respect the dignity of workers and cannot be considered decent. On the contrary, work that cares, that contributes to the restoration of full human dignity, will help to ensure a sustainable future for future generations[30]. And this dimension of care involves, first and foremost, the workers. In other words, a question we can ask ourselves in our daily lives: how does a business, for example, take care of its workers?

The wording and the vocabulary may sound very different, but the basic idea is very close to the one expressed by Leo XIII at the beginning of the Encyclical *Rerum novarum*. From the concern for the miserable conditions of the working class, which are a violation of their human dignity, stems the conclusion that a production system in which "greed" (in *Rerum novarum* terms) or the obsession "with maximizing profits" (with the words of LS, n. 190) crowd out care – and care for workers in the first instance – quickly becomes inhumane.

---

[29] *Ibid.*, n. 534.

[30] Cf. *Care is work, work is care*, Report of "The Future of Work – Labour after *Laudato Si'*" Project, https://futureofwork-labourafterlaudatosi.net/ (appearing as footnote no. 4 in the original text).

> In any case we clearly see, and on this there is general agreement, that some opportune remedy must be found quickly for the misery and wretchedness pressing so unjustly on the majority of the working class: for the ancient workingmen's guilds were abolished in the last century, and no other protective organization took their place. Public institutions and the laws set aside the ancient religion. Hence, by degrees it has come to pass that working men have been surrendered, isolated and helpless, to the hardheartedness of employers and the greed of unchecked competition. The mischief has been increased by rapacious usury, which, although more than once condemned by the Church, is nevertheless, under a different guise, but with like injustice, still practiced by covetous and grasping men. To this must be added that the hiring of labor and the conduct of trade are concentrated in the hands of comparatively few; so that a small number of very rich men have been able to lay upon the teeming masses of the laboring poor a yoke little better than that of slavery itself[31].

Work, care and dignity have been at the core of Catholic Social Teaching since its first steps, and continue to be major drivers of its development. It could not be otherwise, since the world of work is at the core of the transformation today as it was at the end of the XIX century. The publication of *Rerum novarum* in 1891 opened a season of renewed commitment of Catholics within the world of work, which brought to life a variety of approaches, institutions and organizations, such as trade unions, workers' associations and cooperatives. Integral ecology, with the relational notion of work it includes, and the vision of fraternity and social friendship put forward by FT, can have the same mobilizing effect. Some paths for possible developments can already be glimpsed. Here we pick up three:

a) Most religious traditions have developed a rich spirituality of work and a concern for social justice. The promotion of workers' dignity, or, in other words, decent work, can represent a domain in which dialogue among different religious traditions and secular humanistic beliefs may prove particularly fruitful, taking up the call of FT. Actually, that dialogue has already begun, as it is shown by the handbook *Convergences: decent work and social justice in religious traditions*[32], published in 2012 by the ILO in partnership with the World Council of Churches,

[31] Leo XIII, Encyclical letter *Rerum novarum*, 15 maggio 1891, n. 3.

[32] ILO, *Convergences: decent work and social justice in religious traditions. A handbook*, Geneva 2012, retrieved from www.justiceandpeace.va/content/dam/giustiziaepace/pubblicazioni/ILO_Religions_EN_WEB.pdf (consulted on 03/01/2022).

the (then) Pontifical Council for Justice and Peace, and the Islamic Educational, Scientific and Cultural Organization (ISESCO), with contributions of scholars of the Yeshiva University and the European Buddhist Association (EBA).

b) As we have seen, the dialogue among the notion of work proposed by LS and the Decent Work Agenda has allowed identifying an extended scope of the latter. Something similar might involve the vision of social friendship of FT and the practices of social dialogue (even in the technical meaning it has within the world of work), providing stimuli for its renewal.

c) Finally, the much-needed transformation of the economic system needs actors capable of carrying it out: no change can happen without someone pushing for it. The institutions and organizations initiated in response to the publication of *Rerum novarum* performed this very same task when they contributed to shaping a new social model out of the 'classic' liberal capitalism of the XIX century. In today's world, in order to be effective, those actors need to be strongly rooted in a local context, but at the same time very well connected globally, and capable of integrating insights and wisdom coming from the peripheries and the margins. As LS, n. 219, puts it: "self-improvement on the part of individuals will not by itself remedy the extremely complex situation facing our world today. […] Social problems must be addressed by community networks and not simply by the sum of individual good deeds. […] The ecological conversion needed to bring about lasting change is also a community conversion". In this sense, the report *Care is Work, Work is Care* speaks of "transformative global communities". Although this notion needs further investigation, some examples may already be identified. One is the Amazon region: normally considered peripheral, thanks to the 2019 Synod, we discovered that it is the home of a different culture, preserving alternative values to the dominant ones. More recently, Pope Francis expressed the same idea with the phrase "Collective Samaritan", used to describe the popular movements he has met several times during his pontificate, as well as "The protests over the death of George Floyd"[33].

---

[33] Pope Francis, *Video Message on the occasion of the Fourth World Meeting of Popular Movements*, 16 October 2021. For an initial reflection on the innovative potential of the expression "Collective Samaritan", cf. Michael Czerny and Paolo Foglizzo, "Il mondo si vede meglio dalle periferie. Il quarto Incontro mondiale dei movimenti popolari", in *Aggiornamenti Sociali*, 1-2022, pp. 33-40.

There is still a long way to go in order to give human dignity a fuller implementation in the world of work and to explore the many links joining work and care. Integral ecology, together with universal fraternity and social friendship, has the power to set us in motion and to indicate the path to follow. "The Future of Work – Labour after *Laudato Si'*" is warming up for a second phase, with the aim of participating to this great common enterprise.

LAW AND CULTURE IN INTEGRAL ECOLOGY

Francesco Gambino

1. Law as an "effective moderator" of society

A comprehensive conception of environmental problems – which is an integral and integrating vision – requires an in-depth investigation of "the workings of society, its economy, its behaviour patterns, and the ways it grasps reality"[1].

This attentive and unifying examination stems from the need for "integral solutions" that should encompass "the interactions within natural systems themselves and with social systems"[2]. "Everything is interconnected"[3], and "everything is closely interrelated"[4]. In this sense, there are not "two separate crises – environmental and social – but rather one complex crisis which is both social and environmental"[5]. Ecological culture, in which the relation of the human being with the environment is reconsidered, also includes the protection of "the cultural wealth of humanity in the broadest sense"[6]. It is necessary to undertake an interdisciplinary dialogue and "respect the rights of peoples and cultures, recognising that the development of a social group presupposes a historical process which takes place within a cultural context and demands the constant and active involvement of local people *from within their own*

[1] Pope Francis, Encyclical Letter *Laudato Si' on Care for Our Common Home*, 24 May 2015, n. 151.

[2] Pope Francis, *Laudato Si'*, n. 139.

[3] Pope Francis, *Laudato Si'*, n. 138.

[4] Pope Francis, *Laudato Si'*, n. 137.

[5] Pope Francis, *Laudato Si'*, n. 139.

[6] Pope Francis, *Laudato Si'*, n. 143.

DOI 10.32060/C.S.9.113-129

culture"[7]. The theme of the *interactions* among systems, behaviours, ways of knowing reality, which is fundamental in the search for integral solutions to environmental issues, highlights the role of the law in a constantly changing globalised society. Indeed, a society is able to "plan and protect [its] future"[8] if the law acts as an effective moderator, establishing "rules for admissible conduct in the light of the common good"[9].

For its very nature, reflecting on the functioning of society[10], the law takes on a decisive significance in the construction of a juridical culture for the care of the environment. From the perspective of the law *of* or *among* the States of the world, one may explore the *juridical foundations of integral ecology* or, summarizing the concept in an immediate formula, talk of *juridical* or *socio-juridical ecology*.

This is an approach that involves several levels of enquiry.

On the national and local level, the need is felt for a law that is not only punitive but, rather, possesses a promotional function[11]; a law that is not reduced to jargon or standardized measures but is permeable to the complexity of local problems[12]; a law that, inspired by the principle of effectiveness[13], does not remain a dead letter and is applied in the courts of justice.

Society also has a need for a bottom-up law[14]; a law that is formed under the pressure of non-governmental and intermediate associations which compel "governments to develop more rigorous regulations, procedures and controls"[15]. In order to avoid environmental deterioration and damage, it is

---

[7] Pope Francis, *Laudato Si'*, n. 144.

[8] Pope Francis, *Laudato Si'*, n. 177.

[9] Pope Francis, *Laudato Si'*, n. 177.

[10] Cf., in connection with the right to private property, Pope Francis, Encyclical Letter *Fratelli tutti on Fraternity and Social Friendship*, 3 October 2020, n. 20.

[11] Pope Francis, *Laudato Si'*, n. 177.

[12] Pope Francis, *Laudato Si'*, n. 144.

[13] In several passages, the two encyclicals deal with the principle of effectiveness. Moving from the demand for a "healthy, mature and sovereign society", engaged in anticipating environmental risks, *regulatory norms, timely enforcement* (Pope Francis, *Laudato Si'*, n. 177) to the hope that the "legislation and regulations dealing with the environment *will really prove effective*" (Pope Francis, *Laudato Si'*, n. 142). The theme of the *realisation* of rights returns also in Pope Francis, *Fratelli tutti*, nn. 120, 121.

[14] It is that law which, arising from society, is reinforced in the daily experience of social life and achieves "a high degree of effectiveness" (Paolo Grossi, *Ordinamento*, in *Jus*, LIII, 2006. Also in *Paolo Grossi – Maestri del diritto*, ed. Guido Alpa, Rome-Bari: Laterza, 2011, p. 51.

[15] Pope Francis, *Laudato Si'*, n. 179.

necessary for citizens to exercise control over "political power: be it national, regional and municipal"[16].

Some important episodes in the history of environmental law in Italy can be seen in this perspective. Through the initiative of the associations for environmental protection the urgency of caring for the environment reached the courts of justice, fuelling the public debate and the exchange of opinions - which ultimately led to the law that established the Ministry of the Environment, the laws governing environmental damage (l. 8 July 1986, no. 348); and opened the way for the present Environmental Code (d. lgs. 3 April 2006, no. 152). If in Italy today we can talk of environmental juridical culture, it is thanks to that original movement in civil society which launched a broad and shared debate on environmental issues, establishing the basis for political and legislative action.

From another point of view – that of a global society "without a centre and without a summit"[17] – the problems relating to the environment affect the relations between nation States. There is a need for a more robust supranational law to promote new agreements between States, make "international institutions" "stronger and more efficiently organised"[18], and establish "independent bodies empowered to impose sanctions"[19].

In this context, the relationship between world-wide risks and the promise by institutions of controlling them[20] – a long-standing issue for sociologists – was fundamentally and thankfully modified by the rediscovery of the primary role played by socially-aware politics, called upon to challenge the planetary alliance between technology and the economy.

---

[16] Pope Francis, *Laudato Si'*, n. 179.

[17] See along the lines of the thought of Niklas Luhmann, Gunther Teubner and Andreas Fischer-Lescano, "Scontro tra regimi: la vana ricerca di unità nella frammentazione del diritto globale", in *La cultura del diritto nell'epoca della globalizzazione. L'emergere delle costituzioni civili*, trans. and ed. by R. Prandini. Rome: Armando Editore, 2005, p. 159. The paper, with the title *Regime Collisions: the Vain Search for Legal Unity in the Fragmentation of Global Law*, appeared in an English version in 25 *Mich. J. Int'l L.* 999 (2004).

[18] Pope Francis, *Laudato Si'*, n. 175.

[19] Pope Francis, *Laudato Si'*, n. 175.

[20] See the essay by Ulrich Beck, *Un mondo a rischio*, Italian trans. by L. Castoldi. Turin: Einaudi, 2002. Quoting the speech given to the Duma of Moscow in November 2001, the author develops themes and arguments put forward in his book *La società del rischio. Verso una seconda modernità* (1986), Italian trans. and ed. by Walter Privitera. Rome: Carocci Editore, 2000.

2. The cultural side of integral ecology

The observations mentioned – stimulated by the urgency of the environmental problem – raise a basic question that concerns the *cultural side* of integral ecology.

In this problematic scenario, which involves the wider concept of culture and therefore spans over numerous other areas of enquiry, we shall focus on the search for possible interactions between languages, social customs and traditions coming from different social and juridical systems.

Let us start from the empirical fact that language itself creates a distance – the language with which each of us sees the world, thinks and acts. It is the "machine" of language that operates between us and the environment and that (as we are told by anthropological studies) acts as an exosomatic instrument[21]. Each of us, with our education, training, and culture, communicates with others but experiences the world through conceptual, normative, emotional meanings of our own. In the collision of – or interweaving between - symbolic areas, exclusive and excluding meanings can arise. That is because the *sense* of words is always understood through *meanings* that can differ according to *who* is involved in the communication[22]. Complexity – in the workings of linguistic machines – comes from the intrinsic problem of understanding the hidden meaning given by others based on their values or interpretation of reality.

This difficulty is also present in the communication problems existing between different social and juridical systems. Sociological studies explore these issues, as we shall mention.

3. The linguistic structures of law

In this context, marked by the intrinsic difficulty of communicating, which requires participants in the exchange to understand "just enough so that communication can follow"[23], it is necessary to reflect the on the linguistic structures of the law.

---

[21] Carlo Sini, *Inizio*. Milan: Jaca Book, 2016, p. 33.

[22] On the difference between *meaning* and *sense*, cf. Tullio De Mauro, *Guida all'uso delle parole*. Rome-Bari: Laterza, 2019, p. 33.

[23] From the viewpoint of the system in terms of difference, see Niklas Luhmann, *Introduzione alla teoria dei sistemi* (2011), ed.Dirk Baecker; Italian ed. by Stefano Maggiolo. Lecce-Rovato: Pensa Multimedia, 2018, p. 73: "communication takes place only when someone understands approximately or misunderstands, but in any case understands just enough for the communication to continue".

With its own language, the law, aimed at establishing behaviour, setting programmes for the future and making actions assessable[24], affects – or is expected to affect - individual and collective behaviour. Any form of innovation in the world of law or thought pattern that gains ground and is repeated over time to solve juridical and social problems, is determined by the creation of new content within language structures and semantic vessels[25].

The critique of logical neo-positivism and the language analysis which permeated Italian juridical culture in the fifties, "do not affect *the linguistic structures of the law*, as they have come to be established in the modern age and pursuant to the rule of law[26].

From our perspective, we can affirm that the language of the law does not function differently from other languages; and add that – like any other language – it is not identified so much by what it allows us to think as by how it "habitually obliges us to think about particular types of information"[27].

According to studies in linguistics, when a language "forces its speakers to pay attention to certain aspects of the world each time they open their mouths or prick their ears, the result is that these linguistic habits can be turned into mental habits which produce consequences on memory, perception, associations or even practical abilities"[28].

In the same way, the linguistic structures of the law, used by practitioners and borrowed by everyday language, produce new terms, meanings and concepts which become established over time and become part of the culture of a

[24] "Just as we hear and understand interchanges of words, so we see and understand *the juridical language of actions*", Natalino Irti, *Riconoscersi nella parola. Saggio giuridico*, Istituto Italiano per gli Studi Storici. Naples: Il Mulino, 2020, p. 67.

[25] Claudia Roesler, "Legal Concepts from the Standpoint of Innovation and Preservation of Meaning: A Rhetorical Perspective", in *Innovation and Transition in Law. Experiences and Theoretical Settings*, eds by Massimo Meccarelli, Cristiano Paixão, Claudia Roesler. Madrid, Dykinson, p. 127. This paper devotes detailed attention to repetitive structures: Reinhart Koselleck, "Repetitive structures in language and history", in Karin Tilmans, Frank Vanvreee and Jay Winter, eds., *Performing the past*: *memory, history and identity* in *Modern Europe*. Amsterdam: Amsterdam University Press, 2010, pp. 51-66.

[26] Natalino Irti, "I 'cancelli delle parole' (intorno a regole, principi, norme)", in *Un diritto incalcolabile*. Turin: Giappichelli, 2016, p. 76.

[27] On the theme, proposed and elaborated by Wilhelm von Humboldt, of the conditionings of language on the mind, see Guy Deutscher, *La lingua colora il mondo. Come le parole deformano la realtà* (2010), trans. by Enrico Griseri. Turin: Bollati Boringhieri, 2016, p. 175.

[28] By way of reconstruction, see Guy Deutscher, *La lingua colora il mondo*, p. 175.

community in various ways[29]. It is along this path that a new linguistic possibility[30] is converted into a linguistic habit.

4. Juridical concepts, words, ideas

In order to clarify to what extent legal, cultural and linguistic structures influence one another, it is useful to mention here the lively debate over the scientific validity of juridical concepts[31] that occurred in Italy and involved eminent jurists, philosophers, and historians over the ten years between 1935-1945[32].

In what was defined as a "science so bound up with the social sphere and its changes as the law", the main problem was whether it was possible to "crystallise" logical concepts[33]. The practical or scientific character of juridical concepts was not in question, indeed there was agreement on the fact that they constitute arbitrary constructs endowed with a practical function[34]; but the dispute concerned the place and mode of their formation - whether it should be the concrete ground of experiential data or abstract logic, unaffected by contingencies and thus capable of providing consistent solutions.

In my opinion, the difference of opinions on the *point of origin* of juridical concepts – the commanding energy of logic or the persuasive power of experience – does not affect the features of a given juridical culture within a given system. There, the *genesis* of a phenomenon is investigated; here, the *de facto presence* of certain language contents in a community is acknowl-

[29] In this sense, law is "not any kind of manifestation of human culture", but it is "*the structure which has enabled and guided the evolution of culture*", Francesco Romeo, "Idola imperii", in *Diritto frammentato*, eds. Alberto Febbrajo and Francesco Gambino. Milan: Giuffrè, 2013, p. 145.

[30] One thinks, for example, of the greater weight of some words – coming from or taken on by legal language – commonly used in contemporary culture ("consumer", "environment", "user", "sustainable development", "digital identity", etc.).

[31] The debate opened in Italy following the publication of the essay of Salvatore Pugliatti that appeared in 1935 under the title "Logica e dato positivo in rapporto ad alcuni fenomeni giuridici anomali", in *Arch. giur.*, CXIII, 1935, 158-169, now in *La polemica sui concetti giuridici*, ed. Natalino Irti. Milan: Giuffrè, 2004, pp. 5-18, 43-70.

[32] See the writings of Salvatore Pugliatti, Guido Calogero, Arturo Carlo Jemolo, collected by Widar Cesarini Sforza in *La polemica sui concetti giuridici* (Milan 2004), p. 5ff.

[33] Paolo Grossi, *La cultura del civilista italiano. Un profilo storico*. Milano: Giuffrè, 2002, p. 105, n 23.

[34] On the arbitrary and practical character of pseudo concepts, cf. Benedetto Croce, *Filosofia della pratica – Economia ed etica* (1908), 6° ed. Naples: Bibliopolis, 1949, pp. 337-45.

edged, the repetition of particular ways of thinking and acting as well as the stability (from a macro-sociological point of view also) of certain influences exercised over society and over ordinary ways of life is recognised.

In the words of the philosopher of law, one can observe that "the idea does not create the word but, rather, it is the word that creates the idea", since words objectify or model reality[35]. What gives us the illusion of having immobilised concepts (both juridical and non-juridical) that is "the flow of reality and life, is the word, the verbal sign", which is our instrument for expressing and creating ideas[36].

Just as the meaning of things in relation to our life and our goals can be expressed in different ways[37], so our juridical culture[38] – where words, concepts, and linguistic structures are stabilised over time – can express new ways of using juridical phenomena or "juridical things".

Let us go back, in the history of law, to the concept of *obligatio* as expressed in the oldest sources. The abolition of the *nexum,* which took place in 326 B.C., by which the debtor was physically enslaved to his creditor, marks the establishment of the notion of obligation in the modern sense[39]: the shift from the enslavement of the debtor, which was a personal and constrictive measure, to a relationship of an economic character. So an obligation began to lose its material connotation and entered into a new field of establishment and validity[40] – it started to generate new rules for

---

[35] Widar Cesarini Sforza, "Sulla relatività dei concetti giuridici", in *Vecchie e nuove pagine di filosofia, storia e diritto*. I. *Filosofia e teoria generale*. Milan: Giuffrè, 1967, p. 166.

[36] Ibid., p. 166.

[37] With reference to the German term *Zuhandenheit*, used by Martin Heidegger [in *Essere e Tempo* (1927), Italian trans. Chiodi, 17th ed. Milan: Longanesi, § 14, pp. 88-92], one talks of the "usability", in immediate terms, of the meaning of things in relation to our life (Gianni Vattimo, *Introduzione a Heidegger* (1971). Roma Bari: Laterza, 2000, p. 23).

[38] Among the concepts most used in social sciences, the concept of "juridical culture" is able to explain the multiple mutual influences between law and society. Further on, it will be useful in explaining *why*, some legal systems accept claims brought by sectors of society (cf. Alberto Febbrajo, *Sociologia del diritto*. Bologna: Il Mulino, 2009, p. 49). On the relations between social and juridical reality in the analysis of the law being founded on social circumstances, cf. the scholarly pages of Andrea Bixio, L'"immanenza sociale del diritto", in Massimo Severo Giannini, *Le radici del pensiero sociologico-giuridico*. ed. Alberto Febbrajo. Milan: Giuffrè, 2014, pp. 286-95.

[39] Pietro Bonfante, *Istituzioni di diritto romano* (1896), 10th ed.Turin, 1946. Milano: Giappichelli, 1987, pp. 304, 372.

[40] With reference to the studies undertaken by Georges Canguilhem on the "transformations" and on the "shifts" of concepts, see Michel Foucault, *L'archeologia del sapere. Una metodologia per la storia della cultura* (1969), Italian trans. G. Bogliolo. Milan: Bur, 1999, p. 7.

its use, and initiated new theoretical processes in history. This is the first trace of that progressive objectification which, in the full development of juridical culture, was to end up in ensuring the performance of the obligation exclusively through the assets of the debtor.

Today, obligation, which can be expressed by means of the wider concept of contractual relationship, is the object of a broad juridical debate. Having taken different directions on the basis of legal doctrine and jurisprudence, it was rekindled by the Covid 19 pandemic. Given the planet-wide spread of the infection and the regulations issued to contain it, this crisis has produced unexpected and unforeseeable events thereby compromising the stability of juridical and economic relations. Today, according to some trends, the contractual relationship is considered as a *relationship* characterised by *economic, juridical and social elements* that develops over time[41]; in this perspective, it can be modified and remodelled at any moment because of new circumstances and newly arising needs.

Along these very lines, scholars of Latin American law are discussing the project of a *Model Code of Obligations* for Latin America[42], fostering a cultural dialogue between States that differ in terms of traditions, social customs and juridical systems. At the same time, European legal systems consolidate the idea of "relational" contracts[43] – which cannot be isolated from a factual context – placing at the centre of this approach the value of flexibility achieved through our new-found duties of solidarity and inferred from universal legal principles and from a more extensive reading of general provisions (good faith, correctness, etc.).

---

[41] On the level of inter-State relations, one can add that debt – a term which expresses and implies an obligation, the relationship between creditor and debtor – is also a *political* relationship. Indeed, the rights of peoples affect "the way in which this obligation is fulfilled" on the part of poor countries toward rich countries. It "should not compromise" the "existence" and "growth" of the former (Pope Francis, *Fratelli tutti*, n. 126).

[42] See the contributions of European and South American jurists collected in the volumes edited by Simona Cacace and Antonio Saccoccio, under the title, *Europa e AmericaLatina, due continenti, un solo diritto. Unità e specificità del sistema giuridico latinoamericano*. Turin and Valencia: Giappichelli - Editorial Tirant Lo Blanch, 2020.

[43] For an overlook on this theme and a reflection on the contractual relationship conceived as "relational event", perpetually embedded in facts, cf. Massimiliano Granieri, *Il tempo e il contratto. Itinerario storico-comparativo sui contratti di durata*. Milan: Giuffrè, 2007, pp. 54ff.

The sociological[44] or philosophical descriptions[45] of an order founded on the "contract" (and, by virtue of the principle *pacta sunt servanda*, on the programmatic and immobilising effectiveness of the given word), will be integrated or replaced by new views and comprehensive conceptions of reality. In the context of social cooperation in the modern sense, these new approaches will spring from the centrality of the notion of "relationship", which no longer depends on a pre-existing agreement, the "contract" but which, nevertheless, represents the source and justification of this relationship. This being the result of a profound reconsideration of juridical relations, today more exposed than ever to the chaotic and irregular flow of events of our times.

This is the sense in which the language of the law – which always encompasses new words, stands on social reality, programmes for action – can contribute to finding the conditions on which to base a possible interaction between different cultural traditions. The thorny and complex problem of how different social systems may communicate arises at this juncture.

5. The problem of communication between systems

On the one hand, social systems learn, select and translate into their own language the stimuli coming from other systems; on the other hand, however, they can be closed and make themselves impenetrable when these stimuli – or "irritations"[46] – that arrive from outside are deemed incompatible with the laws providing stability to the system.

---

[44] I refer to the macro-sociological hypothesis articulated by Henry Summer Maine (*Ancien* Law, 1861. London and New York: Dent/Dutton, 1972, Italian trans. *Diritto antico*, ed. V. Ferrari. Turin: Giuffrè, 1998), where the contract – in place of *status* – constitutes the new foundation of the conventional system of trade and economic transactions; and, from another perspective – equally aimed at analysing the implications of the change in the mainstay of society from *status* to contract; see the model proposed by Émile Durkheim (*De la division* du travail social. Paris: Presses Universitaires de France; Italian trans., *La divisione del lavoro sociale*. Milan: Edizioni di Comunità, 1962), where there is a description of a society that serves the needs of a plurality of social aggregates.

[45] Here, the reference is to the "pure form of contract", that moved from the *lex mercatoria* to an increasingly broad field of application including international political relations. The scope of the *lex mercatoria*, "in addition to regulating civil society within the various countries", reached out "over international political relations themselves"; turning the "divinised" form of the contract into something of an "*uprooted law*", the "foundation of the only Order possible in this period of ceaseless transformation" (Massimo Cacciari, *Il lavoro dello spirito*. Milan: Adelphi, 2020, p. 18).

[46] On the question of how to make the stimuli and irritations coming from outside circulate in juridical systems, see the pages devoted to the thinking of Gunther Teubner (*Il diritto come sistema autopoietico*. Milan: Giuffrè, 1996, pp. 1ff.) in Alberto Febbrajo, "Dal diritto riflessivo al diritto frammentato. Le tappe del neo-pluralismo teubneriano", in *Diritto frammentato*, eds. Alberto Febbrajo e Francesco Gambino. Milan: Giuffrè, 2013, pp. 183-87.

In this respect, we are not referring so much to the theory of the "operative closure" of the social system[47], with which an internal mechanism can be explained, since the operations performed "are created and remain within the system", being intimately connected with the preceding and future ones produced by the same system[48], but to the theme of the communication between systems (with particular reference, here, to that between juridical systems).

This systemic theoretical approach – in which each system, with its own selection criteria works out structures so that expectations and the evaluation of actions can be formulated – can come useful in comparing juridical systems. Just as a variety of languages and different cultures came into being in the development of communication, so it can be explained – in an open systems theory – how the stimuli coming from outside can produce within the stimulated systems effects, reactions, and substantial variations.

In general, three situations can occur: (a) the first situation is that of *indifference*. When a system interacts with another system – with the norms, principles, and juridical concepts enunciated and applied in it – it may consider it, from its point of view, a body of indistinct and irrelevant facts. In other words, an instance of the mistrust that occurs between foreigners. (b) The second situation gives rise to a *structural modification*. Here the elements of "disturbance and irritation" that impact the system from outside introduce structural modifications in it[49]. It is possible to distinguish between *strong* structural modifications when the system incorporates the external element through political and legislative recognition (*e. g.*, agreements between States, introduction of a provision in the civil code, new laws, etc.); and *weak* structural modification when the recognition takes place through the motivated words of a judge (*e.g.*, the judicial decision draws on principles, arguments, juridical constructs taken from foreign systems). (c) The third situation consists in a *reaction of rejection*[50], aroused by the compelling need to defend one's identity. Where there are operations of selection, there lie criteria for the exclusion of alternative possibilities[51]. In this third hypothesis, there is no room for any contamination.

---

[47] Niklas Luhmann, *Des Recht der Gesellschaft*. Frankfurt a. M.: Suhrkamp, 1993, pp. 34ff.

[48] Alberto Febbrajo, *Sociologia del diritto*, p. 138.

[49] See Niklas Luhmann, *Introduzione alla teoria dei sistemi*, p. 47.

[50] In the systemic perspective, see Alberto Febbrajo, *Sociologia del diritto*, p. 138.

[51] In this sense, it is affirmed, in an anthropo-poietic perspective, that "*making humanity is always also making in/humanity*" (Francesco Remotti, *Prima lezione di antropologia*. Rome-Bari: Laterza, 2000, p. 160).

If we confine our attention to situation (b) – which is the relevant scenario in these pages – we must reflect on the possibility of reducing the external stimuli in something that makes sense within the, so to speak, "irritated" system.

The systemic approaches – recalled here in the comparison between juridical systems – significantly contribute to showing the ways in which systems interact, their deep-set cultural differences, what room is there for communication.

In comparison with Western systems, let us consider the protection of personality rights and of the environment introduced in the new Civil Code of the People's Republic of China, approved in its final version on 28 May 2020 as a result of a historic regulatory process brought about by the reforms undertaken in China over the last two decades[52].

It is not difficult to observe a cultural distance.

Our legal system – referring here to Italian law – was greatly influenced by our Constitution, introduced in 1948, by the *European Convention for the Protection of Human Rights and* Fundamental Freedoms (4 November 1950), by the previous *Universal Declaration of Human Rights, proclaimed by the General Assembly of the United Nations* (10 December 1948) and by jurisprudence and scholarly scientific research.

It is a system in which the exercise of control on the compliance of laws to the Constitution was devolved to the Constitutional Court – established in 1955. A fact that opened the way, in the study of juridical hermeneutics, to assessments based on "evaluative points of view": criteria for evaluation dictated by socially recognised values, to be found in the Constitution or in the civil code[53].

In connection with environmental law, we already mentioned the opinion movements that were set in motion by the associations for environmental protection whose claims made their way into jurisprudence and, to a certain degree, became legal provisions. In liberal democratic systems, founded on the principle of the separation of powers and on the protection of individual rights and freedoms, the pressure of intermediate entities in society on politics[54] produced: the protection of public interest concerning the environment; the right to bring

[52] Guido Samarani, *La Cina contemporanea. Dalla fine dell'Impero a oggi.* Turin: Einaudi, 2017, p. 346.

[53] Luigi Mengoni, *Ermeneutica e dogmatica giuridica.* Milan: Giuffrè, 1996, p. 85.

[54] As has been observed, pressure coming from below is exercised in environmental matters so that governments develop more thorough legislation, procedures and controls (Pope Francis, *Laudato Si'*, n. 179).

a case to court recognised for all the subjects, individuals or associations; the instrument of *class action* that, under United States law represents an effective judicial remedy, and which, according to Italian law, is recognised for the protection of homogeneous individual rights (l. 12 April 2019, n. 31); finally, the extension of the statute of limitations to file for damages, a term that may run from the day when environmental functionality is restored.

In these juridical systems, there are forms, techniques, and instruments of protection, at the centre of public debate and scientific attention, which will undoubtedly drive businesses to adopt strategies of sustainable development in the environmental field[55].

On the other hand, if we were to take on the observation viewpoint of a Chinese jurist, we would come across particular distinctions and novel points of view. We would see "what he sees" and "what he does not see"[56]. His dialogue with an Italian jurist could be complicated to such an extent as to even undermine the "*just enough*" to keep communication possible. Chinese law does not allow for the notion of "subjective right", common in liberal legal systems, that instead has no place in Chinese legal categories in the field of civil law, that are based, rather, on concepts of *power* and of *interest*. In the Chinese system, legal phenomena are intertwined with Confucian social traditions and culture on the one hand, and on the other with the non-independence of the judiciary from political power; in this system there is no authority tasked with checking on laws in terms of their compliance with the Constitution.

6. Points of cultural connection

In the context we have described – characterised by profound differences – the cultural pervasiveness of the recently adopted provisions in Chinese law on the environment will depend on their enforcement by judges in court. It will also depend on the perception – societal and individual – of the role and function that these judges perform in the evolution of Chinese law. This will be the result of their cultural sensitivity, fine-tuned by scientific discourses, academic debates and scholarly meetings with European jurists; of individual efforts to apply rights that

[55] On the responsibility of businesses to respect the environment, see Pope Francis, *Frateli tutti*, n. 122.

[56] The parenthesis is found in a page by Niklas Luhmann, who thus describes the theory of the observation of second order, aimed at capturing profiles of the phenomenology of intersubjectivity (in *Introduzione alla teoria dei sistemi*, pp. 132-33).

would otherwise remain a dead letter; and, finally, will depend on the judges' readiness to share the social claims and ideological trends expressed by Chinese mass organisations, which led the way after the "profound penetration in society of the State party that had characterised the first three decades of the People's Republic of China"[57] had slackened . In this perspective, the reference in the law to the rights of the person, to "freedom and human dignity" (art. 990, par. 2, *Chinese Civil Code*); the general obligation to care for the environment – laid down in arts. 4 and 6 of the *Environmental Protection Law of the People's Republic of China* – which is binding for the State, the citizens, and the entire Chinese community; the introduction in the *Book on Tort Law in the Chinese Civil Code* - in a separate section - of the discipline on "liability for damage to the ecological environment" (arts. 1229-35) represent in themselves legislative initiatives of extraordinary cultural importance.

Just as different cultures produce different legal solutions[58], so contemporary law, accessible to us through technological communication tools and discussed in public debates and in newspapers, penetrates cultures with its persuasive efficacy, generating new habits in thinking and action.

From a systemic point of view, when an "operational connection"[59] is excluded because of the difficulty or impossibility of sharing in the approaches of other systems (for example, the unlikely introduction in Chinese law of class actions in environmental matters), certain elements, tendencies or theoretical pathways are not necessarily prevented from being transposed into the system that tends to reject them. Thus, some tendencies of Chinese jurisprudence that, for deep-seated cultural reasons, keep account of the social effects of judicial decisions[60] can be seen - from the viewpoint of other systems - as referring to a particular argumentative technique, described by the English as "*consequentialist argument*" and by Germans as "argumentation oriented to consequences" (*folgenorientierte Argumentation*). Among several possible solutions, the criterion for choice in terms of juridical effects is based on the practical implications of the rule[61].

---

[57] Guido Samarani, *La Cina contemporanea*, p. 350.

[58] Rodolfo Sacco, *Il diritto muto. Neuroscienze, conoscenza tacita, valori condivisi*. Bologna: Il Mulino, 2015, p. 126.

[59] Niklas Luhmann, *Des Recht der Gesellschaft*, pp. 440ff.

[60] See, on the subject of *pure economic loss*, the pronouncement of Chinese law, *Yong An Construction Company V. Mingzhu Hospital* (Chongqing, 2006), in which – among the decision-making criteria – there are useful arguments offered by European juridical culture.

[61] Luigi Mengoni, *Ermeneutica e dogmatica giuridica*, pp. 91-92.

In line with what expounded in these pages, one could discuss the "cultural connections" that are involved in *transfering*, *transcribing* and *translating* meanings and thought patterns[62] which only practitioners – each in their own role – can forge (in scientific dissertations, in judicial decisions, in public discussions, in newspaper articles, etc.).

Once again, it is the language of the law, though filtered and weakened by the snares of translations, that creates a "bridge" between scholars and practitioners of the law of various origins on the foundation of *what is identical in different cultures, customs, and traditions*: the obvious and clear fact of human nature.

The interdisciplinary perspective of today's jurist thus gets disproportionately wider. The context we have illustrated is one in which comparative law must interact with the history of law; one where sociology must needs resort to anthropological studies.

7. Subjectivity in cognitive processes

In the forms of cultural interaction described above, there emerges the idea of a "tacit dimension" of knowledge, represented by the individual *forma mentis*, the normative convictions, the categories, the creativity and life experience of the subject[63].

In this context, it is useful to once again reflect on the philosophical theme of intentionality and on the role of the subject in cognitive processes, which paves the way for a new *episteme* that can develop in a "dynamic cultural context"[64]. Mankind is capable of articulating an idea only by moving "from his position and from his cognitive and emotional horizons, creating paths in

---

[62] On the history of translation problems, cf. Tullio Gregory, *Translatio linguarum. Traduzioni e storia della cultura*. Florence: Leo S. Olschki, 2016, pp. 16-17: in the hendiadys *transferre aut circumire* translation is seen "as the transfer of a text into a different language from the original, which is closely connected to any *translatio studiorum*, to any passing of civilisation and culture from one geographical, political and linguistic context to another, in order to save legacies which would otherwise be lost".

[63] On the tacit dimension of knowledge, which excludes the existence of experiences that are totally impersonal, see Pope Francis, A*ddress to participants at the international Conference for leaders of Catholic Universities: "New frontiers for university leaders. The future of health and the university ecosystem*, 4 November 2019.

[64] In these terms, Franco Anelli focuses on the horizons of the new *episteme* indicated by Pope Francis (in the inaugural speech to the academic year 2019/2020 at the Catholic University of the Sacred Heart, Milan, 28 November 2019).

which to believe and which, representing his vision of the world, provide him with personal evidence of an idea and a rule for action directed by what he thinks and believes is absolute"[65].

This is an attitude found "in human groups that share the same original tendencies, which are then embodied in common ways of feeling and acting"[66].

When searching for a "common way of feeling and acting" there emerges the great theme of the *Lebenswelt*, the science of the world of life[67]: "something that is 'prior' to any science of nature or of man, something radically vital and active in the actual existence of every human being in his/her quest for life and knowledge[68].

The concepts referred to above ("subjective right", "obligation", "contract", "relationship", "environment", "environmental law", etc.) are the product of abstractions that are indispensable so that different juridical cultures can communicate. With the help of these artificial entities, they can make use of a vast symbolic space in which to find points of contact and, to a certain degree, forms which they can relate to. However, these abstractions cannot be separated from the subject and from his/her intersubjectivity. The *break down (minuere*: to diminish, to divide) necessary for the construction of theoretical practice[69] should not let us lose sight of the path undertaken: moving towards *telos*[70].

That explains the need not to *forget (obliare)*, but, rather, to constantly go back to the original operations with which a science – juridical science included – builds the constructions that are necessary to it[71].

8. Return to the world of doxa

In dealing with the problem of a precategorial foundation of the sciences, there emerges Vico's theme of common sense, which leads us to Husserl's conception of *doxa*[72].

---

[65] Edmund Husserl, *Il destino della filosofia* (1976), Italian trans. Nicola Zippel, Rome: Castelvecchi, 2014, p. 41.

[66] Ibid., pp. 41-42.

[67] Carlo Sini, *Enzo Paci*. Milan: Feltrinelli, 2015, p. 33.

[68] Ibid., p. 33.

[69] Ibid., p. 34.

[70] In connection with the study of structure in the anthropological thought of Lévi-Strauss, cf. Enzo Paci, "Strutturalismo, fonologia e antropologia", in *Il senso delle parole (1963-1974)*, ed. Pier Aldo Rovatti. Milan: Bompiani, 1987, p. 44-45.

[71] Ibid., p. 44.

[72] Carlo Sini, *Enzo Paci*, p. 33.

The reference to *doxa* allows for a more adequate understanding of what "is meant" by the famous expression "world of life"[73]. It is a world that can be compared to the oneness of a "living organism", that is accessible only "if one goes back to its hidden roots, if one pursues, systematically and in all its operations, the life that stirs in it, that pulls it forward, that shapes it"[74].

All the spiritual operations of human beings – whether carried out "in a strictly personal way, or in a cultural dimension" – "are always preceded by a universal operation that is presupposed in any human practice and any pre-scientific life"[75]. We always find ourselves "in a world experienced as prescientific and pre-individual based on the obvious nature of *doxa*, a world that is the premise for any scientific consideration and any human activity"[76].

What is *identical in what is different*, is, in this paper, the basic premise for an interdisciplinary investigation of social and juridical phenomena. Indeed when addressing integral solutions to the environmental question[77], this phenomenon is played out on the anonymous plane - shared by each interlocutor - of "what everybody knows"[78]. In other words, it refers to a "general know-how" which for all time and at every moment "imposes its cogent power and conditions over every other activity or specific knowledge"[79].

It is necessary to be aware that, before the world of science – "before any kind of scientific, philosophical, psychological or sociological consideration" – we find ourselves "in the precategorial world"[80]. Husserl's theme of intersubjectivity, without which scientific objectivity would not be possible, is bound up with the problem (that is significant in anthropological studies) of the relationship between civilisation and culture. With a view to building authentic humanity, all civilisations, including the primitive ones are functionally involved in achieving the *telos* of humanity[81]. In this unifying vision of knowledge, both implicit and explicit, of practices and understandings, the

---

[73] Carlo Sini, "Il piccolo inizio: vita materiale e realtà", in *Nóema,* n. 2, 2011, p. 3.

[74] Edmund Husserl, *La crisi delle scienze europee e la fenomenologia trascendentale* (1959), Italian trans. Filippini. Milan: Il Saggiatore, 2015, p. 140.

[75] Edmund Husserl, *La crisi delle scienze europee e la fenomenologia trascendentale*, p. 139.

[76] Enzo Paci, "Doxa e individuazione nella fenomenologia di Husserl", in *Rivista di filosofia*, 1960, p.161.

[77] Pope Francis, *Laudato Si'*, n. 139.

[78] Carlo Sini, *Il piccolo inizio*: *vita materiale e realtà*, p. 4.

[79] Ibid., p. 4.

[80] Enzo Paci, *Doxa e individuazione nella fenomenologia di Husserl*, p. 160.

[81] Ibid., p. 46.

"techniques and the means that make civilisation possible are not the property of this or that group, of this or that people, but of all the human beings on planet Earth"[82].

[82] Ibid.

LAUDATO SI' AND MANAGING THE COMMONS

Beyond Neo-Realism and Neo-Liberalism

Paul Tang Abomo, SJ

1. Introduction

In the global context, collective or public goods include the natural commons such as the high seas, atmosphere, ozone shield, polar regions, etc. The use of these collective goods highlights the interdependent aspect of human activities and choices. A major risk with common pools is that overuse may result in destruction of the resource. Each individual's use increases the costs or decreases the value to other users. Decisions made by one individual or state actor can considerably affect other individuals or states, which will suffer unanticipated negative consequences as a result of wrong choices made by others. Rules need to be established to manage and govern these commons for the benefits of all. Equally important for maintaining life and enhancing its quality is the use of this common environment. Every activity we engage in whether breathing in and out, growing food, or generating electricity, creates unwanted by-products that affect in one form or another the environment. We want clean air, unspoiled landscapes, and pure water in order to sustain life and health, and indulge our aesthetic sensitivities. The question then becomes how to manage natural global commons to ensure fairness among individual or state users while at the same time protecting and nurturing the same common resource. This paper is an investigation into *Laudato Si's* response to this "tragedy of the commons". I argue that the papal encyclical offers a different perspective that goes beyond neo-realist and neo-liberal institutionalist ap-

DOI 10.32060/C.S.9.131-144

proaches. It calls for an "ecological conversion". The paper is divided in three major parts. The first one sets the stage to the well-known debate on the tragedy of the commons. The second part reviews the solutions that been offered by the neo-realist and neo-liberal institutionalist approaches. Each time I will point out the limits of these approaches and present in the last part the position of the Encyclical and its argument in favor of an ecological conversion.

2. Managing the Commons: setting the debate

A center piece in understanding global governance of collective and public goods is the article of Biologist Garrett Hardin, "The Tragedy of the Commons"[1]. The article tells the story of a group of herders sharing in common a grazing land. Each one finds it individually sustainable to increase the size his own herd, allowing him to sell more in the market and thus make more profits. However, if all herders follow the same individually rational path, the group loses; too many animals graze the area and the quality of the pasture deteriorates, leading to decreased outcome for all. As each person rationally attempts to maximize his own gain, the entire collectivity suffers, and eventually, all individuals suffer. Hardin argues that when personal and communal interests are at odds, overexploitation of resources is inevitable. The same fate, he notes, is likely to befall any shared limited resource, from the atmosphere and oceans to national parks and rivers.

"The Tragedy of the Commons", published in 1968, has become a reference that has subsequently shaped debates on private property, resource management and environmental sustainability. Human collectives, Hardin argues, are spaces of generalized competition, where individuals confront one another because competition between individuals, groups, and species is inherent in biological life. Practically, all resources are likely to be overexploited by some herders who never seem to pay their share of production cost. This "natural" state of conflict is an integral part of human groups. Like ecosystems, societies are battlefields where the struggle for survival rages on: "No activity of man is without competitive use"[2]. And this temptation is universal since abstaining from overexploitation does not have any advantages for the prudent consumer but everyone still incurs the same costs. Therefore, virtually no

[1] Garrett Hardin, "The Tragedy of the Commons", in *Science 162* (December 13, 1968), pp. 1243-1248.

[2] Garrett Hardin, *Nature and Man's Fate*. New York: Rinehart and Co., 1959, p. 253.

benefit is gained from good behavior whereas selfish strategy seems to pay. In other words, individuals with self-interest strategies will overuse the resource necessary for the whole group and will adopt this selfish strategy until the resource falls below the threshold resulting in extinction or depletion. The "tragedy of the commons" has an unavoidable consequence. Common property is incompatible with a sustainable exploitation of resources.

The scope of this tragedy has been extended to global entities such as the Earth, oceans, the atmosphere, etc. in an attempt to find out concerted ways to ensure the sustainability of our common environmental resources such as water, air, and other natural and mineral resources. As Aurelio Peccei puts it, "the entire planet thus provides a typical example of what Garrett Hardin termed the tragedy of the commons – the hopeless fate of something belonging to the community and which everyone tries to exploit more than, or before, the others, without taking the slightest care of it in the common interest"[3]. Does this mean we are doomed to plunder the world's resources and trash our planet?

Even Hardin wasn't entirely pessimistic. He noted that groups can create institutions to manage their communal resources, although these usually fail because of "free-riders" – individuals who try to reap the benefits of cooperation without paying any of the costs. In order to prevent environmental disaster, it is necessary to resort to severe forms of governance that proceed from "mutual coercion, mutually agreed upon by the majority of the people affected"[4]. The coercion is based on a choice between property rights and centralized administration. If we want those resources to last, we must either privatize them or have a higher political and administrative organization control them[5].

3. The neo-realist solution

This approach blames the deterioration of the environment on the insufficient expansion of private property: because these commons are not appropriated and given a trade value, their use is free and limitless, which leads to their

---

[3] Aurelio Peccei, *The Human Quality*. Oxford; New York: Pergamon Press, 1977, p. 76.

[4] Garrett Hardin, "The Tragedy of the Commons", p. 1243.

[5] Garrett Hardin, *Exploring New Ethics for Survival*. Baltimore: Penguin Books, 1973, pp. 109-118; Garrett Hardin, "Extensions of the 'Tragedy of the Commons'", in *Science* 280 (May 1, 1998), pp. 682-683.

deterioration. The proposed solution is to create new property rights. Selling these or trading these property rights will discourage actors from deteriorating the environment and open the way to economic sustainability[6].

3.1 Private ownership and Individual Property

Private allocation of environmental resources and liability for damages could function efficiently within the traditional private market system. If enforceable property rights exist, the possessor of those rights has an incentive to use the resource in such a manner that it provides the highest benefits and is not destroyed. These rights could also be sold for beneficial nondestructive use to those who value them more. Thus an individual pursuing his own interests can be expected to utilize efficiently resources and contribute to the maintenance and sustenance of common resources. Many natural resources such as land, mines, forests, beaches are privately owned and managed. In exchange for this management of the resources, the owner receives payment from users.

The problem with private ownership is that global entities like air, water, and many public land benefits are not easy to capture. Moreover, exclusion or limitation of users, a necessary condition for preservation of an overused common pool, may be virtually impossible and the cost of enforcement may exceed the value of the resource itself. While many examples of efficient resource management through privatization exist, the cost of enforcing property rights, including the right not to be damaged, makes some resources unsuited for relatively unregulated, private management. It may also be very costly or impossible for a private owner to determine the full effects of environmental resource use because their occurrence is distant in either space or time. Private ownership of natural resources leads to the most efficient resource use when there are no third party effects from use, and when users of the resource can easily be charged[7]. These practical limitations could suggest public ownership and management as a credible alternative in preserving environmental resources.

---

[6] Valérie Boisvert, Armelle Caron, and Estienne Rodary, "Privatiser pour conserver? Petits arrangements de la nouvelle économie des ressources avec la réalité", in *Tiers-Monde* 45 (2004), pp. 61-84.

[7] Robert L. Bish, "Environmental Resource Management: Public or Private?" in Garrett Hardin and John A. Baden, and Douglas S. Noonan eds. *Managing the Commons*. San Francisco: Free Publishing, W. H. Freeman, 2017, pp. 74-78.

3.2 Public management

Public management of the common resources means that public officials are in charge of deciding by whom, how, and how much a resource is used, rather than private market transactions. Publicly owned resources are likely to generate high rents and revenue fees that could allow policing, rules enforcement and resource patrolling against free riding. The use of a common resource is often subject to competing claims, each potential user listing all the reasons why he or she, rather than some other user, should be permitted to use the resource. It might also be impossible to identify the value various users place on the resource because preferences cannot be evaluated in terms of market transactions. For example, in the use of common forest resource groups organized around timber or mining interests have different interests than cattle ranchers or individual hikers, campers, and fishermen. Only public officials can manage these competing claims.

The problem here is how laws, rules and regulations are established. Rules generally favor the interests of powerful corporations and exclude other users. Marine users such as commercial fishermen, chambers of commerce, and allied industrial and commercial shoreline users have greater influence on shoreline than regular swimmers, surfers, or other picnickers who would like coastlines to remain in their natural state[8]. These limitations in balancing competing individual and groups interests underline the limits of public management or ownership of the commons. The bottom line is, introducing private rights in what is common is fictionally erasing the problem instead of solving it. When ownership is introduced in the management of common areas, "the resource ceases to be common property and the problem is solved at one stroke"[9].

Today we understand human nature and motivation far better than we did in Hardin's day. In particular, we know that individuals do not always act selfishly but also have some regard for the interests of others and the natural environment. Games such as the prisoner's dilemma and the public goods game demonstrate that under certain conditions people do behave altruistically[10]. Countless success stories attest to the fact that communities can overcome the

---

[8] Robert L. Bish, "Environmental Resource Management: Public or Private?", pp. 80-82.

[9] Partha S. Dasgupta and Geoffry M. Heal, *Economic theory and exhaustible resources*. Cambridge, UK: Cambridge University Press, 1979, p. 77.

[10] Mark van Vugt, "Triumph of the commons", in New Scientist | 22 August 2009, pp. 40-43.

tragedy of the commons without a great deal of coercion. And this is the point made by the neo-liberal institutionalists.

4. The neo-liberal approach: self-governing institutions matter

Neo-institutional scholars have also challenged Hardin's reasoning on the impossibility of managing the commons, using concrete cases to invalidate his conclusion: for centuries, forests, pastures, and fishery resources have been exploited in common without heading for destruction. The publication in 1990 of *Governing the Commons* by Elinor Ostrom – recipient of the Nobel Prize in Economics in 2009 – is an important milestone in that direction. Ostrom and the neo-institutionalists argue that some common environmental resources have been able to survive throughout history despite high levels of uncertainty. This has been made possible because of the existence of strong and robust institutions able to maintain and sustain common resources systems.

Commitment and monitoring are strategically linked in a process that produces private benefits for the monitor and joint benefits for others. The process, Ostrom argues, is more efficient where the "guards" are internal agents, that is, when "the appropriators themselves play a major role in monitoring each other's activities"[11]. As evidence to her claim, she presents historical cases of Swiss and Japanese villages where mountain commons have survived in the midst of a fragile world of avalanches and unpredictable precipitation. She also analyses how large-scale irrigation systems have been kept in order and maintained in the difficult terrain of the Philippines islands. From these cases she identifies a set of underlying design principles and knowledge used to maintain, monitor and sustain these common areas over time.

For centuries, Torbel peasants in Switzerland have tended village cattle on communally owned alpine meadows during summer months. And these common meadows have survived because of strong institutional rules and regulations. The commons were everyone's business. Meadows statutes were voted by all the villagers who also gave legal authority to the local association to manage the meadows. Substantial fines were imposed for any attempt by villagers to appropriate more grazing areas than they had rights to. The association met annually to debate general rules and policies and elect officials. The latter were in charge of hiring the meadows staff, impose fines for misuse

---

[11] Elinor Ostrom, *Governing the Commons. The Evolution of institutions for collective action*, p. 59.

of common property, organize the annual maintenance work, such as building and maintaining roads and paths to the meadows, etc. Overgrazing was prevented by tight controls. Meadows were maintained by investments in weeding and manuring the summer grazing areas. The pattern of sustainability here was deeply rooted in self-governing institutions[12].

In Japan, millions of hectares of forests and uncultivated mountain meadows have been kept and managed in common by thousands of villagers. The environmental conditions of these Japanese villages – Hirano, Nagaike, and Yamanoka – are similar to those of the Swiss village Torbel. Here too the common areas have been regulated by an assembly of elected village officials established to serve the long-term interests of the peasants whose livelihood depends on these commons. Villagers are required to perform collective work for the maintenance of the commons. Each household had to contribute in activities such as annual burning or specific cutting of timber or thatch. Each village devised its own monitoring and sanctioning system. In some villages, local detectives daily patrolled the commons on horseback in groups of two, looking for potential defectors. In others, the task was carried out by males on a rotational basis. It was a requirement for everyone to report violations. Various penalties existed according to the type of violation, ranging from deprivation of their illegal harvest and sizable fines to complete banishment from the village[13]. Here too the common resources system was maintained by self-governing institutions.

A striking similarity in managing the commons is reported in *zanjera* irrigation communities in the Philippines. Zanjeras were born out of the initiative of enterprising tenant farmers banding together to construct an irrigation system on previously non-irrigated land. In return everyone was given rights to a defined portion of the newly irrigated land. Each member of the association was obligated to contribute one day's work during each work season. The members elected an executive team with the task of supervising construction activities, motivating individuals who had to contribute many hours of physically exhausting labor for routine maintenance. There was also a system of incentives and sanctions that help keep up the sustenance and the maintenance of this common and collective good[14]. Here again the system was maintained thanks to self-governing rules and regulations. The point of these three his-

---

[12] Elinor Ostrom, *Governing the Commons*, pp. 62-65.

[13] Elinor Ostrom, *Governing the Commons*, pp. 65-69.

[14] Elinor Ostrom, *Governing the Commons*, pp. 82-88.

torical instances is that common areas and collective goods can be and have been maintained and sustained by robust self-governing institutions. They have relied on collective choice arrangements, a monitoring system guaranteeing accountability and control of actor's behavior, and a system of graduated sanctions against violators operational rules[15].

Hardin believed that such institutions cannot avert the tragedy of the commons without coercion because of the problem of free-riders. In our technologically and culturally complex modern world, common resources systems are managed by institutions, from public companies distributing water to governments limiting air pollution through quotas. In fact, the problem is worse today. The anarchical structure of the international community creates a second-order free-rider problem. Who will patrol the state? This raises the eternal platonic problem as to who will guard the guardians[16]. Former U.S. Vice President Albert Gore fits squarely within this pessimistic vision. His experience led him to the conclusion that the only remedies to the problems before us lie in fundamental changes in our thinking. The solution involves more than mechanical solutions. He cautions that "our children will inherent a wasteland" unless we "dramatically change our civilization and our way of thinking about the relationship between humankind and the earth" [17]. This vision is shared by *Laudato Si'*.

5. *Laudato Si'* and managing the global commons

5.1 Beyond private/public ownership

In an increasingly crowded and interconnected planet, local action and choices have an impact on the planet's natural systems and on our neighbors close and far away. A point of contention has been to establish whether or not a climate or earth policy is a violation of the rights of those who own coal, oil, gas, and other natural resources. *Laudato Si'* argues in favor of the subordination of private property to the "universal destination of goods" (LS, 93). If the earth as a totality is a global common good worthy of protection, these rights should be defined in such a way as to satisfy the social function, that is, the need to serve the common good. The Pope elevates, in the collective

---

[15] Elinor Ostrom, *Governing the Commons*, p. 90.

[16] Plato, *The Republic*, New York: G.P. Putnam's sons, 1980, 403e.

[17] Al Gore, *Earth in the Balance*, Boston: Houghton Mifflin, 1992, pp. 163.

consciousness of humanity, the status of the earth and climate atmosphere to a common global good. He declares the climate and the atmosphere common goods "of all and for all" (LS, 23).

Oceans and other natural goods should also be considered global commons and protected by an appropriate governance system (LS, 174). This position comes as a challenge to the international community because the recognition of the atmosphere and the climate as common goods could have legal consequences: if, for example, nations were threatened, the obligation to protect could result. Some parties to the United Nations Framework Convention on Climate Change feared just that. In the IPCC's Fifth Evaluation Report they refused to recognize climate change as a problem related to a "common good". A footnote states that attributing the concept of "common good" to the climate issue has no implication for a global agreement or for establishing international burden-sharing criteria for climate protection[18].

5.2. Towards an 'ecological conversion'

> Saint Francis of Assisi reminds us that our common home is like a sister with whom we share our life and a beautiful mother who opens her arms to embrace us[19].

The Encyclical invites us to go beyond the misplaced anthropocentric and materialistic conception that considers that creation must be used for the only selfish advantage of human beings. A shift must be made from exploiting and managing common resources systems for the advantage of humanity, to the protection of the creation which we are part of. Throughout the Encyclical, Pope Francis stresses the interconnectedness of all of creation. In managing the environment, we should keep in mind the relationship that exists between human society and the nature which lives in it. Nature cannot be regarded as something separate from ourselves or as a mere setting in which we live. We are part of nature, included in it and thus in constant interaction with it (LS, 139). In fact, it is a completely novel outlook which sees all of creation as interrelated. It is something new and radical. He calls for an "ecological

---

[18] Ottmar Edenhofer O. et al., *Climate Change 2014: Mitigation of Climate Change. Contribution of Working Group III to the Fifth Assessment Report of the Intergovernmental Panel on Climate Change*. Cambridge: Cambridge University Press, 2015, p. 5.

[19] Pope Francis, Encyclical Letter *Laudato Si' on Care for Our Common Home*, 24 May 2015, n. 1.

conversion", wherein man finds his true and just place in relation to all that is around him and thus begins to behave in a manner that is coherent with the common good of all creation.

5.3. Perils in the global commons

> This sister now cries out to us because of the harm we have inflicted on her by our irresponsible use and abuse of the goods with which God has endowed her. We have come to see ourselves as her lords and masters, entitled to plunder her at will[20].

Our common home, our planet is irreplaceable. And it is in peril. From the melting of our polar icecaps to the acidification of our oceans, humans are changing the planet in ways that threaten our own existence. Around the world the poorest populations bear the heaviest burden. Pope Francis calls on humanity to change our own fate. *Laudato Si'* was his rallying cry. It is an urgent appeal to care for and protect our planet. Common air resources are in danger:

> There is also pollution that affects everyone, caused by transport, industrial fumes, substances which contribute to the acidification of soil and water, fertilizers, insecticides, fungicides, herbicides and agrotoxins in general [...] Each year hundreds of millions of tons of waste are generated, much of it non-biodegradable, highly toxic and radioactive, from homes and businesses, from construction and demolition sites, from clinical, electronic and industrial sources. The earth, our home, is beginning to look more and more like an immense pile of filth[21].

Common water resources are also in peril.

> Water supplies used to be relatively constant, but now in many places demand exceeds the sustainable supply, with dramatic consequences in the short and long term. Large cities dependent on significant supplies of water have experienced periods of shortage, and at critical moments these have not always been administered with sufficient oversight and impartiality[22].

The common earth resources are in danger:

> "The earth's resources are also being plundered because of short-sighted approaches to the economy, commerce and production. The loss of forests and

---

[20] Pope Francis, *Laudato Si'*, n. 2.
[21] Pope Francis, *Laudato Si'*, nn. 21 and 22.
[22] Pope Francis, *Laudato Si'*, n. 28.

> woodlands entails the loss of species which may constitute extremely important resources in the future, not only for food but also for curing disease and other uses. Different species contain genes which could be key resources in years ahead for meeting human needs and regulating environmental problems[23].

The encyclical further denounces the false idea that there is an infinite supply of earth's goods and that planet resources are unlimited. This error has led to the planet being squeezed dry beyond every limit (LS, 106). *Laudato Si'* invites the entire humanity to be part of the solution in protecting our global commons from climate change and environmental deterioration. This starts with a new dialogue about how we are shaping the future of our planet (LS, 14). This calls for a change of attitude, from dominion to stewardship.

5.4 From Dominion and exploitation to Stewardship

Dominion and exploitation both refer to the inescapable fact that humans live in relationship to the created order – first as creatures of the created order ourselves, and second, as caretakers of all creation. Both roles are intrinsically intertwined. Humans are continually interacting with creation in our daily activities, from driving to an office, growing food in soil or catching fish in rivers, lakes and oceans. If dominion is responsibility to care for creation, then stewardship describes how we exercise this dominion. We govern creation for the good of humanity and other parts of creation, not in a domineering, selfish and exploitive manner, but by nurturing creation, preserving it. Our dominion is not for selfish exploitation of people or creation. Our understanding of dominion is that it is not unlimited, but is intended to ensure creation functions properly.

The word stewardship is connected with "watch", "guard" and "protect". Humans' stewardship of creation means ensuring that it functions and prospers into the future. The concept denotes careful and close observation – specifically, an attentive and protective tending. This aspect of caring for creation means we are to guard, nurture and protect creation – a good articulation of roles of a steward or servant. Growing vegetables and crops, for example, takes nutrients out of soil. To preserve soil's ability to produce, we must manage the soil and put nutrients back. These activities are inseparable. If we want creation to sustain us, we need to care for creation. If we want creation to feed us, we need to steward creation appropriately. If we are stewards, resources

---

[23] Pope Francis, *Laudato Si'*, n. 32.

at our disposal are not our own. Stewardship is particularly connected with caring appropriately for creation[24].

Humans are not independent actors with regard to creation. Humans find their origins in the soil of the earth. They have no innate ownership of creation. Humanity inhabits creation, but it is not ours. Therefore, our use of creation should preserve its ability to provide – clean water, clean air, sufficient grass and trees, etc. – for the needs of other parts of creation. We cannot simply do whatever we want with earth's resources. Because all creation is connected, both immediate local and long-term global consequences can occur whenever one part of creation is harmed. The earth has complex natural structures, systems and organisms, and when stewarded properly, all work together to provide human sustenance. Wendell Berry writes, "The care of the earth is our most ancient and worthiest and, after all, our most pleasing responsibility. To cherish what remains of it, and to foster its renewal, is our hope"[25]. This responsibility also entails that we care for the others.

5.5. Equal Sharing of the global commons with our contemporaries

In our globally connected world, it is undeniable that our actions and inactions affect the lives of others, both near and far away. As stewards of creation, human beings must ensure it bears enough fruit for all to eat, clean air for all to breathe, pure water for all to drink, etc., for current and for future generations. It becomes a question of justice and accountability. Human activity has the potential to disrupt creation's functioning and provision for others. For example, dumping unwanted waste into a river may 'take away' the waste, but it means the river is unable to provide clean water to others who depend on it for drinking, fishing, irrigation and the like. Or as *Laudato Si'* says,

> The depletion of fishing reserves especially hurts small fishing communities without the means to replace those resources; water pollution particularly affects the poor who cannot buy bottled water; and rises in the sea level mainly affect impoverished coastal populations who have nowhere else to go[26].

---

[24] Jared Hyneman, Christopher Shore "Why Are We Stewards of Creation? World Vision's Biblical Understanding of How We Relate to Creation", in Natural Environment and Climate Issues, May 2013, pp. 16-18. World Vision's Biblical Understanding of How we Relate to Creation_Full.pdf (wvi.org).

[25] Wendell Berry, *The Unsettling of America: Culture & Agriculture*. San Francisco: Counterpoint, 1996.

[26] Pope Francis, *Laudato Si'*, n. 48.

Since the earth provides resources for human needs, these resources must be used in ways that provide well and for all people today, rich and poor as well. It is a matter of justice in the distribution and use of the commons.

5.6 Equal Sharing of the global commons with the future generation

Creation is not the sole preserve of today's generation. Future generations will need the earth to provide for their needs. Stewardship of creation must take into consideration its purpose in providing for other human beings beyond ourselves. Humanity's use of creation must promote – not compromise – the ability of creation to ensure the ability of the planet to provide for humans now and in the future, for current generations and generations to come. Proper stewardship has one eye tending the earth for today and one eye toward tending the earth for future generations. As Pope VI wrote,

> We are the heirs of earlier generations, and we reap benefits from the efforts of our contemporaries; we are under obligation to all men. Therefore, we cannot disregard the welfare of those who will come after us to increase the human family. The reality of human solidarity brings us not only benefits but also obligations[27].

6. Conclusion

In the commons, resources are abundantly available for all with no one having an exclusive monopoly or control over them. It becomes a tragedy when, through competition individuals selfishly exploit the common resource until it is completely depleted and made available to no one. Each individual is tempted to exploit the resource, thus reaping maximum payoffs while sustaining restricted consumption cost. The neo-realist solution to this conflict is privatization or public management of this common resource. But this view obscures the multiple historical examples where human beings have been able to cooperate without coercion. Neo-liberal institutionalists argue that this is the best way to manage and sustain common resource systems is through self-governing institutions. But these views remain anthropocentric as they put creation at the exclusive service of human interests. *Laudato Si'* makes a shift from this materialistic and anthropocentric relation to the natural global

---

[27] Saint Paul VI, *Encyclical Letter Populorum progressio on the Developmen of Peoples*, 26 March 1967, n. 17.

commons and stresses the interconnectedness of all of creation. Nature should not be regarded as something separate from us or as a mere setting in which we live. We are part of nature, included in it and thus in constant interaction with it. This ecological conversion creates a new relation with creation, from dominion to stewardship. This implies new responsibilities. Stewardship calls us to use creation for the sake of human welfare and needs in areas such as farming, fishing, mining, energy generation, engineering, construction, trade and medicine, etc. But, as we provide for human welfare and needs, we are also commanded to care for the earth and all its creatures.

DIALOGUE AS FOUNDATIONAL?

The Role Faith-Based Groups Play in the Environmental Movement in Ireland

Ciara Murphy*

1. Introduction

"The urgent challenge to protect our common home includes a concern to bring the whole human family together to seek a sustainable and integral development, for we know things can change"[1].

With these words Pope Francis is calling on the entire global population to work together to build an ecological movement. In his encyclical letter, *Laudato Si'*, Francis highlights the urgency and scale of the complex crisis that is both environmental and societal. He argues that "everything is closely interrelated. Today's problems call for a vision capable of taking into account every aspect of the global crisis"[2]. Foundational to this process of collaboration is dialogue which develop networks, creating webs of understanding within the ecological movement.

The conversation that is taking place is complex and layered. The dialogue itself weaves across all parts of the environmental movement from engaging people, to building consensus, influencing policies both at national and international levels as well as action on the ground. The complexity which re-

* I owe a huge debt of gratitude to Dr Kevin Hargaden, Martina Madden and Keith Adams for their thoughts and editorial help.

[1] Pope Francis, Encyclical Letter *Laudato Si' on Care for Our Common Home*, 24 May 2015, n. 13.

[2] Pope Francis, *Laudato Si'*, n. 137.

DOI 10.32060/C.S.9.145-183

sults from this dialogue can make the ecological movement seem opaque and undiscernible. Picking apart the threads of this network or networks allows us to gain insights into the function of this movement and facilitates greater understanding of how we can improve and enhance its functioning, including discerning the specific role of faith-based groups.

To facilitate this discussion, I will argue in this chapter that when considering the ecological movement through an integral ecology lens, it might be useful to borrow some of the concepts and terminology of biological ecology. In the same way that in the natural environment complex interactions exist, so too does the ecological movement consist of complex interactions that require precise description. In the case of the environmental and social movements this would invariably manifest as dialogue - words can invite people into the movement, catalyse ecological conversion and lead to action.

Drawing on a wide range of resources including Pope Francis's encyclicals *Laudato Si'* and *Fratelli tutti*, other Catholic teachings, media coverage of Irish and world events, as well as academic research covering sociology, ecology, politics and theology, this chapter explores how faith-based communities, and the catholic church specifically, can enter into dialogue with the ecological movement. The role of these communities within the ecological movement is discussed as well as the opportunities and challenges of this complex collaboration. However, before we enter a discussion about the importance of dialogue in the ecological movement and the role of faith based communities within environmental movement, we first must consider what we mean when we talk about dialogue and what an ecological movement actually is and how it functions. Allowing that this chapter examines the particular circumstances of the Irish ecological movement it is also important attempts are made illuminate the specific circumstances of the movement in Ireland and assess the political, societal and religious context within which it sits[3].

2. Pope Francis on the "foundational" importance of dialogue

Francis is explicit in his challenge to build a diverse and broad movement through dialogue. "We need a conversation which includes everyone, since the environmental challenge we are undergoing, and its human roots, concern and

[3] A detailed account of the legislative, political and societal landscape in which the climate crisis in Ireland can be understood is detailed in David Robbins, Diarmuid Torney, and Pat Brereton (ed.), *Ireland and the Climate Crisis*. London: Palgrave Macmillan 2020.

affect us all"[4]. He recognises the need for dialogue across the entire ecological movement, both within and between different strands of the movement.

Of religious and faith based groups he stresses the imperative for "religions to dialogue among themselves for the sake of protecting nature, defending the poor, and building networks of respect and fraternity"[5]. In saying this he is unequivocal in asking the global faith community to contribute actively in this ecological movement. As this encyclical is inviting "all people of good will"[6] into this conversation, Francis is himself reaching out and inviting open dialogue between religious and non-religious people. In doing so he mandates faith-based communities to do like-wise.

Pope Francis in this encyclical recognises that dialogue and meaningful engagement is fundamental within and between faith-based communities but also in 'secular' society. The siloed and specialised nature of sciences and organisations can hamper holistic and integral ecological solutions to these complex issues. Within each field solutions are sought for problems that are the manifestations or symptoms of wider complex problems and are multifaceted in their cause. Even with ecological movements, different ideologies and world views can inhibit meaningful action and solutions when dialogue is absent[7].

Dialogue is both the process to and the action of ecological conversion. These complex conversations must be evolving and continuous. There exists no defined point at which we can say the conversation has ended and we can stop, constantly working towards something better is the only way in which not to fail[8]. In *Fratelli Tutti,* published five years after *Laudato Si*, Francis again articulates the need for progress through dialogue very clearly: "Approaching,

---

[4] Pope Francis, *Laudato Si'*, n. 14.

[5] Pope Francis, *Laudato Si'*, n. 201.

[6] Pope Francis, *Laudato Si'*, n. 62.

[7] "Dialogue among the various sciences is likewise needed, since each can tend to become enclosed in its own language, while specialization leads to a certain isolation and the absolutization of its own field of knowledge. This prevents us from confronting environmental problems effectively. An open and respectful dialogue is also needed between the various ecological movements, among which ideological conflicts are not infrequently encountered. The gravity of the ecological crisis demands that we all look to the common good, embarking on a path of dialogue which demands patience, self-discipline and generosity, always keeping in mind that "realities are greater than ideas" , Pope Francis, *Laudato Si'*, n. 201.

[8] The risk that dialogue can morph into another form of delay and denial is recognised by Pope Francis. Kevin Hargaden's *Studies in Christian Ethics* essay "An Apocalypse Converted: William Stringfellow and Catholic Social Teaching on Climate Breakdown" (2021) examines how we can steer this conversation away from false solutions.

speaking, listening, looking at, coming to know and understand one another, and to find common ground: all these things are summed up in the one word 'dialogue'. If we want to encounter and help one another, we have to dialogue"[9].

The role dialogue, and by extension networks, play in environmental and social movements cannot be overstated. Networks within and between movements are the glue that keeps diverse members, across every scale (individual, communal and systematic) of the ecological movement together. Laurence Cox, a sociologist at the Maynooth University, argues that to understand social networks:

> we need to put the emphasis above all on the people involved in creating this collective agency, in whatever way, and to ask about the relationships between them. In other words, we look for the networks – formal and clientelistic or informal and radically democratic, with many other shades in between – that connects different kinds of formal organisations and informal groups, parties and trade unions, cultural figures and politicians and even (in some cases) churches, online media, sub-cultures, everyday forms of resistance, popular memories of past revolutions or lifestyles[10].

The development of these networks through dialogue and collaborations between different movements in different areas does not mean conformity and homogeneity within the ecological movement. While the global and Irish ecological movements are intrinsically linked, how this movement manifests itself in different countries and regions will be different. The environmental justice movement would manifest itself differently depending on the societal, political and environmental context; this makes the movement in Ireland different from the rest of Europe and further afield. This holds for smaller scale variations. Even within Ireland, the movement has different focuses depending on regions and urban or rural context. Pope Francis recognised this when he declared: "*There are no uniform recipes*, because each country has its own problems and limitations"[11]. This diversity of movements should allow us to "learn from each other's struggles"[12] but only if we have open dialogue, a conversation that includes everyone.

---

[9] Pope Francis, Encyclical Letter *Fratelli tutti on Fraternity and Social Friendship*, 3 October 2020, n. 198.

[10] Laurence Cox, *Why Social Movements Matter: An Introduction*. Lanham, Maryland: Rowman & Littlefield International, n. d., p. xii.

[11] Pope Francis, *Laudato Si'*, n. 180.

[12] "How Do We Become Able to Win? Learning from Each Other's Struggles", accessed September 4, 2020, https://whysocialmovementsmatter.com/news/2020/7/17/how-do-we-become-able-to-win-learning-from-each-others-struggles.

It is in the context of increased diversity and collaboration, in which faith-based groups must exist in the Irish ecological movement. Developing the niche that these communities, and the Catholic Church in particular, occupy within this ecosystem, the symbiotic relationships they participate in is important work that must be done. It can only be achieved through meaningful dialogue. Francis argues that "Authentic social dialogue involves the ability to respect the other's point of view and to admit that it may include legitimate convictions and concerns" In doing so "it becomes possible to be frank and open about our beliefs, while continuing to discuss, to seek points of contact, and above all, to work and struggle together"[13].

In *Laudato Si'*, Francis calls for this 'authentic social dialogue', a global conversation. This complex dialogue will invariably encourage a diverse and evolving ecological movement. A diverse movement is the only way in which an integral ecology approach towards the "complex crisis which is both social and environmental"[14] can be achieved. There are several useful ways of considering the ecological movement that can help us understand its complexity, as well as aspects of the movement that may need to be further explored in order to make the movement as impactful and relevant as possible. In the same way ecology can be used as a metaphor for religious cultures[15], it can also be used to describe the ecological movement[16].

3. The Ecological Movement within Ireland

3.1 Diversity, Niches and networks within the ecological movement

Ecology is the study of the relationship of organisms between each other and their natural environment. Analysing these complex relationships happens at several scales; from individual organisms, through to communities, ecosystems and all the way up to the interactions within and between biomes. Studying these systems allow the interactions between species, how and where energy flows, how the system reacts to different types of disturbance and the knock-on impacts and feedback loops to be characterised; "it cannot be em-

[13] Pope Francis, *Fratelli tutti*, n. 203.

[14] Pope Francis, *Laudato Si'*, n. 139.

[15] Kevin Hargaden, "Planting Churches or Selling Them? New Competitors for the Metaphors We Use", *Christian Scholar's Review*, 51:2 , 171-188

[16] Mark Engler and Paul Engler, *This is an Uprising How Nonviolent Revolt Is Shaping the Twenty-First Century*. New York, 2016.

phasized enough how everything is interconnected"[17]. These concepts can be adapted to help us understand the functioning of movements. In the same way that healthy ecosystems rely on rich biodiversity to function, where the complexity of the systems confers stability, the same is true for movements.

The diversity of the movement not only provides a plethora of means to become involved, realising the truism that "there are many different spaces of social movement"[18], but also facilitates change from many different angles with many different perspectives. Similarly, there are people who may share the ecological mind set, and the wish for a better future, but would not self-identify (or wish to self-identify) as an activist or environmentalist. A broad range of activities within the movement allows scope to include people within the movement, but without the identifying labels and tags that can be associated with it. Mark and Paul Engler described this succinctly in their book *This is an Uprising*, "When mass mobilizations, established organizations, and alternative communities see themselves as complementary, they can create a movement ecosystem that allows diverse approaches to promoting change to flourish"[19].

The environmental and environmental justice movements originated, in the form in which we recognise them, in USA[20] however it would be a mistake and an injustice to fail to acknowledge environmental movements that originated from within indigenous communities across the world as well as well-established movements within the Global South. Within many indigenous communities there is a tradition and culture of caring for the land[21] resulting in extremely effective environmental and conservation initiatives. The campaign to stop the Keystone pipeline[22] is a prominent example of an indigenous-led movement to stop the expansion of the fossil fuel industry. In New Zealand, the Maori concept of *kaitiakitanga*[23] roughly translated as guardianship or protection of the

[17] Pope Francis, *Laudato Si'*, n. 138.

[18] Laurence Cox, *Why Social Movements Matter: An Introduction,* p. 6.

[19] Mark Engler and Paul Engler, *This is an Uprising How Nonviolent Revolt Is …*, pp. 268.

[20] Ronald L. Sandler and Phaedra. C. Pezzullo (ed.), *Environmental Justice and Environmentalism: The Social Justice Challenge to the Environmental Movement, Urban and Industrial Environments.* Cambridge, Massachusetts: MIT Press 2007.

[21] https://www.ienearth.org/; https://www.nationalgeographic.com/environment/2018/11/can-indigenous-land-stewardship-protect-biodiversity-/

[22] https://www.pri.org/stories/2015-02-19/native-american-tribes-unite-fight-keystone-pipeline-and-government-disrespect

[23] https://teara.govt.nz/en/kaitiakitanga-guardianship-and-conservation

natural environment, influences land management and conservation initiatives within New Zealand. Within the Global South, the environmental movement tend to be associated with protection of the environment on which communities are dependant for their livelihood[24]. These myriad of motivations and reasons for joining the movement is another example of the diversity that is evident, and is vital within the ecological movement. Awareness of these diverse struggles also offers a rich source of knowledge and wisdom as we attempt to evolve the global ecological movement[25].

In ecosystems, different organisms occupy different niches or functions – generally increased biodiversity results in increased complexity and niches. This too is true for social movements. Movements at their core are composed of people; different people are comfortable with different things and have different tastes and preferences. If a movement is comprised solely of protest or policy development or community projects, then while some people will find their niche, many more will not. The more avenues there are to get involved, the more people will get involved. Similarly, none of these activities alone can generate meaningful change – protest alone, without a vision for a better alternative will not change the system in which we live. Likewise, policy suggestions without the pressure for the government to implement those changes will only ever be words on paper[26].

The different elements of the movement which occupy different functions have different capabilities and limitations. The level of organisation and professionalization of some NGOs renders them less agile in some circumstances. This contrasts with some grassroots organisations which tend to be smaller in scale and unhindered by the constraints of grant money and supports. On the other hand grassroots organisations usually have fewer resources and a shorter reach than the larger more structurally organised NGOs[27]. It is when these

---

[24] Raaz Dwivedi, "Environmental Movements in the Global South: Issues of Livelihood and Beyond", in *International Sociology* 16/1-2001 (https://doi.org/10.1177/0268580901016001003).

[25] Brette Clarke, "The Indigenous Environmental Movement in the United States: Transcending Borders in Struggles against Mining, Manufacturing, and the Capitalist State", in *Organisation and Environment* 15/4-2002, (https://doi.org/10.1177/1086026602238170).

[26] "We should not think that political efforts or the force of law will be sufficient to prevent actions which affect the environment because, when the culture itself is corrupt and objective truth and universally valid principles are no longer upheld, then laws can only be seen as arbitrary impositions or obstacles to be avoided". Pope Francis, *Laudato Si'*, n. 123.

[27] Nathalie Berny and Christopher Rootes, "Environmental NGOs at a Crossroads?", in *Environmental Politics* 27/6-2018, pp. 947-972. (https://doi.org/10.1080/09644016.2018.1536293).

disparate parts of the movement symbiotically interact, principally through dialogue, that a functioning ecosystem can emerge. In this way "the whole is greater than the part, but it is also greater than the sum of its parts"[28].

3.2 Ireland's Ecological Movement

The climate and environmental movement in Ireland is a diverse, mature, movement spanning regional, class, and generational divides. The movement has undergone a process of evolution, from an arguably niche group of organisations, where environmental and ecological concern was a core issue almost to the alienation of other questions, to encompassing an ever-broader range of organisations and people with different interests, where ecological issues are often embedded within a broader political vision. This brings together a wide group of organisations that may have differing beliefs and values but with concern for the climate and our surrounding environment as a common link. This process of evolution is continuous, but is still a distance from where it needs to be. It is shaped both by the successes within the movement, its strategic direction, and also the societal and political circumstances in which it exists.

The strands that make up this complex network of organisations include, but are not exclusive to, traditional conservational groups, grassroots-local groups, developmental and social justice groups, faith-based, youth and student groups, political parties, media outlets, academics, groups within particular professions (e.g. Irish Doctors for the Environment and Farming for Nature) and concerned individuals. The groups are organised under several different guises including formal NGOs, quite informal community activism initiatives, or focused project-based groups with different functions ranging from advocacy, legal action, policy development and research, protest, communication, and ecological work including restoration, clean-ups, and community planting. In amongst all of this activity and organisation, coalitions groups[29] and different cross-sectional campaigns[30] function to increase cooperation in the movement and amplify the work that is being done. This rich diversity within the ecological movement is something Francis repeatedly

[28] Pope Francis, Apostolic Exhortation *Evangelii gaudium* on the Proclamation of the Gospel in Today's World, 24 November 2013, n. 235.

[29] These include 'Stop Climate Chaos', 'Sustainable Water Network' (SWAN) and Environmental Pillar.

[30] The 'One Future' campaign preceding the 2020 Irish general elections coordinated a peoples' campaign for faster, fairer climate action. For more see https://www.onefuture.ie/about/.

calls for in *Laudato Si',* arguing that "a consensus should always be reached between the different stakeholders, who can offer a variety of approaches, solutions and alternatives"[31].

Faith-based groups play a large, but unfortunately not fully realised, role in the Irish ecological movement exemplified by the large number of faith-based groups in Ireland's main Climate Coalition, Stop Climate Chaos[32]. These faith-based organisations consisting of grass-root and social justice organisations which have caring for our common home as a core value, are an active, but relatively minor part of the church in Ireland, where Pope Francis' call of ecological conversion has not fully been embraced. Faith-based social justice and development groups that are active within the ecological movement in Ireland – which focus on helping the poor and marginalised[33] – consistently generate excellent research, campaign material, analysis and educational resources ensuring that social justice remains a major consideration of the movement and is not subsumed by other important environmental considerations. A major driver for these development agencies to include environmental and climate justice in their ethos is the realisation that the work in which they were already engaged in is at risk due to the climate crisis, "the warming caused by huge consumption on the part of some rich countries has repercussions on the poorest areas of the world, especially Africa, where a rise in temperature, together with drought, has proved devastating for farming"[34]. This catalyst for involvement in climate justice is not unique to faith-based groups; a large body of research and work exists within Development Studies linking climactic and social destruction[35]. Mary Robinson, the former President of Ireland, former

[31] Pope Francis, *Laudato Si'*, n. 183.

[32] The full list of members can be viewed here: https://www.stopclimatechaos.ie/about/

[33] The Jesuit Centre for Faith and Justice, Tearfund, Christian Aid and Trocaire are among the faith-based groups that produce research and social justice analysis within the Ecological movement in Ireland.

[34] Pope Francis, *Laudato Si'*, n. 151.

[35] Within the Irish context see for example: Orla Kelleher, "A Reflection on the Experience of Climate Justice in Ireland"the Bord na Móna transition already has many of the key ingredients for a Just Transition. Yet to date, the move away from peat extraction in the Midlands has not gotten any of the five key elements of a Just Transition correct. At a minimum the next government should adequately resource the Just Transition Commissioner, urgently raise the Just Transition fund, and provide a multi-stakeholder mediation service to discuss issues currently affecting Bord na Móna workers and future employment opportunities. This could be done by passing into law the Green Party's Just Transition (Workers and Community Environmental Rights, in *Working Notes, Jesuit Centre for Faith and Justice* 86/34-2020, Dublin, pp. 21-30

UN High Commissioner for Human Rights and current Chair of The Elders, has described how the source of her "ecological conversion" lay in how commonly she encountered stories and situations where the changing climate was negatively impacting people's lives and curtailing their access to their rights[36].

Within this diverse collective, where the motives for participation in the ecological movement can vary, groups emerge, at different times, to take leading roles while other members of the movement take a step back in solidarity depending on the circumstances. A striking example of this was the Climate Strike in September 2019, which was predominately led by youth-based groups and student activists, but the organisation of which was supported by climate coalitions, faith-based groups, and trade unions. This alliance resulted in an event where thousands of individuals and groups protested in dozens of locations around the country in solidarity with the teenagers[37].

While the organisations and groups are highly interconnected in terms of areas of interest, not all components of the environmental movement in Ireland are co-ordinated. There is no centralised control hierarchy. There are outlying parts of the movement which are more isolated, but that still see themselves as working towards a better ecological future in Ireland. This flexible set of overlapping alliances has achieved a great deal in terms of making climate and biodiversity mitigation an essential element of the Irish political discourse, but there is still has a long way to go in terms of effecting real change. Central to that work is developing a coherent narrative that compellingly weaves these disparate movements closer together and connects ordinary people's self-understanding with the crisis we face.

One of the main issues of the movement is the perception that it is comprised mainly of urban elite which does not take into account or comprehend issues of rural Ireland. A major factor in this is that the agricultural sector remains an important part of the Irish economy and culture. At the same time farming practices have evolved to become more intensive resulting in increased degradation of the Irish countryside as well as contributing a disproportionate amount to Irelands greenhouse gas emissions[38]. These concerns have be-

---

36 Mary Robinson, *Climate Justice: Hope, Reliance and the Fight for a Sustainable Future*. London: Bloomberg Publisher, 2018, pp. 2-4.

[37] Kevin O'Sullivan Environment and Science Editor, "Friday's Climate Protests in Ireland: When and Where They're on", The Irish Times, accessed August 25, 2020, https://www.irishtimes.com/news/environment/friday-s-climate-protests-in-ireland-when-and-where-they-re-on-1.4023874.

[38] "Ireland's Environment 2020 – An Assessment" *Environmental Protection Agency*: Chapter 13 Available at: http://www.epa.ie/media/EPA-Ireland's-Environment-2020-Chapter13.pdf.

come a focus of criticism by some within the Irish ecological movement. This has produced the impression in certain circles that the ecology movement is 'anti-farmer' and 'anti-rural' and, conversely, that farmers are 'anti-environment'. This perception is not entirely accurate. There are projects, organisations and groups across the country that are working towards a better future in rural Ireland. Typical examples of this are some excellent agri-environmental schemes[39], sustainable agriculture networks[40], individual farmers who engage in organic, ecological or regenerative farming, transition towns, eco villages, rural Green party groups and community ecology groups, including parish-based faith-groups. While not necessarily accurate this perception can be damaging to the overall message and vision of the environmental movement. Work needs to be done to bridge the gaps in the movement as well as invite more people into the ecological dialogue.

3.3 Catalysts within the national and global movement

The evolution of the movement in Ireland has been catalysed and shaped by both increasingly dire scientific reporting[41] and social uprising – both within and outside of the climate movement. In Ireland there have been several notable successes within the environmental movement which encouraged and accelerated the national movement. The ban on fracking in Ireland, signed into law in 2017, was one such achievement. This Bill was the result of a long campaign involving grassroots campaigners, climate coalitions, and NGOs and paved the way for further campaigning against the importation of fracked gas and oil and gas exploration. In July of 2020 the Supreme Court judgement finding in favour of Friends of the Earth in a case known as 'Climate Case Ireland' was another such achievement. This historic judgement quashed the National Mitigation Plan (NMP), a major component of the State's climate change policy. Climate Case Ireland's lawyers successfully argued that this

[39] Caroline Sullivan, "High Nature Value Farmland: Getting Results from Farming for Biodiversity", in *Working Notes, Jesuit Centre for Faith and Justice* 86/34-2020, Dublin, pp. 41-46.

[40] "Home", *Farming for Nature* (blog), April 11, 2018, https://www.farmingfornature.ie/.

[41] Intergovernmental Panel on Climate Change (IPCC), *Special report on the impacts of global warming of 1.5 °C above pre-industrial levels and related global greenhouse gas emission pathways, in the context of strengthening the global response to the threat of climate change, sustainable development, and efforts to eradicate poverty*", 2018. N.d. This report published in October 2018 warned that we had 12 years to limit the consequences of climate change.

plan was inadequate in terms of reducing our emissions by our fair share[42]. More recently, in July 2021 a new Bill, the 'Climate Action and Low Carbon Development (Amendment) Act 2021', was signed into Irish law[43]. While this Bill is the most ambitious Irish commitment to emissions reductions to date, setting out a 51% reduction by 2030 and net-zero by 2050, it still falls short in committing us to doing our fair share in dealing with the climate crisis[44].

Politically there have been advancements, in terms of getting the climate and biodiversity on the agenda. The most recent election in February 2020, saw a record number of Green Party members elected with several other parties evolving their stances on ecological matters as evident in their manifestos[45]. We now have a situation where the Green Party is part of the Government, in coalition with the two major centre-right parties. While this situation has led to major progressions in active and public transport ambitions, a roll-out of a raft of peatland restoration projects as well as the publication of the Climate Bill, this has not come without some considerable negative consequences. As a minor partner in an unpopular Government, during a particularly difficult time to govern (considering the pandemic and the looming recession) there have been decisions made, particularly concerning social justice issues such as wages increases, and concessions made (such as support for the EU-Canada "CETA" trade deal – which they previously opposed) that have harmed the Green Party's support[46]. The merits and disadvantages of the Green Party entering government at such a critical time in the climate crisis will, I am

---

[42] Sadhbh O'Neill, "Carbon Crimes", in *Working Notes, Jesuit Centre for Faith and Justice, Dublin* 32/84-2019 (https://www.jcfj.ie/article/carbon-crimes/); Orla Kelleher, "A Reflection on the Experience of Climate Justice in Ireland", in *Working Notes, Jesuit Centre for Faith and Justice, Dublin* 86/34-2020, pp. 21–30; Orla Kelleher, "The Supreme Court of Ireland's Decision in Friends of the Irish Environment v Government of Ireland ('Climate Case Ireland')", in *EJIL: Talk!* (blog), September 9, 2020, https://www.ejiltalk.org/the-supreme-court-of-irelands-decision-in-friends-of-the-irish-environment-v-government-of-ireland-climate-case-ireland/.

[43] https://www.oireachtas.ie/en/bills/bill/2021/39/

[44] Ciara Murphy, "Green Bill is not Black and White" (Blog) Available at:https://www.jcfj.ie/2021/03/25/green-bill-is-not-black-and-white/

[45] https://greennews.ie/ge2020-manifesto-analysis-climate/

[46] The reception to the activities of the Green Party, both good and bad, can be observed in the Irish media. See for example: https://www.irishtimes.com/news/ireland/irish-news/divisions-in-green-party-over-ceta-latest-complication-for-ryan-1.4440677; https://www.thejournal.ie/neasa-hourigan-green-party-interview-5301891-Dec2020/; https://www.irishtimes.com/news/ireland/irish-news/eamon-ryan-by-2030-the-only-new-cars-will-be-electric-1.4440722; https://www.thejournal.ie/students-nurses-payment-issue-unpaid-placements-5291130-Dec2020/; https://www.jcfj.ie/2020/09/01/agriappeals/

sure, be examined elsewhere and at length when the full consequences are revealed. What is obvious now, in the context of the ecological movement, is the importance of a diverse ecological movement that is active beyond the political sphere. The Green Party, in their current position, cannot be expected to change the entire political and societal system which has resulted in the complex crisis we find ourselves in. A broad and diverse movement, working in every part of society is needed if the hearts and minds of the population are to be ecologically converted.

This broadening of the movement is exemplified by the active involvement of young people. The youth climate activists, who have become engaged through striking from school for climate action, has completely changed the composition and tone of the movement, injecting energy into the movement. In August 2018, the urgency of the climate crisis was starkly highlighted by widespread wild fires and heatwaves across Europe and America. Consequently, the Swedish teenager, Greta Thunberg, decided to strike for climate outside the Swedish Parliament. Emboldened and armed by a scientific report stressing that by 2020 emissions must start on a downward trend[47], she gave her first speech[48] in Stockholm one month later. By the end of the year thousands of students mobilised across the globe in a movement that has come to be known as *Fridays for Future* (FFF). In March 2019, over a million young people across the globe took to the streets to demand climate justice. The empowerment of school students and youth climate activists has enabled wider awareness of the intergenerational aspect of climate justice which is specifically addressed by Pope Francis in *Laudato Si*'"[49]. Thunberg's call for environmental and social justice illustrates that young people "understand *Laudato Si*' with their hearts". The Vatican has been vocal in its praise of Greta's efforts, with Cardinal Turkson declaring that her "protest and her witness brings attention to the great need to be coherent in our care for the environment and also for the people who live on earth"[50]. The youth-led movement

---

[47] Johan Rockström et al., "A Roadmap for Rapid Decarbonization", in *Science* 355/6331-2017, pp. 1269–71, (https://doi.org/10.1126/science.aah3443).

[48] We Don't Have Time, "Greta Thunberg: Our Lives Are in Your Hands", in *Medium*, September 14, 2018, https://medium.com/wedonthavetime/greta-thunberg-our-lives-are-in-your-hands-b5a7b1e24a97.

[49] "The notion of the common good also extends to future generations. The global economic crises have made painfully obvious the detrimental effects of disregarding our common destiny, which cannot exclude those who come after us. We can no longer speak of sustainable development apart from intergenerational solidarity" Pope Francis, *Laudato Si'*, n. 159.

[50] "Vatican Calls Greta Thunberg 'Great Witness' of Church's Environmental Teaching", in

also fostered a shift in the perception of the climate and environmental movement as a more inclusive movement which has more options for engagement that the traditional conservation or environmental NGOs. By initiating conversations that are inherently intergenerational, the FFF movement facilitates a conversation that excludes no one.

The FFF movement was not the only movement that ignited in 2018. The civil disobedience campaign Extinction Rebellion (XR) formally launched on the 31 October with its Declaration of Rebellion[51] outside the Houses of Parliament in the UK. This campaign recruited people across the country who understand the urgency of the ecological crisis, but who were frustrated with the lack of political action. This movement expressed itself in a series of large scale civil disobedience events in November 2018, April 2019 and October 2019 involving thousands of people. This explicitly non-violent civil disobedience movement, while not sparked from within faith-based groups, has resonances with Catholic Social Teaching and a specific Christian branch, Christian Climate Action group, has crystallised[52]. FFF and XR are now firmly established within ecosystem of Irish ecological movement where they have, and continue to play a catalytic role for the movement as a whole.

This step-change in the global ecological movement has emerged with the recognition that the environment in which we live impacts every aspect of our lives in some form or other. Where there is a lack of concern for the natural environment, we will surely find a concomitant lack of concern for social justice. The global ecological movement that has emerged is therefore hugely rooted in social justice. This increased awareness of the interconnecting nature of different environmental and social justice is an important step in developing a truly global and widespread movement. Cox, in his reflections on the evolution of movements argues that "one of the first signs that serious social change is on the cards is when voices within individual movements that argue for alliances with other movements come to the fore, and the possibility of a wider transformation comes to feel like a real possibility"[53]. A constant call

Crux (blog), December 12, 2019, https://cruxnow.com/vatican/2019/12/vatican-calls-greta-thunberg-great-witness-of-churchs-environmental-teaching/.

[51] https://extinctionrebellion.uk/declaration/

[52] CAFOD, "Bridge of Faith: Why Are Extinction Rebellion Taking over a London Bridge?" *CAFOD Blog* (blog), September 20, 2019, https://blog.cafod.org.uk/2019/09/20/bridge-of-faith-why-are-extinction-rebellion-taking-over-a-london-bridge/.

[53] Laurence Cox, *Why Social Movements Matter: An Introduction*, pp. 30

from Pope Francis in *Laudato Si'* is to "come together to take charge of this home which has been entrusted to us"[54], fostering and building associations and collaboration between movements through dialogue as well as inviting people into an ecological conversion is work that everyone, but especially faith based communities, are called to do. What we find is that these new environmental justice movements have served as an impetus to the broader Irish environmental movement and that they have particularly resonated with the faith-based groups which always foregrounded the connection between environmental care and social justice.

3.4 The evolution of political engagement, activism and the role of the Church in Ireland

The ecological movement in Ireland does not exist in a social vacuum and is very much influenced and impacted by other societal and political events. The political, social and religious climate – both in Ireland and globally – has experienced some seismic changes over past few decades. In Ireland, two dominant trajectories must be considered: the decrease in prominence of the Catholic Church, and the end of the long economic boom known as the Celtic Tiger in 2008 and the resultant austerity measures that have shaped the economic and societal environment. Of course, the predominant situation of the present day, the Covid-19 pandemic as well as global conflict, will also impact the ecological movement in both subtle and obvious ways, the latter of which we will also touch on briefly.

At the time of writing, the devastating war in Ukraine is in its first weeks. We cannot but acknowledge the suffering and the destruction that this conflict, and indeed any war causes. Francis is unequivocal in *Laudato Si'* that "sin is manifest in all its destructive power in wars"[55]. Pope Francis asserts that "the mindset which leaves no room for sincere concern for the environment is the same mindset which lacks concern for the inclusion of the most vulnerable members of society"[56]. At this early stage we do not know what the long term consequences will be of this war, what is certain is that the senseless slaughter, destruction of people's livelihoods, homes and even their towns will ripple through every facet of our lives with implications on how we react

---

[54] Pope Francis, *Laudato Si'*, n. 224.
[55] Pope Francis, *Laudato Si'*, n. 66
[56] Pope Francis, *Laudato Si'*, n. 196

to the climate crisis. A lot of wars, in one way or another, are related to access to energy, particularly fossil fuels.

> Iraq, Syria, Nigeria, South Sudan, Ukraine, the East and South China Seas: wherever you look, the world is aflame with new or intensifying conflicts. At first glance, these upheavals appear to be independent events, driven by their own unique and idiosyncratic circumstances. But look more closely and they share several key characteristics – notably, a witch's brew of ethnic, religious, and national antagonisms that have been stirred to the boiling point by a fixation on energy[57].

Within the Irish, and indeed the global context, the response to this war may precipitate a faster turning away from fossil fuel energy, as the insecurity and destruction it presents become more obvious. Alternatively, and more negatively the recognition that Ireland's, and indeed Europe's overreliance on Russian gas may result in an impulse reaction of investing in further exploration and fossil fuel infrastructure thus prolonging our dependence in fossil fuels. There is recognition within the Irish climate movement that this crisis must not derail climate action momentum[58]. In fact active forms of solidarity from within the climate movement have already formed with global Friday for Future marches, protesting the war, taking place early in the year[59].

When discussing *Laudato Si'* and how the principles of integral ecology are played out in Ireland, it is impossible to separate it from the recent history of the church in Ireland and the current context that exists. For over more than century and a half, the Catholic Church in Ireland held exceptional power. Church institutions bore the primary responsibility to educate the young and tend to the sick. As it stands, fully 90% of Irish children are educated in Catholic schools[60]. Signs of the disintegration of this concentration of power, observed as far back as the 1970s[61], were catalysed by the devastating series of scandals came to light in recent decades. Since the turn of the millennium,

---

[57] Michael T Klare, "Twenty-first-century energy wars", *Le Monde diplomatique* (Blog) July 2014, https://mondediplo.com/openpage/twenty-first-century-energy-wars

[58] See for example: Daly et al, "Ireland must play its part in ending Europe's dependence on Russian fossil fuels" March 2022. https://hannahdaly.ie/2022-03-08-oped-energy-security/

[59] Paul Krantz, "We Don't Know What Will Happen Tomorrow", *New Internationalist* March 2022 https://newint.org/features/2022/03/04/ukraine-war-climate-fridays-future

[60] https://www.education.ie/en/Publications/Statistics/Data-on-Individual-Schools/enrolments/statistical-bulletin-enrolments-september-2019-preliminary-results.pdf

[61] Michael Paul Gallagher, "Atheism Irish Style", in *The Furrow* 25/4-1974, pp. 183-192.

decades of abuse and exploitation carried out within Catholic institutions[62], have been exposed, precipitating collective turning away from the church[63], hastening the transition to a largely post-Catholic country[64]. The result of this can be seen in reduced church attendance numbers and in a wider rejection of Catholic or Christian perspectives on important issues. The results of the marriage equality referendum in 2015 and the repeal of the 8th amendment to the Irish constitution in 2018 are examples of this; where the Irish public overwhelmingly voted against the church's teachings. It is important to note that while ecclesial influence in such matters has waned, the Catholic Church still plays a huge role in Irish people's lives, even if this influence is highest among the older generations[65].

In parallel to this tidal change of the Church's position in Ireland, the economic situation in Ireland was undergoing a major upheaval. After undergoing a sustained period of economic growth from the 1990s onwards[66], the years since the global financial crisis, specifically from 2008, have seen massive changes within the Irish society, where social injustices of inequality, a flagging health-care system, and homelessness are at crisis point[67]. The doubling down of the neo-liberal policies of deregulation, reduced welfare and privatisation[68] created a reality where the prospects of housing security is out of

---

[62] The anger Irish citizens feel towards this exploitation and abuse is immense, and the pain that is felt by its survivors continues to the present day. In 2019 an Irish journalist published a book recounting some of the stories of the women and children who were victims of this exploitation. Caelainn Hogan, "Republic of Shame: Stories from Ireland's Institutions for 'Fallen Women'", n.d., 231.

[63] Different starting points for the decline can be identified. In 1987, Tom Inglis was able to publish his seminal sociological account of the Catholic church's cultural and political power, *Moral Monopoly*, which was subtitled the "Rise and fall" of the Catholic church in Ireland. Tom Inglis, *The Moral Monopoly*. Dublin: Gill & Macmillan 1987.

[64] Sat, 25 Aug, and 2018 - 01:00, "Losing Our Religion: Can Francis Offer New Hope and Energy to Church?" Irish Examiner, August 25, 2018, https://www.irishexaminer.com/opinion/commentanalysis/arid-30864433.html.

[65] "Religion - Religious Change - CSO - Central Statistics Office" (CSO), accessed September 8, 2020, https://www.cso.ie/en/releasesandpublications/ep/p-cp8iter/p8iter/p8rrc/.

[66] In one year alone, 1994, the Irish economy grew by almost a tenth, 50,000 people were added to the work force, and unemployment plummeted. Peadar Kirby, *Celtic Tiger in Collapse*. London: Basingstoke Palgrave Macmillan, 2010, p. 33.

[67] A detailed theological account can be found in Kevin Hargaden, *Theological Ethics in a Neoliberal Age,* Eugene OR: Cascade, 2018.

[68] "Murphy - 2014 - Ireland Celtic Tiger in Austerity—Explaining Iris.Pdf", accessed August 14, 2020, http://mural.maynoothuniversity.ie/9055/1/MM-Ireland-2014.pdf.

reach for a large proportion of the population, along with depressed incomes and pensions for public sector workers. These economic and societal circumstances, alongside the reality that for today's youth the climate and biodiversity will manifest itself in an increasingly hostile future, has precipitated a wave of youth interested in environmental and social justice matters.

While austerity politics created the conditions for a generation of youth to be disillusioned with business-as-usual politics there were also several specific events which catalysed the politicisation of young people in Ireland. The Marriage Equality and Repeal the 8th (which would effectively remove a ban on abortion in Ireland) referenda, in 2015 and 2018 respectively, saw huge youth voter registration and turn out, which suggests that young people are becoming more engaged with politics[69]. While these referenda were undoubtedly contentious, in particular the Repeal the 8th campaign, in terms of representing a break in Ireland from Catholic social teachings, it is important to consider these events and what learnings can be achieved from these campaigns.

Both of these referenda were won by creating movements that positioned themselves on the side of social justice and human rights[70]. While acknowledging objections to whether these positions do in fact instantiate social justice or human rights, it is critical to note these campaigns successfully activated a huge proportion of the general population across all age and socio-economic groups, but in particular youth, on that premise. So while these citizens took a position diametrically opposed to that of the church, many did so with the motivation of supporting social justice. Consequently, there exists a quandary whereby there is significant potential space for dialogue around social justice concerns between these youth activists and the Church, even though there are obvious fundamental differences in application and perhaps significant philosophical disputes.

In fact, anecdotal evidence suggests that activists who were involved in the referenda were primed for engagement in climate justice work. Campaigners

---

[69] "CAREY - An Examination of Young People's Relationships Wit.Pdf", n.d.; https://usi.ie/featured/usi-welcomes-large-youth-turnout-in-referendum/

[70] The most interesting movements that have arisen in response to these referenda grant that the successful campaigns were built on a vision of justice and as such they can be engaged with by Catholic activists on those grounds. For example, "The Minimise Project", an initiative led by young pro-life Catholics who work in conjunction with secular pro-life advocates and seek overlapping spaces to productively work with pro-choice movements so as to reduce the rate of abortion: https://theminimiseproject.ie/about/our-vision-mission-values/

who cut their teeth in the referenda often extended their activism to specific climate justice campaigns and in those contexts found themselves working alongside faith-based groups[71]. The successful political engagements of the referenda also functioned to show what is possible with mass mobilisation, paving the way for future engagement in environmental justice activism[72] and political activity[73].

There is a positive link between youth engagement in politics and activity into adulthood[74], albeit the level of activity can depend on a number of factors including the persistence of activist networks, availability of opportunities, initial level of engagement and burnout[75]. One of the consequences of such a large political mobilisation around the referenda has been a spill-over into environmental activism. This phenomenon of increased youth engagement around specific political issues is not unique to Ireland but is also experienced internationally. Increased youth engagement during specific political events (e.g. Brexit) illustrate that when the issues are perceived to be relevant to young people[76] they will become more engaged and active[77].

---

[71] The "One Future" campaign in the run up of the 2020 General Election is an example of where faith based groups and campaigners who previously worked on the referenda in 2015 and 2018 collaborated.

[72] "Senator Lynn Boylan on Climate Justice, Both at Home and Abroad - Green News Ireland", accessed July 24, 2020, https://greennews.ie/lynn-boylan-climate-justice-interview/.

[73] A striking example of this is a statement (on twitter) from a young Irish TD (member of parliament) Holly Cairns stating that the referenda were the catalyst needed to enter politics – "I never thought I'd join a political party. I had a light bulb moment after the marriage equality and repeal referenda – knocking on doors and asking for votes works! You can make a difference and effect the positive change you want to see" (https://twitter.com/HollyCairnsTD/status/1297963649739427842).

[74] Maor Shani, Daniel Horn and Klaus Boehnke, "Developmental Trajectories of Political Engagement from Adolescence to Mid-Adulthood: A Review with Empirical Underpinnings from the German Peace Movement", in Nikola Balvin and Daniel J. Christie, ed., *Children and Peace: From Research to Action, Peace Psychology Book Series*. Springer International Publishing, 2020, (https://doi.org/10.1007/978-3-030-22176-8_17),\\uc0\\u8221{} in {\\i{}Children and Peace: From Research to Action}, ed. Nikola Balvin and Daniel J. Christie, Peace Psychology Book Series (Cham: Springer International Publishing, 2020 pp. 271–90.

[75] Eugene Nulman, "Political Engagement Trajectories of Youth Activists Following Recruitment into High-Intensity Mobilisation", in *Journal of Youth Studies*, August 19, 2019, pp. 1–19, https://doi.org/10.1080/13676261.2019.1655535.

[76] Darren Lilleker, *EU Referendum Analysis 2016: Media, Voters and the Campaign : Early Reflections from Leading UK Academics*, 2016.

[77] "Brexit Is Going to Happen. What Is a Young 'Remainer' to Do?", in *Christian Science Monitor*, December 13, 2019, https://www.csmonitor.com/World/Europe/2019/1213/Brexit-is-going-to-happen.-What-is-a-young-remainer-to-do.

While the referenda have indirectly impacted the environmental movement through an increased number of experienced and enthusiastic activists, there is another direct impact on the ecological movement. The success of the 'Yes Equality' campaign in the marriage referendum and the 'Together for Yes' campaign in the Repeal the 8th referendum provided a template which other political campaigns could follow[78]. The 'One Future' campaign, a people-led movement which called for 'Faster, Fairer Climate Action' before the 2020 General Election was modelled on the experiences of these successful campaigns. A coalition of organisations within the climate movement developed eight socially just climate demands. The breath of the demands represents the range of interests of the coalition. The focus on environmental justice alongside emission and biodiversity demands, illustrates that this campaign attempted a sort of 'Integral Ecology' approach which recognised that we cannot have one without the other.

I want a government that will:

Reduce Ireland's greenhouse gas emissions by at least 8% every year

Support the rural economy as farms reduce emissions and become more sustainable

Ban new fossil fuel projects

End peat and coal burning for electricity and ensure fair treatment of workers

Deliver an ambitious State-backed home insulation programme

Advocate internationally for faster and fairer climate action and justice

Restore and protect nature and wildlife

Ensure affordable and accessible public transport for all

Enable and support community-owned renewable energy projects

OneFuture.ie **#ClimateVote2020**

[78] Susan Parker, "The Path to Marriage Equality In Ireland", n.d., p. 21.

'One Future' functioned to unify environmentally minded individuals and organisations across Ireland, increase awareness and engagement with environmental justice issues, and aimed to make climate and the environment an election issue. A key component of this campaign was a network of local support-groups equipped with branded materials and information which were widely disseminated. Groups were encouraged to go into their community and campaign for the climate, without reference to particular political parties or electoral candidates, urging voters to challenge their representatives with the eight climate asks which were central to the campaign. This evolution in the tactics could herald an ever more engaged and inclusive environmental movement.

When a seismic event, like a global pandemic, occurs, there is no aspect of society that remains untouched. This holds true for social movements. From early in the pandemic, links have been identified not only between the causes of the ecological emergency and the emergence of the novel coronavirus but also the response to these emergencies[79]. Covid-19 has also highlighted that everything is indeed interconnected[80] - everything from public health, sports, childcare, transport infrastructure, housing, economics and how we work and experience our environments have all been impacted by the pandemic. The war in Ukraine similarly highlighted how interconnected everything is. Our reliance on fossil fuels can both precipitate and fund wars. Global politics and conflict can have a direct impact on peoples' ability to heat themselves and their families. In this way campaigning for peace, which the church has a long history of, can be viewed as another manifestation of climate justice advocacy. Acknowledging the interconnection between these crisis allows for imaging solutions and policies for recovery that addresses both simultaneously. The calls to "Build Back Better" from across the ecological and social justice movements[81] demand that how we recover from this pandemic should result in a fairer, more equal and ecological society. In Ireland, the ecological movement have shifted and intensified their demands for change post Covid-19 with a prominent example being

[79] Ciara Murphy, 2020 "We are living through two global emergencies" (Blog) Available at: https://www.jcfj.ie/2020/03/19/globalemergencies/

[80] https://www.jcfj.ie/2020/05/18/climatejustice/

[81] These organisations include Trocaire https://www.trocaire.org/build-back-better/, Ecojesuit https://www.ecojesuit.com/wp-content/uploads/2020/07/Ecojesuit-Commitment_new-normal_EN.pdf, TASC https://www.tasc.ie/blog/2020/06/30/build-back-better-or-back-to-business-as-usual/

the huge investments in active and public transport[82] and accelerated plans for cycle lanes and green ways across the country.

4. Evaluation of the Ecological Movement from a Catholic Perspective

Understanding the ecological movement from a catholic perspective can facilitate greater collaboration and dialogue. Examining the various ways in which faith-based organisations can become involved in the ecological movement, alongside the text of *Laudato Si'* can reveal the necessity of diversity within the ecological movement and the importance of fostering networks, through dialogue and cooperation.

4.1 Individual/community and system change – the scale of the movement

The question of how we effect change is ever present. There is an ongoing debate about the importance of individual and community change verses systemic change, with advantages and disadvantages associated with every avenue of action. Effective movements for change can be considered the same way we observe interactions at different scales in ecology, where examination of these functions is at the individual, community, and ecosystem scale.

Change can only start from within ourselves. Recognising the ecological issues that surround us, and how our behaviour and the society in which we live contribute to this crisis, is the first step to making real change. Trying to change our individual behaviour and understanding how these changes can simplify and improve our lives is the natural second step. On this topic Pope Francis has much to say:

> There is a nobility in the duty to care for creation through little daily actions, and it is wonderful how education can bring about real changes in lifestyle. Education in environmental responsibility can encourage ways of acting which directly and significantly affect the world around us, such as avoiding the use of plastic and paper, reducing water consumption, separating refuse, cooking only what can reasonably be consumed, showing care for other living beings, using public transport or car-pooling, planting trees, turning off unnecessary lights, or any number of other practices. All of these reflect a

---

[82] https://www.irishtimes.com/news/environment/climate-and-the-budget-carbon-tax-hike-and-funding-for-sustainable-transport-1.4380128.

> generous and worthy creativity which brings out the best in human beings. Reusing something instead of immediately discarding it, when done for the right reasons, can be an act of love which expresses our own dignity[83].

The responsibility to make these changes can be especially pertinent for individuals depending on circumstances[84]: more affluent people usually to have a higher carbon footprint[85] and tend to be insulated from the worst of the impacts. "The warming caused by huge consumption on the part of some rich countries has repercussions on the poorest areas of the world"[86].

But it is a dreadful mistake to imagine we can solve the current ecological crisis with individual action alone. Individual responsibility for climate change was the mainstay neoliberal policy for years. Many of the ideas and concepts that furnish our ecological imagination were in fact shaped to keep our thinking on the individual level. The fossil fuel company BP consciously popularised the 'Carbon footprint' concept as a means of deflecting responsibility to the consumer, away from the producers[87].

Individual reductions are also inherently limited in terms what is possible; simply existing within the systems that have evolved within our society constrains our possible impact for change. Our economic system and culture often incentivises carbon-intensive lifestyles and generates barriers to living a low carbon life. A common example of this can be seen in the housing and transportation systems in Ireland. There is an unfortunate situation where people often live in areas without adequate public transport or active transport infrastructure resulting in high private car ownership and usage. This in turn leads to more infrastructure being built to accommodate these cars which further disincentives people from cycling and walking on unsafe and busy roads. An individual's desire to reduce their carbon footprint is limited to what is practicable. While

---

[83] Pope Francis, *Laudato Si'*, n. 211,

[84] "Who Is Really to Blame for Climate Change? - BBC Future", accessed September 8, 2020, https://www.bbc.com/future/article/20200618-climate-change-who-is-to-blame-and-why-does-it-matter.

[85] Timothy Gore, "Extreme Carbon Inequality Why the Paris Climate Deal Must Put the Poorest, Lowest Emitting and Most Vulnerable People First" (Koninklijke Brill NV), accessed September 8, 2020, https://doi.org/10.1163/2210-7975_HRD-9824-2015053.2020, https://doi.org/10.1163/2210-7975_HRD-9824-2015053.","plainCitation":"Timothy Gore, "EXTREME CARBON INEQUALITY Why the Paris Climate Deal Must Put the Poorest, Lowest Emitting and Most Vulnerable People First" (Koninklijke Brill NV

[86] Pope Francis, *Laudato Si'*, n. 52.

[87] Julie Doyle, "Where Has All The Oil Gone? Bp Branding And The Discursive Elimination Of Climate Change", n.d., 24.

individual change is usually what one has the most control over – and changing diet, driving less, cutting down flying, and so on might make a big difference in an individuals' emissions – it remains but a grain of sand on the beach.

Allowing the conversation to be framed primarily in terms of individual action is also problematic because of how it shields our vision from seeing clearly: larger emitters are overwhelmingly responsible for the crisis. Individualist accounts fail because it obstructs very wealthy individuals, very intensive polluting corporations and institutions and those in political power from accepting their moral responsibility. Relatively few companies are responsible for the vast majority of greenhouse gas emissions yet "many of those who possess more resources and economic or political power seem mostly to be concerned with masking the problems or concealing their symptoms"[88] Any reductions an individual achieves will be dwarfed into meaninglessness by these companies – which are not abstract entities but are collections of humans with the same moral responsibility to Care for Our Common Home.

Again here Pope Francis recognises these issues:

> self-improvement on the part of individuals will not by itself remedy the extremely complex situation facing our world today. Isolated individuals can lose their ability and freedom to escape the utilitarian mindset, and end up prey to an unethical consumerism bereft of social or ecological awareness. Social problems must be addressed by community networks and not simply by the sum of individual good deeds. This task "will make such tremendous demands of man that he could never achieve it by individual initiative or even by the united effort of men bred in an individualistic way. The work of dominating the world calls for a union of skills and a unity of achievement that can only grow from quite a different attitude". The ecological conversion needed to bring about lasting change is also a community conversion[89].

Individual changes and actions can precipitate the wish for more – either by observing the benefits of the change or conversely experiencing the barriers and limitations of individual actions. Individual actions can lead to more community-based initiatives and actions including community gardens, cycle lanes, sharing services which can make a huge difference to a region or community. In this way, community level projects and activity is vitally important

---

[88] "New Report Shows Just 100 Companies Are Source of over 70% of Emissions - CDP", accessed September 8, 2020, https://www.cdp.net/en/articles/media/new-report-shows-just-100-companies-are-source-of-over-70-of-emissions.

[89] Pope Francis, *Laudato Si'*, n. 219.

within the ecological movement, this is where work on the ground gets done, allows communities to create change, and improve their own situation.

> The existing world order proves powerless to assume its responsibilities, local individuals and groups can make a real difference. They are able to instil a greater sense of responsibility, a strong sense of community, a readiness to protect others, a spirit of creativity and a deep love for the land[90].

While the benefit of these projects are undeniable for the respective communities, to scale to regional or national systemic change requires extensive meaningful dialogue to establish consensus and political will to support and promote community action[91]. Again, this is a challenge fruitfully considered from the perspectives offered by Catholic Social Teaching: a shared vision for a better future, developed through dialogue, is needed before we can start to build and re-build.

Although these projects have more of an impact than individual change, the scale of the ecological problem can make these actions, when carried out alone, seem insignificant. The opposite end of the scale is widespread system change or replacement. We need to change the current system, which has resulted in inequality and ecological destruction, into one where living and thriving within our planetary boundaries is possible[92]. "We are faced not with two separate crises, one environmental and the other social, but rather with one complex crisis which is both social and environmental. Strategies for a solution demand an integrated approach to combating poverty, restoring dignity to the excluded, and at the same time protecting nature"[93]. This feat can seem like an intractable problem which calls for nothing short of a global ecological conversion. The individual voice can feel weak and inconsequential in the face of this mammoth task. Finding the balance between these seemingly separate avenues is something every movement and ecologically minded individual is striving for.

---

[90] Pope Francis, *Laudato Si'*, n. 179.

[91] "CLM Environmental Justice Roundtable Agenda & Briefing Paper.Pdf", n.d.

[92] Kate Raworth, an economist based in Oxford has pioneered a new economic model, coined "Doughnut Economics", which takes into these planetary boundaries alongside critical human development. See: Kate Raworth *Doughnut Economics – Seven ways to think like a 21st-Century Economist.* London: Random House, 2017.

[93] Pope Francis, *Laudato Si'*, n. 139.

4.2 Dialoguing Towards Systemic Change

The ecological movement will only be able to implement profound integral ecological solutions, addressing the source of the problem and not just its symptoms, when they develop approaches to change which encompass all levels of ecological participation and action. Individual, collective and systematic change act symbiotically to facilitate and drive change[94], one without the other will not succeed. Once again, the ecological metaphor, laden as it is with systemic interaction, is illuminating. Our efforts, for example, to refuse single use plastic, reduce private car use, and conserve our environment need to be supported by legislative changes which facilitate and incentivise change at a higher level than individual actions can achieve. With those high-level changes in place, the individual or communal action becomes easier to achieve. In this way "political and institutional frameworks do not exist simply to avoid bad practice, but also to promote best practice, to stimulate creativity in seeking new solutions and to encourage individual or group initiatives"[95]. Changes in funding, administration burdens, and decision-making processes could remove barriers for community projects allowing more work to be done. Involvement in community projects can also result in longer term change than an individual basis affords, especially where there is support network readily available. We can conceive of a maturing ecological movement which constitutes a sort of cornucopia of hope and change, insisting powerfully that governments become instruments of positive environmental policy. Real change will accelerate when those in power make it their priority or recognise that it is an electoral priority.

Throughout all this process, from individual to community to system change, communication and dialogue and ecological communication is central to all action. The Integral Ecology approach that Pope Francis is asking us to adopt "calls for openness to categories which transcend the language of mathematics and biology, and take us to the heart of what it is to be human"[96]. This complex dialogue facilitates the transition to each stage and without meaningful dialogue it will be impossible to conceive holistic solutions to this complex problem.

---

[94] Author Jussi, "Individual Change or System Change Is Not the Right Question" in *Jussi Pasanen* (blog), April 10, 2019, https://www.jussipasanen.com/individual-change-or-system-change-is-not-the-right-question/.

[95] Pope Francis, *Laudato Si'*, n. 177.

[96] Pope Francis, *Laudato Si'*, nn. 1, 11

4.3 Role of Faith based groups in the Environmental movement

"Now, faced as we are with global environmental deterioration, I wish to address every person living on this planet.... I would like to enter into dialogue with all people about our common home"[97].

The church has a long history of advocating for social justice issues. The increasing understanding of the devastating impact on ecological destruction on social justice makes caring for our common home a natural extension of this catholic tradition. There is also a rich tradition of considering foundational and philosophical principles within the Church, which remains relevant and important when in dialogue with 'secular' organisations and individuals. Before we discuss in more detail how the role of the Catholic Church in Ireland could evolve and what, if anything, makes faith-based groups unique in the environmental movement, we need to acknowledge the ambiguity that exists when referencing secular and religious groups and concepts.

4.3.1 Collaboration Blurs the Secular/Religious Borders

The faith-based groups that are active within the ecological movement in Ireland are usually easily defined groups which have, at their core, the desire to serve God through their work. However, within these groups there also exist projects, such as ecological centres, that are openly accessible to those of all faiths and none, blurring the lines between 'religious' and 'secular'. While the faith based groups and institutions can self-identify as belonging to a particular religion and stem from specific beliefs and practices, the concept of religious or secular motives to care for our common home is more ambiguous.

That ambiguity ought to prompt us to question the wisdom in accepting any stark demarcation between 'secular' activism and 'faith-based' initiatives, especially if they conceptually arrange the secular as the default. The American theologian William Cavanaugh has written extensively about the complications involved in this approach. In dialogue with scholars of secularity like Talal Asad and José Casanova, he has gone so far as to say that the concept of defined religious and secular ideologies and institutions is "arbitrary and unsustainable on either theoretical or empirical grounds"[98].

---

[97] Pope Francis, *Laudato Si'*, nn. 1, 3.

[98] William T. Cavanaugh, *The Myth of Religious* Violence. Oxford: Oxford University Press, 2009, p. 16.

When we consider these questions in terms of caring for our common home, we see the wisdom of Francis' openness to collaboration with people of goodwill. The concept of social and environmental justice, including intergenerational justice are articulated by Pope Francis in *Laudato Si'* in distinctively Catholic terms, but these concepts are not the sole preserve of Christian or religious groups. Similarly, spiritual connection with nature, "taking time to recover a serene harmony with creation"[99] and the idea that the natural environment has intrinsic value and must be protected is observed not only in Catholic social teaching but is also widely found in environmental ethics research and within the ethos of the environmental movement. Indigenous communities across the globe also hold the protection of the natural environment deeply within their faith and culture. From a theological perspective, integral ecology is an approach that is inherently permeable. "Faith-based" groups need not ask for special permission to be included in the activities of the otherwise modern, liberal, secular activism movements.

Recognising that the borders are blurred is therefore a step towards clarity. There is clearly a distinctive set of commitments, practices, and even aesthetic preferences that might differentiate Catholic environmental activism from other approaches, but all approaches are concerned with meaningful moral goods and are embedded in traditions. Paying attention to those concrete real-world differences is much more useful than developing abstract, generalised theories. Understanding the strengths and weaknesses within specific communities and groups as a means to identify potential collaboration is an important component in this global dialogue, where similar concepts can be discussed by different groups coming from slightly different angles or using different language which might resonate differently with different groups. Faith based groups might practice 'creation care' which to the eyes of someone outside this group might appear to be environmentalism. Those studying *Laudato Si'* might talk about Integral ecology where different academics would use terminology like 'holistic ecology' or 'systems thinking'. These differences matter. It cannot all be boiled down, reduced to some common activity. But there is permeability between the concepts, such that even when there is not full alignment in meaning or perfect harmony in practices or values, a complex dialogue and fruitful dialogue can yet emerge.

---

[99] Pope Francis, *Laudato Si'*, n. 125.

So: while faith-based groups are not the only groups involved in the examination of the moral and ethical justifications for environmental conservation, there exists rich resources within the Christian tradition which can inform, contribute, and strengthen the overall message of the wider movement. The field of eco-theology has the potential to be able to articulate *why* - ethically, philosophically as well as theologically - we need to do everything in our power to protect and conserve our natural environment. As the Irish Jesuit, Gerard Whelan SJ has pondered: "might religion offer help to resolve the difficulty that nations find in arriving at consensus on ethical principles that underlie environmental policy?"[100].

On contemplating this possibility, I now ponder "how might religion fulfil this role, specifically in Ireland?" As mentioned earlier in the chapter, the perception of the Catholic Church in Ireland has undergone a complete overhaul in the past few decades with a pervasive distrust of the Institute. Consequently, as messengers for the protection of our Common Home, Pope Francis and the Catholic Church could be viewed as barriers to communication with non-faith based groups rather than beneficial to the overall dissemination of the message. To achieve meaningful dialogue across these perceived barriers we must concentrate on the commonalities rather than the differences and focus on the message over the messenger. While different terms might be used by different people, the overarching goal of protecting our Common Home is shared among everyone within the ecological movement. Communication that is based on dialogue rather than top-down instruction is more likely to persuade and engage.

4.3.2. The Faith-Based Element in the Irish Environmental Movement

In the ecosystem of the ecological movement faith-based groups fulfil another important niche function – vocalising hope. Pope Francis in his Encyclical Letter *Fratelli tutti* illustrated the many paths to hope and asks to "realize that our lives are interwoven with and sustained by ordinary people valiantly shaping the decisive events of our shared history"[101], Francis invites

[100] Gerard Whelan, "'Everything Is Interconnected': Laudato Si', an Ethical Framework to Guide Environmental Policy?", in *Working Notes, Jesuit Centre for Faith and Justice, Dublin* 34/86-2020, pp. 9–20.

[101] Pope Francis, *Fratelli tutti*, n. 54.

> everyone to renewed hope, for hope "speaks to us of something deeply rooted in every human heart, independently of our circumstances and historical conditioning. Hope speaks to us of a thirst, an aspiration, a longing for a life of fulfilment, a desire to achieve great things, things that fill our heart and lift our spirit to lofty realities like truth, goodness and beauty, justice and love… Hope is bold; it can look beyond personal convenience, the petty securities and compensations which limit our horizon, and it can open us up to grand ideals that make life more beautiful and worthwhile". Let us continue, then, to advance along the paths of hope[102].

The climate and biodiversity crisis can induce climate grief, despair and feelings of hopelessness. Within the ecological movement activists have sometime been working tirelessly for decades. The psychological toll of standing in the role of prophet, often seeing predictions of doom repeated for years suddenly take shape in real-time, and still without recognition, can be immense. The situation is so serious that it can seem like the solutions that we know exist simply won't be implemented because of greed and apathy towards our common home. Despair appears a rational response. Burn-out is a predictable consequence. The faith-based groups have a resource to hand here that is often unformed or unavailable to secular reason. Even when there are no grounds for optimism, the church can inhabit hope. This has a revolutionary potential for environmentalist groups; to be able to stare unflinchingly at the dark consequences of our present commitments and yet still go on. Awe and simple delight in God's good creation runs through the teachings of the church, her practices, and then spill over into activism[103]. This hope can generate a vision and narrative of environmental action marked by a confidence and generosity towards others that is not common in environmental groups that are not faith-based[104]. Francis articulated very clearly the importance of positive, constructive, yet realistic narratives to environmentalism and how we frame our approach to activism, "if we approach nature and the environment without this openness to awe and wonder, if we no longer speak the language of fraternity and beauty in our relationship with the world, our attitude will be that of masters, consumers, ruthless exploiters, unable to set limits on their

102 Pope Francis, *Fratelli tutti*, n. 55.

103 Elizabeth Bomberg - A. Hague, "Faith-Based Climate Action in Christian Congregations: Mobilisation and Spiritual Resources", in *Local Environment* 23/5-2018), pp. 582–96, (https://doi.org/10.1080/13549839.2018.1449822).

104 Adams et al., "Manifesto for a Green New Deal".

immediate needs. By contrast, if we feel intimately united with all that exists, then sobriety and care will well up spontaneously"[105]. The contrast between the sobriety of hope and the despair induced by unmet optimism is much more stark than any conceptual distinction between secular and religious activism, but it is a concrete expression of the unique contribution of "faith-based" groups to Irish environmentalism.

This well of hope that faith-based groups possess is continuously present in the various other ways in which the church contributes to this global struggle. Intellectual contributions, with *Laudato Si'* being the most notable example, differing perspectives and faithful action to make the world a better place including a preferential option for the poor are all infused with this hope.

4.3.3 Scale matters

As we explored previously, ecology and ecological movements exist and operates at different scales. This is also true for the religious belonging and identity that informs environmental activism. The individual, community, and institutional sense of belonging within a religion all exist concurrently with extensive interaction between scales. This, however, does not necessarily mean that values will be conferred or promoted in the same way throughout[106].

The structure of religious institutions is complex, consequently disseminating ecological values throughout its entirety is not as simple as might be imagined. The Catholic Church is hierarchical, but it is also a global institution with more than a billion members spread across every continent, in every country, practically in every region on the planet. Consequently, within specific regions and cultures, external norms and values affect and influence what teachings are disseminated and how. In an organisation so large, it can take years or decades for common teachings and values to become embedded[107].

In Ireland the message of Integral Ecology has not yet spread across the entire church family. This could be due partly to the waning of the churches influence in Ireland, competition for parish attention and resources, or due to the organisational structure of the church and the recalcitrance within the organ-

---

[105] Pope Francis, *Laudato Si'*, nn. 1, 11.

[106] Although focusing particularly on British churches – ecumenically broad at that – the studies by Ives and Kidwell explore this diversity well. For example: Christopher D. Ives - Jeremy Kidwell, "Religion and Social Values for Sustainability", in *Sustainability Science* 14/5-2019, pp. 1355–1362, (https://doi.org/10.1007/s11625-019-00657-0).

[107] Ives and Kidwell, "Religion and Social Values for Sustainability".

isation to preach and inculcate its stated values. Despite this, faith-based and religious groups contribute greatly to the environmental movement in Ireland. The involvement of the church in Ireland's ecological movement concentrates mainly at grassroots level, as well as social justice centres and development organisations advocating for environmental justice action and policies. At the grassroots level there are many *Laudato Si'* and ecological groups working within parishes to improve the sustainability of the parish and parishioners' lives[108]. Religious congregations and communities have also been very active in terms of setting up ecological centres and organic farms[109], fostering a culture of integral ecological within their surrounding communities. These examples highlight what would be possible in the Catholic Church in Ireland if the integral ecology message was ingrained more fully in the teaching of the church throughout the entire Church. The full potential of faith-based involvement in ecological activism has not been realised, either in terms of political and cultural influence or in terms of faith formation and mission. What is happening at the parish level is usually grown from the grassroots, taking its cue directly from the Pope but with relatively little sustained attention on ecological conversion coming from within the Catholic institutions in Ireland.

The network of parishes within Ireland covers every square inch of the island. Through this complex web of thousands of grassroots communities, the church has a unique ability to impact wider society. The church's role within the ecological movement in Ireland has, so far, been extensive, and yet still is underutilising its potential in guiding people towards an ecological conversion, with all the social, environmental, and spiritual benefits that follow. If caring for our common home was a top priority across the catholic church in Ireland the ecological groups that are dotted across the country would be the norm and not the exception. The activism that does exist remains significant. It embodies a distinctive Catholic commitment, while at the same time serving as an example of the fertile collaboration that can exist when religiously-informed actors engage in otherwise 'secular' activities. While it is a large-

[108] Many of these are members of the Eco-congregation network - https://www.ecocongregationireland.com/

[109] Examples of these faith-based ecological centres include: "An Tairseach", a farm which produces an abundance of organic vegetables as well as organic beef and pork from its livestock, a wildlife conservation area and an ecological education centre which is run by the Dominican Sisters; The Congregation of the Sisters of Mercy have several eco-justice initiatives in Ireland including community gardens and education centres.

ly grassroots response to *Laudato Si'* and the generations of eco-theology that preceded Francis' encyclical, it is grass growing healthy and strong, in a way that encourages not just hope, but even some optimism.

4.4 Challenges and opportunities of continued integration and collaboration between groups

With a phenomenon as diverse and complex as the environmental movement, it is inevitable that tensions will emerge at times between factions. Groups that have different core values or visions of what a better environmental future might be, groups answerable to different constituencies, or pursuing different priorities may find themselves in opposition. In the Irish context there are several examples of this, including tension between urban and rural based communities, industries that are either impacted by or impact the ecological health of our country, as well as groups which may clash when it comes to different perspectives on reproductive health and human rights. Drawing, once again, on ecological systems as a metaphor for the ecological movement we identify that dialogue, as is the common theme across this entire chapter, is at the root of both the challenges and opportunities that exist within the movement. The challenges that are faced can be overcome by turning towards and relying on dialogue and conversation to guide our actions.

Strengthening the networks between the different niches that make up the ecological movement is needed to encourage, cultivate and intensify these symbiotic relationships. This work needs to be done to strengthen both the bonds within distinct groups, such as faith based communities, as well as between groups, such as between faith-based groups and regenerative agriculture groups. These types of networking, 'bonding' and 'bridging' respectively, can only be achieved through dialogue and are vitally important for healthy functioning systems. Bonding within groups creates strong ties within the groups actively keeping people engaged in the work but can create insular, exclusionary groups if practiced without actively practicing other networking activity. Bridging between these groups helps to create a more diverse movement overall. Building strong bridges which connects different groups, who may have different priorities and values, enhances the symbiotic interactions between groups. Pope Francis in his encyclical *Fratelli tutti* argues that "we, as a people, should be passionate about meeting others, seeking points of contact, building bridges, planning a project that includes everyone"[110]. Francis also describes how our mutual concern our natural environment can facilitate these connections across groups, he also argues that this same concern needs to be shown to every person if we are to maintain the intrinsic connection between environmentalism and environmental and social justice:

> When we speak of the need to care for our common home, our planet, we appeal to that spark of universal consciousness and mutual concern that may still be present in people's hearts. Those who enjoy a surplus of water s choose to conserve it for the sake of the greater human family have attained a moral stature that allows them to look beyond themselves and the group to which they belong. How marvellously human! The same attitude is demanded if we are to recognize the rights of all people, even those born beyond our own borders[111].

Networking across the ecological movement demands work and effort in actively reaching out to, and dialoguing with, people we would not normally engage with. Failing to do this will result in fragmentation and splintering of the movement which can only reduce momentum in demanding and implementing real change. Just as in fully functioning ecosystems, if these modes of interaction cease or are interrupted the functioning of the system is lost. The process of the conversation must be continuous.

While the importance of cultivating connections between different groups cannot be overstated it would be naive to imply that this is a simple process. There are many barriers to generating these symbiotic relationships which can easily be observed in the relationship between traditional environmental groups and the environmental justice groups. There is much discussion about the disparate mission and values of the environmental and environmental justice movements. Traditional Environmental organisations tend to be larger organisations, mainly concentrating on preservation of scenic areas or specific species. Environmental justice organisations on the other hand, which tend to be at the grass roots level and consequentially more informal, concentrate on how environmental destruction impacts people. While there are commonalities between these two movements, the relationship between the two is fractioned, with conceptual and cultural differences existing. Charges against the larger environmental groups include racism, classism and sexism while

[110] Pope Francis, *Fratelli tutti*, n. 216.
[111] Pope Francis, *Fratelli tutti*, n. 117.

the some in the environmental movement fear that the environmental justice movement can directly conflict with environmental protection[112].

While these issues are very real there is a strong argument that "in the long run environmental justice, environmental sustainability, and regional justice will either be accomplished together or not at all"[113]. This echoes the teachings of Pope Francis in *Laudato Si'* that "the human environment and the natural environment deteriorate together; we cannot adequately combat environmental degradation unless we attend to causes related to human and social degradation". The role that faith based groups and the Catholic Church can play in reinforcing the link between these two movements is significant. Viewing these issues through an integral ecology lens allows the bridge to be built in protecting species, ecosystems and scenic areas for their intrinsic value and protecting the dignity of humans through environmental justice action. The tradition of the Church in promoting a 'preferential option for the poor' strengthens the role it plays within the movement where the continual representation of the marginalised should balance any tendency within the more traditional environmental organisations to prioritise excessive 'wilderness' conservation over social justice issues[114]. While it is undoubtedly true that for healthy ecosystems and biodiversity to survive there must be space in the world free from exploitation, the delicate balance to address the complex crisis that is both social and environmental will need to be found[115].

As we demonstrated above in relation to consequences of the referenda results, faith-based groups advocating for climate and environmental justice will, without exception, be working alongside people they fundamentally disagree with on certain topics. From a theological perspective Christians practice hospitality for those who hold different opinions. The onus is on Christians, by dint of their own profession, to make space so that "enemies" can be

[112] Sandler and Pezzullo (ed.), *Environmental Justice and Environmentalism: The Social Justice Challenge to the Environmental Movement*, 2007.

[113] Sandler and Pezzullo (ed.), *Environmental Justice and Environmentalism*, 2007, p. 16.

[114] There is a very real risk of 'land grabs' from local and indigenous communities in the name of conservation. See Tom Blomley et al. 'Land Grabbing': Is Conservation Part of the Problem or the Solution? International Institute for Environment and Development, 2013, (www.jstor.org/stable/resrep01526). This misguided and counter-productive phenomenon contributes to the polarising discussion of conservation, rewilding and social justice.

[115] Raworth, who pioneered the doughnut economics model, might be a fruitful dialogue partner when grappling with this complex issue. Her model aims to articulate the balance that needs to be found between planetary boundaries and critical human development. See Raworth, *Doughnut Economics – Seven ways to think like a 21st-Century Economist.*

loved and become friends. "An open and respectful dialogue is also needed between the various ecological movements, among which ideological conflicts are not infrequently encountered"[116].

It is possible that the strong interest in and commitment towards justice found among Irish young people, who may not identify as 'religious', may find its origin in the deeply embedded influence of Catholic Social Teaching on the Irish imagination. What is certain however is that if the church in Ireland is to realise its ambition of catalysing a broad ecological conversion it will need to collaborate generously with all people outside of reach if its own institution.

Pope Francis, on publishing *Laudato Si'*, led by example in this regard by inviting Naomi Klein a "secular Jewish feminist" to speak at the press conference launching *Laudato Si',* followed by attendance at a two-day symposium to explore this encyclical. Recognising allies in looking towards climate and environmental justice is vital if we are to convert billions of people to care for our common home. In her book, "*On Fire: the burning case for a green new deal*", Klein recounts the experience of visiting the Vatican but also highlights some of the challenges that apply in this collaboration. From the church's side, as mentioned before, there is opposition to working with people who may not share core values on abortion and reproduction rights. For the 'secular' advocates, this hesitation can be reversed, and there is also distrust of the church for its failing to respond to abuses that have been perpetuated within the institution for decades. While acknowledging these challenges are important, they must not be allowed to stall the vital movement for climate and environmental justice and the important work of understanding what faithfulness looks like in the Anthropocene. Again, we see the importance on focusing on the common message, and our common goal, of protection of our natural environment in fostering dialogue and climate action.

This challenge is not only observed between 'secular' and faith-based groups but is also clearly seen between some factions of rural and urban Ireland, as well as those who work the land such as foresters and farmers. The animosity between environmentalists and the agricultural sector in Ireland is particularly problematic. Those working in the agricultural sector have a complicated relationship to climate breakdown and environmental destruction. Intensive agriculture, which has been incentivised by successive Irish governments, is a major cause of greenhouse gas emissions, water pollution, and biodiversity

---

[116] Pope Francis, *Laudato Si',* n. 201.

loss. On the flip side, this sector is also one of the most vulnerable to climate breakdown with changing weather patterns an intrinsic risk to both animal welfare and yield. The challenge to bridge this gap is a difficult one, but within the context of our country a vital one. Here integral ecology teachings from *Laudato Si'* could be used to find a narrative and a vision of the future of agriculture that protects both farmer and the natural environment. Indeed the Jesuit Centre of Faith and Justice, in its *Working Notes* journal has attempted to create this narrative[117] as well as highlighting the possible schemes that could be utilised to make this vision a reality[118]. By articulating possible alternative ways forward, the church in Ireland could play a role in bridging urban and rural communities and engaging in a conversation on creation care that is inclusive and non-confrontational.

Religious organisations have certain characteristics which allows them to be proficient mobilisers and action takers[119]. Pope Francis in *Laudato Si'* challenges the church to define and characterise creation care as integral component of practicing our faith. He argues that "living our vocation to be protectors of God's handiwork is essential to a life of virtue; it is not an optional or a secondary aspect of our Christian experience"[120]. If this assertion became a central part of church teaching the sporadic grassroots ecology groups that have sprung up across the country could be ubiquitous across every diocese, catalysing a rapid ecological transition. The pre-existing social interactions and parish networks could facilitate a broad emergence of 'creation care' volunteers who may not otherwise associate with environmentalism. Experiments in adult faith-formation can be expansively explored while engaged in such activities.

While the church must accept the challenge of becoming a partner in the environmental movement, it also need to be recognised that a balance must be found which allows the church to participate as equal partners with whom ideas and concepts can be debated and not simply as a 'spiritual resource'[121] to be utilised. The counter side of the balance is for the church to enter into the complex dialogue of the ecological movement, not only as a leading participant, but also in solidarity with the other members of this diverse collec-

[117] Adams et al., "Manifesto for a Green New Deal".
[118] O'Sullivan, "High Nature Value Farmland".
[119] Bomberg - Hague, "Faith-Based Climate Action in Christian Congregations".
[120] Pope Francis, *Laudato Si'*, n. 217.
[121] Bomberg - Hague, "Faith-Based Climate Action in Christian Congregations".

tive working towards a common goal with mutual respect. We must always remember that "Solidarity means much more than engaging in sporadic acts of generosity. It means thinking and acting in terms of community"[122]. In the Irish situation, where the power and influence of the church has changed so significantly in recent decades, this has the potential to be both an opportunity to engage with people who may be disaffected by the Church but also a challenge to accept this altered role.

5. Conclusion

The diversity within the ecological movement, in terms of who participates, the paths that can be taken to action and ways of being involved, are intrinsic to the health of the movement. A vital component for this diversity to function effectively is the networks that link all the different components, and as we have seen in this chapter, dialogue is foundational to this. It is an important foundation for deepening our notion of integral ecology. Reflection on "foundational principles" for integral ecology that do not emerge from a practice of dialogue and real-world engagement in the complexities of the ecological movement in different countries may prove limited and require careful interpretation.

Dialogue is the process by which environmental action is developed and implemented but it is also intrinsic in the functioning of the networks which allows the environmental and social movements to function. Dialogue does not imply abandoning the wisdom accumulated in Catholic social and moral teaching over centuries, but, rather, a kind of rediscovery of its relevance to debates occurring in the secular world—always remembering that Catholic teaching must itself develop as it learns from others in the dialogue process. Without dialogue each part of the movement separates, depriving it from the energy and life blood that feeds it. *Laudato Si'* and *Fratelli tutti* both call for meaningful dialogue to create a better future. We must answer this call as we, as faith based communities, fulfil our role in Caring for Our Common Home.

---

[122] Pope Francis, *Fratelli tutti*, n. 116.

THE THROWAWAY CULTURE AND TECHNOCRACY

Results of the Absence of a Metaphysical Vision

Gaetano Piccolo, SJ

1. Introduction: some questions raised by *Laudato Si'*

A rabbinic story, repeated by L. Pareyson in *Filosofia della libertà*, tells how annoyed the letter *aleph* was with God for not choosing it with which to begin the text of *Genesis* but beginning instead with *beth*, that is, with the second letter of the Hebrew alphabet[1]: God explains the appropriateness of beginning the account of the creation with the letter *beth* on the grounds the shape of this alphabetical character which is closed on its right side, that is, from the place where one begins to read in Hebrew. This was to say that man can only know from then on. Man is placed before a limit which characterises his existence. It is not man's role to generate creation, but it is he who receives it as a gift, he who is asked to cultivate and guard this garden in which he gratuitously finds himself[2].

This limit, which cannot be ignored, opens up two possible outcomes, depending on the way it is perceived. It can be the occasion of unbearable humiliation or the occasion of a valuable identity. In the first case, man seeks to trespass across the limit or even tries to destroy it. He finds himself filled with the possibility of evil but, at the same time, he is unable to handle it. In the second case, the limit allows him to define himself as the non-creator, it en-

[1] Luigi Pareyson, *Filosofia della libertà*. Genova: Il Melangolo 1991, pp. 19ss.

[2] Gen 2,8.

DOI 10.32060/C.S.9.183-195

ables him to escape from the delusions of suicidal omnipotence and puts him back in his rightful place. It enables him to look at things not as a possession but as gifts that he has received. In other words, it prepares him for gratitude. Man thus finds himself living between the limit and the gift, between the recognition of his identity and his relationship with what appear to be only things (but which, on close examination, are revealed to be gifts).

In this paper, I would like to investigate the foundation lying beneath man's relation with everything else, namely, with other beings and with things. I would like to argue that the study of this relationship, that is, of human activity within creation, requires an in-depth analysis of the type of relationship which precedes this activity. I would like, therefore, to lay my analysis on those metaphysical foundations which are unavoidable if there is to be a healthy relationship between man and the creation. This investigation is intended to contribute in some way to a greater understanding of two points in the encyclical, *Laudato Si'*: the first point which I would like to study concerns what Pope Francis calls 'the throwaway culture' (in particular, sections 20-27)[3]; the second concerns the deviation which results from the 'technocratic paradigm' (especially sections 101-114).

When we think about it, these two themes are also involved in the debate to which the pandemic of Covid-19 should give rise[4]: at those times when the hospitals were most congested, we were aware that the doctors found themselves faced with a grave choice, namely, that of deciding who to save when the available means, such as the number of beds in intensive care units, were not enough. In fact, whoever has been a patient or accompanied their own families when undergoing medical treatment knows all too well that, even before the pandemic, patients were being sorted out on the basis of a scale of

---

[3] This theme was taken up again later in the encyclical, *Fratelli tutti*, specifically in sections18-21. In particular, Pope Francis writes: "Some parts of our human family, it appears, can be readily sacrificed for the sake of others considered worthy of a carefree existence. Ultimately, persons are no longer seen as a paramount value to be cared for and respected, especially when they are poor and disabled, 'not yet useful' – like the unborn, or 'no longer needed' – like the elderly. We have grown indifferent to all kinds of wastefulness, starting with the waste of food, which is deplorable in the extreme", Pope Francis, Encyclical Letter *Fratelli tutti on Fraternity and Social Friendship*, 3 October 2020, n.18.

[4] Paradoxically, it is precisely through a harrowing experience like the pandemic that we can be more fully aware of and recognise the intimate bond which exists between ourselves and the creation: "a worldwide tragedy like the Covid-19 pandemic momentarily revived the sense that we are a global community, all in the same boat, where one person's problems are the problems of all", Pope Francis, *Fratelli tutti*, n.32.

eligibility. On this basis of priorities, those who are old or affected by other pathologies are considered 'untreatable', that is, they are already condemned. The pandemic has simply made this process public. Thus, there are contexts in which this throwaway culture is all too evident.

On the other hand, however, the months of this emergency, and especially the early months, exposed us to a total subjection to the opinions of experts, virologists, epidemiologists and technico-scientific committees which actually, when put to the test, were shown on several occasions to be inadequate and incompetent. We found ourselves disorientated in the face of opinions which were not simply different but wholly contradictory. With greater humility, science should have recognised its own lack of preparation and inadequacy when it came to the task of assessing methods and making forecasts. The economic, social and even religious life of entire peoples was subjected completely to the opinion of technicians who, inevitably, were not *au fait* with many particular aspects of the various areas for which they were proposing measures.

However, Pope Francis's encyclical, *Laudato Si'*, warns about what he describes as the 'mysterious network of relations'[5]. Indeed, there is "a technology – writes Pope Francis - which, linked to business interests, is presented as being the only way of solving these problems"[6]. If we ignore the complexity of relations, we "risk solving one problem by creating others"[7]. The throwaway culture is also the product of this slapdash and imprudent way of tackling problems: whatever seems useless, or creates problems, or requires investment which exceeds the profit is immediately jettisoned or eliminated, thus transforming the earth into an immense rubbish dump, not only in the material sense but also as an image of those who are considered human rejects[8]. Thus, the basic question we are trying to answer is, what leads us to treat things and people in this way.

As stated above, a possible answer is to be found in those paragraphs where Pope Francis reflects on technocracy, claiming that it is necessary to seek "the human origins of the ecological crisis"[9]. We must focus, the Pontiff continues,

---

[5] Pope Francis, Encyclical Letter *Laudato Si' on Care for Our Common Home*, 24 May 2015, n. 20.

[6] Ibid.

[7] Ibid.

[8] "The throwaway culture affects the excluded just as quickly as it reduces things to rubbish", Pope Francis, *Laudato Si'*, n. 21.

[9] Pope Francis, *Laudato Si'*, n. 101.

"on the dominant technocratic paradigm and the place of human beings and of human action in the world"[10]. We could describe the first dimension of the problem as a formative one since the encyclical argues that modern man has not been educated in the right use of power, "because our immense technological development has not been accompanied by a development in human responsibility, values and conscience"[11].

More generally, Pope Francis sees the basic problem in the way in which human beings have adopted technology and its development "according to an undifferentiated and one-dimensional paradigm"[12]. In this paradigm, there is a concept of the subject as a being able to understand and so to possess external objects. However, this possession does not seem to be inscribed in the reality of things so much as in the way in which human beings have decided to treat things. Indeed, what we have can be taken away from us at any moment in our lives without our permission being sought. Certainly, various types of contract have been invented to guarantee the defence of property within human relations, but, outside these relations, time, death and health do not respect those contracts which human intelligence has produced in abundance.

Thus, this subject who wants to possess things has developed a technique of "possession, mastery and transformation"[13]. Such an attitude reveals the conviction that one is "in the presence of something formless, completely open to manipulation". The expression 'something formless' recalls the text of Genesis 1: "the earth was without form and void"[14]. It is as if man desires to remake the creation. He sets himself over the earth as if there had been no creation, as if there was no starting point, and decides, autonomously, without God, what to do with everything that he encounters. In this way, a technology conceived thus expresses a logic of dominion[15].

2. The essence of technology

To understand how we have come to this supremacy of the technocratic paradigm and to attempt to go beyond it, we shall have to go back and ask ourselves about the essence of technology. When we speak of the essence of

---

[10] Ibid.

[11] Pope Francis, *Laudato Si'*, n. 105.

[12] Pope Francis, *Laudato Si'*, n. 106.

[13] Pope Francis, *Laudato Si'*, n. 106.

[14] Gen 1,2.

[15] Cf Pope Francis, *Laudato Si'*, n. 107.

something, we are referring to its nature; we are trying, that is, to answer the question *what is it?*

Heidegger tackled this question in *The Question concerning Technology*, a text contained in the collection *Vorträge und Aufsätze*. The first observation which springs from the question itself is that the essence of technology cannot be technology itself just as the essence of a tree is not a tree. When we ask what technology is, there are usually two replies:

a. Technology is a means in view of certain ends.
b. Technology is a human activity[16].

It is easy to acknowledge that these two definitions are connected:

> According to ancient doctrine, the essence of a thing is considered to be what the thing is. We ask the question concerning technology when we ask what it is. Everyone knows the two statements that answer our question. One says: Technology is a means to an end. The other says: Technology is a human activity. The two definitions of technology belong together. For to posit ends and procure and utilize the means to them is a human activity. The manufacture and utilization of equipment, tools, and machines, the manufactured and used things themselves, and the needs and ends that they serve, all belong to what technology is. The whole complex of these contrivances is technology. Technology itself is a contrivance, or, in Latin, an *instrumentum*. The current conception of technology, according to which it is a means and a human activity, can therefore be called the instrumental and anthropological definition of technology[17].

If we think of technology in an instrumental sense, it has to do with causes, in the sense that it causes something. Obviously, when we are speaking of causes, we are referring to the Aristotelian doctrine of the four causes, that is, to take up again, as Aristotle himself says, a doctrine well known to his predecessors[18]. The cause is *responsible* for the object, inasmuch as, if the object is what it is, that is due to its causes. In fact, the causes make the thing *happen* (*casus*, from the Latin *cedere*), bring about that it occurs, is made present, in other words, they unveil it. Therefore, the causes concern the truth of the thing[19].

---

[16] Martin Heidegger, *The Question Concerning Technology and Other Essays*. New York-London: Garland, 1977, p. 4.

[17] Ibid., pp. 4-5.

[18] Aristotle, *Metaphysics*, I, 983 b 1-5.

[19] Cf Martin Heidegger, *The Question Concerning Technology and Other Essays*, pp. 7-10.

In the light of these considerations, Heidegger considers technology "allows us to see what is not produced by itself"[20]. Continuing in this direction, *technè* ends up being connected with *episteme*, that is, knowing and disclosure. The essence of technology is indicated by Heidegger with the term 'Gestell'. Here we find the essence of technology in something which is not technical[21]. *Gestell* is system or structure (from the German, *stellen*, 'to place'). In modern technology, the *Gestell* finds its most complete expression. In this system or mechanism, there is a confrontation between man as an *animal rationale* on the one hand, and, on the other, the being considered as his foundation and placed at his disposition. In Heidegger's vision, the *Gestell* is characterised by a shared belonging of man and being in the imperative and oppositional form of the "reciprocal provocation" and "calculating machination"[22].

Precisely because technology in its essence reveals the relationship of man with the being, it has to be considered as the final outcome of the journey of metaphysics, that is, the height of the forgetfulness of the sense of being which is illusorily reduced to *what exists* in such a way that it is manipulable. In fact, being is rendered available for every kind of calculation and manipulation in the moment in which it is fixed and stabilised, uprooted from its essence as event. Here, we understand the reason for Heidigger's inclusion of this reflection on technology within his paper on the *Principle of Identity*.

The principle of identity is usually understood as a law, a principle or an assertion about the meaningless identity of being with itself. Within, being is thought of as something stable, immutable and constantly present. This is the reason why being is considered the foundation of *what exists*. This stable foundation also makes possible the work of science which is thus able to concentrate clearly on an object that has to be known[23]. Thus, the principle of identity becomes the secure foundation for the construction of a metaphysic which Heidegger says can only be onto-logical at this point (for example, in the sense of Hegelian logic), inasmuch as, if we presuppose that being is the stable, general and primary basis of *what exists* (understood with Aristotle as *hypocheimenon*), it claims that thought is foundational. Moreover, at the same time, this metaphysic can only be theo-logical, since, if we presuppose that

---

[20] Ibid, p. 10.

[21] Ibid., pp- 15-16.

[22] Martin Heidegger, *Identità e differenza.* Milano: Adelphi, 2013, pp. 10-11.

[23] Ibid., p.10.

being is the external, supreme and ultimate foundation of *what exists* (as *causa prima* and *causa sui*), it claims that thought arrives at the concept of God[24].

It is at this point that we discover Heidegger's explosive insight: "onto-theology is the first form of technology, technology is the final form of onto-theology"[25]. Our interest here is to show that there is actually a profound relationship between technology and metaphysical vision: in other words, technology is the expression of a certain way of understanding the relation with things. To take the stability of things as fundamental renders them open to manipulation and opens the way to the possibility of violence[26]: it is a current way of looking at things which no longer sees in *what exists* its dynamic as event and connection with the other beings. Once isolated and defined, we can make of *what exists* all that we want[27]. To rethink technology and avoid its metaphysical deviation, Heidegger seems to think, what we have to rethink is the principle of identity which is not an empty repetition of itself (*A*=*A*) but tells of the relation *with* itself. The principle of identity tells of the relation which *all that exists* has with itself. *All that exists* is in itself relation[28].

The principle of identity has been adopted in the metaphysical tradition as a guarantee of the unity of *what exists* with itself which enables science. Only in this way has scientific research succeeded to ensure the possibility of working. Yet, even if science is not disposed to listen, its own working says something of *what exists*. The experience of the pandemic has shown that, despite all its progress, reality always remains a mystery. Science works by continually appropriating more of reality to itself but without ever reaching the potential foundation of this mystery[29]. Thus, Heidegger argues, metaphysics has always thought that identity belongs to being. In other words, if there is being, then there is identity as its property. Taking up a famous affirmation attributed to Parmenides, "thinking and being are the same", However, Heidegger inverts the relation, claiming that the being belongs to the identity: where there is a relation with itself, then there is being.

---

[24] Ibid., 13.

[25] Ibid., p. 15.

[26] Pope Francis also takes up the theme of violence in the Encyclical, *Fratelli tutti*, within a particular discussion on the relations between religions. Cf Pope Francis, *Fratelli tutti*, n. 281ff.

[27] According to Heidegger, at this point it is necessary to bring man out of the inheritance of the West, that is, from the nihilism of technology. Heidegger finds a way in remembering (*vordenken*) and remembering again (*andenken*). Ibid., p. 16.

[28] Ibid., pp. 28-29.

[29] Ibid., pp. 31-32.

Thus, the age of technology is the age of metaphysics, that is, of the human subject counter posed to the world. The risk run by ecology is that of being put in the same perspective as technology: even though one is seeking to use things in a better, healthier way, one is actually putting oneself in the perspective of using things[30]. That is why Heidegger has been described as "the true and most important philosopher of ecology. In fact, it is he who has explained that *what exists* is of itself that which refers to human use and consumption"[31]. The human subject is convinced that he is such to the degree in which he places *what exists* under his complete control. For this reason, as we have tried to show, following the thought of Heidegger, technology can be considered the final point of human destiny which is the metaphysic, that is, the forgetting of being. The being is thus destined to remain outside this metaphysic understood as technology because it is never allowed to be reduced to *what exists*.

If technology is the system which makes up modern culture, it is easy to imagine that it affects and forms human beings in a certain way. Thus, the anthropological crisis of our time arises from this metaphysical vision which dispossesses the being from its reflection and focuses on *what exists* in a delirium of total control which, inevitably, is open to the possibility of violence. Tecnique (or *technology* as G. Anders says, using more recent language) is not actually a neutral instrument in human hands. It affects the form and the meaning of human occupations: "The culture of the mass-media possesses an ontological ambiguity which enables the overlapping of reality and its representation"[32]. The aberration lies in the representation in the sense that the spectator consumes the image without needing to take part in the event which it represents. Someone can consume the world while remaining at home. In this way, there is less capacity for compassion but also that for perceiving otherness. Everything becomes simultaneously familiar and indifferent. The individual identity is weakened in favour of a transformation of the individual spectator into mass-man[33].

---

[30] On this subject, cf. the reflection of Cimatti in Felice Cimatti, *Cose. Per una filosofia del reale*, Bollati Boringhieri, Torino 2018, 47.

[31] Idem, 45.

[32] Günther Anders, *Il mondo dopo l'uomo. Tecnica e violenza*. Milano: Mimesis, 2008, p. 8.

[33] Anders has coined the expression "popular hermit" to indicate the person who sits isolated in front of his radio or television, and, today, we would obviously say in front of his computer screen. This is someone who receives the same auditory and visual "food" as the others. Cf Günther Anders, *Il mondo dopo l'uomo*, p. 93.

Technology's risk lies in the human incapacity to control what is produced on account of the disparity between feeling and doing. Anders describes as *Promoethean disparity* this difference between the ability to feel and that of doing, that is, the human inability to foresee the consequences of the unlimited possibility of production. The infinite possibilities available to man as producer are not sufficient to raise him from a feeling of inferiority described as *Promethean shame*. Precisely because what man produces is unmeasurably superior to him, he has no control over it[34]. The frustration arises, therefore, from the tension between the pretence of control over *what exists* and the reality which throws this claim into question. Inevitably, this frustration then generates anger and violence.

3. The human being freed from the supremacy of technology

We return now to following Heidegger's reflection in order to clarify the destiny of humanity in this era of the metaphysic as technology. In my opinion, Heidegger's analysis explains the origin of the throwaway culture and the technocratic paradigm addressed by Pope Francis. Moving on from this diagnosis, however, I would like to propose alternative ways which can lead towards a more authentic and integral existence.

This attempt to go further was perhaps already implicit in the inaugural lecture, published later with the title, *What is Metaphysics?*, which Heidegger delivered on 24 July, 1929, in the University of Freiburg, where he had been invited after five years of teaching at Marburg, to succeed Husserl. When the work was reprinted, an important *Postscript* was added, and the fifth edition of 1949 saw the appearance of an *Introduction*. Despite its being one of Heidegger's minor works, he returned to it constantly.

Heidegger begins his enquiry precisely by starting off from the public he is addressing, namely, men of science. It is exactly this condition, Heidegger argues, to be paradoxical in appearance because "man, *who exists* among others, does science"[35]. Since science wishes to control its object, its focus is on *something* excluding the nothing from its enquiry. Science considers the nothing as something with which it does not wish to be concerned but falls thereby into a paradox which logic, by itself, cannot overcome: if the nothing is something, it must equally be part of the interests of science. Taking for

[34] Ibid., 10-11.

[35] Martin Heidegger, *Che cos'è metafisica?* Milano: Adelphi, 2001, p. 40.

granted that there is something implies the inability to marvel before that which it is. The object which we study is taken for granted. It is no longer taken as a gift which could also not be there. Precisely for this reason, science can run down the slippery slope of manipulation and violence with regard to its object. On the contrary, distress brings us to face the nothing because we daily have the experience that, from one moment to the other, what there is could no longer be there. This way of doing science, which no longer focuses on the background on which *all that exists* is grasped, is at the root of the technocratic paradigm which we have analysed[36].

This attitude of science with regard to *what exists* could be described in a certain sense as inhuman in that, as Heidegger was to say in his *Letter on Humanism*, "man is not the master of *what exists*. Man is the pastor of the being"[37]. Man is the one who enables the meaning of things to emerge, thanks to reflection and thought. When, on the other hand, he concentrates on the objects, thinking he can make something from them, then he betrays his own identity. With the modern period, man's focus has become definitively that of technology, a focus, namely, which tends to transform *all that exists* into a thing: "The technology spoken of by Heidegger is not an accessory feature of humanity, it is its distinctive characteristic. Technology has to do neither with ethics nor with economics but rather with anthropology. It is the human animal who treats the world as a collection of things at his disposition"[38].

The question is anthropological because, although man is also a thing, by contrast with the other things, he is not (or ought not to be) one that simply lives. Man is called to make something of his life: "Every human being chooses, that is, he chooses the life that he can live"[39]. Man is man, therefore, when he has this possibility. To remove from the human being this possibility of choosing what to make of his own life means to deprive him of his dignity. By contrast with man, the herring is not an existent because it will never know it is a herring. In Heidegger's language, we could say that the herring never experiences anguish. By contrast, the first characteristic of the existent with the world is precisely this anguish. Whereas the herring is afraid in the face of the attack of the seal which eats vast quantities of herrings, the existent is

---

[36] Ibid., pp. 38; 46; 49; 50.

[37] Martin Heidegger, *Lettera sull'umanesimo*. Milan: Adelphi, 2013[10], pp. 56 and 71.

[38] Felice Cimatti, *Cose. Per una filosofia del reale*. Torino: Bollati Boringhieri, 2018, p. 44.

[39] Ibid., p. 49.

in anguish in the face of the world as world. This anguish is not directed at a specific object as happens in the case of the herring's fear before the seal. Before the world, the existent finds himself with his freedom to be able to choose. It is precisely this condition which is revealed by his anguish[40]. Thus, the existent does not identify himself with the world in which he stands; he is aware of the difference. The existent never confuses himself with the world. The world is not his home as the sea is for the herring. However, the existent can transform his existence into something which it is not yet. The existent is able to imagine himself elsewhere and for this reason contributes to the progress of existence.

Thus, man is free to take his farewell from the world he has contributed to construct because, by contrast with the animal, he has no relationship of belonging to it. This difference between man and the animal was also identified by Max Scheler in a conference entitled "The position of man in the cosmos", which was held in 1927 but published in the following year. Whereas the animal is enclosed in its own environment and condemned to react there in an impulsive-instinctive way - wrote Scheler – man is distinguished by his ability to say no to reality, to open himself to the world, to take his distance from reality and, precisely because of this position, to grasp it fully. Only man, who is endowed with spirit and self-knowledge (that is, he is not simply conscious as an animal can be) is able to make free decisions. Man is thus capable of emancipating himself from the original tendencies imposed on him by life[41].

4. Conclusion: a new humanism

From what we have learned in this brief study, it appears urgent to educate ourselves for a new relationship with things and with others, First and foremost, we must escape from the logic of use: it is not a question of using better but of understanding that we are not called, through our very existence, to use so much as to live relation. Our every interaction with the world constitutes an act which defines our identity. Our choices define us. Only if we start from this perspective, one which is actually more realistic, is it possible to avoid the manipulation of the world which easily results in violence, understanding by violence every action which is imposed on the other. If we move from use to

[40] Ibid., p 50.

[41] Günther Anders, *Patologia della libertà. Saggio sulla non-identificazione*. Salerno: Orthotes Editrice, 2015, p. 12.

relation, this withdraws us from a false view of reality: the world is not at our disposal; rather, it is the space which enables us to give meaning to existence. It is above all science, and so also, technology, which needs to get beyond this manipulative view of its object.

This leads to a second aspect which we must bear in mind: the relationship between wonder and anguish. If we try to anaesthetise man so that he does not experience anguish, he is also deprived of the possibility of wonder. Everything is taken for granted. It is already clear what I can do. Within a relation, on the contrary, this direction of the action is never given at the start. Each time, we are called to wonder that the world is there, to consider that it could also not be there, to be aware that what there is can also end, and that, as has happened with the Covid-19 pandemic, everyday life can also be turned upside down.

To return to addressing our fundamental question about the meaning of things involves certainly the affective task of giving up control, recognising one's own ignorance, and restoring to the world its mysterious character. Only in this way shall we stand more satisfactorily within reality, renouncing constructing it artificially simply in order to enjoy a little more peace. A peace which, sooner or later, and very easily, can be changed to reveal an unintended face.

Every relationship with things and with others represents a precious and irreplaceable occasion to ask oneself and to find the meaning of existence because it is always placed as the possibility of a choice (and not of a use). Every alterity constitutes an appeal to relation: on the contrary, waste is a denial of relation. It is as if one is saying of this existent, I do not wish to know it. It, therefore, represents the betrayal of one's own nature: I do not take responsibility for *what exists*. But it is precisely here that we find the paradox because, as Lévinas already reminded us, responsibility for the other is incumbent on me. I do not choose it; it is infinite:

> But the relation with a past that is on the hither side of every present and every re-presentable, for not belonging to the order of presence, is included in the extraordinary and everyday event of my responsibility for the faults or the misfortune of others, in my responsibility that answers for the freedom of another, in the astonishing human fraternity in which fraternity, conceived with Cain's sober coldness, would not by itself explain the responsibility between separated beings it calls for. The freedom of another could never begin in my freedom, that is,

abide in the same present, be contemporary, be representable to me. The responsibility for the other cannot have begun in my commitment, in my decision. The unlimited responsibility in which I find myself comes from the hither side of my freedom, from a "prior to every memory", an "ulterior to every accomplishment", from the non-present par exellence, the non-original, the anarchical, prior to or beyond essence. The responsibility for the other is the locus in which is situated the null-site of subjectivity, where the privilege of the question "Where?" no longer holds. The time of the *said* and the *essence* there lets the pre-original saying be heard, answers to transcendence, to a dia-chrony, to the irreducible divergency that opens here between the non-present and every representable divergency, which in its own way – a way to be clarified – makes a sign to the responsible one[42].

[42] Emmanuel Lévinas, *Otherwise than being or beyond essence*. Pittsburgh: Duquesne University Press, 2000[3], pp. 10-11. We cannot do less than note that in this text we find various themes which are dear to Pope Francis and are taken up not only in *Laudato Si'* but also in *Fratelli tutti*, namely, the themes of fraternity/brotherhood, waste and responsibility.

NO ORGANISM IS AN ISLAND: THE PHILOSOPHICAL CONTEXT REGARDING LIFE AND ENVIRONMENT

Louis Caruana, SJ

The encyclical *Laudato Si'* constitutes an important addition to Catholic social teaching[1]. Many commentators have analyzed its social and political contributions from a theological viewpoint and have produced useful results about the origin of its main ideas and about their implications[2]. One perspective however remains somewhat underexplored. Few studies have explored the presuppositions that are primarily philosophical. Few have surveyed the conceptual history that serves as the background to its entire message. To appreciate the full impact of this major contribution of the Church, it is becoming increasing evident that we need to explore this philosophical background. We need to uncover its ontological and epistemological commitments with respect to complex debates in both philosophy and science, debates that have been laboriously hammered into shape in the course of history. For instance, it would be quite irresponsible, and perhaps even counterproductive, to summarize the encyclical's message by merely using clichés like "everything is connected to everything else" without referring to the underlying philosophical debates supporting this statement. A full appreciation of the encyclical's

[1] Pope Francis, Encyclical Letter *Laudato Si' on Care for Our Common Home*, 24 May 2015, n. 15.

[2] Overviews of such contributions are many. See for instance the special issue of Paul G. Crowley, ed. *Theological Studies* [special issue], vol. 77/2 - 2016; and Humberto Miguel Yañez SJ, ed., *Laudato si'. Linee di lettura interdisciplinari per la cura della casa comune*, Rome: Gregorian & Biblical Press, 2017.

DOI 10.32060/C.S.9.197-220

message, therefore, requires us to position the document and its arguments within the long narrative of conceptual and scientific history of which it forms a significant part. Given its rich and interdisciplinary content, many of its arguments could not be presented except in a summary form despite their complex background. It is obvious therefore that if the conceptual and historical background is explored carefully, the encyclical's message will stand out in a clearer light.

The philosophical debates we are referring to here are, of course, many. They include debates concerning reductive materialism, intrinsic and instrumental value, the ontological category of relation, the common good, and many others. This paper will focus on one of these only. It will concentrate on the debate concerning the very idea of environment and the way life is intelligible in terms of relations. In its simplest, primary meaning, the word "environment" refers to the surroundings of the thing that we are considering, the environment here being conceived somewhat as their container. We ourselves however, as human individuals, can be considered an environment with respect to the enormous number of bacteria living within us. And so also all other living creatures[3]. Although the distinction between a living thing and its environment is very often clear, life itself cannot be understood exclusively in terms of the dispositions within the organism taken in isolation. Life is a function of both the organism and its environment. The boundary between them is porous. These preliminary remarks already indicate how the very concept of environment is certainly not simple. An awareness of the philosophical and scientific background in this area, and more detailed inquiry, are therefore indispensable for a deeper appreciation of the message of *Laudato Si'*. The guiding question for this paper will be the following. If we determine how the encyclical's arguments regarding life, the environment, and ecology, are situated within the history of ideas, what new insights become available, especially as regards the encyclical's original contribution? The first section of the paper will deal with the idea of environment as this is presented in the encyclical. This presentation is followed by three sections on the history of the concept: a section on early views, namely those that flourished roughly up to the early nineteenth century, a section on views that see nature as a guided process, and another section on views that see nature as an unguided pro-

[3] The concept of *milieu intérieur* was the key concept for the physiologist Claude Bernard (1813-1878).

cess. The fifth section will explore how ecological explanations and human sociology became mutually entangled, as it were, each discipline borrowing metaphors from the other, with consequences both positive and negative. The concluding section will then situate the message of *Laudato Si'* within this long line of inquiry so as to bring out the encyclical's specific contribution.

1. The idea of environment in *Laudato Si'*

It should not come as a surprise that the concept of environment plays a major role within the encyclical. It is present in various ways in different sections, often where creatures are mentioned and presented as living within it, surrounded by it, and sustained by it. Human beings of course are no exception. Rather than enumerate all the appearances of this concept, in all its various forms, within the entire document, what follows will focus on the most significant paragraphs only, namely on those that expose a fundamental philosophical debate concerning the very idea of environment. We find these paragraphs in section four of Chapter Two, which is entitled "The mystery of Creation", and in section one of Chapter Four, which is entitled "Integral Ecology"[4].

The first of these refers to the somewhat undervalued metaphysical category of relation for the correct understanding of being in general. Chapter two, section four, expounds the theme of the harmony of creation and starts by highlighting the balance there should be between the importance of human beings and that of other creatures: "each creature has its own purpose. None is superfluous" (para. 84). Here, the main point is that the finality of creatures is not necessarily associated with, or dependent upon, the needs of human beings. The dynamism evident in our universe, in all its levels of complexity, suggests that, to understand it correctly, we need to see all creatures as enjoying a degree of integrity and autonomy. We need to see them as constituting a fellowship of travelers, each one engaged on a journey of self-fulfillment together with us humans. Up to this point in the encyclical, there is no direct reference to the idea of

[4] As regards textual details, one might be interested to know that the word "environment" and its derivatives, like "environmental", appear 150 times in the encyclical. At times, the encyclical qualifies "environment" by adjectives like "natural", "social", or "human". The adjective "environmental" occurs most often in the expression "environmental degradation". In para. 95 we find an explicit attempt at a definition: "the natural environment is a collective good". The most direct treatment, however, of the very idea of environment occurs in para. 139. These textual considerations in this footnote refer to the English version. One may recall that the encyclical was released as a simultaneous publication of several official translations. The Latin version uses "ambitus" for the word "environment".

environment. The unexpressed assumption is apparently that the environment is the overarching container within which creatures live and flourish as they find their way back to the Creator from whom they had issued. The paragraphs immediately following paragraph 84 are crucial because they highlight the importance of creatures as constituting a society, a fact that should urge us to see ourselves "in relation to all other creatures". The idea of a fabric of crisscrossing relations starts emerging. In paragraph 86, this idea becomes more explicit in the expression "the universe as a whole, in all its manifold relationships". We notice therefore how the universe is not just the collection of all creatures. It encompasses rather this collection of all creatures together with the relations between them. The text then quotes the important statement from the Catechism of the Catholic Church, "creatures exist only in dependence on each other, to complete each other, in the service of each other"[5].

The main section dealing with the very idea of environment is in Chapter Four, entitled "Integral Ecology". The inter-relatedness highlighted above becomes the focus of the argument with a special emphasis on the inevitable links between the natural, social, and economic dimensions. The starting point of this proposal lies at the biological level. The encyclical acknowledges that humanity is learning more and more about the intricacy of biological inter-relatedness: "Just as the different aspects of the planet – physical, chemical and biological – are interrelated, so too living species are part of a network" (para. 138). This inspires us therefore to extend our view beyond the biological sphere. It demands that we reach out beyond the merely material aspects and become aware of another network that is immaterial, namely the network of social, economic, and cultural inter-relatedness. This other network is typically human, constituting a distinct covering layer over the biosphere. The following paragraph, para. 139, therefore, defines the environment as the relationship "existing between nature and the society which lives in it". Natural reality and human reality are not two distinct and separate things but one complex whole. The encyclical therefore exposes the limitations of seeing the environment as a mere container in which creatures are allegedly meant to flourish. Nature is not "a mere setting in which we live" (para. 139). The emphasis on the importance of the philosophical category of relation obliges us to consider the essence of humanity not in terms of specific attributes of the

---

[5] Catholic Church, *Catechism of the Catholic Church.* Vatican City: Libreria Editrice Vaticana, 1994, n. 340.

human individual, each individual taken as an isolated unit, but in terms of the network of relations that individuals have with the environment in which they live and flourish, in other words, with the environment in a broad sense. Separating the organism from its environment will lead to distorted understanding.

The encyclical therefore defends the idea that organism and environment need to be seen, in some special way, as one whole that enjoys a dynamic balanced state. Moreover, groups of organisms are woven together among themselves by mutual relations, forming ecosystems. These ecosystems are themselves connected to each other by further mutual relations at a higher level, thus constituting other larger ecosystems. We are far therefore from that idea that organisms are mere occupants of their environment, seen as an inert container. The environment of some organisms is itself constituted by organisms. Ecosystems themselves need other ecosystems at higher level as their environment. Due to this intricacy, we need to acknowledge the value of nature not only in relation to our human needs but also in relation to itself.

> Ongoing research should also give us a better understanding of how different creatures relate to one another in making up the larger units which today we term "ecosystems". We take these systems into account not only to determine how best to use them, but also because they have an intrinsic value independent of their usefulness. Each organism, as a creature of God, is good and admirable in itself; the same is true of the harmonious ensemble of organisms existing in a defined space and functioning as a system[6].

The philosophical assumption behind this position is quite clear. Since we need to take a broader view of living things, since we need to zoom out and see the organism as part of the ecosystem, and then to zoom out even further and see that ecosystem as part of a larger ecosystem, we are obliged to revise our simple notion of environment as inert container. We need to avoid trying to understand life by focusing exclusively on the organism. We need to be aware of the limitations of models of biological and ecological explanation that are satisfied with micro-explanations dealing with the "within" of the organism but neglect the importance of relations at higher levels.

This completes our overview of the encyclical's position regarding the environment. What can we learn if we compare these views with the various ideas that emerged and were discussed, analyzed, evaluated, quantified, verified, or falsified in the course of many centuries of inquiry and speculation about the environment?

---

[6] Pope Francis, *Laudato Si'*, n. 140.

2. The emergence of the very idea of environment

It could be argued that early religious and philosophical cosmologies were all, in a sense, an effort to describe and understand the surroundings of human life. Consider, for instance, how the cosmologies of Ancient Greece dealt primarily with the fundamental nature of the universe. Plato in his *Timaeus* left us an influential narrative that highlights the order and beauty manifested by the entire world. What is ordered and beautiful here refers to the environment understood in the broadest possible sense, extending from our immediate surroundings to the distant stars. Plato's proposal, whether intended literally or metaphorically, highlights how the material universe is not a chaos but a cosmos, an ordered whole that manifests goodness and beauty as intended by the craftsman who produced it. The intricate relations holding the various parts of the universe together are of the same order as those that occur within a living organism. Plato therefore takes the universe itself as one enormous living thing, whose parts include all the living species we know of, even ourselves[7]. In such a description of the living universe, we do not find the use of direct personal attributes. Plato does not say that the universe loves us or cares for us. In this respect, his idea differs from what we find in the first line of *Laudato Si'*, "our common home is like a sister with whom we share our life and a beautiful mother who opens her arms to embrace us". There is no indication that Plato envisaged the universe in line with this metaphor, as a caring person. His view however did point in this direction. His student Aristotle continued more on less on these lines. He did not agree with Plato's idea of a definite beginning of the world, but he nevertheless accepted Plato's insistence on the major features of the universe, namely order and beauty. The basic unit of Aristotle's cosmological understanding was the concept of nature, which he took to be the inner principle of change or motion[8]. For him, to understand an entity's change or motion, we need to refer to that entity's nature and also to any intervening circumstances, which themselves are explainable in terms

---

[7] His exact wording is fascinating, bringing together the concept of universe, life, intelligence, and God. See, Plato *Timaeus* 30b "τὸν κόσμον ζῷον ἔμψυχον ἔννουν τε τῇ ἀληθείᾳ διὰ τὴν τοῦ θεοῦ γενέσθαι πρόνοιαν" which B. Jowett translates as "the world came into being – a living creature truly endowed with soul and intelligence by the providence of God". See Plato, *Timaeus*. Translated by Benjamin Jowett in Edith Hamilton and Huntington Cairns. ed. *The Collected Dialogues of Plato*. New Jersey: Princeton University Press, 1982, p. 1163.

[8] Aristotle, *Physica*. Translated by R. P. Hardie and R. K. Gaye, in *The Basic Works of Aristotle*, edited by Richard McKeon. New York: Random House, 1941, p. 236.

of the nature of other entities. All motion or change is either according to the entity's nature or contrary to it. In the latter case, the motion is called violent. For Aristotle, the good of the universe resides in both its cause, the Unmoved Mover, and in itself. This latter kind of good is evident in the way the universe is ordered. What kind of order are we talking about here? Aristotle uses the analogy of an army. We say that an army is good when it is ordered and when that order corresponds to what its leader wants. The universe is like that. It manifests goodness "as an army does; for its good is found both in its order and in its leader, and more in the latter; for he does not depend on the order but it depends on him. And all things are ordered together somehow, but not all alike – both fishes and fowls, and plants; and the world is not such that one thing has nothing to do with another, but they are connected"[9].

These foundational ideas were developed further within a Christian setting by St. Thomas Aquinas who not only accepted nature's order and beauty but also accepted that nature is not all rosy and attractive. It has a dark side to it: elements of conflict, violence, and contraries. How are we to understand this dark side? For some philosophers, the occurrence of contraries in nature revealed environmental disorder. Predation, for instance, apparently indicated an intrinsically conflictual universe, a kind of battleground where the principle of good wrestles against the principle of evil. Aquinas, resorting to the revealed doctrine of Creation, resolved these problems by fixing as his point of departure the truth that God is the origin of all there is. The contraries in nature, therefore, are only apparent. "The very order of things created by God manifests the unity of the world [...] and whatever things come from God, have relation of order to each other, and to God Himself"[10]. The kind of order he had in mind included what for us appears as conflictual, and moreover, it was not the kind of order established by mere multiplicity, as in the case of many similar bricks constituting a house. It was rather the kind of order that is evident in coordinated diversity, as we see in the case of different organs constituting a living being. "The good and the best of the universe consists in the order of its parts to one another, which cannot

---

[9] Aristotle, *Metaphysica*. Translated by William Davis Ross, in *The Basic Works of Aristotle*, edited by Richard McKeon. New York: Random House, 1941, p. 886.

[10] Thomas Aquinas, *Summa Theologiae* I, Q 47, art. 3: "ipse ordo in rebus sic a Deo creatis existens, unitatem mundi manifestat. [...] Quaecumque autem sunt a Deo, ordinem habent ad invicem et ad ipsum Deum" (my translation). (https://www.corpusthomisticum.org/iopera.html) accessed 18/11/2020.

be without distinction"[11]. Difference is a richness. The fact that the environment contains enormous diversity is neither a mistake nor the result of chance. It shows the wisdom of the Creator. The perfection of the universe in fact consists in the order among parts precisely when these parts differ from one another. "There is a certain affinity and order of one with the other. For plants are for the sake of animals, and animals are for the sake of human beings. And that all are ordered to one another is evident from the fact that all are ordered to one end at the same time"[12]. Aquinas returns to Aristotle's two images, the universe as a house and the universe as an army, but adds his own insight. He adds the overarching truth that the origin of everything resides within God's will. For Aquinas, the house analogy highlights the order and complementarity between parts in a static way. The army analogy, on the contrary, highlights the order that is dynamic. It highlights how the interaction between the parts, despite their differences, is a kind of communication with the whole, each part enjoying a specific degree of participation within an overall project. For the universe as a whole, the project is the will of the Creator[13].

The early seventeenth century started introducing significant changes as these medieval views were questioned, modified, or abandoned by some prominent philosophers. Galileo Galilei and René Descartes engaged in a radical revision of the key concepts they grew up with, and launched a new view according to which the quantifiable properties of the environment, properties

[11] Thomas Aquinas, *Summa Contra Gentiles*, II, c 39: "bonum et optimum universi consistit in ordine partium eius ad invicem, qui sine distinctione esse non potest" (my translation). (https://www.corpusthomisticum.org/iopera.html) accessed 20/11/2020.

[12] Thomas Aquinas, *Sententia Metaphysicae*, lib. 12, I, 12 n. 6 "est aliqua affinitas et ordo unius ad alterum. Plantae enim sunt propter animalia, et animalia sunt propter homines. Et quod omnia sint ordinata adinvicem, patet ex hoc, quod omnia simul ordinantur ad unum finem" (my translation). (https://www.corpusthomisticum.org/iopera.html) accessed 6/11/2020.

[13] Further analysis here should include also another analogy used by Aristotle. Apart from the house and the army, Aristotle sometimes resorts to the analogy of a household, which is not the building as such but the social unit with its different levels. "For all [things] are ordered together to one end, but it is as in a house, where the freemen are least at liberty to act at random, but all things or most things are already ordained for them, while the slaves and the animals do little for the common good, and for the most part live at random; for this is the sort of principle that constitutes the nature of each" (*Metaphysica* Bk XII, chap. 10, 1075a 20-23; p. 886). Aquinas draws inspiration from this Aristotelian text to understand the ordered hierarchy between creatures. This point is of secondary importance as regards the scope of this paper. For more information, see Oliva Blanchette, *The Perfection of the Universe According to Aquinas: A Teleological Cosmology*. PA: Penn State University Press, 1992, pp.12-19.

like size, shape, and motion, were given priority, while other attributes, like heat, weight, and color, were considered either reducible to the other quantifiable attributes or else irrelevant. The fact that nature manifested an impressive order was not denied. What was denied was the assumption that this order, which included the innumerable dependency relations between entities, was the result of the specific nature of each entity. Instead of this assumption, the new project adopted a purely mechanistic worldview according to which the only way to understand the order of nature was to refer to the properties of quantity and extension. All material entities, living or non-living, were considered nothing more than combinations of corpuscles or atoms, and each of these corpuscles or atoms were seen as following the laws of nature with mathematical precision. The world manifested the order of an extremely sophisticated machine. The fascination that this mechanistic view of nature, together with its technological potential, generated was considerable. Nevertheless, the reaction against it was not long in coming. It took the form of what historians now call the Romantic movement, which including the works of prominent figures like W. Wordsworth, F. Schelling, and J. W. von Goethe. By highlighting the limitations of the mechanistic philosophy's basic method, namely analysis and experimentation, these authors generated interest in organismic philosophies and sought to understand the environment in terms of holism or vitalism. This meant a return to the previous mode of appreciating nature, namely the mode of seeing it not just as a passive source of raw material, waiting to be exploited, still less as an enemy of human flourishing, but, on the contrary, as a major work of art exhibiting goodness, beauty, and harmony[14].

Against this conceptual background, the general features of the science of ecology emerged slowly during the nineteenth and twentieth centuries. A full and responsible account of the tentative beginnings of new ideas is not easy. To facilitate our task, the emergence of ecology will be presented by highlighting two main trends. These were evident from its earliest stages, one giving priority to the idea of ecological harmony and the other to the opposite: ecological randomness and discord.

---

[14] The basic project of the Romantics has re-emerged in current phenomenological approaches to ecology. These approaches, which will not be mentioned here because of lack of space, apply Edmund Husserl's method to reveal the limitations of a technology-dominated mentality and to suggest novel ways of understanding our experience of the world. See for instance, S. Charles Brown and Ted Toadvine, *Eco-Phenomenology: Back to the Earth Itself.* New York: SUNY Press, 2003.

3. The environment as a guided dynamism

Prominent among the early ecologists who defended a harmony view of the environment were Alexander von Humboldt (1769-1859), Henry D. Thoreau (1817-1862), and Frederick Clements (1874-1945)[15]. The first one was a leading figure of Romanticism, and he engaged in empirical work to defend the idea that nature is a beautiful, unified whole. Like his contemporary Charles Darwin (1809-1882), he embarked on long expeditions in the New World and drew inspiration from the rich flora and fauna he encountered. He interpreted what he observed as an optimistic view of nature, in the sense that nature, for him, was clearly driven forward by collaboration and mutual support among the various species. The environment was the canvas on which all forces and counterforces are brought together, in a meaningful whole, forming the complex but harmonious picture we call life[16].

Henry Thoreau, like Humboldt, sought to build his understanding of nature on empirical grounds, and became a keen observer of the various kinds of natural environment within the United States of America. His observations eventually convinced him that people needed to disengage from an excessively anthropocentric view of nature and to adopt a new perspective, the one we now would call a biocentric view[17]. He accepted of course that the view we have is always our view. We cannot see things from elsewhere. Nevertheless, he argued that only if we downplay the centrality of the human viewpoint could we have access to the rich relations that constitute the environment. For him therefore, being human is nothing more than the place where we stand as we observe. It should not be valued more than that. On these grounds, he de-

---

[15] For the history of ecology from the 18th century onwards, useful sources include Donald Worster, *Nature's Economy: a history of ecological ideas*. Cambridge: Cambridge University Press, 1994; Paul Warde, Libby Robin and Sverker Sörlin, *The Environment: a history of the idea*. Baltimore: Johns Hopkins University Press, 2018; S. Etienne Benson, *Surroundings: A History of Environments and Environmentalisms*. Chicago: University of Chicago Press, 2020.

[16] See for instance Alexander von Humboldt, *Cosmos: a sketch of the physical description of the universe*. Translated by E. C. Otté. New York: Harper & Brothers Publishers, 1877. He describes his project as "an attempt to delineate nature in all its vivid animation and exalted grandeur, and to trace the *stable* amid the vacillating, ever-recurring alternation of physical metamorphosis", Vol. I, p. xii.

[17] See, for instance, Henry David Thoreau, "Walking" in *The portable Thoreau*, edited by Carle Bode. New York: Penguin, 1982, p. 621: "Here is this vast, savage, howling mother of ours, Nature, lying all around, with such beauty, and such affection for her children, as the leopard; and yet we are so early weaned from her breast to society, to that culture which is exclusively an interaction of man on man [...]".

fended an ethic of perception. For him, this meant that we are obliged, in the moral sense, to perceive properly, correctly, and completely. For instance, the child's appreciation of the flowering meadow is rich in simple forms of appreciation. Unfortunately, these forms of appreciation shrivel away as the child grows up. They become inaccessible to the adult, whose perception is dulled and diminished by instrumental thinking, selective memory, and misplaced attention. For Thoreau, this change in adult perception is morally significant. If humans want to perceive the truth and the beauty of nature, they need to purify themselves first. An ascetic effort is needed. This aspect is reminiscent of Plato's myth of the cave in which the prisoner can only leave the shadows and arrive at the light through effort and determination. For Thoreau, the light corresponds to the appreciation of the harmony and the beauty of nature.

Further developments on this arcadian view of nature are found in the work of Frederick Clements. His main contribution was very influential. He proposed that, within any given environment, the various processes of nature work in successive stages to arrive at a climax state. This idea depended upon the work of other ecologists whose focus was on inter-species dependence, collaboration, and symbiosis, rather than on competition between species. The unity between the various species was for them a reflection of the internal unity and effective collaboration we find within the individual living body. Cohabiting species, therefore, are best described as forming a community. Clements and his peers argued that, for any given environment, organisms progress slowly but steadily to establish the most diverse, well-balanced, and self-perpetuating society of organisms. Clements was particularly interested not just in the flux of nature but in the way that flux was ordered. His reasoning started from what we see within the individual organism. We all accept that the changes we see within the individual organism are not aimless but stages within a growth, an ordered movement towards a final point of arrival. Clements enlarged this view. He shifted the idea of finality from the single organism to the entire environment, claiming that the changes we observe regarding the various species, their flourishing, their competing, their migration, their extinction, and so forth, are linked together to form a kind of string of life with a final end in view. He called this end the climax stage that is supported or allowed by the profile of that particular environment. Nature's drive, therefore, is not aimless. It is rather a slow and steady movement towards a climax stage that remains relatively stable due to the homeostatic potential of nature itself[18]. These reflections, which Clements derived mainly from his plant ecological studies, led him to restore the ancient idea, now articulated in empirical terms, that the entire world is one huge living thing. He was convinced that, when the final climax configuration of a given environment is reached, that configuration is indeed a collective organic entity in its own right.

At this point of our historical overview of ecology, it is interesting to note that the idea of environmental harmony and equilibrium discussed so far corresponds directly to some key concepts in *Laudato Si'*. For instance, when discussing biodiversity, the encyclical refers to the intricate balance discernible within any given environment, a balance that depends not only on organisms that are easily observable but also on those too tiny to see: "Some less numerous species, although generally unseen, nonetheless play a critical role in maintaining the equilibrium of a particular place" (34). In paragraph 84, as indicated above, the encyclical mentions how each organism has its own purpose within the overall harmony of nature. The encyclical does not see natural contraries as an objection to the idea of harmony. On the contrary, faith in a loving and merciful Creator supports the conviction that the enormous diversity we see in nature, despite its conflictual values as we see in predation, is perfectly intelligible with reference to divine goodness. On this point, the encyclical quotes St. Thomas Aquinas who argued that God allowed multiplicity and diversity of living things so that "what was wanting to one in the representation of the divine goodness might be supplied by another"[19].

During the early twentieth century, this harmony view of ecology gained support not only from empirical studies but also from some philosophical works of that period. The most influential of these was probably the work of Alfred North Whitehead, who, as a proponent of organicism, inspired ecologists to explore the interrelatedness of nature in terms of organic wholes nested within each other at successive levels, forming thereby various hierarchies. For Whitehead, "actuality is through and through togetherness", and "nature is a theatre for the interrelations of activities"[20]. By such typical affirmations, he meant that any

---

[18] For an overview, see Frederic E. Clements, "Nature and Structure of the Climax" (original 1936), in *Dynamics of Vegetation*, edited by Berton Wendell Allred and Edith S. Clements. New York: H. W. Wilson Company, 1949, pp. 119-160.

[19] St. Thomas Aquinas, *Summa Theologiae* I, Q 47, art. 1, quoted in para. 86 of *Laudato Si'*: "quod deest uni ad repraesentandam divinam bonitatem, suppleatur ex alia".

[20] The first quote is from Alfred North Whitehead, *Science in the Modern World*. New York: Free

given entity is intelligible, and thereby real, not in isolation but in relation to its neighboring entities, and ultimately in relation to the universe as a whole. Now, we often assume that an individual organism is quite distinct from its surroundings and can therefore be studied as an independent unit. That individual organism however may form part of a larger whole, about which we know little or nothing at all. If even the lifecycles of simple creatures we can readily perceive, like butterflies and beetles, can be bewildering, how complex then would be the larger ecosystems we cannot readily perceive. Ecologists wanted to face such natural complexity and, with Whitehead's support, felt philosophically supported in their proposal that the higher-level unit was, in a sense, an organism as well. The harmony of nature reveals itself as a kind of evolution that is emergent, in the sense that wholes that appear at higher and higher levels of organization are more than the mere sum of their parts.

Another philosopher who inspired ecologists was Jakub von Uexküll who worked mainly as a biologist but wrote an influential philosophical book with the title *A foray into the world of animals and humans* (1934)[21]. To study the phenomenon of life in its various forms and interrelations, Uexküll used the word *Umwelt*, which means surrounding-world or milieu[22]. Each kind of animal, even if very simple, has specific perceptual possibilities and these possibilities determine that animal's world. For Uexküll, even the simplest, primordial kind of sentience is perception, and every lifeform is therefore a subject. The organism's ways of registering the surroundings, as it perceives and as it thereby constitutes its *Umwelt*, are brought together into a single project. The fact that the organism survives through time shows that it and its *Umwelt* are well tuned to each other. A simple organism is well tuned to its simple *Umwelt*; a complex organism to its complex *Umwelt*.

> Subject and object are interconnected with each other and form an orderly whole. […] All animal subjects, from the simplest to the most complex, are

Press, 1967, p. 174; the second from his *Modes of Thought*, p. 140 (Alfred North Whitehead, *Modes of Thought*. Toronto: Free Press, 1938).

[21] First published in Germany in 1934, *Streifzüge durch die Umwelten von Tieren und Menschen*; English translation: Jakob von Uexküll, *A Foray in the Worlds of Animals and Humans*, translated by J. D. O'Neil. Minneapolis and London: University of Minnesota Press, 2010. This English edition includes the paper "A theory of meaning". Throughout this book, Uexküll justifies his arguments by many interesting biological observations.

[22] This notion had a decisive influence on a number of key twentieth century thinkers like Martin Heidegger and Gilles Deleuze.

> inserted into their environments to the same degree of perfection. The simple animal has a simple environment; the multiform animal has an environment just as richly articulated as it is [i.e. as the animal is][23].

Notice how, for Uexküll, we should not visualize the organism as choosing some features of the world while neglecting others. From its viewpoint, there is no choosing. For any given organism, there are no features of the world except those that it can perceive. Moreover, those features it can perceive are the determinants for its existence[24]. For him, the harmony in nature is like music. Life is a matter of "being in tune with". For instance, maple keys are not just blown around by the wind[25]. They are not passive, like, say, clouds. On the contrary, they "use" the wind. Their form is attuned for the wind. The point here is not that maple keys use the wind deliberately. Nor is it that evolution can never explain this phenomenon. The point is rather that, irrespective of how the phenomenon arose, we should acknowledge this basic feature of life, namely one thing "using" another, one thing being in tune with another[26].

A third philosopher we may mention who inspired ecological holism and the idea of a harmonious nature was Georges Canguilhem especially through his book *La connaissance de la vie* (1952). In this work, he presents an original exploration of the very idea of environment or milieu, with reference not only to biology but also to technology. He argues that, all along the rapid growth of the natural sciences and technology, "the problem of the relation between machine and organism has been studied, in general, in one direction only"[27]. By this he means that we have sought to explain the organism in terms of the machine, never the other way round. We thereby have never considered the specific richness of life, as such, a real question[28]. We saw it only as a question

---

[23] J. von Uexküll, *A Foray in the Worlds of Animals and Humans*, pp. 49-50.

[24] He writes, "Every subject spins out, like the spider's threads, its relations to certain qualities of things and weaves them into a solid web, which carries its existence". J. von Uexküll, *A Foray in the Worlds of Animals and Humans*, p. 53.

[25] Maple keys are maple seeds attached to a kind of wing that rotates as it drops, carrying the seed across a considerable distance before it hits the ground.

[26] Uexküll's link between the idea of meaning and that of use finds an analogue not only in Ludwig Wittgenstein's later writings but also in Martin Heidegger's ideas, especially his study of *Zuhandenheit*.

[27] "Le problème des rapports de la machine e de l'organisme n'a été généralement étudié qu'à sens unique". Georges Canguilhem, *La connaissance de la vie*. Paris: Hachette, 1952, p. 124 (my translation).

[28] I am assuming that a question becomes real for a specific scientific community in a specific

to be dismantled, broken down into pieces that could allegedly be explained mechanically. The organism in fact became a complex system of push-pull relations, a system that enjoys a state of internal equilibrium until it deteriorates. This one-way explanatory method, reducing all life to mechanics, assumes that, in the universe, the stable position is death, not life. Life becomes a kind of precarious superstructure inevitably destined to return sooner or later to the state of inanimate matter. For Canguilhem, a serious distortion is at work here. For a correct understanding, we need to recall that machines are human products. They do what humans do. The realm of technology is anthropomorphic. Therefore, the starting point for explaining the natural environment should be life. The machine is nothing more than an extension of the human, not the other way round. He arrives thereby to the interesting conclusion that technology, understood rightly as an extension of humanity life, is a biological phenomenon. The starting point needs to be life. To understand nature correctly, therefore, we need to understand how the environment on which an organism depends is structured and organized by that organism itself, according to what is valuable to that organism. If we disregard the network of values, if we try to produce a value-less description of nature, what we end up with will necessarily be a caricature. Living things are centers of perception and organization, each one carrying within itself a project for life. To arrive at a comprehensive view of nature, in so far as this is possible, we need therefore to add up, as it were, all these values together to see nature as a harmonious whole.

There is certainly much more to say about each of these philosophical approaches mentioned briefly so far, but even this quick overview is enough to show how they were instrumental in inspiring and supporting the ecologists who defended a harmonious nature[29]. Within such a picture, where do human beings fit in? Do humans fit in at all? The early empirical ecologists rarely referred to the humans dwelling in the areas they were studying. They attended to some sector of the flora or fauna living there, and little else. Eventually, they broadened their interest and began to include tribal or indigenous human cultures as participants

time when that community has reasons to believe that the question is relevant. For more on real questions within the context of environmental studies, see Louis Caruana, "Questions concerning Science, Theology and the Environment", *Gregorianum* 79/1-1998, pp. 149-161.

[29] For lack of space, this overview did not refer to the way metaphors regarding harmony, balance, and equilibrium affected empirical research. On this point, see Kim Cuddington, "The 'Balance of Nature' Metaphor and Equilibrium in Population Ecology", *Biology and Philosophy* 16-2001, pp. 463-479.

within harmonious biotic communities. This was relatively easy because tribal lifestyles, even when involving tools, hunting and agriculture, merged well with other life forms, and everything could be pictured within the general harmonious model of the environment. As opposed to this, recent human lifestyles including massive technological intrusions were, of course, totally different. Should we consider recent humans in this sense part of the environment or not? For some ecologists, even nowadays, current human populations remain clearly distinct from the natural environment. Massive technology is an aberration. It is an unacceptable means of disruption and unjust exploitation of the environment. This position, highlighting the distance, the incongruence, between humans and nature, does not go unnoticed in *Laudato Si'*. The encyclical describes it as the position according to which "men and women and all their interventions on the planet as no more than a threat, jeopardizing the global ecosystem"[30]. There are other ecologists, however, who strive to retain humans within their overall account of nature. They accept that technology can be destructive, but they argue that, whatever happens, nature will always readjust things to ensure life. For them, global homeostasis is part of the equation. A kind of invisible hand will ensure that the climax state of a given environment will always be reached, but there is no guarantee that this climax state will include humans. To express this idea, James Lovelock introduced the Gaia hypothesis in 1969, claiming that the Earth acts like a superorganism that will eventually counterbalance any excessive human intrusion by an appropriate counterforce to ensure the survival of life on the planet. This position corresponds somewhat to what is expressed in paragraph 60 of *Laudato Si'*, but it is not the same. The encyclical mentions the misguidedness of "those who doggedly uphold the myth of progress and tell us that ecological problems will solve themselves simply with the application of new technology and without any need for ethical considerations or deep change". The encyclical here does refer to a kind of homeostasis. It holds that technology is beneficial but brings collateral damage; and this collateral damage will then bring more technology. This kind of homeostasis however depends on humans. It is a kind of homeostasis that allegedly functions because of the human technological ingenuity. As opposed to this, the homeostasis endorsed by the Gaia hypothesis supporters is much broader. It allegedly functions because of the nature of the universe itself. Both kinds, however, minimize the urgency of thinking ahead and of evaluating our technological footprint responsibly.

---

[30] Pope Francis, *Laudato Si'*, n. 60.

4. The environment as a blind dynamism

We turn our attention now to the opposite trend within the emergence of ecological thinking. Even while discussing the harmony model of the environment, we seem obliged, especially by the use of the idea of homeostasis, to resort to the vocabulary of tension, friction, disturbance and repair, force and counterforce, action and reaction. It should not come as a surprise therefore that, despite the appeal of the harmonious model, a diametrically opposed model is possible as well, a model that is conflictual, random, and blind. The emergence of such a model is precisely what we see during the twentieth century. As indicated previously, Humboldt had used empirical data that he had carefully collected from his explorations in the New World. The same kind of data served Charles Darwin in the opposite way. He proposed natural selection as one of the mechanisms whereby the environment filters off some lifeforms while leaving others to propagate. His followers emphasized the predominance of natural selection and produced a violence-saturated picture of nature in which domination and survival constituted the main engine of change. These Darwinians contributed to ecology a new set of key concepts: competition, aggression, climatic disruption, scarcity, invasion, survival, extinction, and so forth. Described in these terms, the environment becomes a stage on which actors are surrounded by terror, constant danger, and menacing powers: "Nature, red in tooth and claw"[31]. The peaceful meadow is not peaceful at all. It hides organic discord and deadly warfare. According to the harmonious view described previously, violence and intrusion were a lamentable attribute of human technocrats only. On this view now, the violence and discord constitute the essential characteristic of all nature, at all levels.

With the idea of finality out of the way, this view of ecology aligned itself more and more with the explanations used by physicists rather than with those of biologists. Its protagonists in fact became increasingly interested in measurable, empirical details. The idea of a food-chain became important to map a given environment in terms of relations constituting a kind of vertical hierarchy within that environment. Studies revealed how such chains have a well-defined structure, related to the needs and predatory capabilities of organisms at various levels. The corresponding idea of ecosystem, a direct derivative from the idea of system in physics, was employed to bypass the vague

[31] Alfred Tennyson, "In Memoriam", Canto LVI, in Alfred Tennyson, *The Works of Alfred Lord Tennyson*. Hertfordshire UK: Wordsworth Editions Ltd., 1994, p. 315.

idea of organic wholes and to explore the environment in terms of a network of exchange-relations involving transfer of energy and of chemicals like water, nitrogen, and phosphorus[32]. Eventually, the previous idea of an organic community, which was analogous in some way to human social reality, was abandoned also. Instead of a community, or a community of communities, the environment became a collection of atomic configurations relating dynamically with other atomic configurations. On this new account, all constituents of the environment, whether animate or inanimate, ranging from rocks to higher animals, fall under the same explanation. Ecology becomes therefore more like a subdiscipline of physics than of biology. The mathematization of ecology introduced also other terms deriving from economics: energy capture, energy cost, energy budget, energy investment, and so on. Food chains came to be seen as systems of producers and consumers. The community analogy was replaced by the analogy of individualistic liberal economy. Nature became the stage on which individualistic species made their way through thick or thin, as best they could, without any kind of guidance from an alleged overall finality or plan. The idea of an environment having a climax stage, which was a central idea in the harmony view described above, has no role within this model. Even the idea of disturbing the environment therefore loses its importance. The word disturbance has meaning primarily with respect to the idea of normality. If a given environment, as a whole, does not have a way of flourishing that could be called normal or optimal, no change imposed from elsewhere could really be called a disturbance. There is no criterion anymore by which we could decide whether to call a change a disturbance or not.

5. Ecology and human society mirroring each other

It should be clear by now that the trajectories in ecological thinking described in the preceding sections, one harmonious and guided, the other conflictual and blind, are both associated with analogies deriving from human political existence, the first one inspired by communitarian ideals while the second by individualism. As already mentioned, within the explanatory picture of the environment, human existence, with all its social, political, cultural complexity, was at first considered largely irrelevant. Nevertheless, the

---

[32] For a comprehensive overview of the development of the idea of ecosystem, see, Frank B. Golley, *A History of the Ecosystem Concept in Ecology: More than the Sum of the Parts*. New Haven: Yale University Press, 1993.

exchange of metaphors and analogies between the human and the non-human spheres was already happening even before ecology started to include human existence. There was in fact something like two-way conceptual traffic: ideas from ecology triggered new political insights and ideas from politics triggered new ecological insights in return[33]. Expressions like social evolution, survival of the fittest, profit maximization, and competition started being used for both human and non-human environments, thus blurring the boundary between these two camps. As regards explanation, the dynamics of non-human nature and that of human cultural history became parts of one continuous fabric. Ecologists and environmentalists eventually became convinced that the non-human environment is disturbed not only by factors like droughts, climate change, parasites, or earthquakes. It is disturbed also by typically human factors like colonization, market takeovers, terrorism, and social media. Nature and humans are in the same boat. Notice how such inclusion of human reality within the broader picture of nature does not depend on whether we adopt one of the above ecological trends or the other. Such inclusion is possible in both models. On the one hand, those who see traces of finality in single organisms see it also in the way species collaborate and see it also in the overall drive of the entire universe, humans included. On the other hand, those who see only conflict and competition between individual organisms see this also between species, and also within human society.

Moreover, in more recent ecological thinking, while the explanatory boundary between human nature and non-human nature disappears, so also is disappearing the explanatory boundary between organism and environment. Recent studies are showing how the organism and its environment are like two sides of the same coin. They are intricately related to each other and dependent on each other to such an extent that we cannot really say that life resides primarily within the organism. We need to accept that life is a phenomenon revealed by both organism and environment taken together. We need to say, for

[33] Studying politics in relation to human biology is certainly not just a post-Darwinian phenomenon. Many famous philosophers had developed political views based on biology before Darwin, e.g., Aristotle, Aquinas, Hegel, and others. With the advent of the theory of evolution however, the interest in seeing politics as a biological phenomenon flourished in new ways. For early views, see for instance Herbert Spencer, "The social organism", *Westminster Review* 73-1860, pp. 90-121; Jakob von Uexküll, *Staatsbiologie (Anatomie-Physiologie-Pathologie des Staates)*. Berlin: Gebrüder Paetel, 1920. For a useful overview, see Henry John McCloskey, "The State as an Organism, as a Person, and as an End in Itself", *The Philosophical Review* 72/3-1963, pp. 306-326, doi: 10.2307/2183166, accessed 18/12/2020.

instance, that life is present in both the seedling and the soil, taken together. This holds for all life, from the smallest protozoa to human beings. All life is symbiosis, with inanimate matter included. Ecology has now moved beyond the original simple idea that the environment is a kind of inert container of organisms, a kind of niche with the conditions that permit a given species to flourish. Ecology has moved beyond the simple idea that the environment is a collection of niches or "houses" waiting passively for some species to occupy. The picture scientists have now revealed is much more complex. In the mid-1970s, prominent biologists like Richard Lewontin and Richard Levins started to highlight the need to revise the kind of Darwinism "that represents the environment as a preexistent element of nature formed by autonomous forces, as a kind of theatrical stage on which the organisms play out their lives"[34]. They realized that assuming an asymmetrical relation between a living thing and its environment is like seeing half of the picture only. In reality, the organism and its environment affect each other. There is interaction in both directions. The environment influences the organism, and the organism influences the environment. Just as the predator population affects the prey population and vice versa, so also the soil affects the seeds, and the seeds affect the soil. Humans affect the biosphere, and the biosphere affects humans. Organisms and the environment form one evolving whole[35].

6. The specific contribution of *Laudato Si'*

To recapitulate therefore: we are faced with two possible pictures, both of which seem reasonably plausible. On the one hand, we could proceed with the assumption that life is characterized by an intrinsic order that manifests itself not only at the level of single organisms but also at higher levels, including even the entire biosphere itself. On this view, in any given environment, the

[34] Richard Lewontin and Richard Levins, "Organism and environment", *Capitalism Nature Socialism* 8/2-1997, p. 96. doi: 10.1080/10455759709358737.

[35] The co-determination happening between the seed and the soil is explained clearly in Richard Levins and Richard Lewontin, "Dialectics and Reductionism in Ecology", *Synthese* 43-1980, p. 49: "the seedling is the 'environment' of the soil in that the soil undergoes lasting evolutionary changes of great magnitude as a direct consequence of the activity of the plants growing in it, and in turn feeds back on the conditions of existence of the organisms". On this point, see also Trevor Pearce, "The origins and development of the idea of organism-environment interaction", in *Entangled Life: Organism and Environment in the Biological and Social Sciences*, edited by Gillina Barker, Eric Desjardins and Trevor Pearce. Dordrecht: Springer, 2014, pp. 13-32, doi 10.1007/978-94-007-7067-6 2.

various organic forces and counterforces are characterized more by collaboration than by mutual destruction, they balance each other out, and they eventually determine an overall, sustainable optimal state. This holds, or should hold, for all of nature, humans included. On the other hand, we could proceed with the contrary assumption that life is nothing more than the unpredictable result of piecemeal struggles for survival, involving instances of random disturbances of momentary equilibrium situations, and natural elimination of unsustainable organisms, with no overall guide, no finality, and no clear sense of flourishing or progress. Given these possible pictures of ecology, what is the role of humans? The two trends within current ecological thinking work with different assumptions but both recognize that climate, terrain, flora, fauna, human social reality, culture, and politics are all interconnected. Human ecology affects non-human ecology and vice versa. The key concepts we use to explain one explain the other also.

This constitutes the background for *Laudato Si'*. We are now in a position to reply to the original question that we set ourselves at the start of this paper. What is the encyclical's original contribution in this specific area? The answer can be presented in three main points.

First, we need to refer to the important adjective "integral". This word derives from the Latin *tangere* and literally means untouched, undivided, or complete. When applied to ecology, it refers to the kind of ecology that leaves nothing out, ecology as it should be[36]. The previous paragraphs show that *Laudato Si'* is not the first study that highlights the fact that non-human ecology is inseparable from human ecology with all its social, cultural, and political dimensions. During several decades before the encyclical, some philosophers and ecologists were highlighting the importance of including human reality within the broader context of life on the planet. In fact, as we saw, these studies often used the same keywords and expressions for describing non-human reality as for human reality. The idea of an integral ecology therefore was indeed already present. Pope Francis however added a new emphasis, especially as regards the kinds of causal connections that characterize life

[36] It is interesting to note that previous significant uses of "integral" in this sense include Gustavo Gutierrez's idea of integral liberation in chapter 9 of his *Teología de la Liberación. Perspectivas* (Gustavo Gutierrez, *Teología de la Liberación. Perspectivas*. Lima: CEP, 1971), and Jacques Maritain's idea of integral humanism in his *Humanisme intégral* (Jacques Maritain, *Humanisme intégral: problèmes temporels et spirituals d'une nouvelle chrétienté*. Paris: Aubier - Éditions Montaigne, 1936).

on the planet. Before *Laudato Si'*, many scientists and environmentalists had shown how behind major ecological imbalances there are sometimes human causes, and that, vice-versa, ecological imbalances sometimes cause human suffering. The Pope agreed. He not only agreed however with this causal interdependence[37]. He went further by uncovering causal connections in areas where few others had dared to explore. He explained how degradation can be seriously contagious and can therefore jump from one level to another. The environment is not immune to the degradation we see in human society, and, in the opposite sense, human society is not immune to environmental degradation. We need to consider also personal or moral degradation, which refers to egoism, indifference to the suffering of others, lack of foresight and responsibility regarding future generations, indifference to the unnecessary suffering of animals, and similar negative attitudes within the human individual. Human society, culture, and politics are not immune to personal degradation, and the individual person is not immune to social, cultural, and political degradation. The ecology defended in this encyclical, therefore, is integral in a more robust sense than had been envisaged before. It is integral because it includes all these causal connections, down to personal choices embedded within the individual's heart.

The second area one can mention regarding the originality of *Laudato Si'* concerns the link that it establishes between ecological concerns and Christian faith. Christians should support the kind of ecology that is integral not only because science is unveiling the holistic richness of nature. They should support it also because the oneness of nature and the mandate to care for creation constitute an essential part of the Biblical message. So, the encyclical did not merely reiterate that the biosphere includes the human phenomenon. It did not just restate that ecology is intimately connected with human social and political realities. It did not merely highlight again the dangers of a blinkered, utilitarian, mechanistic worldview. It went further. It produced a comprehensive and consistent picture in which the previous achievements regarding ecology gain added support by being seen within the broader horizon of religious faith. It produced a unified declaration in which the scientific, social, cultural, and

[37] He clearly disagrees with those who deny the evidence and insist that there are no human causes of global warming and its consequences. He also disagrees with those who seek to dedramatize the situation by claiming that the evidence is unclear and that the current problems are temporary because they are merely part of a recurring natural cycle.

personal dimensions of ecology can be seen together within the framework of creation's divine origin and destiny. Within this picture, degradation, conflict, and individualism are not just aberrations, miscalculations, or simple errors. They are dimensions of sin. Flourishing, collaboration, and communion are dimensions of grace. By highlighting these links, these resonances, between the material and the spiritual, the encyclical retrieves and enhances some fundamental insights that are already briefly sketched in the Old Testament, for instance the following text from the Book of the Prophet Hosea, which is surprisingly missing within *Laudato Si'* even though it conveys this unified message brilliantly. "Swearing, lying, and murder, and stealing and adultery break out; bloodshed follows bloodshed. Therefore the land mourns, and all who live in it languish; together with the wild animals and the birds of the air, even the fish of the sea are perishing"[38].

Another area of originality concerns the encyclical's balanced view of anthropocentrism. These last decades, we have seen various heated philosophical debates regarding anthropocentrism, many of which arose as a direct result of starting to see human reality as an important variable for ecological understanding in all its complexity. *Laudato Si'* avoids extreme positions and points towards a responsible understanding of the role of human beings. On the one hand, it refutes positions that adopt an arrogant disregard for the state of the biosphere and that adopt any form of shortsighted exploitation. On the other hand, it refutes also the opposite positions that disregard or devalue humanity in favor of biocentrism, of cosmo-centrism, and of no centrism at all. The encyclical defends a realistic position that acknowledges the specific gifts with which the Creator has endowed *Homo sapiens*. Hence, it does not consider humans a kind of jarring, cosmic mistake. It does not take human rationality as a kind of disease that makes this species disrupt the ecosystems in which it multiplies without control. It does not see human social and cultural refinement as a kind of self-destructive, life-asphyxiating excrescence on the planet. On the contrary, the encyclical adopts a balanced approach. Without neglecting the seriousness of ecological sin, it outlines and supports a scientifically informed role for humans within creation, a role that is appreciative, far-sighted, responsible, supporting, humble, and joyful. The overall

[38] Hosea 4:2-3 (NRSVCE). Notice the conjunction "therefore" at the beginning of verse 3, highlighting the fact that human immoral living and environmental degradation are not just correlated phenomena but that the former is the cause of the latter.

approach here is founded on Christian faith, and, as such, it cannot but support the first trend mentioned above. In other words, it is in line with the idea that the environment, understood in the broadest sense, is a harmonious reality. The contrary forces we see, the contingencies, random mutations, struggles for survival, even predation itself, are not senseless. They are all part of the handwork of an intelligent and loving Creator. This is a theologically inspired view. Those who limit themselves to science and philosophy will point out that the discussions regarding finality in nature have not yet reached any definite conclusion. Nevertheless, current developments in these disciplines seem in favour of the idea that finality is an emergent property, in the sense that, although random changes and statistical averaging may predominate at micro-levels, some form of finality or goal directionality starts to become increasingly relevant as we consider higher levels of complexity. It starts to become relevant especially when we consider animals and obviously humans. Since humans are part of the environment, human intentionality, deliberation, and responsibility are an essential part of the environment as well. In this sense, therefore, it may not be difficult to bridge the gap between the faith-based approach of *Laudato Si'* and views that are founded exclusively on science and philosophy.

The three areas of the encyclical's originality mentioned above are only a part of its overall message. There are certainly many more aspects to be appreciated. This paper has focused on one issue only, with the hope that the encyclical's message regarding life and environment comes across in clearer light now that the reader can see the encyclical not in isolation but in continuity with other ideas, now that the reader can see it as part of the long struggle that humanity has engaged in since earliest times, the struggle to understand itself and its surroundings correctly, and to behave and flourish within it accordingly.

THE FUNDAMENTALS OF INTEGRAL ECOLOGY

A Philosophical Reflection

Lorella Congiunti

In this study, we propose to offer some reflections on the fundamentals of the integral ecology of Pope Francis, favouring the point of view of philosophy. The path is complex. We shall begin by giving some hints on the thought of Pope Francis, retracing the fundamental principles which attain maturity in his Encyclicals but which sink into his previous journey. Having acquired these interpretative elements, we shall move into the analysis of the text of the *Laudato Si'*[1], extracting what appears fundamental in it and the philosophical sources to which it refers. With this, we shall try to draw the peculiar principles of integral ecology within the wider system of thought of Pope Francis[2]. Finally, we shall try to briefly conclude on the value of this thought for dealing with the present and planning the future.

1. The Thought of Pope Francis

As has been repeatedly noted, it is improper to affirm that the thought of Pope Francis has a "uniquely pastoral" depth, and that it is instead important

---

* For the English translation, I thank my friend and colleague, Prof. Johnson U. Ozioko.

[1] Pope Francis, Encyclical Letter *Laudato Si' on Care for Our Common Home*, 24 May 2015 (hereinafter referred to as LS).

[2] We decided to give ample space to the thought of Bergoglio/Francis, for this reason his texts are cited in an ample and diffuse manner.

DOI 10.32060/C.S.9.221-252

to understand his "theological method"[3]. His thought establishes itself in his Encyclicals, with a precise cultural depth:

> Despite what is commonly thought, the more one reads his encyclicals, his speeches or homilies, the more one has the impression that Francis knows the world of the intellectuals and that he has solid convictions on the role of culture in the contemporary society[4].

The thought expressed in the Encyclicals and in the speeches of Pope Francis sinks into the thought of Jorge Maria Bergoglio, which is a thought developed and cultivated over time. Behind the deliberately "simple" language of Pope Francis, as Borghesi writes, "lies a rich and original thought process, derived from the Jesuit school and nourished not only by Argentine teachers but also, and above all, by European ones"[5].

Bergoglio's education took place in the Colegio Máximo San José in the city of San Miguel in the province of Buenos Aires from 1958 to 1971, where he completed his studies in philosophy and theology. These are the years of the Second Vatican Council; the decree on priestly formation *Optatam Totius*[6] of 1965 requests the revision of ecclesiastical studies, so that philosophical and theological studies be "more suitably aligned and that they harmoniously work toward opening more and more the minds of the students to the mystery of Christ" (OT, n. 14). This renewal is ensured through the valorisation of the traditional heritage, but with a greater rootedness in the Holy Scriptures and a stronger bond with contemporary challenges. For the teaching of philosophy, there is maintained "a solid and coherent knowledge of man, the world, and of God, relying on a philosophical patrimony which is perennially valid and taking into account the philosophical investigations of later ages" (OT, n. 15), but with specific attention to "the necessary connection between philosophy and the true problems of life", and to "the links between the subject-matter of philosophy and the mysteries of salvation" (OT, n. 15). For theology, it is recommended to take "particular care in the study of the Bible" and that "biblical themes are proposed first of all" but the reference to "the guidance

[3] See for example, Walter Kasper, *Pope Francis' Revolution of Tenderness and Love*. New York: Paulist Press, 2015.

[4] Agostino Giovagnoli, "Francesco sfida gli intellettuali", in *Avvenire*, 3 November 2015.

[5] Massimo Borghesi, *The Mind of Pope Francis. Jorge Mario Bergoglio's Intellectual Journey*, transl. B. Hudock, Collegeville MN: Liturgical Press, 2018, p. XIX.

[6] Second Vatican Ecumenical Council, Decree on Priestly Training *Optatam Totius, 28* October 1965 (hereinafter referred to as OT).

of St. Thomas" remains (OT, n. 16). Bergoglio experienced a transition from a certain dried up education based on less dynamic manuals to the search for methods and contents which help the human formation and the opening up to the mystery of Jesus Christ.

He himself recalls: "Unfortunately, I studied philosophy from textbooks that came from decadent Thomism"[7]; this consideration is similar to what Benedict XVI wrote, recalling his philosophical formation in the Seminary of Frisinga in the years 1945-1947, where Arnold Wilmsen was teaching rigid Neo-Scholastic Thomism, presenting the thought of St. Thomas as closed, impersonal and prepackaged[8].

Starting from the second half of the twentieth century, Thomism itself lived experiences of intense rereading:

> attempts of comparison that the tradition which refers back to Aquinas intended to undertake with modern and contemporary thought and which attest to the integrating capacity of Thomistic reflection with respect to themes initially born outside its own sphere: the dialogue between Thomism and *phenomenology*, between Thomism and *analytic philosophy*, between Thomism and *post-secular* reason, thus showing a fruitfulness of the reference to St. Thomas not only on the level of thought, but even more on that of *thinking*, of a style of thought that is an expression of the Catholic *Denkform*[9].

Bergoglio exudes this change.

Among his Argentine teachers who represent the dynamism of the moment, it is appropriate to remember Ismael Quiles, a Jesuit of Spanish origin, who spent his entire life in Argentina and who taught the history of philosophy and metaphysics beginning from the 1940s[10]. In the years in which Bergoglio gravitated to San Miguel, Quiles elaborated the development of his thought, which he himself divided into three phases: the rational, fundamentally Thomistic stage, with attention to Suarez, from 1938 to 1948; the in-sistential stage in which he

[7] Pope Francis and Antonio Spadaro, *My Door is always open. A Conversation on Faith, Hope, and the Church in a Time of Change,* trans. S. Whiteside. London: Bloomsbury, 2013, p. 20.

[8] See Joseph Ratzinger, *Aus meinem Leben*: München DVA, 2005.

[9] Daniele Aucone, "Tomismo dialettico. Teoria polare e sfida della complessità nella formazione culturale di J.M. Bergoglio", in *Domenicani*, https://www.dominicanes.it/predicazione/meditazioni/1407-tomismo-dialettico.html.

[10] See Lorella Congiunti, *Soggettività ed ontologia. Introduzione alla filosofia "in-sistenziale" di Ismael Quiles*. Ardea: Abelardo, 1996; Lorella Congiunti, *Sostanza e in-sistenza,* in *Persona y Acción del RP Ismael Quiles SJ. Homenaje de la Fundación "Ser y Saber" en su 30 aniversario*. Buenos Aires: Ediciones Universidad del Salvador, 2010, pp. 187-207.

assumes some traits of the phenomenological method and reacts to existentialist philosophy by valorising, from an ontological point of view, the inner dimension of the person and the movement of *ensimismamiento* until 1960, and finally the attempt at synthesis between Western and Eastern thought.

Pope Francis in *Evangelii Gaudium*[11] cites the pedagogical work of Quiles for its reference to interiority[12], and in the preface to a collective book dedicated to Quiles in 2010 he recalled having known him and defined him[13].

Borghesi, who conducted the most documented and in-depth analysis of Bergoglio's thought, notes that in the years of study "his standard references were the Jesuit intellectuals of the French sphere: Henri de Lubac, Gaston Fessard, Michel de Certeau. Some were representatives of the Lyon school. These are his teachers"[14].

These same authors are the basis of the thought of the Uruguayan Alberto Methol Ferré, who greatly influenced Bergoglio, also through the magazines "Vispera" and "Nexo" which he read assiduously[15]. Methol Ferrè is an original exponent of "dialectical Thomism": "Methol Ferré found, in Thomas's thought, and particularly in its hylomorphic vision of reality based on matter-form tension, what he recognized to be the correct form of dialectic thought"[16]. Ferrè was lay collaborator of the Latin American Episcopal Council in the years between the General Conference of Puebla in 1979 and that of Santo Domingo 1992, and precisely in this context Bergoglio and Ferré met and spent time with each other, and Ferré "was very influential on Bergoglio's

---

[11] Pope Francis, Apostolic Exhortation *Evangelii gaudium* on the Proclamation of the Gospel in Today's World, 24 November 2013 (hereinafter referred to as EG).

[12] "But if we look more closely at these biblical texts, we find that the locus of this reconciliation of differences is within ourselves, in our own lives, ever threatened as they are by fragmentation and breakdown" EG, n. 229. Cf. Ismael Quiles, *Filosofia de la educación personalista*. Buenos Aires: Depalma, 1981, pp. 46-53.

[13] "*Padre y maestro en época convulsionadas, Quilescual pedagogo supoarcercarse al ministerio del hombre y ayudar al hombremismo a entronarse dentro de sì. Frente al drama contemporáneo de una cultura pragmática y relativista, Quiles nos presenta una reflexión plena di sabiduría*". Jorge M. Bergoglio, *Prólogo,* in *Persona y Acción del RP Ismael Quiles SJ, Homenaje de la Fundación "Ser y Saber" en su 30 aniversario*. Buenos Aires: Ediciones Universidad del Salvador, 2010, p. 9. Cfr Jorge M. Bergoglio, *Insistencialismo y hombreactual en Reflexiones en Esperanza*. Buenos Aires: USAL, 1992, p. 349. La *Fundación Ser y Saber,* dedicated to the studies and diffusion of the thought of Quiles is born in 1979 with the authorization of Bergoglio, at that time the provincial of the Companions of Jesus.

[14] Massino Borghesi, *The Mind of Pope Francis,* pp. XIX-XX.

[15] *Ibid.*, pp. XIX -XX.

[16] *Ibid.*, p. 91.

thought, particularly with respect to his understanding of the relationship between the universal and the local in ecclesiology"[17]. The leitmotif of this formation and of the re-elaboration Bergoglio made of it is the "reconciliation" understood in a theoretical sense as the ability to grasp the unity of opposites. Borghesi notes again:

> *Bergoglio's entire system of thought is one of reconciliation* -not an irenic, optimistic, naively progressivist thinking, but rather a dramatic thinking, marked by a tension, that, having matured during the course of his Ignatian studies in the 1960s[18], finds its first formulation in the 1970s, in the tragic context of an Argentina divided between a right-wing military and left-wing revolutionaries. It is a contrast that marks both the Church and the Society of Jesus[19].

This thought of "reconciliation" appears as the fruit of a thought which, starting from solid foundations, is able to face the challenges offered by contemporary reality, in the light of the Gospel message.

It has to do with a complex movement of thought which will continue to characterize the thought of Pope Francis. It should not be equivocated with an irenist or pacifying movement, or even with a merely dialectical *Aufhebung*, but it is a matter of realistically grasping the tensions that run through reality, at every level, to bring them back to their unity: "a theoretical/practical/religious attempt to propose an antinomian unity, an agonic solution achieved by way of the contrast. It is, therefore, a dialectical view in which reconciliation is not entrusted, as in Hegel, to philosophical speculation, but to the Mystery that acts in history"[20].

Bergoglio's philosophical training continued in Germany, starting in 1986, where he came into contact with the theoretical thought of Romano Guardini whose spiritual writings he already knew. He elaborates the project of his doctoral thesis precisely on the thought of Romano Guardini; the thesis will never be concluded but Guardini's thought touches him deeply and will be one of the recurring sources of his thought. We will see that in LS itself, Guardini is mentioned seven times.

Thanks also to the studies on Guardini, a philosophy that is based on the "*symphony of opposites*" is gradually and decisively defined: "It is a philos-

---

[17] Ethna Regan, "The Bergoglian Principle's: Pope Francis' DialecticalApproach to PoliticalTheology", in *Religions*, 2019, 10, 670; doi: 10.3390/rel10120670, p. 3.

[18] See Gaston Fessard, *La dialetiquedes "Exercices spirituels" de saint Ignace de Loyola*. Paris: Aubier, 1956.

[19] Massimo Borghesi, *The Mind of Pope Francis,* p. XXIV.

[20] *Ibid.*, p. XXIV

ophy that is situated firmly in the flow of Catholic tradition, understood as a *coincidentia oppositorum*, as expressed in the work of Johann Adam Möhler, Erich Przywara, Romano Guardini, Henri de Lubac[21].

It is an education that sinks into the traditional path of philosophical and theological studies, but evolves in the context of religious and social problems, having as a recurring point of reference the "*coincidentia oppositorum*". For example, Przywara, who is one of Bergoglio's reference authors, speaks of "a cosmic '*coincidentia oppositorum*', as Nicholas of Cusa ultimately concludes from Aquinas's principle"[22].

The valorisation of unity without suppressing differences appears constant, which is gradually deepened and applied in different areas: from Jesuit spirituality to social and political issues, to ecclesiastical issues. It is a dynamic, polar, universal, convergent thought.

At the foundation of this approach, there is certainly a metaphysical vision, sometimes unexpressed, which derives from the Aristotelian-Thomist tradition, variously interpreted. Nature as a connected totality, which we will study later as the central notion of LS, must be understood in this theoretical context.

2. The Method and the Principles

Bergoglio would carry such ideas to his teaching as pontiff. These are social principles but applicable to every level of reality, insofar as it has to do with a true and real social and political ontology illuminated by the Christian faith: the tensions of reality are grasped and, without being suppressed, are resolved in unity. They are principles that allow us to "build" as they sink into reality, interpreted in the light of the "pillars of the Church's social doctrine" (EG, n. 221).

The "constant tensions present in every social reality" (EG, n. 22) are between fullness and limitation, unity and diversity, ideas and reality, globaliza-

[21] *Ibid.*

[22] Thus Przywara writes: "Metaphysics is methodologically 'polymorphous' -indeed, 'polymorphous in infinitum' (just as, for Thomas Aquinas, the creaturely is an '*infinitum potentia*'). It is only in the form of such a polymorphous metaphysics (in keeping with the 'other and other' that obtains among the various regions of being proper to the 'physica') that this metaphysics becomes the ultimate 'organon' for the 'Meta' of the divine –that which, according to Thomas, can be experienced only as what is '*simpliciter et uniformiter*' with regard to the antithetical polymorphousness of the creaturely universe, and which for this reason is experienced within the medium of a cosmic '*coincidentia oppositorum*', as Nicholas of Cusa ultimately concludes from Aquinas's principle": Erich Przywara, *Between Metaphysics and Christianity* (1958) in Id., *Analogia Entis. Metaphysics. Original Structure and Universal Rhythm*. Grand Rapids, Mi: Wm. B. Eerdmans Publishing Co., 2014, p. 528.

tion and localization: these are couples which concern contemporary social reality, such as globalization and localization, but which also concern social reality as such, or even more so reality as created reality. Everything is crossed by this tension between the parts and the whole, between limit and fullness, which for the human being, capable of conceiving ideas and acting, becomes tension between idea and reality.

The originality of the Bergoglio-Francis' method is not to ignore the tension or even to hypostatize it, but to resolve it, not however with a dialectical movement, but by grasping its intrinsic unitary movement: the solution is unity that maintains diversity.

From the solution of this tension, four principles are born: *Time is greater than space; Unity is greater than conflict; Realities are greater than ideas; The whole is greater than the part.*

First of all, "*Time is greater than space*" (EG, n. 223): the first principle set out in *EG* means accepting the tension between fullness and limitation, without crystallizing the processes in the present. The time extends beyond the present: the time has roots in the past and prospects for the future. The fullness lies in the process of touching the limits and overcomes them.

In this attention to time "Francis draw on a rich tradition of discernment from the Ignatian tradition to the Second Vatican Council's concept of 'reading the signs of the time'"[23].

The constructive aspect lies in the recognition that processes are more important than the possession of spaces. For this principle, the EG refers directly to Romano Guardini: "The only measure for properly evaluating an age is to ask to what extent it fosters the development and attainment of a full and authentically meaningful human existence, in accordance with the peculiar character and the capacities of that age" (EG, n. 224)[24].

The principle of the superiority of time is present in *Lumen Fidei*[25] in connection with hope: "Space hardens processes, whereas time propels towards the future and encourages us to go forward in hope" (LF, n. 57). And the same prin-

[23] Ethna Regan, *The Bergoglian Principles*, p. 5.

[24] In the footnote he refers to Romano Guardini, *Das Ende der Neuzeit*, Würzburg, 1965; English translation: *The End of the Modern World:* Wilmington: Del. ISI Books, Intercollegiate Studies Institute, 1998, pp. 30-31.

[25] Pope Francis, Encyclical Letter *Lumen Fidei* on Faith, 29 June 2013 (hereinafter referred to as LF).

ciple is expressed twice in *Amoris Laetitia*[26] in reference to the journey of the Magisterium (AL, n. 3) and to the educational process of children (AL, n. 261).

The second principle is "*Unity is greater than conflict*" (EG, n. 228); this principle constitutes a third way between ignoring conflicts and remaining prisoners of it: "it is the best way to deal with conflict. It is the willingness to face conflict head on, to resolve it and to make it a link in the chain of a new process" (EG, n. 227). This is a particularly important principle, which explains how thought works, a real option of method: "a resolution which takes place on a higher plane and preserves what is valid and useful on both sides" (EG, n. 228). The principle is founded on the Gospel and is of the Christological order: "This principle [...] reminds us that Christ has made all things one in himself" (EG, n. 229).

The third principle appears to be more philosophical: "*Realities are greater than ideas*" (EG, n. 233). It starts from the constant tension between ideas and realities, insofar as "Realities simply are, whereas ideas are worked out". Hence the need for a "continuous dialogue between the two". It is an affirmation of philosophical realism to which are opposed the extremisms of "angelic forms of purity, dictatorships of relativism, empty rhetoric, objectives more ideal than real, brands of ahistorical fundamentalism, ethical systems bereft of kindness, intellectual discourse bereft of wisdom" (EG, n. 231) and also "ineffectual forms of idealism and nominalism" (EG, n. 232). The condemnation of idealism (as the reduction of reality to the idea) and nominalism (as the negation of any realism of the idea), which historically are the extreme opposites in the question of universals[27], means the implicit affirmation of moderate realism[28] which essentially consists in founding universal ideas in individual reality, because "*contra factum non valet argumentum*": in fact reality is greater than idea.

---

[26] Pope Francis, Post-synodal Apostolic Exhortation *Amoris Laetitia* on Love in the Family, 19 March 2016 (hereinafter referred to as AL).

[27] As Alain de Libera notes, although the problem of universals is typical of the medieval debate that contrasts and unites Platonism and Aristotelianism starting from late antiquity, "the debate still continues today": Alain de Libera, *La querelle des universaux de Platon à la fin du Moyen Age*. Paris: éditions du Seuil, 1996.

[28] Neglecting the rich philosophical tradition of "universal" concepts, capable of affirming what is essential starting from the individual, tradition is part of Bergoglio/Francis' training and thought, can lead to enormous misunderstandings, such as affirming that: "the priority of reality over ideas delegitimizes human rights" Margaret MacLeisch Mott, *Love the prisoner, Ban the Substance: Pope Francis and the War on Drugs*, in Alynna Lyon, Christine. A. Gustafson and Paul C. Manuel (eds), *Pope Francis as a Global Actor: Where Politcs and Theology Meet*. Cham: Palgrave Macmillan, 2018, 81-92, pp. 81-92.

Finally "*The whole is greater than the part*" (EG, n. 235). It is, in some ways, the most advanced principle because it takes up the challenge of globalization, solving it without reducing it. The principle of unity which is superior to conflict is completed by this definition of the "whole" that "is greater than the part, but it is also greater than the sum of its parts". This principle is an invitation not to be obsessed with details and "to broaden our horizons". This movement is also linked to the priority of time over space, in the invitation "to sink our roots deeper into the fertile soil and history of our native place" (EG, n. 235).

The polyhedron model is extremely significant in this regard: "Here our model is not the sphere, which is no greater than its parts, where every point is equidistant from the centre, and there are no differences between them. Instead, it is the polyhedron, which reflects the convergence of all its parts, each of which preserves its distinctiveness" (EG, n. 236).

The geometric image is interesting and not obvious, already used by Bergoglio in his interreligious dialogues with Rabbi Skorka[29].

Even in the case of this principle, the ultimate foundation is evangelical: "The Gospel has an intrinsic principle of totality: it will always remain good news until it has been proclaimed to all people, until it has healed and strengthened every aspect of humanity, until it has brought all men and women together at table in God's kingdom" (EG, n. 237).

These same principles have already been fully expressed by Bergoglio in the Opening Address to the Provincial Congregation of the Jesuits in Buenos Aires in 1974, therefore almost forty years before EG, "in the context of the violent dialectical opposition that divided both the church and society in Argentina in the 1970s. This was not a facile effort to find a 'middle ground' between the two opposing sides. Rather, it attempted to offer a different point of view, one that, without ignoring tensions and problems, avoided the fragmentation of the people's unity"[30]. In fact, Bergoglio writes at the conclusion of his speech: "The fundamental criteria for carrying out these processes and which must inspire our work are: unity is superior to conflict; the whole is superior to the part; time is superior to space"[31]. One is still missing; in fact,

---

[29] See Jorge M. Bergoglio and Abraham Skorka, *Sobre el Cielo y la Tierra*. Buenos Aires: Editorial Sudamericana, 2010.

[30] Borghesi, *The Mind of Pope Francis,* p. 57.

[31] Jorge M. Bergoglio, *Una institución que vive su carisma* 18 February 1974, in *Boletín de Espiritualidad*, 55.

the principle of the superiority of reality over idea was introduced in 1980[32], but it is in some ways implicit; as Borghesi writes

> The first two–"unity is superior to conflict" and "the whole is superior to the parts"- are criteria of synthesis and are intended to foster social and political peace. […] The third principle– "time is superior to space"- indicates the method for arriving at synthesis, that is, the importance accorded to processes over the desire for domination that calls for the occupation of spaces. The fourth principle, at this point only present potentially-"reality is superior to ideas"- was intended to oppose the abstractness of ideologies and, subsequently, to oppose gnosis in general[33].

In the opening speech of the Faculty of Theology and Philosophy in San Miguel in 1989, addressing the complexity of postmodern reality, Bergoglio proposes the method of the "inner overflow":

> I refer to the *method* of thinking: method of approach, understanding and description of the situation of today's man, of societies, and therefore, a method for elaborating a political anthropology. […] A style of thinking that implies a phenomenological method with a dialectical nuance. In this sense, it is necessary to strip the term "dialectical" of any Hegelian resonance, and simply see in it *the expression of a mutual interaction of reality*. It is a dynamic dialectic between life and thought in which rhythm is decisive, in which mobility does not distract, but rather concentrates and intensifies attention. It could be called *sineidetic thinking*, in which the details must be studied as a function of the whole[34].

The reference to Romano Guardini, defined as a "prophet of postmodernity" and his theory of polar oppositions is explicit[35].

---

[32] See, Jorge M. Bergoglio, *Formación permanente y reconciliatión.*

[33] Massimo Borghesi, *The Mind of Pope Francis,* p. 59.

[34] "In this way of thinking, *unity* and *rhythm* are essential. The *rhythm* comes from the application of the phenomenological method; unity is a "transcendental" being of the oppositions of the dynamic and living being. This method requires from the mind a *sineidetic tension* necessary to see the parts as a function of the whole and the whole as a function of the parts, with the awareness that in every vital whole (and the politico-social fact is a vital whole) it is impossible to separate the parts from the whole and vice versa, for the simple reason that it is not possible to understand a part without at the same time grasping the whole to which it is bound, and so to the opposite. It is a work of togetherness, in tension, sineidetic" Jorge M. Bergoglio, *Necesidad de una Antropología Política. Un Problema Pastoral,* in "Stromata", January-June 1989, pp. 173-189.

[35] See Romano Guardini, *Der Gegensatz Versuche zu einerPhilosophie des Lebendig-Konkreten.* Mainz: Matthias GrünewaldVerlag, 1955.

The method of the innerflow remains in the terminology of Pope Francis. The term "desbordar" is recurring, for example, in *Querida Amazonia*[36]: "On the contrary, solutions are found by 'overflow', [*por "desborde"*] that is, by transcending the contraposition that limits our vision and recognizing a greater gift that God is offering" (QA, n. 105). We note, however, that in the same passage, a few lines below "an overflowing fountain" image is spoken of which figuratively renders the *desbordar*, but which does not recur at a terminological level in the other linguistic versions of QA analysed[37].

The term "overflowing" occurs twice in LS and always in reference to love (nos. 231 and 236) and through LS it also appears in FT (FT, n. 181 cites EG, n. 231): precisely the love of God is overflowing:

> The phenomenon of *desborde* that the Pope has presented as the main analogy is the action of God in our life and in our history: 'It is proper to the Heart of God to overflow with mercy'. Francis points to this as God's greatest gift (cf. QA, n. 105). Both overflow as a way of acting synodally and overflow as a practical method of thinking are founded in it. They are a way and a method that try to adapt to this ever greater gift of God, to this 'vital environment' that the overflowing of God's love and mercy never cease to create and offer[38].

The overflowing method overcomes the objectivist analytical method, which does not work with the sciences of the spirit, and recovers the breadth of the intellect in its full value as "the capacity of reason and at the same time understanding"[39].

---

[36] Pope Francis, Post-synodal Apostolic Exhortation *Querida Amazonia* to the People of God and to all Persons of Good Will, 2 February 2020 (hereinafter referred to as QA).

[37] In the Spanish version, we read "*fuente generosa*" and analogically in Italian "*fonte generosa*") (QA, n. 105).

[38] Reference is made to Pope Francis, *Homily in Holy Chrism Mass*, 24 March 2016.

[39] "The analytic-objectivist method, borrowed from the natural sciences, when it was applied to the sciences of the spirit. Having to face the political problem, an overcoming of rationalism is therefore crucial, but not as a veneration of a romantic returnism, nor of a presumed affective knowledge of reality, but by recognizing its full value of the intellect: the ability to reason and to understand, that is to say the ability of the *vision* (*Anschuung*) of things" Jorge M. Bergoglio, *Necesidad de una Antropología Política*. We remember that the ability to reason and to understand are aspects of a single faculty. St. Thomas already wrote: "*In homine eadem potentia est ratio et intellectus*" Thomas Aquinas, *Summa Theologiae,* Transl. Fathers of the English Dominican Province, 5 vols.,: Christian Classics, Westminster, Md. 1981, I, q. 79, a.8, *co*.

In this perspective, the mere contrast between objective and subjective is overcome, avoiding falling from pure objectivism to subjectivist relativism, and the category of the "super objective" is proposed:

> "The social being is found in the context of the *collective*, which does not mean a 'gregarious whole', a *unitas accumulationis* as the scholastics would say, but an *organic whole*; which is why it is not opposed to the individual, but rather integrates it. *Objective* and *personal* are not opposite terms, since the objective-subjective duality is ontologically incomplete if not downright confused. The sphere of *super-objective* being is, in the sense of reality, strictly "objective". Consequently, if a reality is alive, and therefore organic, it can be *collective, objective and personal* at the same time[40].

In Bergoglio's thought, the four principles, the methodology for overflow and sineidetics, the super-objective perspective, as well as the polar tension, are part of the philosophical sphere, but their full foundation is of a theological order. With this Bergoglio seems to restore unity in diversity also to reason and faith, philosophy and theology. In fact, Bergoglio's thought is never presented as exclusively philosophical and, like any good theological thought, it is never devoid of philosophical elaboration.

To understand the relationship between philosophy and theology according to Pope Francis, it is important to carefully read the apostolic constitution on the ecclesiastical universities and Faculties, *Veritatis Gaudium*[41], in particular the six paragraphs of the preface which "summarize the pontiff's vision of what theology should be"[42]. Re-launching the perspective of the OT, VG asks academic institutions to assume "a radical shift": "to offer the decisive contribution of leaven, salt and light of the Gospel of Jesus Christ and the living Tradition of the Church, which is ever open to new situations and ideas" (VG, n. 3).

---

[40] Bergoglio, *Necesidad de una Antropología Política*, pp. 173-189.

[41] Pope Francis, Apostolic Constitution *Veritatis Gaudium* on Ecclesiastical Universities and Faculties, 8 December 2017 (hereinafter referred to as VG).

[42] Gerard Whelan, "Il metodo teologico e pastorale di papa Francesco" in Fabrizio Mandreoli (ed.), *La teologia di papa Francesco. Fonti, metodo, orizzonte e conseguenze*. Bologne: EDB, 2019, p. 90. For a reflection on *Veritatis Gaudium,* cf. Pino Di Luccio and José Luis Narvaja, "*Veritatis Gaudium* e rinnovamento degli studi ecclesiastici" in *La Civiltà Cattolica,* 4053 (2019), vol. II, 272-283; Giuseppe Tanzella Nitti, *Unità del sapere e transdisciplinarità. Per una lettura della* Veritatis Gaudium, in "Osservatore Romano", 12 March 2020; John-Okoria Ibhakewanlan (ed.), *Catholic University Education Today*. Bloomington (IN): Author House, 2020.

This renewal "against this vast new horizon now opening before us" must follow some fundamental criteria "for a renewal and revival of the contribution of ecclesiastical studies to a Church of missionary outreach" (VG, n. 4).

The first criterion is "that of contemplation and the presentation of a spiritual, intellectual and existential introduction to the heart of the kerygma, namely the ever fresh and attractive good news of the Gospel of Jesus Christ, which continues to take flesh in the life of the Church and of humanity". Hence a special task in the formation of Christian culture, namely "discovering in the whole of creation the Trinitarian imprint that makes the cosmos in which we live a "network of relations" in which "it is proper to every living being to tend towards other things" (VG, n. 4a, quoting LS, n. 240).

The second criterion, connected to the first, concerns dialogue, a broad dialogue that is "an intrinsic requirement for experiencing in community the joy of the Truth and appreciating more fully its meaning and practical implications" (EG, n. 4b) for the purpose of the formation of a culture of encounter. The third criterion, which derives directly from the second, consists of "inter-disciplinary and cross-disciplinary approaches carried out with wisdom and creativity in the light of Revelation". The fundamental characteristic of ecclesiastical studies must be "the vital intellectual principle of the unity in difference of knowledge and respect for its multiple, correlated and convergent expressions" (EG, n. 4c). This implies flexibility, organicity and dynamism. The interdisciplinary approach has the "weak" form of "multidisciplinarity" which shows the different aspects of the same phenomenon and the "strong" form of cross-disciplinarity, whose purpose is "situating and stimulating all disciplines against the backdrop of the Light and Life offered by the Wisdom streaming from God's Revelation" (VG, n. 4c).

The last criterion concerns "the urgent need for "networking" between those institutions worldwide that cultivate and promote ecclesiastical studies, in order to set up suitable channels of cooperation also with academic institutions in the different countries and with those inspired by different cultural and religious traditions" (VG, n. 4d).

This renewal of studies in fact proposes again the necessary dialogue of philosophy and theology, re-launching it in the context of the unitary dynamism of knowledge. It seems that the four principles of the EG are fully reflected in these four criteria of the VG, valorising processes, unity and totality even in the context of studies.

3. Analysis of *Laudato Si'*[43]

In this section we want to conduct a textual analysis of *Laudato Si'*, to look for what is defined as "fundamental" or "principle" in it. The analysis is conducted not only on the English translation, which is the language in which this article is written, but also on the Spanish text, which is the language of Pope Francis, on the Latin text which remains the ecclesiastical reference language, and on the Italian translation, which is affirmed as one of the languages of communication of the Church[44], all the more important in the case of *Laudato Si'* and also of *Fratelli tutti*[45], whose title is in Italian (and not in Latin as it is generally appropriate, taking the first two words) as in both cases they are quotations from St. Francis of Assisi. It might seem useless, but it is a duty to respect the text, and also to highlight the problematic nature of the translation process.

First of all, let's look for what is "fundamental" in LS, and therefore the recurrences of the term. I am aware that this terminological investigation does not exhaust the search for foundations, but it is a good start to investigate the theoretical structure of LS, while maintaining fidelity to the text.

First of all, rights recur as "fundamental": the right to drinking water is a fundamental human right (LS, nos. 30 and 185); the disadvantaged have fundamental rights to be taken into account from an ecological perspective (LS, n. 93); fundamental human rights are ordered to the integral development of

---

[43] See Lorella Congiunti, *Per una rinnovata comprensione della natura alla luce della Laudato Si'. Considerazioni filosofiche,* in Simone Rosati – Javier Belda Iniesta (ed.) *Le risorse naturali come patrimonio collettivo dell'umanità. Percorsi interdisciplinari tra scienza e Fede,* in "Bollettino - Società Tarquiniense di Arte e Storia", 2020, 55-72.

[44] On the fact that Latin remains the official language for the liturgy and the Magisterium, while Italian is affirmed as the universal language of "exchange and communication" in the Church, cf. Franco Pierno, "Tra universalità e compromessi locali. Il Vaticano e la lingua italiana" in Massimo Arcangeli (a cura di), *L'italiano nella Chiesa fra passato e presente*. Tourin: Allemandi, 2011, 173-89. Msgr. Paolo Rizzi recently said that Italian is "in a certain sense, the official language of the Church, as it is the most used in mass meetings of the faithful from various continents, especially when the Pope is present in Rome or in other countries. Certainly, it is the one usually and widely used in the various areas and at every level of the entire Roman Curia" and he also specified the criteria of "curial Italian": in the drafting of the internal practices and the official texts of the Roman Curia, the Italian allows a literary production that combines some important qualities: naturalness, elegance and harmony, avoiding sophistication, rhetoric, haughtiness and neglect. With this balanced combination, the text is clear and linear, and the distinction is obtained in simplicity, that is, *nobilis simplicitas*, which characterizes the style of curial Italian" Paolo Rizzi, "L'italiano internazionale della Chiesa cattolica: la voce della Curia romana", in *Accademia della Crusca*, 29 ottobre 2018.

[45] Pope Francis, Encyclical Letter *Fratelli tutti* on Fraternity and Social Friendship, 3 October 2020 (hereinafter referred to as FT).

the human person (LS, n. 157); many people are deprived of basic human rights (LS, n. 158).

In the English text, "fundamental" and "basic" alternate; in the same way, the Spanish text oscillates between "*fundamental*" and "*básico*"; in Italian "*fundamentale*" is always used; in the Latin text we find "*fundamentalis*" but more often "*praecipuus*". They are terms that are equivalent in their constructive function: they describe that which is fundamental, basic and that which is taken before the others, according to the etymology of "*praecipuus*" which is composed of "*prae*" and "*capere*".

The second term that often appears combined with the adjective fundamental is "relationship": "human life is grounded in three fundamental and closely intertwined relationships: with God, with our neighbour and with the earth itself" (LS, n. 66); relations with the environment cannot be healed if fundamental human relations are not healed (LS, n.119).

Some problems and consequently some attitudes appear to be defined as "fundamental".

The "basic problem" consists in having developed technology "*according to an undifferentiated and one-dimensional paradigm*" (LS, n. 106). In this case, the term "*primaria quaestio*" was chosen in Latin, i.e. the problem that comes first.

Reflecting the fundamental problem, there is a fundamental solution, and it consists in seeking "comprehensive solutions which consider the interactions within natural systems themselves and with social systems" (LS, n. 139). In this case, the English preferred the adjective "essential" and the Latin the expression "*magni momenti*".

Furthermore, recognizing the dignity of the poor is a fundamental ethical duty "an ethical imperative essential for effectively attaining the common good" (LS, n. 158).

As for the fundamental attitude expressed in number 208, the versions are partially different. In Spanish we read: "*La actitud básica de autotrascenderse, rompiendo la conciencia aislada y la autorreferencialidad, es la raíz que hace posible todo cuidado de los demás y del medio ambiente, y que hace brotar la reacción moral de considerar el impacto que provoca cada acción y cada decisión personal fuera de uno mismo*" (LS, n. 208). This passage is literally similar in Italian and Latin, while in English it presents some differences in form: "Disinterested concern for others, and the rejection of every form of self-centeredness and self-absorption, are essential if we truly wish to care for our brothers and sisters and for the natural environment. These attitudes also attune us to the moral imperative of assessing the impact of our every action and personal decision on the world around us" (LS, n. 208). It is noted that there is no self-transcendence and also the reference to "*raíz*", which in some ways is absorbed by "essential" which translates "*básica*".

LS also speaks of "foundations" of social life destroyed by moral degradation, causing the rise of violence and cruelty (LS, n. 229)[46] Above all, God's paternal love is defined as "fundamental": God's love is the fundamental cause of creation (LS, n. 77). Moreover, the term chosen has some interesting variables in the various language versions: God's love is "*móvil*", "*ratio*", "*ragione*", "moving force" of the created. Besides, the Paternity of God is a "fondamental" truth (*dato* and *quiddam* in other linguistic versions) (LS, n. 96). The Father is the ultimate source of everything, the loving and communicative foundation of all that exists (LS, n. 238).

It seems possible to conclude that rights and relationships are fundamental in LS; in the ecological question, there are fundamental problems and fundamental attitudes. The love of God the Father is the foundation of everything. Tracing what is literally fundamental in the Encyclical allows us to truly grasp its fundamental, profoundly clear skeleton: there are rights that concern the balance given by God the Father with love, and when this order of justice is violated in its foundations, it is fundamental to compensate it. The track of the words leads us to the truth of the thought expressed.

Let us now look for the principles expressed in LS. First of all, it should be pointed out that the "necessity" of the principles is asserted. It is in fact emphasized that without stable principles, relativism would dominate and there would be no possibility of limiting unjust attitudes:

> In the absence of objective truths or sound principles other than the satisfaction of our own desires and immediate needs, what limits can be placed on human trafficking, organized crime, the drug trade, commerce in blood diamonds and the fur of endangered species? Is it not the same relativistic logic which justifies buying the organs of the poor for resale or use in experimentation, or

---

[46] Even in this case, the English chooses a partially different form: "When the foundations of social life are corroded, what ensues are battles over conflicting interests, new forms of violence and brutality, and obstacles to the growth of a genuine culture of care for the environment", while in Spanish: "*Esa destrucción de todo fundamento de la vida social termina enfrentándonos unos con otros para preservar los propios intereses, provoca el surgimiento de nuevas formas de violencia y crueldad e impide el desarrollo de una verdadera cultura del cuidado del ambiente*" (LS, n. 229).

> eliminating children because they are not what their parents wanted? This same "use and throwaway" logic generates so much waste, because of the disordered desire to consume more than what is really necessary. We should not think that political efforts or the force of law will be sufficient to prevent actions which affect the environment because, when the culture itself is corrupt and objective truth and universally valid principles are no longer upheld, then laws can only be seen as arbitrary impositions or obstacles to be avoided[47].

The double use of the term "principles" in this passage shows the importance of fundamental truths to support the ethical system. Without immutable principles the same integral ecology would be unspeakable. The principles on which the encyclical is built are of an ethical nature, that is, normative judgments based on objective truths. It is clarified in LS that ethical principles are learned from reason, and can "reappear in different guise and find expression in a variety of languages, including religious language" and that it would be "quite simplistic to think that ethical principles present themselves purely in the abstract, detached from any context" (LS, n. 199).

The work of great politics should be based on these principles: "True statecraft is manifest when, in difficult times, we uphold high principles and think of the long-term common good" (LS, n. 178). We note, however, that while in the Spanish, Italian and Latin versions we speak of "*grandeza política*", "*grandezza politica*", "*politica magnitudo*", the English version prefers "true statecraft" instead of the literal "political greatness". The choice of "statecraft" appears interesting considering how, for Bergoglio, politics can be assimilated to poetry, arts, *techne*:

> Politicians must be poets in the Greek meaning of the term, they must be creative and move from a pragmatic balance to a fruitful creativity: this is what wears them down, How can they do it? how can it be done? What is the right way? Politicians are, by virtue of this creativity, fundamentally architects: this is why I have defined them as poets, potters[48].

The link between politics and ethical principles is explained in these terms: "Politics must not be subject to the economy, nor should the economy be subject to the dictates of an efficiency-driven paradigm of technocracy" (LS,

---

[47] Pope Francis, *Laudato Si'*, n. 123.

[48] Jorge M. Bergoglio, *Ciclo di Formación y ReflexiónPolítica,*1 June 2004 in Marco Gallo, *El pensamiento social y político de Bergoglio y papa Francis*co. Salta: Ediciones Universidad Católica de Salta, 2018, p. 329.

n. 180), observing that precisely the financial crisis of 2007-2008 is an opportunity to be "more attentive to ethical principles" (LS, n. 189). Concern is then expressed "when some ecological movements defend the integrity of the environment, rightly demanding that certain limits be imposed on scientific research, they sometimes fail to apply those same principles to human life" and equally "when technology disregards the great ethical principles, it ends up considering any practice whatsoever as licit" (LS, n. 136).

Among the great principles emerges "the principle of the common good" to which chapter IV on "Integral Ecology" is dedicated: "An integral ecology is inseparable from the notion of the common good, a central and unifying principle of social ethics" (LS, n. 156).

The common good is one of the fundamental principles of the Social Doctrine of the Church, it is central to the Magisterium and is "at the heart of the Bergoglian Principles"[49].

In LS, the "common good" is mentioned 30 times, and is defined as "the sum of those conditions of social life which allow social groups and their individual members relatively thorough and ready access to their own fulfilment" (LS, n. 156) with reference to *Gaudium et Spes*[50].

The common good is "the unifying principle of social ethics" (LS, n. 126), concerns respect for the human person as such (LS, n. 157) and is extended to future generations (LS, n. 159). In the contemporary globalized world, "the principle of the common good immediately becomes, logically and inevitably, a summons to solidarity and a preferential option for the poorest of our brothers and sisters" (LS, n. 58).

In the context of integral ecology, the common good is enriched with specifications: the climate is defined as a common good (LS, n. 23[51]) "the creation of jobs as an essential part of its service to the common good" (LS, n. 129). The principle of the common good is strictly connected to the principle of subsidiarity, another fundamental principle of the Social Doctrine of the Church and the pivot of integral ecology itself (LS, nos. 157, 196), and also "The principle of the subordination of private property to the universal des-

[49] Ethna Regan, *The Bergoglian principles*, p. 4.

[50] Second Vatican Ecumenical Council, Pastoral Constitution *Gaudium et Spes* on the Church in the Modern World, 7 December 1965, n. 26.

[51] See United States Conference of Catholic Bishops, *Global Climate Change: A Plea for Dialogue, Prudence and the Common Good,* 15 June 2001.

tination of goods" which, with reference to *Laborem exercens*[52], is "the first principle of the whole ethical and social order" (LS, n. 93, citing LE, n. 19).

Another principle invoked is that of precaution, when we are faced with an ecological threat not yet supported by full scientific certainties (LS, n. 186). There are also principles in particular matters such as those concerning forests (LS, n. 167). There are also economic principles such as the principle of maximization of profits which, not being a first principle, if isolated from other considerations, leads to "a misunderstanding of the very concept of the economy" (LS, n. 195).

The term "principle" is also used to indicate the beginning and the cause. In this regard, the polysemanticity of the term is interesting in n. 244 where Saint Basil the Great is cited: "if the world has a beginning and if it has been created, we must enquire who gave it this beginning, and who was its Creator" (LS, n. 244). While the English uses the term "*beginning*" twice analogically to the Latin "*principium*", the Italian and the Spanish versions instead distinguish "*principio*" and "*inizio/inicio*" where the Greek text uses in both occurrences the term "*archè*"[53].

In LS, the foundation of everything is the Trinity that created the world "as a single divine principle, but each one of them performs this common work in accordance with his own personal property. Consequently, 'when we contemplate with wonder the universe in all its grandeur and beauty, we must praise the whole Trinity'"[54].

The profoundly ethical structure of LS is confirmed, which is a development of the Social Doctrine of the Church. All ecology is based on objective truths and fundamental principles, of which the central one is the principle of the common good. Principles of subsidiarity and subordination of private property are linked to it.

This perspective is inserted into a broad thought, which looks at reality in all its complexity from above and knows how to bring it back to its principles and its First Principle.

This view is further illuminated by the four principles of the EG, which in LS are placed in the context of integral ecology.

---

52 Saint John Paul II, Encyclical Letter *Laborem Exercens* on Human Work on the ninetieth anniversary of *Rerum Novarum*, 14 September 1981 (hereinafter referred to as LE).

53 Basil the Great, *Exegetic Homilies: on the Hexameron*. Catholic University of America Press, Washington D.C., 1963, 1, 2, 6 *PG* 29, 8.

54 Saint John Paul II, *Catechesis* 2 August 2000, 4.

The principle of unity above conflict, LS refers specifically to the conflict between economics and politics, between "financial gain" and "increasing power". In this type of "conflicts or spurious agreements [...] the last thing either party is concerned about is caring for the environment and protecting those who are most vulnerable. Here too, we see how true it is that "unity is greater than conflict" (LS, n. 198).

The principle of the superiority of time manifests itself in the myopia of a policy committed only to immediate results, in the search for an easy consensus on the part of "consumerist sectors of the population", thus "we forget that 'time is greater than space' that we are always more effective when we generate processes rather than holding on to positions of power" (LS, n. 178).

The superiority of reality over the idea is manifested in "a number of symptoms which point to what is wrong, such as environmental degradation, anxiety, a loss of the purpose of life and of community living" (LS, n. 110); the horizon of a fragmented knowledge, dominated by technique without ethical vision and the absolutization of the scientific view lead to the loss of the vision of the complexity of reality.

But, it is especially the fourth principle that is central: "There is an interrelation between ecosystems and between the various spheres of social interaction, demonstrating yet again that "the whole is greater than the part" (LS, n. 141).

As already noted, "This fourth principle [...] is discussed in a radically broader context in *Laudato Si'*. [...] The whole in question is Earth, our common home, and this becomes a *locus theologicus*"[55]. The term "whole" is recurrent in LS, with several uses: "society as a whole", "the whole, human and social", "the whole of humanity", but above all "the natural world as a whole, "nature as a whole", "the universe as a whole in all its manifold relationship", "creation as a whole". Nature as a whole includes the human being who is its center.

Nature is presented in LS as a connected totality: "everything in the world is connected" (LS, n. 16); "all creatures are connected" (LS, n. 42); "everything is connected" (LS, nos. 117 and 138). This definition is so peculiar and characteristic that it is conceptually prevalent for understanding nature within integral ecology.

In fact, nature in LS is configured as a complex universe, a totality not only material, in which everything is in relation and communication, constitutively open to transcendence, to God: "In this universe, shaped by open and inter-

---

[55] Ethna Regan, *Four Bergoglian Principles*, p. 12.

communicating systems, we can discern countless forms of relationship and participation. This leads us to think of the whole as open to God's transcendence, within which it develops" (LS, n. 79). The valorisation of the fourth principle in the context of LS allows us to better understand its synthetic and inclusive value compared to the other three principles. The connection is expressed in the relationships that hold together the complexity of nature, so much so that the universe as a whole expresses the inexhaustible perfection of God: "The universe as a whole, in all its manifold relationships, shows forth the inexhaustible riches of God" (LS, n. 86).

Precisely the valorisation of the connection is more understandable if read within the framework of the overflowing method. As we have seen, the term occurs only twice in the encyclical, but indicates a track that is structural and fundamental. In fact, "Love, overflowing with small gestures of mutual care, is also civic and political, and it makes itself felt in every action that seeks to build a better world" (LS, n. 231). Love, therefore, is a totality that overflows into small gestures of a civic and political nature. Love therefore holds great projects and daily actions together. Love is what holds together the whole of reality and should motivate every gesture and every action of the human community. The four principles seem, then, all integrated into the overflowing love, which seems to hold together the ethical motivation of LS.

4. The Sources of the *Laudato Si'*

To better understand the thought of LS in the context of Bergoglio/Francis' thought, it is important to evaluate the sources that are not directly magisterial, that is, which authors are cited.

This is an important factor, also considering the particular role that Bergoglio attributed to the classics. In 1978, addressing the Provincial Congregation of the Jesuits, he affirmed:

> Neither one thing, nor the other: neither traditionalists nor utopians. ... resorting to the 'classic', which is very different from the easy recourse to what is 'traditional', to empty traditionalism that is concerned only with maintaining peace ... but that of sepulchres. Speaking of 'classics', we are referring to those strong moments of experience and religious and cultural reflection that make history because, in some way, they touch the irreversible events of the journey of a people, of the Church, of the Christian. It is a question of always having before our eyes the fundamental nucleus that constitutes and identi-

> fies us (cf. Heb 10:32ff; 13:7ff.,) to be able to fulfil, without deviating from our identity, the steps that the concrete historical situations demand of us. We are inspired by the 'classics' to carry on these two attitudes, apparently antinomic, which however reflect our way of being. Memory of the past and courage to open new spaces to God. The 'classics' have had the strength to find synthesis in moments of conflict. These are not easy 'compromises' or cheap 'irenisms'. These are the syntheses that, without denying the contrary elements that cannot be composed in crises, refer them to a higher plane and resolve it there, through a mysterious journey of understanding and fidelity to what is perennial in history. For this reason the 'classics' have this double virtue of being faithful to history and inspiring new paths to take[56].

Therefore, just turning to the classics is a method of finding the synthesis that resolves conflicts. The classics seem to have the same function as the roots, in the awareness that time is superior to space: they maintain fidelity to the past and open to the future.

Also in the light of this reflection, it seems important to consider the sources explicitly cited in LS.

First of all, we note that references to Saint Basil appear. In relation to the universe in which "Every creature is thus the object of the Father's tenderness, who gives it its place in the world" (LS, n. 77) is expressly written "Saint Basil the Great described the Creator as 'goodness without measure'"[57]. Then Saint Basil is also mentioned at the end of LS "we journey through this land seeking God, for 'if the world has a beginning and if it has been created, we must enquire who gave it this beginning, and who was its Creator'" (LS, n. 244)[58].

Therefore, Saint Basil is mentioned in the context of the creatural totality that depends on the creative love of the Father.

Saint Bonaventure is mentioned twice as a biographer and "disciple" of Saint Francis (LS, nos. 11 and 66) and twice as theologian, and precisely with regard to the possibility of encountering God in everything: "The ideal is not only to pass from the exterior to the interior to discover the action of God in the soul, but also to discover God in all things. Saint Bonaventure teaches us that 'contemplation deepens the more we feel the working of God's grace within our hearts, and the better we learn to encounter God in creatures out-

[56] Jorge M. Bergoglio, *Apertura de la Congregación Provincial*, 8 February 1978, in *Meditaciones para religiosos*. Bilbao: Mensajero, 2014.

[57] Reference is made to S. Basil the Great, *Hom. in Hexaemeron*, 1, 2, 10: *PG* 29, 9.

[58] As we earlier analysed, here S. Basil the Great, *Hom. in Hexaemeron*, 1, 2, 6: *PG* 29, 8 is cited.

side ourselves'" (LS, n. 233)[59]; and in relation to the Trinitarian structure of creation: "Saint Bonaventure went so far as to say that human beings, before sin, were able to see how each creature 'testifies that God is three'. The reflection of the Trinity was there to be recognized in nature 'when that book was open to man and our eyes had not yet become darkened'. The Franciscan saint teaches us that *each creature bears in itself a specifically Trinitarian structure*, so real that it could be readily contemplated if only the human gaze were not so partial, dark and fragile. In this way, he points out to us the challenge of trying to read reality in a Trinitarian key" (LS, n. 239)[60]. The passages dedicated to St. Bonaventure are undoubtedly important and on central themes; Di Maio, in his interesting study on Bonaventure's "traces" and "ideas" in LS, warns that "we must not forget that the respective contexts, vocabularies and problems are profoundly different"[61]; however, he finds interesting references. For example, he believes that the principle of totality can find a founding point in the Bonaventure concept of "*resolutio pieno*"[62].

Dante is also cited in n. 77: "Dante Alighieri spoke of "the love which moves the sun and the stars"[63]. The universe narrated by Dante Alighieri moved by the love of God agrees with the image of totality traced in LS.

Two other sources of LS, used in a certain sense as "classics" by Pope Francis, are also Teilhard de Chardin and Juan Carlos Scannone.

In relation to the becoming of the universe and to Jesus Christ understood as the fullness of God, a reference is made in a footnote to the Jesuit Teilhard de Chardin: "The ultimate destiny of the universe is in the fullness of God, which has already been attained by the risen Christ, the measure of the maturity of all things" (LS, n. 83) and in the footnote "In this horizon we can set the contribution of Fr. Teilhard de Chardin", some passages of the magisterium in which Teilhard de Chardin is mentioned are cited[64].

---

[59] Reference is made to S. Bonaventure, *In II Sent.*, 23, 2, 3. Cf. Bonaventure, *Opera omnia,* Vols. I-IX, The Fathers of the Collegii S. Bonaventura (eds.), Quaracchi, Florence 1882–1902

[60] Reference is made to Id., *Quaest. disp. de Myst. Trinitatis*, 1, 2, concl.

[61] Andrea Di Maio, "Tracce e spunti bonaventuriani nella *Laudato Si'*", in *Antonianum*, XCI (2016), p. 830.

[62] *Ibid.*, 835; reference is made to S. Bonaventure, *PerfEv.* 2.1 e 2.3 co.; *Hex.* 5.

[63] Reference is made to Dante Alighieri, *Divina Commedia,* Transl. Langdon C. Cambridge: Harvard University Press, 1918, Canto XXXIII, 145.

[64] Saint Paul VI, *Address in a Chemical and Pharmaceutical Plant,* 24 February 1966; Saint John Paul II, *Letter to the Reverend George Coyne,* 1 June 1988; Benedict XVI, *Homily for the Celebration of Vespers in Aosta*, 24 July 2009.

There is a short reference in the footnote to the Jesuit Juan Carlos Scannone: "Some authors have emphasized the values frequently found, for example, in the *villas*, *chabolas* or *favelas* of Latin America" (LS, note 117)[65].

Both references, albeit brief, seem perfectly consistent with the context of a dynamic and Christocentric vision of the cosmos regarding Teilhard de Chardin and with the context of a strong social commitment regarding Scannone.

The classical authors cited most often in LS are St. Thomas Aquinas and Romano Guardini.

St. Thomas Aquinas is quoted several times: in n. 86 three quotations from the *Summa Theologiae*, in n. 80 a quotation from the *Summa theologiae* and one from *Commentary on the Physics* of Aristotle, and note 171 of number 240 refers to three places in the *Summa Theologiae.*

Nature as a total expression of God the creator is explained by referring to the thought of Thomas:

> The universe as a whole, in all its manifold relationships, shows forth the inexhaustible riches of God. Saint Thomas Aquinas wisely noted that multiplicity and variety 'come from the intention of the first agent' who willed that 'what was wanting to one in the representation of the divine goodness might be supplied by another', inasmuch as God's goodness 'could not be represented fittingly by any one creature'. Hence we need to grasp the variety of things in their multiple relationships. We understand better the importance and meaning of each creature if we contemplate it within the entirety of God's plan[66].

All the quotations in this passage come from question 47 of Part I of the *Summa Theologiae*[67], which is dedicated to the plurality and distinction of things, and serve to argue that the complexity and complementarity of creatures in the totality of their relationships are an expression of divine goodness and express God's plan.

The link between God and universal relations is explained with reference to three passages of the *Summa Theologiae*: I, q. 11, art. 3; q. 21, art. 1, ad 3; q. 47, art. 3, dedicated to the unity of God and the justice of God in relation

---

[65] In the footnote, reference is made to Juan Carlos Scannone, «La irrupción del pobre y la lógica de la gratitud», en Juan Carlos Scannone and Marcelo Perine (eds.), *Irrupción del pobre y que hacer filosófico. Hacia una nueva racionalidad.* Buenos Aires: Editorial Bonum, 1993, 225-230.

[66] Pope Francis, *Laudato Si'*, n. 86.

[67] He cites precisely S. Thomas Aquinas, *Summa Theologiae* I, q. 47, art. 1, co; art. 2, ad 1; art. 3, co.

to the natural order, and to the unity of the world which manifests itself in the order of things.

On the other hand, nature understood as the internal essence of things themselves, is explained in LS with reference to the commentary on Aristotle's *Physics* by St. Thomas Aquinas: "Nature is nothing other than a certain kind of art, namely God's art, impressed upon things, whereby those things are moved to a determinate end. It is as if a shipbuilder were able to give timbers the wherewithal to move themselves to take the form of a ship" (LS, n. 80)[68]. Therefore, nature is the rational principle that God the creator wrote in things themselves, and that moves them to the end: as if the shipbuilder could give the wood the ability to take the shape of the ship.

We can note that the complexity of the notion of "nature" presented in LS as a connected totality and as an internal principle of dynamism rests conceptually in the Aristotelian-Thomist tradition; in fact, the references to St. Thomas appear to provide the conceptual framework for understanding this notion[69].

It seems that the dynamic spirit of Tommasian thought is being recovered, saving the manuals that had not satisfied the young Bergoglio, recovering "the genius of Thomas Aquinas"[70], "the depth and originality of St. Thomas Aquinas"[71], perhaps also thanks to the mediation of the dialectical Thomism of Ferrel.

Also in EG the reference to the *Summa Theologiae* is relative to the variety and unity of reality, and to the link between things and the Creator:

> Saint Thomas Aquinas noted that the multiplicity and variety 'were the intention of the first agent', who wished that 'what each individual thing lacked in order to reflect the divine goodness would be made up for by other things', since the Creator's goodness 'could not be fittingly reflected by just one creature'. Consequently, we need to grasp the variety of things in their multiple relationships. By analogy, we need to listen to and complement one another in our partial reception of reality and the Gospel[72].

---

[68] Reference is made to Thomas Aquinas, *In octo libros Physicorum Aristotelis expositio*, transl. Blackwell R. J. - Spath R.J. -Thirlkel W.E. New Haven: Yale University Press, 1963, lib. II, lectio 14.

[69] For a discussion of this point, see Serge-Thomas Bonino and Guido Mazzotta, *Dio creatore e la creazione come casa comune. Prospettive Tomiste*. Città del Vaticano: Urbaniana University Press, 2018.

[70] Pope Francis and Antonio Spadaro, *My Door is Always Open*, p. 20.

[71] Jorge M. Bergoglio, *Prefación*, in Enrico C. Bianchi, *Pobres en este mundoricos en la fe. La fe de lo spobres en América Latina según Rafael Tello*. Buenos Aires: Agape libros, 2012.

[72] Pope Francis, *Evangelii Gaudium*, note 44. Reference is made to: S. Thomas Aquinas, *Summa-Theologiae*, I, q. 47, a. 1; a. 2, ad 1; a. 3.

In the Encyclical Letter *Fratelli Tutti*, the reference to Saint Thomas is related to the movement of God's love:

> Saint Thomas Aquinas sought to describe the love made possible by God's grace as a movement outwards towards another, whereby we consider 'the beloved as somehow united to ourselves'. Our affection for others makes us freely desire to seek their good. All this originates in a sense of esteem, an appreciation of the value of the other. This is ultimately the idea behind the word 'charity': those who are loved are 'dear' to me; 'they are considered of great value'. And 'the love whereby someone becomes pleasing (*grata*) to another is the reason why the latter bestows something on him freely (*gratis*)'[73].

As Borghesi observes: "A Thomistic character is one of the constants of Bergoglio's thinking; this is where his gnoseological-metaphysical realism and his appreciation of the tangible world find their source"[74].

It is precisely in relation to the complex unity of nature that the reference to Thomas' realism appears relevant.

In LS there are also numerous quotations and references to Romano Guardini relating to the notion of modernity and the connected anthropological crisis.

As we have already highlighted, Guardini is mentioned seven times, always with reference to his work, *The End of the Modern World*.

The analysis of modern man in the face of the growth of power brought about by progress (LS, n. 105) is taken from Guardini: "contemporary man has not been trained to use power well"[75]; "The risk is growing day by day that man will not use his power as he should"[76]. Guardini therefore offers the key to interpreting the anthropological roots of the ecological crisis, that is the affirmation of the univocal paradigm of technoscience, which ultimately refers to the problem of "power" (LS, n. 108): "in the most radical sense of the term power is its motive – a lordship over all"[77].

The point of view that absolutizes technology as a question of power, transforms the whole of nature into "an insensate order": "the technological mind sees nature as an insensate order, as a cold body of facts, as a mere 'given', as an object of utility, as raw material to be hammered into useful shape; it views

[73] Pope Francs, *Fratelli tutti*, n. 93. Reference is made to: S. Thomas Aquinas, *Summa Theologiae* II-II, q. 27, art. 2, co; I-II, q. 26, a. 3, co.; q. 110, a. 1, co.

[74] Massimo Borghesi, *The Mind of Pope Francis*, p. 2.

[75] Romano Guardini, *The End of the Modern World*, p. 82.

[76] *Ibid.*, p. 83.

[77] *Ibid.*, p. 56.

the cosmos similarly as a mere 'space' into which objects can be thrown with complete indifference".[78]

Not only is the diagnosis taken from Guardini, but also the cure; the ecological conversion requires the contribution of the community, and "will make such tremendous demands of man that he could never achieve it by individual initiative or even by the united effort of men bred in an individualistic way. The work of dominating the world calls for a union of skills and a unity of achievement that can only grow from quite a different attitude"[79].

The prevailing sources of LS are, therefore, St. Thomas Aquinas for the metaphysical vision of nature as a unitary complexity and Romano Guardini for the analysis of modern anthropocentrism.

Finally, however, it should be noted that the most important "classic" in LS is St. Francis. Not only is the title a true quotation from the *Canticle of Creatures*, which allows us to understand what the vision on the nature of integral ecology is, but also references to St. Francis in LS are precisely in terms of "reconciliation".

> As a result, the originally harmonious relationship between human beings and nature became conflictual (cf. *Gen* 3:17-19). It is significant that the harmony which Saint Francis of Assisi experienced with all creatures was seen as a healing of that rupture. Saint Bonaventure held that, through universal reconciliation with every creature, Saint Francis in some way returned to the state of original innocence[80].

Ecological reconciliation takes the form of a personal conversion: "In calling to mind the figure of Saint Francis of Assisi, we come to realize that a healthy relationship with creation is one dimension of overall personal conversion, which entails the recognition of our errors, sins, faults and failures, and leads to heartfelt repentance and desire to change" (LS, n. 218).

In LS, the thought of reconciliation concerns man's relationship with creation, however, in a particular dynamic: man himself is part of the unity of creation and he is the one who thinks about creation and must take care of it.

Moreover, the relationship between man and nature is precisely the peculiarity of integral ecology which is fundamentally and essentially human ecology.

---

[78] *Ibid.*, p. 55.

[79] *Ibid.*, pp. 55-56.

[80] Pope Francis, *Laudato* Si', n. 66. Reference is made to Bonaventure, *The Major Legend of Saint Francis*, in *Francis of Assisi: Early Documents*, vol. 2. New York-London-Manila: New City Press, 2000, p. 586.

5. Peculiarity of LS: There is no Ecology without Anthropology

The thought of reconciliation and dynamic unity therefore finds a peculiar development in LS. In my opinion, a characteristic principle of LS can be deduced in the statement: "There can be no ecology without an adequate anthropology (LS, n. 118).

This principle seems to be an application of the principle of totality superior to conflict, in the spirit of a polar dynamic: in fact, integral ecology sees the whole as including both the environment and man, without opposing and without identifying them, in a unity in which the differences are not cancelled but integrated.

The need for a new anthropology is already present in Bergoglio's thought, as appears very clearly in the aforementioned speech in San Miguel in 1989 entitled "The need for a political anthropology"; it is placed in the context of the Argentine situation in reference to the return of the democratic political system, but offers a far-reaching reflection on who is the "politician" and which anthropology is required in the face of contemporary challenges.

In this discourse he denounces the fracture of the semantic, empirical, theoretical order, that is, the partial views that prevent understanding of who the politician is; it would require a vision capable of grasping all the tensions, conceiving politics in relation to the person and aimed at the common good. In an attempt to avoid reducing tension to an unstable equilibrium or cancelling it by joining only one of the poles, the solution is found when "politics rediscovers a deeper and less instrumental and immediatist meaning, in a classical and Christian conception, as that which - in the words of Aristotle – "makes man better". The tension in terms of bonum is between the common good and the particular good, which configures the political tension as such"[81].

The method required for a political anthropology involves "turning our gaze to the roots of man himself, of the political animal, sketching out *some traits* that I consider important in reference to today's man, to the real man to whom we announce the Gospel"[82]. In this speech, the sources used by Bergoglio are the speech of John Paul II at the plenary meeting of the Secretariat for non-believers on March 5, 1988 and the texts of the "prophet of postmodernity" Romano Guardini.

For an adequate anthropology, an appropriate method is needed that is not reduced simply to "quantitative reason". The "reduction of ethics and politics to physics" is denounced, a sentence which is also identical to what is written

---

[81] Jorge M. Bergoglio, *Necesidad de una Antropología Política.*

[82] *Ibid.*

in the FT: "What is now happening, and drawing us into a perverse and barren way of thinking, is the reduction of ethics and politics to physics. Good and evil no longer exist in themselves; there is only a calculus of benefits and burdens" (LS, n. 210).

Instead, "An anthropology that wants to guide the overcoming of the crisis must be dialectical: strictly personal and at the same time supportive", must place the uniqueness of the person at the centre, his being unique, inalienable, indispensable, irreplaceable. Such an anthropology "cannot evade the confrontation between the person and the Person who transcends him and who gives him a foundation in that same transcendence"[83].

Here we find the premises of what is stated in LS, where anthropology is internal to the ecological discourse itself, within an interdisciplinary, never reductionist methodology, illuminated by Faith.

Ecology and anthropology in LS are built together: insofar as nature is configured as the house of man and man as the guardian of nature. Integral ecology is in fact above all a human ecology.

Benedict XVI in *Caritas in veritate*[84] already stressed the importance of keeping together the concept of human ecology and that of environmental ecology: "It is contradictory to insist that future generations respect the natural environment when our educational systems and laws do not help them to respect themselves" (CiV, n. 51). In this context, Benedict XVI uses the image of the book of nature which LS proposes again and cites in n. 6:

> The book of nature is one and indivisible: it takes in not only the environment but also life, sexuality, marriage, the family, social relations: in a word, integral human development. Our duties towards the environment are linked to our duties towards the human person, considered in himself and in relation to others. It would be wrong to uphold one set of duties while trampling on the other. Herein lies a grave contradiction in our mentality and practice today: one which demeans the person, disrupts the environment and damages society[85].

Integral ecology is therefore a human ecology, and needs a correct anthropology. The search for this anthropology passes through the identification of the human roots of the ecological crisis to which the entire chapter 3 is dedicated.

---

[83] *Ibid.*

[84] Benedict XVI, Encyclical Letter *Caritas in Veritate* on Integral Human Development in Charity and Truth, 29 June 2009 (hereinafter referred to as CiV).

[85] Benedict XVI, *Caritas in Veritate*, n. 51.

In this context, not anthropocentrism as such is condemned, but modern, deviant, excessive anthropocentrism, which "has paradoxically ended up prizing technical thought over reality" (LS, n. 115). It therefore seems that precisely the techno-centric thought is an example of an idea that does not compare with the superiority of reality.

This "excessive anthropocentrism" wears out what is common and social relations (LS, n. 116).

However, "A misguided anthropocentrism need not necessarily yield to 'biocentrism', for that would entail adding yet another imbalance, failing to solve present problems and adding new ones. Human beings cannot be expected to feel responsibility for the world unless, at the same time, their unique capacities of knowledge, will, freedom and responsibility are recognized and valued" (LS, n. 118).

It therefore seems that the tension between deviant anthropocentrism and biocentrism must find its solution in correct recognition of the role of man in nature, from which human responsibility towards the environment follows.

The human person has peculiarities of reason and science, he can make use of the vegetable and animal world when it is needed for his life, but his power is limited.

Every human person has an "inalienable worth" (LS, n. 136), and "Christian thought sees human beings as possessing a particular dignity above other creatures; it thus inculcates esteem for each person and respect for others". (LS, n. 119).

An adequate anthropology highlights the relationships of the human person with other people, with the natural environment, with God: "A correct relationship with the created world demands that we not weaken this social dimension of openness to others, much less the transcendent dimension of our openness to the 'Thou' of God" (LS, n. 119).

In LS, the right relationship between man and nature is configured in terms of work correctly understood (LS, n. 125). Work as a correct relationship implies taking care of the existing and producing fruit, according to the biblical warning: "The biblical texts are to be read in their context, with an appropriate hermeneutic, recognizing that they tell us to 'till and keep' the garden of the world (cf. *Gen* 2:15). 'Tilling' refers to cultivating, ploughing or working, while 'keeping' means caring, protecting, overseeing and preserving" (LS, nn. 67 and also 124). Work also means human development and personal fulfilment (LS, n. 128), investment in people (LS, n. 128), valorisation of the spiritual sense of manual work (LS, n. 126).

In these terms, work itself seems a means of grasping the reconciliation of the tensions of reality.

Integral ecology therefore requires a complex, synthetic, reconciling method that is able to grasp all the dimensions of reality and arrive at the essence of the human person: “an integral ecology calls for openness to categories which transcend the language of mathematics and biology, and take us to the heart of what it is to be human” (LS, n. 11).

6. Concluding remarks looking towards the future

The richness of Bergoglio’s thought helps to understand the fundamentals of LS, grasping the peculiarity of integral ecology.

The thought of Pope Francis is a thought of the present, which has roots in the past and leans towards the future. It is a thought that captures tensions to resolve them without cancelling them, that knows how to think of multiplicity in the complexity of totality.

It is a thought that is constitutively open to action, because it arises from confrontation with reality, and moreover: “What calls us to action are realities illuminated by reason” (EG, n. 232).

The reason that can illuminate reality is a broad reason, capable of reasoning and understanding, of grasping the essence of man and of opening up to transcendence, of dialoguing in unity with the Faith.

The denunciation of reductionism seems to me to be one of the most fruitful aspects of the contemporary debate.

For example, the Coronavirus pandemic that has hit the whole world since the end of 2019, emphasizing previous injustices and imbalances, often arouses opposing reductionisms: rationalism and spiritualism oppose each other in always partial visions that claim to be definitive; scientism and scepticism scramble in claiming that the sciences give absolute truths or in doubting every assertion.

The remedy, on the other hand, lies in grasping the tension and always seeking the solution in unity, addressing a complex method in which each specialization constitutes one face of the polyhedron.

In concrete terms, praying and wearing protective devices are not choices that oppose each other, but actions that interpenetrate. Solidarity must devise gestures appropriate to the situations, because even the request for a hug in the face of the risk of contagion can mean the claim that the idea is superior to reality.

True solidarity places the concrete person at the centre with his non-quantifiable value. For this reason LS is naturally fulfilled in FT. Again St. Francis is a teacher of reconciliation, with his invitation "for a love that transcends the barriers of geography and distance" (FT, n. 1).

Recourse to the "classics" intended as solutions of reconciliation constitutes a suggestion of great usefulness to escape both traditionalism and progressivism.

It is a method of understanding and action that is based on a metaphysical, ontological vision, which is committed to grasping reality, and does not seek abstract unity, but seeks truth.

This philosophical realism in the thought of Bergoglio/Francis is constantly nourished and illuminated by Revelation: Jesus Christ is the foundation and the meaning of every unity, yesterday, today, always.

THE DWELLING OF MEN

Towards an anthropology adequate for integral ecology

Jakub Gorczyca, SJ

"There can be no ecology without an adequate anthropology"[1]. If *Laudato Si'* (LS) understands "ecology" as that discipline which "studies the relationship between living organisms and the environment in which they develop" (LS 138), we must immediately add that, in the perspective of the encyclical, the subject is "integral ecology". This qualification results from the vision of reality in which "everything is interconnected". Therefore, according to the encyclical, the care of nature and of human relations with nature must proceed hand in hand with the care of human relations themselves, whether at the interpersonal level or in the social, cultural, political and economic areas. Still further: "Learning to accept our body, to care for it and to respect its fullest meaning, is an essential element of any genuine human ecology" (LS 155). It is not surprising, therefore, that the ecological crisis, whose multiple symptoms are amply described in the first chapter of the papal text, concerns in equal measure the world of nature and the world of human relations; both "deteriorate together" (LS 48). "We are not faced with two separate crises, one environmental and the other social, but rather with one complex crisis which is both social and environmental" (LS 139). However, as it is easy to realise, the deterioration of the natural environment originates from human roots (cf. LS 101). Consequently, the remedy for this deterioration has to be found in the area of human relations and actions lived authentically and, on the theoretical

[1] Pope Francis, Encyclical Letter *Laudato Si' on Care for Our Common Home*, 24 May 2015, n. 118. Hereafter LS.

DOI 10.32060/C.S.9.253-271

level, in an anthropology that is able to think of the human being in a way that is radically different from that which has led to the present crisis. In fact, according to the encyclical, the root of this crisis is to be found in a "misguided anthropocentrism" that, in its turn, "leads to a misguided lifestyle" (LS 122), characterised by individualism, moral relativism, consumerism, technocratic pragmatism, and the "throwaway culture which affects the excluded just as it quickly reduces things to rubbish" (LS 22). This anthropocentrism, which is also labelled "despotic", because it does not respect subhuman beings, "has ended up prizing technical thought over reality" (LS 115). In this way, we have reached a situation of "a constant schizophrenia wherein a technocracy which sees no intrinsic value in lesser beings coexists with the other extreme, which sees no special value in human beings" (LS 118).

At this point, the nucleus of our argument materialises: what kind of philosophical anthropology – for it is this that we are speaking of – can be suitable for integral ecology? In the light of the assumptions set out above, a preliminary answer can only direct us to an anthropology that takes account of the relational dimension of the human being and towards a starting point for our reflection in which this dimension emerges with clarity. In our opinion, the privileged starting point, thanks to which it is possible to clearly grasp the truth about the human being as essentially relational and existing in relation, is the ethical experience. The description and the subsequent interpretation of this experience should thus enable us to discover the basis of an anthropology that is not only adequate for integral ecology but adequate *tout court*, that is, one that conforms to human reality. In advancing our hypothesis, we are guided by the conviction that the phenomenon of the moral good, given originally in the ethical experience, manifests humanity, that is, the very essence of the human being in its multiple dimensions.

A model of the ethical experience that clearly demonstrates the moral good is found in the parable of the Good Samaritan that Jesus told in answer to the question "And who is my neighbour?"; a question posed by a doctor of the Law in connection with obedience to the commandment of love (cf. Lk 10:29). In Jesus' parable, the moral good is realised in the human behaviour that expresses the love of neighbour. In the context of philosophical reflection, it is appropriate to recall that Pope Francis made extensive use of the parable of the Good Samaritan in the encyclical *Fratelli tutti* (FT), seeing in it a story that can be understood universally, that is to say, by "all people of good will, regardless of their religious convictions"[2]. According to the Pope, therefore, this parable is able to appeal to everyone and, we can add, in different historical circumstances and in various ways. We too will allow ourselves to be addressed by this story in order to draw out its ethical and anthropological significance[3].

1. The event of the encounter

"The human person, with his or her inalienable rights, is by nature open to relationship. Implanted deep within us is the call to transcend ourselves through an encounter with others" (FT 111). In reflecting on the parable (cf. Lk 10:25-37), our very first question is: How and under what conditions did the encounter between the Samaritan and the injured man take place?

The man, who lay "half dead" at the side of the road, attacked and wounded by brigands, entered into the "field of vision" of the Samaritan, who, "passing by, saw him". In this connection, we could distinguish two different ways of seeing: one that results in knowledge and the other that establishes an encounter. The former can be expressed in theoretical judgements; in the Samaritan's situation, this way of seeing and knowing objectively could have been expressed in judgements: "he is a foreigner", "he is a rich man", "this bleeding man is impure". The second way of seeing, that of the Samaritan, was not the subject of a merely cognitive intention. The Samaritan's gaze was filled with compassion. As the first reaction to the wounded man, his compassion could certainly not be classified a kind of involuntary emotional reaction (such as could have been a reaction of fear or disgust). With moving concern, the Samaritan "bound up his wounds" and, "setting him on his own beast and brought him to an inn, and took care of him".

In this context, we could observe the difference between "meeting someone" and "meeting with someone": in the event of an encounter with someone, there is a form of reciprocity that is more or less symmetrical. It can be very rudimentary and consist of a simple meeting of kindly glances or an exchange of greetings; it can take the form of a verbal dialogue or, even, of a communicative silence[4]. In the encounter with someone, therefore, not only is

[2] Pope Francis, Encyclical Letter *Fratelli tutti on Fraternity and Social Friendship*, 3 October 2020, n. 56. Hereafter FT.

[3] In the reflections that follow, we shall draw on some pages from our *Essere per l'altro. Fondamenti di etica filosofica*, 2nd ed. Rome: G&BPress, 2016.

[4] Cf. Martin Buber, "Dialogo", in Martin Buber. *Il principio dialogico e altri saggi*. San Paolo: Cinisello Balsamo, 1993, ed. or. "Zwiesprache", in Martin Buber, *Das dialogische Prinzip*, 5th ed. Heidelberg: Lambert Schneider, 1984, pp. 187-88.

the other present to the subject but also the subject is present to the other. In the encounter with someone, there takes place a reciprocal intermingling of presences and mutual communication. Simply "encountering someone" does not necessarily imply a symmetrical reciprocity: maybe someone encountered me, became aware of my presence, without my being aware of him ("I was encountered"), or else that I myself encountered someone without his becoming aware of my presence.

However, in order to speak of the event of encounter, whatever its mode, the presence of the other must *involve*, in some way, a subject at least. Indeed, I can be aware of the presence of someone or perceive that he is in the same space as I am and, notwithstanding, remain ethically indifferent, as one remains indifferent when passing by a stone. The Samaritan did not remain indifferent to the lot of the man who, all of a sudden, became present to him and whose presence engaged him; responding to his call, "he went to him". However, we could ask ourselves, how a man reduced to the end of life, abandoned "half dead" on the street, perhaps totally unaware of what was happening around him, would be able to call the Samaritan and seek his help just as, for example, a beggar sitting on the steps of a church seeks it? Certainly, such a call would have been wholly impossible. In encountering the other, the Samaritan found himself in the presence of a face; and the face itself speaks, seeks help silently.

Although the mode of the encounter of the other and with the other can vary according to the individual characteristics of the people involved and according to the circumstances of time and place, nonetheless their ethical dimension remains as an essential constant. The experience of the presence of the other, *ethically demanding* of the subject, is the original mode of the manifestation of being "someone" and not "something". In the event of the encounter, the subject discovers his ethical bond with the other and becomes aware of his responsibility for him.

2. The face and responsibility

In introducing the concepts of the "face" and of "responsibility" in the descriptions of encounter, we are referring to some among the best-known ideas of the philosophical thinker Emmanuel Lévinas. In its ethical significance, which is imposed without presuppositions as absolutely original, the face does not coincide with a specific part of the human body whose physiognomy can be perceived, described, characterised, or classified. Therefore, it can be argued, as the French philosopher of Jewish origin does, that "the best way of encountering the other is not to note even the colour of his eyes! (...) The relation to the face is directly ethical"[5].

If we express it in a paradoxical way, it is not incorrect to say that the face cannot be seen. "Seeing", that is, the cognitive approach to reality – which we have just mentioned – "thematises" the object and seeks the adjustment between the judgement and what is perceived because in this way it can be understood and classified. However, such a theoretical approach misses the mark and cannot not miss it before the face, if the face is encountered in an authentic way. That does not mean that the face is presented as "irrational": the impossibility of understanding the face, that is, its insignificance in the cognitive order, means that the significance of the face is of a different order from the theoretical because it belongs, precisely, to the ethical order. Not only does it belong there. According to Lévinas, the face establishes, opens the ethical order and, together with it, the whole perspective of the first philosophy.

The presence of the neighbour's face immediately establishes an ethical relationship with the subject, that is, a social relationship. In other words, for the subject, the closeness of the face does not mean entering into a relationship of reciprocal intimacy "I-Thou" or, on the contrary, opening up an objectifying knowledge. The face does not allow itself to be "captured" or let itself be described "objectively", because it is "abstract", that is to say, it comes from beyond every context – aesthetic, cultural, historical, and also...ontological – which could enable it to be identified and signify "something". Being "invisible" in this sense, the face signifies without a context, it expresses itself all by itself. The way in which the face "appears" to the subject can be described as "epiphanic" and "enigmatic". Its way of making itself present is not "figurative" but "eloquent".

The face is a word that imposes on the subject the first commandment: "you shall not kill me!" Although expressed with absolute authority, this prohibition of violence is not authoritarian; it originates in the helpless nakedness of the face, its extreme poverty and weakness. In the face, the other is present as radically fragile, alone and exposed to suffering and death. The face "commands" by praying, by seeking, as we sometimes hear in the expression: "You are kindly requested to go...". Thus, the face expresses both the essential poverty of

---

[5] Emmanuel Lévinas, *Etica e Infinito. Dialoghi con Philippe Nemo*. Rome: Città Nuova, 1984, ed. or. Éthique et Infini. Dialogues avec Philippe Nemo. Paris: Fayard, 1982, pp. 99-101.

the neighbour as well as his authority as master. In seeking, the poor – symbolised in the biblical figures of the widow, orphan, and stranger – commands and teaches the subject how he ought to act so that his freedom is not overbearing or despotic. In other words, the helpless face of the neighbour reveals the meaning of humanity of man. The authority of this teaching springs from a different order from that of the natural world where the struggle for survival often applies: in this sense, the authority of the neighbour is literally "extra-ordinary". In the face that does not allow itself to be understood and yet is not irrational is signified the Infinite and Goodness. Following Lévinas, we could also say that in the face of the neighbour is signified the Other of Infinite Goodness.

Infinitely separate and immediately close, by demanding in a prayer a response, the other awakens the subject to his unique responsibility. Indeed, the warning "Thou shalt not kill" is not only a prohibition of murder, the forbidding of the commission of any evil against him: fundamentally, expressed in positive terms, it is a call to responsibility in view of the good of the other.

"The encounter with the other represents immediately my responsibility for him. Responsibility for the neighbour, which undoubtedly is the austere name for what is called the love of neighbour…"[6]. Encountering the other in the immediacy of responsibility for him, replying to his call-prayer, signifies loving him as a neighbour, that is, "making oneself a neighbour" in the gradual but never definitive process of coming close to him.

Responsibility for the other, encountered in his dignified poverty, appears as a more fundamental ethical structure in comparison with the responsibility of the subject for his individual acts, the frequent concern of various treatises on morality. The original meaning of the whole of the moral life springs effectively from the relations in which the subject is called to respond to the other and for the other. This responsibility – unlimited, like the demands of love of the neighbour – is not the result of the free choice of the subject. On the contrary: one can say that the subject has been chosen and called to this responsibility. But chosen and called by whom? Chosen and called by the Goodness signified in the face of the neighbour. It is precisely this Goodness which arouses in the subject the desire to take care of him. Such a responsibility, which arrives at the "substitution" of the other and becoming his "hostage",

[6] Emmanuel Lévinas, "Filosofia, giustizia e amore", in Emmanuel Lévinas, *Tra noi. Saggi sul pensare-all'altro*. Milan: Jaca Book, 1998, ed. or. "Philosophie, justice et amour", in Emmanuel Lévinas, *Entre nous. Essais sur le penser-à-l'autre*. Paris: Grassset, 1991, p. 137.

constitutes the human subject as a subject and as human. Another human subject than the ethical one does not exist. Equally, we can say that responsibility for the neighbour is constitutive of the very humanity of man.

From Lévinas' perspective, the relation of responsibility is originally asymmetrical and, in a certain way, unidirectional. In disinterested service, the subject is responsible for the other without aiming at a reward, without any parity of rights and duties or any reciprocity. However, in the human world, there also exists the "third", and "he too is my other, my neighbour. Therefore, I am interested to know which of the two goes forward: is not the one, perhaps, the persecutor of the other? Should men who are incomparable be compared? Therefore, in taking on oneself the fate of the other, justice here has the priority: I have to judge where I must take on responsibility first of all"[7]. The demand of justice makes itself alive together with the plurality of the "unique" and the incomparable who must yet become "equal" in some way in the inevitable comparison that equity demands. Here, according to Lévinas, is the first violence in the bosom of society. In fact, the other is not only my neighbour; he is also my fellow citizen and the fellow citizen of others. Thus, in the manifold relations between citizens, protected and regulated by the institutions of the State, the initial asymmetry of responsibility becomes counterbalanced in some way by the egalitarian order of law.

It seems, therefore, that the structure of *fundamental responsibility*, namely, the dynamic of the love of neighbour – which we have tried to describe following the thought inspired by biblical wisdom – can be summarily expressed in some key terms. One can speak, in fact, of the *respect* for the dignity of the other (non-violence), of *acceptance* (hospitality), of *concern* for his good (benevolence), and of *service* (charity). It goes without saying that all these features of charitable relations, elements of responsibility, are interdependent and, in practice, overlap one another in a single movement of love of the neighbour.

3. The goodness in love of the neighbour

Through acts of love for the neighbour, human goodness enters the world. In the Gospel parable, it appears as that quality which characterises the behaviour of the Samaritan and his very being. He has become good because he did not remain indifferent to the man in need. He responded to his call by

---

[7] Ibid., p. 138.

taking care of him. His response and responsibility materialise in personal involvement, beginning with the look of compassion, through the gift of time, the expenditure of energy, the change of travel plans, the use of various material goods, ending up with concern for the future of the unfortunate person.

What appears to be essential in moral goodness – to the degree in which we can know it on the basis of the situation described – is the wish to achieve the good of the other in a disinterested way. "The Samaritan who stopped along the way departed without expecting any recognition or gratitude" (FT 79). When such a disinterested intention is lacking, the help offered has to be interpreted differently, as, for example, an intervention made to earn the recognition of the man who was saved, the result of a utilitarian calculation, an act of self-gratification or vainglory. Thus, the decisive element for the "generation" of the moral good seems to be the free and disinterested *gift of self* to the other.

Recounting and listening to the story of the Samaritan as that of a good person, we are aware that the goodness that "pertains" to his behaviour characterises the Samaritan in an absolute sense, that is, as a human being. We describe him as "good" because he behaved in the way in which everyone finding himself in the same situation ought to behave: he did what is appropriate for human beings as such. We could also express it thus: moral goodness distinguishes a person in the essential manifestation of his humanity; in yet other terms, moral goodness reveals the truth of being and acting properly human. Whereas we recognise the model of authentic humanity in the man who is morally virtuous, no one becomes the ideal of humanity through possessing, for example, a sharp intelligence or an artistic talent.

We have said that everyone ought to behave as the Samaritan behaved. Indeed, the indifference of the priest and the Levite appears as morally reprehensible. But what is the source of this unconditional duty that conditions a person's authenticity? Undoubtedly, the duty of help is born in the presence of the unfortunate man, springs from his personal being. In fact, like no other being on earth, the person demands that, where he is concerned, the good be done and the evil avoided. The response of love, the only adequate response to such a categorical need, will be expressed, as we have seen, in an attitude of respect, acceptance, and concern for the good of the other: in short, in the service of the other. In other words, the objective need to behave in a certain way and not in another is not the result of the wish to reach a certain objective (e.g., something pleasant or useful for the subject); it stems, not primarily from the law promulgated by a civil or religious authority, but rather from the understanding of the *truth* that the person must be unconditionally respected, that is, loved in a disinterested way. Therefore, the foundation of moral duty coincides with the dignity of the person.

Not a few "men of good will", agnostics and atheists, stop at this foundation, considering it sufficiently intelligible and definitively final. In other words, in order to justify the absolute nature of moral duties, it would be enough to recognise the dignity of the human being that consists principally in his rationality. By virtue of his rational nature, man has to be considered superior to all the other beings. However, we could ask ourselves, if we found human superiority on the empirical difference in comparison with other beings who are nonrational or endowed with "inferior" rationality, does this perhaps not signify that we are maintaining a superiority that is only one of degree? And is the resulting dignity, which is, in fact, relative, suitable for justifying absolute duties?

Certainly, in order to *know* about moral demands that are categorically binding, the presence of the being of the person is sufficient. However, if the ultimate *ontological* foundation of his dignity were identical with his empirical superiority in relation to other beings, nothing could guarantee the absolute nature and the universal permanence of the moral duties in his regard. That this conclusion is not mistaken or imaginary, it can receive confirmation indirectly from some ethical theories according to which, in situations of conflict, the "right" to life of certain animals (e.g., of a healthy adult chimpanzee) would take precedence over the respective right of some human beings (e.g., the elderly and gravely ill).

However, even if we accept the assumption that essentially – and not only through the difference in degree of a natural quality – the human being is superior to all the other beings, it is possible to raise further questions. In fact, although eminent in his dignity, every human person is contingent, fallible, and mortal. Can such a being really constitute the ultimate source of absolute duties? Therefore, if, despite all the existential and moral imperfections, the absolute need to behave in conformity with the demands of the person's dignity is sensed, then we should probably think of another foundation for moral duties; otherwise, their absolute nature would not be sufficiently intelligible. To think of such a non-contingent foundation means going down the path of metaphysics and affirming as necessary the existence of the personal Absolute of Being and Goodness. In other words, being free and self-aware, the human person possesses his incomparable dignity by virtue of his constitutive relational openness to that Absolute which we have previously called by the name of the Other of Infinite Goodness.

4. The gift and gratitude

In our descriptions of responsibility, the moral good, and unconditional duty, the goodness of the being of the human person appeared as that property of his which is manifested with conspicuous evidence in the ethical experience. This same goodness can be revealed again by another way, that is to say, through the experience of *gift*.

Like every living being, every human lives thanks to the resources of the natural world. When he recognises in the elements of the world the "goods" that are useful to him, he can observe that the majority of the goods that are needed to maintain him in life do not originate directly from nature but, rather, have been produced and/or supplied by other people. Many of these goods have been given to him freely. If he opens himself to the theological perspective, man is also able to come to the vision of the world of nature as a great gift of the Creator (cf. LS 76, 220). Moreover, through the gift of these goods, the presence of the other can be revealed to him as beneficent and the other himself, in his being, as a good for him. However, the other, the neighbour, is a good that is radically different from everything that comes from the world of nature – directly or indirectly, that is, through work and the use of technology. He is a good that cannot be "assimilated" or used like any other vital good. Every other vital good appears as such in that it corresponds to the needs of the bodily subject. The usual approach to the "vital" goods coming from nature is that of using them and, so, destroying them. Precisely in this process of consumption, their good or lack of good is often verified. The right approach to natural goods, achieved with the aid of tools and technologies, is regulated chiefly by needs and pragmatic reasoning. On the other hand, the right and fundamental approach to the other person is – as we have already glimpsed – completely different: the other is properly encountered and received as a good when he is received as an *absolute gift,* namely, as a good that is not correlated with consumerist needs and dominion of the subject. To wish to use the other in the function of needs would be equivalent to debasing him to the level of things to be consumed and the means of sustenance.

Thus, the just relation of the subject with the good that he encounters in the person of the other is not the same as the gratification of a consumerist need. It is a relation animated by the desire for the Infinite (to return once again to the vocabulary of Lévinas). This desire does not exclude the need of the other; however, it will be a "non-consumerist" need because it is filled, precisely, with the desire for Transcendence. In the face of the beneficent presence of the other, the subject can exclaim: "It is good that you are here!". This expression contains his acknowledgment according to a double sense of the word: affirmation of his being and gratitude for his being. In fact, the very being of the other is given as good and beneficent, and the other is close, neighbouring, as a *good that is desired* and a *gift that is free*. In this way, too, all the goods offered by the other can be recognised and received as gifts. In fact, it is unthinkable that something should come from the goodness of the neighbour which is not good and which only appears to be a free gift. In their turn, the individual gifts received could strengthen, if there was need of it, the conviction of the subject that the being of the other – beyond all his imperfections – is a gift for him (cf. FT 143).

The act of giving requires a response of *gratitude*. This requirement of gratitude, however fragile it is from the ethical point of view, belongs to the intrinsic understanding of the relation of "giving – receiving". In fact, ingratitude or the "simple failure" to acknowledge appear deplorable. Certainly, the very act of accepting the gift can already include a gesture of gratitude towards the donor. Knowing how to accept well is to recognise the other in his goodness; it gives him the possibility of giving himself and also offering gifts to others. Thanking is giving.

In the experience of being accepted and loved – and the helpful closeness of others is a reassuring testimony of this – the subject can come to the awareness of being himself a good. He knows, indeed, that evil as such can be neither desired nor loved. His limitations and faults notwithstanding, his being among men manifests itself as fundamentally good. It is recognised as a good that remains "substantially" through all the "bad" vicissitudes of his life. His being-good and his possibilities for giving are given to him in order to be set in motion by the other and actualised as a gift for the other.

While we are asking ourselves about the subjective conditions that make possible the gift to the other, it seems clear that the subject who gives and gives himself has to be able to dispose of himself, that is to say, he must be able to be free and to know the significance of the situation in which he finds himself. In this sense, the subject must be ontologically "autonomous". However, although this condition is necessary, it is not sufficient because, if someone regards his own being as an evil, hates himself, and rejects his own existence, he will not be able to give himself to others since only a good can be offered as an authentic gift. Therefore, only by respecting himself and accepting himself as a good can the subject give himself to others to the point of sacrificing his own life. Only one that respects himself as a being that is fundamentally good can practise *kenotic* responsibility for others; only in *kenosis* for others do people accept themselves fully as gift.

Thus, what is called "love of oneself" in philosophical-theological language appears as a condition of possibility for love for others. That does not mean to say that it is necessary to love oneself "before", "then" loving others in a chronological sense, or that love for others can be "inferred" from the subject's love for himself. However, if there is an absence of this attitude of love of self, lived more or less explicitly, the sacrifice of oneself for others could be interpreted as a hidden rejection or even hatred of oneself. The authentic "love of oneself", therefore, has nothing in common with egoism. In fact, the sacrifice involved in giving to the other and gratitude for the possibility of giving protect the subject from every kind of self-sufficient closing in on himself and so from the falsehood of narcissism and from moral guilt.

We notice to aspects of the human reality. On the one hand, there is the radical ontological and ethical fragility of human goods. On the other hand, there is the existence in the human world of goodness and generous self-giving. Arising from this, we can conceive of another being who cannot not be and who cannot not be good: we can think, therefore, of the Being of Infinite Goodness. Such an Infinite Goodness who gives himself in the being of persons cannot consist in something less than personal being. If he is the absolute foundation of being and the ultimate source of every gift, then, in the response of gratitude to the other, the subject addresses himself implicitly to the Other of Infinite Good. The response of gratitude to this Other can even take an explicit form, thus becoming one of the most fundamental of religious acts, that is to say, the act of prayer.

5. The person as gift

Constitutively open to the Other of Infinite Goodness, the human being is an autonomous being. His existential autonomy means that he is free to think, to decide, and to act. In other words, insofar as he is autonomous, the person "possesses himself" and is capable of determining himself and transcending himself. He is autonomous also in the ontological sense. That is, as a "substantial" being, the human person exists "in himself" (not as an "accident" of another) and "for himself" (not as a "means" serving the purpose of another being).

As an autonomous being, the human person is not a mere "part" of the world of nature, or of society, or of the State in which he lives; he does not allow himself to be absorbed and dissolved in a "totality". By virtue of his ontological consistency and his continuity of identity over time, the person also transcends the totality of his own acts. These are, certainly, "his", but he is not reduced to their "totality". Thus, to describe the human person through the metaphysical category of "substance" does not mean to attribute to him a being that is static and closed in on itself, compact and impenetrable, to which the concept of the "monad without windows" is fitting, but it is completely the opposite: in fact, the other ontological dimension of the person, equally fundamental, is its *relationality*.

On the basis our awareness of ethical experience and it reflective explanation, we must now reiterate that the human person is a being that is constitutively "related" to the other. If we express this truth in yet a different way, we can say that the human person is a "relational substance". That means that the relationality, or the relating to another personal being, belongs to the intrinsic substantial constitution of the person. This relationality is not an add-on to the personal substance "from outside", as happens in the case of some accidents (including that of "relation"), but is intrinsically inherent in the essence of the personal substance. If this is the case, what can be identified and called an act of effective "openness to the other" is possible on the basis of this constitutive dimension of the person. In other words, the person is able to open himself to the other thanks to his essential orientation to personal alterity. If, on the other hand, we observe a real "closing" of someone in the face of the other, then we must equally suppose that this closing, too, is possible on the basis of his substantially constitutive dimension of relationality. In fact, the person can give himself and gives himself or, on the contrary, refuses to give himself, as essentially related to the other.

The person can, therefore, be defined as the "spiritual subject who is autonomous gift". In this context, the term "subject" does not refer primarily to the epistemological sense but rather to the ontological: it designates the relational substantiality of the person. His "spirituality" signifies his freedom and openness to the truth, that is, his "inner reality". If we have ascertained that the being of the person is a gift from the Other of Infinite Goodness, then we must now add that it is not this in the mode of an object or as "something". "Something", belonging to "someone", can undergo a change of owner, can be given away, modified, even destroyed. The person, on the other hand, presents as a free being, as a subject capable of self-giving: of love and responsibility for the other.

However, to exist as autonomous gift and to depend in being on the being of the Other of Infinite Goodness implies the presence of this Other in the gift itself; otherwise, the gift would be only "an object", existing as a "something". Therefore, it seems legitimate to conclude that the generosity of the Other of Infinite Goodness does not abandon the gift of being present in the human person.

In affirming, as we have done a short while ago, that man is a person insofar as he is a spiritual subject in relation to the other, we must not "forget" that his being a person includes his bodily nature. The spirit as soul and the matter organised into a living body form the unity which – following a venerable philosophic tradition – can be called "substantial". Although – as is obvious in the light of ordinary experience – not all the bodily phenomena manifest the personal being as such, the human body itself, by participating in the being of the person and his dignity, is personal and, so, also relational. In fact, it constitutes an essential element of the subject-person. Consequently, we can now clarify that the *human person* is the *corporal-spiritual subject who is autonomous gift*.

6. Values and virtues in the light of the person

In reflecting on the ethical experience, we were able to affirm that the person achieves the moral good in the act of giving to the other. So, if his being gift belongs to the nature of the person, we can equally assert that the source of the moral good is found in the free fulfilment of the personal nature of the man. To be more precise: the essence of the moral good, that is, the constitutive structure of ethical value, is identified with action that conforms to the demands of the personal nature of the man.

In making this formula explicit, we can say, therefore, that the human person as a being is good and subsists as gift. Sometimes, by virtue of a free decision, he acts according to the demands of his nature (acting in conformity with the *truth* of his own being), fulfils himself in being gift to others and becomes morally good. This reveals that the moral good is founded on the ontological good of the *person who is gift*. Therefore, in the perspective opened up by the ethical experience of the encounter and of responsibility, the ancient sapiential sentence that exhorts: "Become what you are!", can be rendered with the maxim: "Live as gift!".

Living as a gift will be translated principally in the demand to respect the character of gift of every being. In fact, every contingent being, inasmuch as it exists thanks to the generosity of the Other of Infinite Goodness, reveals its character as gift. Without doubt, in a manner different and differentiated, the character of gift of the world of nature (animals, plants, minerals, etc.) has to be respected as also that of the autonomous gift of the human person. The subject himself is not excluded from this need to respect the person in his character as gift: so the human person becomes the "object" of ethical concern for himself. This concern will be expressed in the grateful acceptance of his own being and will begin – not necessarily in the chronological sense – with not harming himself.

Respecting the being of the human person in his nature as gift means respecting him in all the elements of his spiritual-psychical-bodily structure. In the concrete, this respect demands having to take account of the different goods that are *for* the person and that are indicated by this same structure. The term "indicated" does not actually mean that these goods are spontaneously and immediately known as such. Even though such a kind of immediate and spontaneous knowledge can be identified in many cases, in others, in order to identify in what the objective good of the person consists (and, still more, to identify the good he needs *hic et nunc*), it will be necessary to dialogue carefully with the person himself and with others and, not rarely, with the help of the experts in some sciences. Realised, protected and promoted, the goods of the person become, in the context of human action, the means through which love of the neighbour is expressed and given concrete manifestation. Even the more ordinary and common needs of the neighbour – which, as *his*, are no longer ordinary and common – and the corresponding goods are, in fact, entrusted to the responsibility of the subject, and, as such, become vehicles of his love and factors that co-determine his goodness.

In the ethical sense, according to the types of relations in which it is realised, love assumes different forms, and nothing forbids us from calling these forms *moral values*. We can, therefore, understand moral values as the various expressions of the single goodness of love, that is to say, as the different modes of realising the moral goodness of man. The moral values exist as a multiform quality of the human acts and attitudes that are actuated both in interpersonal and social relations as also in dealing with the subhuman beings and the Other of Infinite Goodness. For example, the values of justice and truth refer to the correctness and goodness of the relations that are essentially different from those in which the values of compassion and patience are realised – as ideals of morally good behaviour. The actions, relations, and respective moral values that, for example, make up part of the sphere of international commerce differ essentially from those that apply within the family (between father and son), the hospital (between the patient and the doctor), or the university (between professor and student). Consequently, the moral correctness and goodness of such relations will be structured ontologically in ways that are essentially different and will also carry different names (for example: "international justice", "solidarity", "trustworthiness", "honesty", "trust", etc.). Thus, it becomes clear that the ontological constitution of the moral value is *relational*, more precisely, *interpersonal.*

If it is true that the moral values have their foundation in the human action that conforms to the demands of the individual's personal nature, namely, in the free fulfilment of being gift, then we can also say that, by "incarnating" values in himself, the morally virtuous person realises goodness in various modes of his being "with" and "for" others. In still other words, the virtues appear as different forms of the human perfection that consists in being gift. Giving oneself in the attitude of acceptance and service signifies loving the other and, indirectly, loving one's own being, that is, respecting it and accepting it as a gift from the Other of Infinite Goodness. In this perspective of the self-giving of the person, we can note that every moral virtue is a different expression of the one love.

Earlier, we defined the human person as a "corporal-spiritual subject who is gift". "To be gift" can be more or less actualised, more or less perfect, that is to say, more or less close to the ideal of humanity. The degree of perfection of being gift depends on the freedom of the person, on his decisions, and on his actual conduct. Let us say, therefore, that the person can and should be "always better" or "ever more" gift for the other. To be gift is gift and, at the same time, the task of the freedom that knows itself and recognises responsibility for the other.

Through just decisions and good acts, the person himself becomes morally good, gaining new ways of being gift. In this way, his character and personal being are constituted. Different attitudes and acts, expressions of his being gift, are "deposited" over time in his being and constitute in it a new capacity for self-giving. We call such capacities, formed through practices that enable and incline the person to act in gift-mode, almost in confirmation of the principle *bonum diffusivum sui*, "moral virtues". Thus, moral virtue appears as *a stable and freely constituted mode of being gift*. We believe that this conception of virtue also corresponds to that demand of which the encyclical *Fratelli tutti* speaks when it says that "… Yet if the acts of the various moral virtues are to be rightly directed, one needs to take into account the extent to which they foster openness and union with others. That is made possible by the charity that God infuses. Without charity, we may perhaps possess only apparent virtues, incapable of sustaining life in common" (FT 91).

7. The culture of encounter and the civilisation of love

Involved in the drama of human existence, both in the interpersonal and socio-political spheres, also at the international level, the man who is ethically virtuous becomes a "beautiful" person. His interior integration and the harmony between his being and acting, between his words and his deeds, between the affections, the will, and the reason are manifested as beautiful and attractive. The roots of this metaphysical beauty are found in the dynamic union of truth and goodness in his very being. Inasmuch as he is "true" in his helpful openness to the other, the virtuous man is "intelligible"; in that he is "good" in his disinterested giving of himself to the other, he is "desirable"; in that he is "beautiful" in his being, which is true and good, he is "admirable". His presence radiates an interior light that has nothing in common with the "worldly" lustre spread by the vain, proud man, greedy for power and celebrity. Indeed, the splendour of the spiritual beauty of the virtuous man comes, as from a secret source, from his *humility*. Entering the world of the moral virtues softly together with Christianity, this virtue is not autonomous in the sense of possessing its specific "object". It accompanies the love of neighbour in its different expressions. Put another way, love itself is humble and, so, beautiful and admirable.

> Love, overflowing with small gestures of mutual care, is also civic and political, and it makes itself felt in every action that seeks to build a better world. Love for society and commitment to the common good are outstanding expressions of charity (…). Social love is the key to authentic development[8].

In fact, the love that is established between persons on the level of relations cannot not concern social, political, and economic relations – in different degrees and modes according to the people and the circumstances. If it were

---

[8] Pope Francis, *Laudato Si'*, n. 231.

otherwise, it would be a question, not of love, but rather of the relationships of a selfish self-reference.

> "Social love" makes it possible to advance towards a civilization of love, to which all of us can feel called. Charity, with its impulse to universality, is capable of building a new world. No mere sentiment, it is the best means of discovering effective paths of development for everyone[9].

Adopting the statement of the Second Vatican Council, according to which human beings cannot be fulfilled or find their own fullness "except in the sincere gift of themselves", the encyclical *Fratelli tutti* adds that, in order to recognise the real truth about themselves, people must encounter others and communicate with them. In this connection, it cites a short passage from Gabriel Marcel (cfr. FT 87) who, as is well known, devoted some exquisite philosophical pages to the analysis of interpersonal relations. Taking our cue from Pope Francis' text, we too would like to reconnect in this context – and in this historical moment when humanity is undergoing manifold negative impacts not only from the ecological crisis but also from the pandemic – with the thought of the French philosopher.

When, in the tempestuous period of the Second World War, Marcel spoke of the "perilous situation of ethical values" – seeing that "the forces of destruction that have been unleashed everywhere" were tending "towards all the incarnate values" present "in a family or in a school, in a hospital or in a museum or in a church" –, he discovered the roots of this barbarism in the repudiation of the real (more exactly, of its "sacred" character) and in the "collapse of religious beliefs"[10]. As a remedy for this evil and malaise, he postulated, among other things, the raising of human consciousness to "natural" love – or, as he called it "perichristian" love—and the fulfilling of the task of "reawakening in the other the awareness of what he is, of his divine sonship (...), through the love which is its witness"[11].

The Christian thinker had no illusions about the immediate practical efficacy of philosophical discourses in this matter. His hope for the reawakening of religious faith and the renewal of customs was founded on the reality of

---

[9] Pope Francis, *Fratelli tutti*, n. 183.

[10] Gabriel Marcel, *Homo viator. Prolegomeni ad una metafisica della speranza*. Roma: Borla, 1980, ed. or. *Homo viator. Prolégomènes à une métaphysique de l'espérance*. Paris: Aubier, 1944, pp. 188 and 187.

[11] Ibid., p. 186.

fraternal life that was authentically human within small communities. He was persuaded that men were capable of "relearning how to live when they are placed in real conditions and a light from the height illuminates the group that they form from among themselves and with the things on which they draw for their subsistence". Moreover, he added: "Everything leads to the belief that the guarantee of success of enterprises like these is bound up with the humility of their origins and of their initial goals"[12]. Undoubtedly, the spreading of the civilisation of love in the globalised world – subject for the most part to the technocratic mentality and the mechanisms of the free market – must also primarily draw on the fruitfulness of the culture of encounter (cf. FT 215), of the love experienced "in one's own home" and the depth of respect for the face of the neighbour. As the encyclical *Fratelli tutti* opportunely reminds us, "Service always looks to [the] faces [of the brothers], touches their flesh, senses their closeness, and, even, in some cases, "suffers" that closeness and tries to help them" (FT 115).

The development of the culture of encounter and dialogue, the promotion of the civilisation of love and life according to the demands of integral ecology, cannot do without the education of consciences. Indeed, it is beginning with the conscience – that secret place where the voice of the truth about the good echoes and calls the subject to continual conversion – that the human *ethos* is built, namely, the *dwelling* where brothers and sisters live in justice and reciprocal love, in harmony with the gift of nature and in communion with the Giver. It is our hope that these reflections on the anthropology, that is intended to be adequate for integral ecology, can contribute in some way to the building of such a dwelling.

---

[12] Ibid., p. 190 [the translation has been slightly modified – J.G.] .

ROBERT DORAN AND THE VISION OF POPE FRANCIS

Gerard Whelan, SJ

This contribution explores the philosophical foundations of the notion of integral ecology found in *Laudato Si'*. It explores two aspects of this issue: the philosophical foundations of Pope Francis as he proposes this notion, and how these foundations might be deepened. It traces two major philosophical influences on the Pope: romantic philosophy and a more traditionally Catholic realistic metaphysics. It suggests that the philosophical foundations of integral ecology, as articulated by the Pope, remain less than systematically elaborated and proposes the critical realist thought of Robert Doran, a disciple of Bernard Lonergan, as a resource improve on this.

1. Foundations of the Notion of Integral Ecology in Pope Francis

Before identifying the philosophical undercurrents of *Laudato Si'*, it can help to outline the general structure of its argument.

1.1 The Structure of Argument of *Laudato Si'*

This encyclical employs an inductive method of "See", "Judge", and "Act", a method that was made familiar in the document of Vatican II, *Gaudium et Spes*. In his Introduction, Francis makes two points clear: that he is addressing all people of good will, so his argument will be primarily philosophical, and not theological; and that his approach to the ecological crisis will be holistic. He presents the figure of St. Francis of Assisi as setting the tone for the holistic message of the encyclical claiming that this saint, "shows us just

DOI 10.32060/C.S.9.273-316

how inseparable the bond is between concern for nature, justice for the poor, commitment to society, and interior peace"[1]. Introducing the key notion of the encyclical, "integral ecology", he states: "Saint Francis is the example par excellence of care for the vulnerable and of an integral ecology lived out joyfully and authentically"[2].

In Chapter 1, the Pope employs the data of natural science to outline the main dimensions of the ecological crisis: climate change, the reduction of biodiversity, and the reduced access for humans to clean water. Giving evidence of his holistic approach, he immediately adds comments that draw on the social sciences. He states, "today, however, we have to realize that a truly ecological approach *always* becomes a social approach; it must integrate questions of justice in debates on the environment, so as to hear *both the cry of the earth and the cry of the poor*"[3].

Chapters 2, 3, and 4 broadly conform to the step "Judge". Chapter 2 constitutes a digression from the promise made in the Introduction to treat the ecological problem without explicit appeal to Christian theology. Francis defends this by pointing out that religion motivates so many people in the world that it must necessarily play a role helping to shift political culture to accept policies that express a concern for the environment. Chapters 3, returns to a philosophical key and again focuses on questions of culture. It identifies one primary cause of the ecological crisis: a "technocratic paradigm" in culture. The Pope describes this as an approach to life that "exalts the concept of a subject who, using logical and rational procedures, progressively approaches and gains control over an external object"[4]. He notes that this paradigm stresses the use of scientific method, "which in itself is already a technique of possession, mastery and transformation", without the constraint of counterbalancing, ethical, considerations. He adds:

> Human beings and material objects no longer extend a friendly hand to one another; the relationship has become confrontational. This has made it easy to accept the idea of infinite or unlimited growth, which proves so attractive to economists, financiers and experts in technology. It is based on the lie that there is an infinite supply of the earth's goods[5].

---

[1] Pope Francis, Encyclical Letter *Laudato Si' on Care for Our Common Home*, 24 May 2015, nn. 1-11.

[2] Pope Francis, *Laudato Si'*, n. 10.

[3] Pope Francis, *Laudato Si'*, n. 49 (emphasis in the original).

[4] Pope Francis, *Laudato Si'*, n. 106.

[5] Pope Francis, *Laudato Si'*, n. 106.

In Chapter 4, entitled "Integral Ecology", the Pope proposes an alternative to the technocratic paradigm. He asserts: "An integral ecology includes taking time to recover a serene harmony with creation, reflecting on our lifestyle and our ideals, and contemplating the Creator"[6]. He describes this notion as multi-faceted:

> A distinctive way of looking at things, a way of thinking, policies, an educational programme, a lifestyle and a spirituality which together generate resistance to the assault of the technocratic paradigm[7].

The last two chapters of the encyclical conform to the step "Act" and offer practical proposals for how a culture of integral ecology might be promoted. Chapter 5, "Lines of Approach and Action", appeals for processes of dialogue to be established at different levels: within the international community; at the level of national and local politics; for "transparency in decision-making"; between economics and politics, and between religions and science. Chapter 6, "Ecological Spirituality and Education", stresses that, in order to be capable of authentic dialogue, citizens need to be virtuous. Pope Francis states: "Many things have to change course, but it is we human beings above all who need to change"; he adds: "A great cultural, spiritual and educational challenge stands before us, and it will demand that we set out on the long path of renewal"[8]. As this is the concluding chapter, the Pope reverts to a theological mode of speaking. He introduces a notion of "ecological conversion", which he explains in terms of Christian spirituality. Here he makes explicit reference to a Christian vision of the end of the world:

> At the end, we will find ourselves face to face with the infinite beauty of God (cf. *1 Cor* 13:12), and be able to read with admiration and happiness the mystery of the universe, which with us will share in unending plenitude[9].

He explains how such an eschatological vision influences a notion of stewardship of the created world in the present: "In the meantime, we come together to take charge of this home which has been entrusted to us, knowing that all the good which exists here will be taken up into the heavenly feast"[10].

---

[6] Pope Francis, *Laudato Si'*, n. 225.
[7] Pope Francis, *Laudato Si'*, n. 111
[8] Pope Francis, *Laudato Si'*, n. 106.
[9] Pope Francis, *Laudato Si'*, n. 243.
[10] Pope Francis, *Laudato Si'*, n. 244.

He explains how such beliefs inspire a sacramental vision of the universe: "The Sacraments are a privileged way in which nature is taken up by God to become a means of mediating supernatural life"[11], adding, "it is in the Eucharist that all that has been created finds its greatest exaltation"[12].

1.2. *Laudato Si'* and Romantic Philosophy

Much of the above argument finds a resonance in romantic philosophy. One commentary explains that romanticism has been described as less of a coherent, single, philosophy than "an attitude or intellectual orientation" that is associated with "Western civilization over a period from the late 18th to the mid-19th century", and adds that, "the movement finds expression in 'works of literature, painting, music, architecture, criticism, and historiography'". To the degree that one can describe romanticism as a philosophy, the commentary continues:

> It was also to some extent a reaction against the Enlightenment and against 18th-century rationalism and physical materialism in general. Romanticism emphasized the individual, the subjective, the irrational, the imaginative, the personal, the spontaneous, the emotional, the visionary, and the transcendental[13].

Commentators have studied how, from the early nineteenth century, Catholic thinkers, in Germany and elsewhere, began to engage with romantic thought finding that it "sailed along the coastline of sacramentality and mysticism"[14]. Two aspects of *Laudato Si'* are particularly resonant with romantic thought. First, identifying the technocratic paradigm as a central cause of the ecological crisis has much in common with romantic reaction against the rationalism of the Enlightenment. Second, the manner in which Pope Francis speaks of integral ecology echoes romantic themes such as, the beauty of nature, spirituality, community, culture, and educating the young.

---

[11] Pope Francis, *Laudato Si'*, n. 235.

[12] Pope Francis, *Laudato Si'*, n. 236.

[13] Encyclopedia Britanica, "Romanticism", https://www.britannica.com/art/Romanticism, lines 7-11 accessed on March, 2022. Authorities on Romanticism who write within the Anglo-American tradition include, Isaiah Berlin, *The Crooked Timber of Humanity: Chapters in the History of Ideas*. London: Henry Hardy, John Murray, ed., 1990; and his former doctoral student Charles Taylor *A Secular Age*. Cambridge. MA: Belknap Press of Harvard University Press, 2007.

[14] Thomas F. O'Meara, *Romantic Idealism and Roman Catholicism: Schelling and the Theologians*. South Bend, IN: University of Notre Dame Press, 1982, 5.

The romantic attributes of *Laudato Si'* are similar to those found elsewhere in the teaching of Pope Francis. Commentators point to the origins of this in his cultural and educational background. They note how a romantic nationalism characterised the culture of Argentina in general, producing a political philosophy of "Peronism". They also note how a version of liberation theology emerged in that country after Vatican II, that differed from the version found in most other Latin American countries in that it was particularly influenced by romantic philosophy. This school of thought gained the name, a "theology of the people" (*la teologia del pueblo*)[15]. A founder-figure of this school was Lucio Gera, a diocesan priest and an occasional lecturer of the young Jorge Bergoglio. Gera had studied theology in Europe in the 1950s where he was influenced by thinkers of the nineteenth Century School of Tübingen, who had attempted to relate Catholic theology to the romantic movement and to idealist philosophers. Walter Kasper notes that an early founder of the German romantic movement was Johann Gottfried Herder[16] and adds:

> The parallels between the romantic teaching of the early Tübingen theologians about the national spirit, which goes back to Johann Gottfried Herder (1744-1803), and the Argentine theology of the people are striking and clearly not accidental[17].

Other commentators note that another source by which the romanticism of the School of Tübingen influenced Pope Francis was through the thought of Romano Guardini, a scholar of the early twentieth century who had been both a student and a professor in Tübingen[18].

---

[15] Resources for an account of romanticism, a theology of the people, and their influence on Pope Francis include the following: Rafael Luciani, *Pope Francis and the Theology of the People*, translated by P. Berryman. Maryknoll, N.Y.: Orbis Books, 2017; Juan Carlos Scannone, "Pope Francis and the Theology of the People", in *Theological Studies* 77/1-2016, pp. 118-135; Austen Ivereigh, *The Great Reformer: Francis and the Making of a Radical Pope*. London: Allen & Unwin, 2014.

[16] See, Vicki A. Spencer, *Herder's Political Thought: A Study of Language, Culture, and Community*. Toronto: University of Toronto Press, 2012; Sonia Sikka, "Heidegger's Concept of Volk", in *The Philosophical Forum* XXVI/2-1994.

[17] Walter Kasper, *Pope Francis' Revolution of Tenderness and Love*. NJ: Mahwah, 2015, p. 18.

[18] See, Massimo Borghesi, "Polarity Model: The Influences of Gaston Fessard and Romano Guardini on Jorge Mario Bergoglio", in Brian Y. Lee and Thomas L. Knoebel, ed., *Discovering Pope Francis. The Roots of Jorge Mario Bergoglio's Thinking*. Collegeville, MN: Liturgical Press, 2019, pp. 93-113. See also, Massimo Borghesi, *The Mind of Pope Francis: Jorge Mario Bergoglio's Intellectual Journey*, trans. Barry Hudock. Collegeville, MN: Liturgical press, 2018.

1.3. Pastoral Action and an Ontology of Polarity

Romano Guardini is quoted six times in *Laudato Si'*, more than any other thinker. This did not come as a surprise to those who were familiar with the intellectual biography of the Pope. In the 1980s, Bergoglio travelled to Germany and began work on a doctoral dissertation on the thought of Guardini. He produced a draft of the dissertation, entitled, "Polar Opposition as Structure of Daily Thought and of Christian Proclamation"[19], which, for various reasons, he did not publish. The notion of polar opposition in the dissertation is based on the thought of Guardini as expressed in the book, *Gegensatz: Versuche zu einer Philosophie des Lebendig-Konkreten* ("The Principle of Opposition: Toward a Philosophy of the Living-Concrete")[20].

Guardini was a priest of Italian-German heritage who lived out his life in Germany in the first half of the early twentieth century[21]. He was critical of many aspects of modernity, but sought, nevertheless, to help Catholics learn how to speak to this culture and to influence it. He recognized that the current attachment of Catholic thought to neo-scholasticism was reducing its credibility in the minds of educated Europeans. He was a prodigious commentator on current art and literature, always making explicit that he related to such art-work as a Christian and a Catholic. This helped him become a respected public intellectual, at least until the rise to power of the Nazi party, which marginalised him[22]. He was aware of the problem in modern thought of relating subjectivity to objectivity and attempted to resolve this challenge in an early work in philosophy. In this book, he speaks of the "the polarities of personal existence", of which the polarity between subjectivity and objectivity is central. He distinguishes his notion of polarities from contradictory oppositions, such as those between good and evil. He explains that, polarities constitute "a peculiar relation, in which two elements exclude each other and are nevertheless bound together, indeed ... presuppose each other"[23]. He suggested that one cannot resolve these tensions with some univocal insight, but must rather just try to act as "living-concrete persons" who maintain a healthy bal-

[19] Massimo Borghesi, *The Polarity Model,* 105.

[20] Romano Guardini, *Der Gegensatz.*

[21] See Robert A. Krieg, S.S.C., *Romano Guardini: A Precursor of Vatican II.* South Bend, IN: University of Notre Dame Press, 1997.

[22] Robert A. Krieg, *Romano Guardini,* 115-136.

[23] Romano Guardini, *Der Gegensatz,* p. 16, cited in Krieg, *Romano Guardini,* p. 15.

ance between these polarities. Guardini claimed this "theory of opposition" between polarities was his distinct contribution to ontological philosophy and he referred to this principle throughout his life.

Already from an early stage of his priestly life, Bergoglio had articulated "four pastoral principles", in which polarity-thinking is evident. These principles are introduced in *Evangelii Gaudium* and references to them recur regularly in the teaching of the Pope, including in *Laudato Si'*[24]. The first of the pastoral principles of Bergoglio is "time is greater than space". Here he speaks of the importance of patience in trying to address the "processes of people-building" and adds that those who "opt for space", instead of time, try to short-circuit this process and impose solutions on situations that are not the fruit of the insight that comes from dialogue and deliberation. He suggests that those who opt for space usually do so for selfish reasons. This pastoral principle is in direct continuity with what Guardini has to say about the "living concrete".

Bergoglio's second principle is "unity prevails over conflict". Here he discusses the appropriate manner of engaging with conflict, insisting, "conflict cannot be ignored or concealed" when it arises. He suggests that we commit ourselves to "face conflict head on, to resolve it and to make it a link in the chain of a new process". His third principle, "realities are more important than ideas", expresses his distrust of abstract ideologies. His fourth and final principle is "the whole is greater than the parts". Here he speaks of the importance of not over-emphasizing the importance of the particular processes of experience and dialogue in which one finds oneself, but, rather, of recognizing that one is part of a wider and complex process. He sates, "an innate tension also exists between globalization and localization" and proposes that we pay attention to both. He recommends paying attention to the global "so as to avoid narrowness and banality" and to the local "which keeps our feet on the ground"[25].

In addition to the influence of romantic philosophy on the notion of polarity and dialogue in Pope Francis, one can recognize the influence of the *Spiritual Exercises of St. Ignatius of Loyola*. In an interview given shortly after his

[24] Pope Francis, Apostolic Exhortation *Evangelii gaudium* on the Proclamation of the Gospel in Today's World, 24 November 2013, nnn. 224-234; Pope Francis, *Laudato Si'*, n. 178, 141. Scannone suggests that these four principles had already been established as central to the thought of Bergoglio when he was Jesuit provincial during the 1970s. See Juan Carlos Scannone, "Pope Francis", pp. 127-130; also, Austen Ivereigh, *The Great Reformer*, pp. 140-143.

[25] These quotations are taken from Pope Francis, *Evangelii Gaudium,* nn. 224-234.

election[26], Francis explains how closely related is his notion of dialogue to the principle of discernment of spirits that characterizes a Jesuit way of life. He also points out how this practice involves embracing various polarities. He describes how St. Ignatius of Loyola states, "*non coerceri a maximo, sed contineri a minimo divinum est (*not to be limited by the greatest and yet to be contained in the tiniest—this is the divine)". He explains that discernment of spirits requires us "to hear the things of God from God's 'point of view'", and adds, "according to St. Ignatius, great principles must be embodied in the circumstances of place, time and people". He adds that this effort produces a constant tension as the Jesuit seeks to understand the situation and decide on what action needs to be taken within it. He insists, "the Jesuit must be a person whose thought is incomplete, in the sense of open-ended thinking". This open-endedness involves pursuing a dialogical path recognizing a polarity of goals one is trying to achieve, and being confident that the insight into what action to take will arrive as one seeks to engage pastorally in concrete situations.

Biographers have stressed that this broadly romantic interpretation of the Ignatian vision has guided many of the decisions made by Jorge Bergoglio throughout his life. They point out how, as a Jesuit Provincial, Bergoglio challenged Marxist tendencies in his fellow Jesuits, seeing in these a tendency toward abstract and universalising thought that ends up imposing inappropriate solutions on situations. Next, they note that when he became a bishop he engaged in provocative disagreement with the neo-liberal political philosophy of President Menem and other leaders of Argentinian governments. Here they note a paradox. While neo-liberalism might seem to be the polar opposite of Marxism, from a romantic perspective, both philosophies are expressions of Enlightenment rationalism; both are unwilling to see how different situations might, at the very least, require adaptations of the general principles they seek to apply[27].

Some biographers suggest that, as Bergoglio rose to the position of Cardinal, he became increasingly aware of another intellectual opponent. These point out that, as Cardinal, Bergoglio had increasing engagement with the affairs of the Church of Latin America and witnessed the way the Vatican was interacting with it. They suggest that he became concerned about the kind of

---

[26] "Interview with Pope Francis", by Fr Antonio Spadaro SJ, 2013 September 2013, http://www.vatican.va/content/francesco/en/speeches/2013/september/documents/papa-francesco_20130921_intervista-spadaro.html.

[27] Juan Carlos Scannone, "Pope Francis and the Theology of the People", pp. 121-123.

theological thinking that was governing decision-making by the Vatican[28]. He found it to be clericalist, moralizing, centralizing, and lacking in respect for the prerogatives of local churches. They suggest that, for obvious reasons, Cardinal Bergoglio did not express his opinion in public on this matter until he became pope. However, this policy changed when he came to write *Evangelii Gaudium*, an apostolic exhortation that some regard as a manifesto for his papacy[29]. Some see a romantic denunciation of rationalistic ideological tendencies in the critique of "spiritual worldliness" offered in this letter:

> This insidious worldliness is evident in a number of attitudes which appear opposed, yet all have the same pretence of "taking over the space of the Church". In some people we see an ostentatious preoccupation for the liturgy, for doctrine and for the Church's prestige, but without any concern that the Gospel have a real impact on God's faithful people and the concrete needs of the present time[30].

Finally, commentators recognize the characteristic stress on polarity and dialogue in the calls made by Pope Francis for a "synodal Church", a theme that came to characterize his later pontificate. He explains that synodality involves sincere listening and dialogue, especially with lay members of church, so as to help guide the mission of the church to "the world in which we live, and which we are called to love and serve, even with its contradictions"[31].

1.4 Weaknesses of Romantic Philosophy

Given the influence of romantic thought on Pope Francis, it is worth exploring criticisms of such an approach, which are not lacking. Some critics focus on the difficulty romanticism has in engaging with the strengths of rationalism: an ability to engage with modern science, and a concern for the validity of truth-claims. One scholar claims that romanticism, at least in its earliest expression, so disrupted traditional notions of rationality and ethics, that it led to "something like the melting away of the very notion of objective truth"[32]. Another identifies in romanticism, "formulations of the primacy of

---

[28] Austen Ivereigh, *The Great Reformer*, pp. 241-252, 294-301.

[29] Austen Ivereigh, *The Great Reformer*, pp. 210-216.

[30] Pope Francis, *Evangelii Gaudium*, n. 95.

[31] Pope Francis, "Address during Ceremony Commemorating the 50th Anniversary of the Institution of the Synod of Bishops", 17 October 2015, https://www.vatican.va/content/francesco/en/speeches/2015/october/documents/papa-francesco_20151017_50-anniversario-sinodo.html.

[32] Isaiah Berlin, *The Crooked Timber*, p. 57.

aesthetics that may seem, initially, to erase any distinction between what is 'real' and what is 'poetic'"[33]. Many note that romanticism is an amorphous phenomenon that is difficult to analyse. The earliest expressions of romanticism began to decline around 1815, with the defeat of Napoleon. However, the spirit of romantic protest against a modernization driven by rationalism constantly recurs. Some see it in the emergence of political populism in the twenty-first century. Others find in romanticism a precursor of "the scepticism the romantics raised about first principles and about systematicity"[34].

Within Catholic theology, many have studied the role that romantic philosophy played in *la nouvelle théologie* that would so influence Vatican II. Some who are supportive of the general direction of the Council nevertheless recognize that romantic philosophy leaves a legacy of unclear thinking. One describes how many German theologians of this school were "conceptually fuzzy" and employed theological terminology that "eschews clear distinctions and precise definitions"[35]. This commentator points out that this resulted in some such thinkers becoming drawn, uncritically, into the attractive symbolic world of Nazism[36].

Some commentators extend this criticism of romantic thought in general to the notion of polarity proposed by theologians of the Tübingen School. One commentator acknowledges that Guardini considered a theory of opposition to be, "crucial for his writings"[37], but suggests that Guardini does not employ this principle in a consistent and coherent manner. He points out that Guardini is not a systematic thinker and often employs the theory of opposition loosely. Next, he suggests that, regarding the key polarity between the subjective and the objective, Guardini does not follow his own advice of seeking a wise *via media* between the poles. Rather, claims this commentator, Guardini favours the subjective. He explains that, as a result, "Guardini was not spontaneously inclined to see the good in modernization"[38]. He offers reasons for this including the fact that Guardini held a nostalgic "intellectual attraction to the middle

---

[33] Keren Gorodeisky, "19th Century Romantic Aesthetics", in *The Stanford Encyclopedia of Philosophy*, Fall 2016, edited by E. N. Zalta, see https://plato.stanford.edu/entries/aesthetics-19th-romantic/, accessed March 2021.

[34] Keren Gorodeisky, "19th Century Romantic Aesthetics".

[35] Robert A. Krieg, C.S.C. *Karl Adam: Catholicism in German Culture*. South Bend, IN: University of Notre Dame Press, 1992, p. 163.

[36] Robert A. Krieg, *Karl Adam: Catholicism in German Culture*, p. 171.

[37] Robert A. Krieg, *Romano Guardini,* p. 16.

[38] Robert A. Krieg, *Romano Guardini,* p. 167.

ages" and a personality characterised by "melancholy"[39]. Finally, he notes that Guardini, towards the end of his life, was unable to respond positively to some of the innovations of the Second Vatican Council[40].

Other commentators note that a notion of polarity has roots in romantic-idealist Catholic thought that pre-dates Guardini[41]. One commentator traces the notion of polarity to the thought of a founding figure of the School of Tübingen, Johan Sebastian Von Drey, who, in turn, based this notion of polarity on the thought of the romantic-idealist philosopher Schelling[42]. This commentator makes a criticism of Drey that is similar to the one that Krieg makes of Guardini. He notes that Drey wished to employ a notion of polar opposition as a central philosophical notion for a fundamental theology but adds:

> It must finally be asked, however, whether Drey succeeded in this purpose. Did he, in fact, come down on one or the other side of the opposition? Or did he simply oscillate between the two, without being able to integrate them?[43]

Such criticisms of Guardini and Drey raise the following questions: Might the influence of romantic philosophy weaken the ability of *Laudato Si'* to present a truly integral vision, one that includes clear and precise ideas regarding what integral ecology might mean? To be more specific, might the philosophical tools employed by Pope Francis be insufficient to engage with the task for which he calls: the careful mediation of practical and theoretical insights, including a dialogue between science, ethics, and religion? Some commentators have suggested that such weaknesses exist. One Catholic theologian asks of *Laudato Si'*, "whether the Pope has gone too far in this chapter in favour of a particular reading of modernity?" Noting the manner in which Pope relies on the vision of Romano Guardini, this commentator states:

---

[39] Robert A. Krieg, *Romano Guardini,* p. 165.

[40] Robert A. Krieg, *Romano Guardini,* p. 20. Regarding Guardini's difficulty in relating the subjective to the objective, Krieg points out that Guardini had difficulty relating speculative theology (a subjective activity) to church doctrines (an legacy of objective truth-statements). He states: "Guardini's theological method manifests an important omission: it did not lead him to develop his ideas explicitly in relation to church doctrines" (Ibid.).

[41] Cardinal Walter Kasper wrote his doctoral dissertation on the early Tubingen School and was an important figure in encouraging others to study the influence of this school on Vatican II. See Kristin M. Colberg and Robert A. Krieg, *The Theology of Walter Kasper: Speaking Truth in Love.* Collegeville, MN: Liturgical Press, 2014.

[42] Wayne L. Fehr, *The Birth of the Catholic Tübingen School: the Dogmatics of Johann Sabastian Drey, AAR Academy Series*[37]. Chicago CA: Scholars Press, 1981.

[43] Wayne L. Fehr, *The Birth of the Catholic Tübingen School*, p. 259.

> Some will say that the Pope is asking a little too much of the contemporary reader if, in order to accept the central message of the encyclical, he or she must also accept a whole analysis of modernity made mainly in Germany 60 years ago[44].

Another such voice is the William Nordhaus, a Nobel prize winner in economics, and one of the founders of "climate economics". Nordhaus wrote a review of *Laudato Si'*[45] that was largely favourable but which suggests that Pope Francis is over-fond of making criticisms of the capitalist system without entering into much detail regarding just how to relate ethics to the technicalities of economics. He states that Pope Francis exhibits a reluctance to recognize the role that, "markets, particularly market-based environmental policies such as carbon pricing, must play if countries are to make substantial progress in slowing global warming"[46]. He adds: "One senses the struggle of an ancient institution, immersed in its doctrine and history, slowly and incompletely adapting to modern science"[47]. Nordhaus does not discuss the philosophical foundations of the encyclical, but his critique is consistent with the criticism that rationalists make of romantics.

1.5 Metaphysical Realism in *Laudato Si'*

If questions arise about the influence of romanticism on Pope Francis, a number of commentators recognize that Pope Francis also draws on another philosophical tradition, one that can broadly be described as metaphysical realism. Massimo Borghesi notes that three of the pastoral principles outlined by Pope Francis have direct links to Guardini. However, he asserts that, "the second polar pair, idea-reality, does not have a parallel in the Guardinian set of categories"[48]. He traces how Bergoglio had begun to explore the question of realism in philosophy in his years prior to being Pope. While still in Argentina, Bergoglio had explored the notion that, "what unites is reality illuminat-

---

[44] René Micallef, "Laudato Sì e la sua metodologia: un critico ascolto delle scienze empiriche?", in Humberto Miguel Yáñez, ed., Laudato Si': Linee de lettura interdisciplinari per la cura della casa comune. Roma: G&B Press, 2017, p.41 (translation into English provided by current author).

[45] William D. Nordhaus, "The Pope and the Markets", in *The New York Review*, October 8, 2021. Access to this review is available at: https://www.nybooks.com/articles/2015/10/08/pope-and-market/, accessed March 2022.

[46] William D. Nordhaus, "The Pope and the Markets".

[47] William D. Nordhaus, "The Pope and the Markets".

[48] Massimo Borghesi, "Polarity Model", p. 108.

ed by reason, idea, and the intuitive perception [underlying them]"[49]. Borghesi notes how this desire for philosophical realism in Bergoglio is based on a belief that we know reality by an "intuitive perception", and that, subsequent to this intuition, we think through the intelligibility of what it is whose reality we have encountered. Bergoglio states:

> Reality is. Idea is elaborated, induced. [Idea] is instrumental in the understanding, perceiving, and engaging with reality. Between the two there must be dialogue—a dialogue between reality and the explanation of it that I produce. This represents another bipolar tension[50].

Borghesi describes such realism as a "meeting point with Thomism"[51]. However, Parts 2 and 3 of this chapter will raise questions about such a statement, suggesting that the realism involved here is not Thomist but rather Scotist. However, there is no doubting the truth of Borghesi's claim that, with such statements, Bergoglio is attempting, "the broadening of the Guardinian framework"[52].

Commentators also trace other influences on Bergoglio's thought that are more clearly Thomist. One such commentator is Juan Carlos Scannone, a philosopher and former teacher of the young Bergoglio. Scannone stresses that the Theology of the People was propounded by a school of thinkers (including himself) and not only by Lucio Gera. He notes that one theologian who exercised an important influence on Bergoglio was Raphael Tello who was expert in the thought of Thomas[53]. He describes how Tello sharpened Bergoglio's awareness that a problem with rationalist ideologies such as liberalism or Marxism is that they have a tendency toward deductive, abstract, thinking. He traces the origins of such an approach in the thought of Plato. By contrast, Tello stresses how Aquinas relies on an Aristotelian notion of insight into

---

[49] Massimo Borghesi, "Polarity Model", p. 108, citing Bergoglio, *Nosotros como ciudadanos, nosotros como pueblo. Hacia un bicentenario en justiciar y solidaridat 2010-2016 (Buenos Aieres: Claretiana, 2011).*

[50] Pope Francis, *Noi Come cittadini noi come popolo. Verso un bicentenario in giustizia e solidarietà*. Milano: Jaca Book, 2013, 565, cited in Massimo Borghesi, "Polarity Model", p. 108.

[51] Massimo Borghesi, "The Polarity Model: The Influences of Gaston Fessard and Romano Guardini on Jorge Mario Bergoglio", in Brian Y. Lee and Thomas L. Knoebel, ed., *Discovering Pope Francis: The Roots of Jorge Mario Bergoglio's Thinking*. Collegeville, MN: Liturgical Press, 2019, pp. 93-113.

[52] Massimo Borghesi, "The Polarity Model …", p. 109.

[53] See, Juan Carlos Scannone, "Pope Francis and the Theology of the People", pp. 119-120. See also, Allan Figueroa Deck, SJ, Chapter 4: "Raphael Tello and the Faithful People of God", in *Francis, Bishop of Rome: The Gospel for the Third Millennium*. Mahwah, NJ: Paulist Press, 2016.

phantasm that is inductive and concerned with concrete instances of being. Drawing on Aquinas, and criticizing all forms of nominalism and rationalism, Tello avers that the object of Christian care is "not an abstract person, considered as some idea or conception" but rather "the entire human person, in all his or her dimensions—eternal, temporal, spiritual and corporal, individual and communitarian"[54].

On some questions, Tello achieves a blending of a Thomistic approach with themes that are also important for romantic philosophy. This occurs when he discusses the theme of a preferential option for the poor. On this issue, he first differentiates his position from a Marxist one, which he claims exhibits a Platonic tendency to prefer the abstract over the concrete. In contrast to general theorizing about "the working class", he insists on the importance of listening to the poor and finding in their thinking and general attitude a guide for theological reflection. He asserts that a reason to do this is that that even the uneducated poor have a practical wisdom that is close to that which Aquinas praises[55].

There are various points in *Laudato Si'* where one witnesses a rigor of thought and a comfort with metaphysical assertions that can be considered more Thomist than romantic. Three examples demonstrate this. A first is straightforward. Chapter One of *Laudato Si'* exhibits an engagement with natural science that would be foreign to a document that was guided entirely by romantic thought. Second, the critique of the technocratic paradigm offered in Chapter Three also has Thomistic tendencies. While this is also fully consistent with a romantic critique of rationalist reductionism, the Pope employs a degree of subtlety in rejecting biocentric alternatives to the technocratic paradigm that goes beyond romantic generalizations:

> There can be no ecology without an adequate anthropology. When the human person is considered as simply one being among others, the product of chance or physical determinism, then "our overall sense of responsibility wanes". A misguided anthropocentrism need not necessarily yield to "biocentrism", for that would entail adding yet another imbalance, failing to solve present problems and adding new ones. Human beings cannot be expected to feel responsibility for the world unless, at the same time, their unique capacities of knowledge, will, freedom and responsibility are recognized and valued[56].

---

[54] Allan Figueroa Deck, SJ, *Francis Bishop of Rome*, p. 49.
[55] See Allan Figueroa Deck, SJ, *Francis Bishop of Rome*, p. 49 and 54.
[56] Pope Francis, *Laudato Si'*, n. 118.

This position has been described as representing a "moderately anthropocentric"[57] argument and one can recognize in it a notion of a hierarchy of being in which humans find themselves at the apex, which is typical of Thomistic metaphysics. In fact, Pope Francis deepens this perspective by stretching the bounds of what is intended to be a primarily philosophical argument and introducing a "sacramental" dimension into his vision of the natural universe:

> Developing the created world in a prudent way is the best way of caring for it, as this means that we ourselves become the instrument used by God to bring out the potential which he inscribed in things[58].

A third Thomist—or at least realistic—characteristic in *Laudato Si'* is evident when the Pope makes comments on sexual ethics. Here he recalls that his predecessor, Pope Benedict XVI, liked to employ a notion of "human ecology". While *Laudato Si'* prefers to adopt a wider notion of integral ecology, Pope Francis stresses that this latter notion includes the former. Quoting Pope Benedict XVI, he states that, "the relationship between human life and the moral law, which is inscribed in our nature and is necessary for the creation of a more dignified environment"[59]. He adds:

> The acceptance of our bodies as God's gift is vital for welcoming and accepting the entire world as a gift from the Father and our common home, whereas thinking that we enjoy absolute power over our own bodies turns, often subtly, into thinking that we enjoy absolute power over creation[60].

---

[57] Michael S. Northcott offers an overview of the debates between biocentric and "moderately anthropocentric" arguments within ecological ethics in, *The Environment and Christian Ethics*. Cambridge: Cambridge University Press, 1996, pp. 113-123.

[58] Pope Francis, *Laudato Si'*, nn. 1 and 124. Pope Francis will later conclude the encyclical with a similarly sacramental insight: "We take charge of this home which has been entrusted to us, knowing that all the good which exists here will be taken up into the heavenly feast. In union with all creatures, we journey through this land seeking God" (Pope Francis, *Laudato Si'*, n. 244).

[59] Pope Francis, *Laudato Si'*, n. 155, quoting Pope Benedict XVI, *Address to the German Bundestag*, Berlin, 22 September 2011: AAS 103 (2011), 668.

[60] Pope Francis, *Laudato Si'*, n. 155. Pope Francis goes so far as to implicitly criticise "gender ideology": "Valuing one's own body in its femininity or masculinity is necessary if I am going to be able to recognize myself in an encounter with someone who is different. In this way we can joyfully accept the specific gifts of another man or woman, the work of God the Creator, and find mutual enrichment. It is not a healthy attitude which would seek "to cancel out sexual difference because it no longer knows how to confront it", see Pope Francis, *Laudato Si'*, n. 155.

1.6 The Significance of *Fratelli Tutti*

In 2020, five years after the publication of *Laudato Si'*, Pope Francis published another social encyclical, *Fratelli Tutti*. Much can be said about this wide-ranging document. However, in the context of this current discussion, it is sufficient to note that, in this document, the realistic aspect of the thought of Pope Francis comes to the fore. The reason for this is that the Pope recognizes in current culture a tendency toward fragmentation that is caused by an increasing tendency to employ merely emotive means of making truth claims[61]. He identifies a key cause of this deterioration being the "illusion of communication"[62] provided by new technologies of social communication. He states that, communication that employs these devices often leads to a situation where "exchanges are merely parallel monologues"[63], and concludes, "social aggression has found unparalleled room for expansion through computers and mobile devices"[64]. Related to this phenomenon, he identifies a deterioration in political discourse which is conducted, "appealing to the basest and most selfish inclinations of certain sectors of the population"[65].

Over against increasing social aggression, Pope Francis explores strategies to build social friendship. As always, he looks to education and dialogue as a way forward. This time, however, he stresses the importance of intellectual clarity in dialogue, as well as a presupposition that we are all oriented to the truth and capable of finding it through dialogue. As he explores the relationship between reasoning processes and the ability to affirm truth, he identifies the need for "greater interdisciplinary communication"[66] that is "enriched and illuminated by clear thinking, rational arguments, a variety of perspectives and the contribution of different fields of knowledge and points of view"[67]. He insists that one important perspective is the ethical and that this cannot "exclude the conviction that it is possible to arrive at certain fundamental

---

[61] Pope Francis, Encyclical Letter *Fratelli tutti on Fraternity and Social Friendship*, 3 October 2020, n. 11.

[62] Pope Francis, *Fratelli Tutti*, nn. 42-50.

[63] Pope Francis, *Fratelli Tutti*, n. 200.

[64] Pope Francis, *Fratelli Tutti*, n. 44.

[65] Pope Francis, *Fratelli Tutti*, n. 160. This encyclical was written before the invasion of the Ukraine by Russia, but this tragic event has given a new significance to the Pope's denunciation of social aggression. Indeed, the use of social media as a remote instrument of war by Russia illustrates the ambiguous value of this technology.

[66] Pope Francis, *Fratelli Tutti*, n. 204.

[67] Pope Francis, *Fratelli Tutti*, n. 211.

truths always to be upheld"[68]. The title of Chapter 7 is "Paths of Renewed Encounter", and the first section of this is entitled, "Starting Anew From the Truth"[69]. Here he explores how entering into dialogue does not have to imply a descent into relativism. In that section, and elsewhere, he speaks of how religion can help all people that they share a notion that our authentic selves are oriented toward certain moral truths. Concerning such truths, he makes a metaphysical claim, "we realise that they rise above consensus; they transcend our concrete situations and remain non-negotiable"[70]. Expanding on this point, he states:

> If something always serves the good functioning of society, is it not because, lying beyond it, there is an enduring truth accessible to the intellect? Inherent in the nature of human beings and society there exist certain basic structures to support our development and survival. Certain requirements thus ensue, and these can be discovered through dialogue, even though, strictly speaking, they are not created by consensus. The fact that certain rules are indispensable for the very life of society is a sign that they are good in and of themselves. There is no need, then, to oppose the interests of society, consensus and the reality of objective truth[71].

With statements such as these, Pope Francis seems to be moving beyond a notion of realism as the result of intuitive perception. In this approach, reason does not arrive as a second step after a reality has already been intuited. Rather, an orientation to truth, as an ultimate destination, seems to be implicit in the reasoning process itself. As will be suggested in Part 3, such an approach has firm roots in the thought of Thomas Aquinas, and is close also to the critical realism of Bernard Lonergan and Robert Doran.

1.7 Evaluating the Philosophical Foundations of Pope Francis

Having studied the echoes of both romanticism and philosophical realism in Pope Francis, it becomes clear that it is not fair to attribute to Pope Francis any simplistic version of romantic philosophy. Rather, he is also a thinker who employs a traditional, Catholic, orientation toward truth statements that are based on a realist metaphysics. Walter Kasper states that the true novelty

---

[68] Pope Francis, *Fratelli Tutti*, n. 211.
[69] Pope Francis, *Fratelli Tutti*, n. 226-227.
[70] Pope Francis, *Fratelli Tutti*, n. 211.
[71] Pope Francis, *Fratelli Tutti*, n. 212.

of Pope Francis lies, by comparison with previous popes, lies at the level of "differences not in truths of faith, but certainly in style, in methodological starting point, and in emphasis"[72].

Nevertheless, the above analysis does not completely resolve questions that arise regarding the philosophical foundations of the Argentinian Pope. The question now becomes: granting that Pope Francis wants to propose a vision that draws both on romantic and realist resources, does he do so in a manner that is philosophically coherent? The above discussion has adverted to a fact that will be explained at greater length below, that the realistic philosophical resources employed by the Pope seem to be both Scotist and Thomist— systems that are not compatible with each other. Consequently, one notices a certain eclecticism in the philosophical approaches invoked by Francis. Part 2 proposes that the thought of Robert Doran provides a resource to support the basic project of Pope Francis in a way brings coherence and explanatory depth.

2. Critical Realist Foundations for Integral Ecology

Robert Doran (1945-2021) was a Jesuit from the USA who was deeply influenced by the thought of his mentor and fellow Jesuit from Canada, Bernard Lonergan (1904-1984). Both were motivated by a pastoral concern for the poor[73]. While neither tended to address the ecological crisis, their thought can be easily extended to do this. Lonergan was born near Montreal and was deeply affected by witnessing the poverty in this city that resulted from the Great Depression in the 1930s. He also lived in Italy during the later years of that decade and witnessed the rise of fascism and the outbreak of the Second World War. He conceived a life-project of developing a philosophy of history to help inform a theology that could help avoid such evils in the future. He conceived of this project as a response to the insight of Karl Marx who had developed a philosophy of history—and then a theory of economics—as a key step in his plan to redirect the course of history[74]. Doran grew up in Milwaukee and is a product of the Jesuit thinking about its mission in the years after Vatican II, a mission that speaks of "the service of faith and the promotion of justice". His closeness to Lonergan was indicated by the fact that the older

[72] Walter Kasper, *Pope Francis' Revolution of Tenderness and Love,* p. 10.

[73] Gerard Whelan, *Redeeming History: Social Concern in Bernard Lonergan and Robert Doran.* Rome: G&B Press, 2013.

[74] Gerard Whelan, *Redeeming History*, pp. 26-37.

man invited him to be co-editor of the 25-volume *Collected Works of Bernard Lonergan*, a project to which Doran devoted much of his life. However, Doran's own work was innovative. His major work, *Theology and the Dialectics of History*[75], complements Lonergan's work by extending it to engage both with the thought of psychoanalysts such as Carl Jung, and with various liberation theologians of Latin America. Arguably, the ways in which Doran expands Lonergan's thought makes his thought especially applicable to the notion of integral ecology.

2.1 Lonergan's Epistemology and Metaphysics

Lonergan is known for two masterworks: *Insight: A Theory of Human Understanding* (1957), and *Method in Theology* (1972)[76]. The first half of Lonergan's 800-page *Insight*, is devoted to cognitional theory and epistemology. Explaining the invitation to "intellectual conversion", he states:

> Thoroughly understand what it is to understand, and not only will you understand the broad lines of all there is to be understood but also you will possess a fixed base, and invariant pattern, opening upon all further developments of understanding[77].

The second half of *Insight* is devoted to metaphysics, i.e. "all further developments of understanding", mentioned in the above quotation. Here, Lonergan states that intellectual conversion reveals that, "being is the object of the pure desire to know"[78] and that knowing succeeds in attaining objectivity. He then explores the consequences of the correspondence theory of knowing and being that he has invited his readers to establish for themselves. He suggests that there is an isomorphism between the structure of our knowing and the structure of being. He states, "we also must recognize in heuristic structure itself a clue

---

[75] Robert M. Doran, Theology and the Dialectics of History. Toronto: University of Toronto Press, 1990. For practical reasons, the discussion of the thought of Doran, below, limits itself to this 700-page work. I should be noted that Doran proceeded, in his later years, to work intensively in the field of systematic theology.

[76] Bernard Lonergan, *Insight: A Study of Human Understanding*, in Frederick E. Crowe and Robert M. Doran, ed., *Collected Works of Bernard Lonergan, Volume 3*. Toronto: University of Toronto Press, fifth edition, revised and augmented 1992 (1957); Bernard Lonergan, *Method in Theology*, in Robert M. Doran and John D. Dadosky, ed., *Collected Works of Bernard Lonergan, Volume 14*. Toronto: University of Toronto Press, 2017 (1972).

[77] Bernard Lonergan, *Insight*, p. 22.

[78] Bernard Lonergan, *Insight*, p. 372.

to the nature of the universe proportionate to our capacities to know"[79] One recognizes an echo of Aquinas when he asserts that an elementary example of this isomorphism is to be found between the cognitional levels of experience, understanding, and judgement, and the metaphysical elements of matter, form, and existence. However, Lonergan adds that because knowing is a dynamic and cumulative process, involving a series of "higher viewpoints", and there must be an analogous upward tendency, or "finality", in being.

A central characteristic of the metaphysics of Lonergan is to explain that the "actual order of the universe"[80] is characterised by "emergent probability"[81]. He outlines how modern science employs both classical methods and statistical methods to gain insight into reality. Following the principle of isomorphism, he explains that there must be a corresponding quality of randomness and directedness in being itself. He offers an evolutionary vision of a universe "conditioned series of schemes of occurrence"[82], were the emergence of schemes of a higher level of being are conditioned by the prior emergence schemes at a lower level. He explains that the resulting universe of proportionate being (i.e. being that is proportionate to our ability to know it), "results from the combination of the conditioned series of schemes with their respective probabilities of emergence and survival"[83]. He states that the notion of emergent probability has "remarkable potentialities of explanation" and, as examples, speaks of its applicability to disciplines such as evolutionary biology[84] and then begins to explain how it can also be applied to the study of, and evaluation of, human affairs.

The final chapters of *Insight* are devoted to discussing the human moment in the emergent process. Lonergan identifies how humans are unique in being able to exercise freedom and to create new instances of being, such as culture and social structures. He explains that, in an ideal world, humans would always employ their freedom authentically and would create a line of progress in history. Here technological discovery would always lead to economic and political consequences best suited to serve the common good. Similarly, culture would pro-

[79] Bernard Lonergan, *Insight*, p. 475.

[80] Bernard Lonergan, *Insight*, p. 719.

[81] Lonergan introduces his notion of emergent probability in Chapters 2 and 4 of *Insight*, pp. 57-92 and 126-162.

[82] Bernard Lonergan, *Insight*, pp. 144-145.

[83] Bernard Lonergan, *Insight*, p. 145.

[84] For a discussion of Charles Darwin, see Bernard Lonergan, *Insight*, pp. 154-162.

vide the ideas and values to guide this process, educating members of society to behave authentically. Both before and after the authoring of *Insight*, Lonergan, devoted considerable periods of time working on a theory of macroeconomics that illustrated how the dynamics of emergent probability are present in an economic system and how respecting this can help avoid of economic depression[85].

Next, Lonergan notes that, because of the proclivity of humans to abuse their freedom, there exists also a principle of decline in history. He employs the metaphor of vector analysis from algebra to speak of two "vectors" in history: progress and decline[86]. He speaks of various forms of bias that lead to decline. He speaks of a relatively innocent bias, one of misunderstanding the nature of human history, that nevertheless produced inadequate theories of economics which, in turn, led to poor political decision-making and the Great Depression of the 1930s. More sinister, according to him is the "general bias" which glorifies common sense knowing and mocks the work of intellectuals who seek a theoretical understanding of the social situation. He then describes how "the repeated exclusion of timely and fruitful ideas involves a cumulative departure from coherence"[87]. He then speaks of how general bias appeals to a notion of "reality" to imply that a situation of decline must be accepted as inevitable. Writing shortly after the Second World War, he states:

> Reality is the economic development, the military equipment, and the political dominance of the all-inclusive state. Its ends justify all means. Its means include not merely every technique of indoctrination and propaganda, every tactic of economic and diplomatic pressure, every device for breaking down the moral conscience and exploiting the secret affects of civilized man, but also the terrorism of a political police, of prisons and torture, of concentrations camps, of transported or extirpated minorities, and of total war[88].

Lonergan next speculates on how an emergently probable universe might be capable of receiving a "divine solution for the problem of evil"[89]. He describes how a moral impotence in human living gives decline an advantage over progress. Anticipating the theological work that he will produce subsequent to *In-*

---

[85] See, Bernard Lonergan, *Macroeconomic Dynamics: an Essay in Circulation Analysis,* in Frederick E. Crowe and Robert M. Doran, ed., *Collected Works of Bernard Lonergan, Volume 15*. Toronto: University of Toronto Press, 2005.

[86] Bernard Lonergan, *Insight*, pp. 232-269.

[87] Bernard Lonergan, *Insight*, p. 145.

[88] Bernard Lonergan, *Insight*, p. 257.

[89] Bernard Lonergan, *Insight*, pp. 715-718.

sight, he now speculates philosophically about the nature of the intervention that such a God might exercise so as to address the problem of evil in history. He states that such an intervention would constitute a third vector, or heuristic category, by which one can study human affairs. He calls this heuristic category, redemption, and explains how it must have the capacity to reverse decline and promote progress. He explores some the other necessary characteristics of the vector of redemption, claiming that one would be a respect for the laws of an emergently probable universe that God had already created, including the freedom to choose what is unintelligible, irrational, and evil[90].

Lonergan describes how the redemptive vector in history will necessarily be comprised of communities who are dedicated to a response to decline that seeks to overcome bias, above all the general bias that rejects intelligent, rational, and responsible reflection on society. He borrows the term "cosmopolis" from Kant to speak of the service such a community can render to a culture:

> What is both unnecessary and disastrous is the exaltation of the practical, the supremacy of the state, the cult of the class. What is necessary is a cosmopolis that is neither class nor state that stands above all their claims, that cuts them down to size, that is founded on the native detachment and disinterestedness of every intelligence, that commands man's first allegiance, that implements itself primarily through that allegiance, that is too universal to be bribed, too impalpable to be forced, too effective to be ignored[91].

Two concluding points can now be made regarding the epistemology and metaphysics that underpins Lonergan's notion of cosmopolis. First, employing the Kantian term, cosmopolis, indicates why Lonergan accepts the term "critical realism" as describing his philosophy. He had studied the thought of Thomas Aquinas at length and valued the cognitional theory and realistic metaphysics found there[92]. However, he also believed that the thought of this medieval genius needed to be transposed into a modern key. This transposition would hinge on an invitation to intellectual conversion which, in turn, is based on the turn to the subject, a "critical" step initiated by Descartes and Kant[93]. Secondly, for all

---

[90] Bernard Lonergan, *Insight*, p. 719.

[91] Bernard Lonergan, Insight, p. 263.

[92] Bernard Lonergan, *Verbum: Word and Idea in Aquinas* in Frederick E. Crowe and Robert M. Doran, ed., *Collected Works of Bernard Lonergan, Vol. 2*. Toronto: University of Toronto Press, 2005.

[93] Lonergan makes the remarkable statement about the cognitional theory of Aquinas, "as an ontology of knowing it is satisfactory; as an epistemology it is null and void"; see Bernard Lonergan, *Verbum: Word and Idea in Aquinas*, 86.

his respect for these philosophers, Lonergan criticises both Descartes and Kant for submitting to a cognitional theory that was biased. In fact, he locates the origin of this bias in the philosophy of the medieval thinker, Duns Scotus. This theory holds that knowing is analogous to "taking a good look". According to this approach, one knows a thing by an immediate act of "intuition" of its reality. Only subsequently, does one achieve acts of insight into its intelligible form. Lonergan insists that such "naïve realism" is misguided[94] and will consistently produce distorted cultural and social ideas. He insists the structure of cognition is how Aquinas describes it: we first gain direct insight into the intelligible form of a thing, and only subsequently, affirm whether or not our insight is true. For Lonergan, a large part of the task of cosmopolis involves distinguishing between authentic and inauthentic processes of knowing that lead to ideas and values that become the constitutive meaning of culture.

2.2 Lonergan, Conversion and Theological Method

Lonergan's thought continues to develop after *Insight* and, in spite of the title, *Method in Theology* includes significant philosophical developments. After completing *Insight*, Lonergan read widely in post-Kantian authors, many of which were influenced by a heritage of romantic philosophy. He described the authors he was now discovering as "German historicists", who were writing in the fields of phenomenology, hermeneutical philosophy, and existentialism. He learnt from these philosophers in a variety of ways. He recognized that they brought a "historical consciousness" to philosophy, a theme that is related to emergent probability and which had always been close to his heart. He also appreciated how they addressed the role of affectivity in decision-making, something he had neglected. When Lonergan writes *Method in Theology*, he is able to expand on the epistemology of *Insight* and to outline more clearly than before how Christian theology can be an instrument of cosmopolis in history.

At the level of epistemology, the key development in *Method in Theology* concerns ethics. Lonergan explains how subjectivity includes a fourth level, concerned with decision-making, that builds on the previous three, characterized by experience, understanding, and judgment. Now Lonergan describes how we move beyond a third level of consciousness when we no longer mere-

---

[94] Bernard Lonergan, *Insight*, pp. 365-366, 391-392, 396-398.

ly judge the facts of situation but begin to evaluate them and discern how we should respond. He explains how a first step in process of the fourth level of consciousness involves an affective response to the facts we have affirmed as true. He avers that we discern values in feelings, explaining:

> Feelings that are intentional responses regard two main classes of objects: on the one hand, the agreeable or disagreeable, the satisfying or dissatisfying; on the other hand, values, whether the ontic value of persons or the qualitive value of beauty, understanding, truth, virtuous acts, noble deeds[95].

He suggests that our intentional responses reveal a "scale of values", explaining, "we may distinguish, vital, social, cultural, personal and religious values in an ascending order"[96]. He describes how responding to values in this way is partly natural to us, but also needs to be cultivated in the young by an educational process. He next explains that we advance from discerning value to making a judgement of the particular value that is at issue in the situation we are evaluating. This involves bringing the process of discerning value to a close with a judgement of value that has an objective quality, one that is related to the ability for objective judgment at the third level of consciousness. He next describes how a judgment of value proceeds to a decision to act.

As with his analysis in *Insight*, Lonergan adds to this expanded account of the authentic person a description of how the individual can fall into inauthenticity. He describes how bias manifests itself at the fourth level of consciousness. He suggests that it can begin with a failure to discern feelings appropriately. This will often involve affirming vital values, or perhaps vital and social values, at the expense of cultural, personal, and religious values. This results in judgements of value that prompt selfish decision-making. Next, he describes the phenomenon of rash decision-making where the individual advances to action believing that he or she is motivated by noble motives, without, in fact, having given sufficient time to discerning what are really the values the values at issue in a situation. In addition, Lonergan notes that, even when we have made an authentic judgment of value, we can still be indecisive and fail to act as we should, perhaps because we foresee negative consequences for us if we behave authentically[97].

Based on this expanded philosophy, Lonergan proceeds to revisit what he had said in *Insight* about the divine solution to the problem of evil. Now he

---

[95] Bernard Lonergan, *Method in Theology*, p. 32.
[96] Bernard Lonergan, *Method in Theology*, p. 32.
[97] Bernard Lonergan, *Method in Theology*, p. 38.

speaks of religious conversion and explains how it is registered, primarily, at the fourth level of consciousness. He begins by exploring how we respond to religious values, which he has identified as occupying a place at the top of the scale of values. He accepts that, in principle, it is possible to discern religious value in a philosophical way as being a logical necessity if personal values are real and true. However, he claims that, in practice, there is unlikely to happen unless the mind is assisted by a supernatural source. Revisiting his discussion of the supernatural at the end of *Insight*, he describes, now in an existentialist way, how a moment comes in a person's life when he or she is confronted with a sense of ineffable mystery. He describes how individuals encounter a "mysterious other" that represents "an apprehension of transcendent value"[98]. Concerning this apprehension he adds:

> That thrust to unrestrictedness may be objectified as a clouded revelation of absolute intelligence intelligibility, absolute truth and reality, absolute goodness and holiness. With that objectification there recurs the question of God in a new form. For now it is primarily a question of decision. Will I love him in return, or will I refuse?[99]

Lonergan then states that if we decide to love God in return there occurs a supernatural event constituted by "'the love of God flooding our hearts through the Holy Spirit given to us' (*Romans*, 5:5)"[100]. He calls this event religious conversion. He speaks of how such religious experience transforms us into a "being in love" where one is filled by the gift of divine love. He explains, "all love is self-surrender, but being in love with God is being in love without limits or qualifications or conditions or reservations"[101]. He speaks of religious conversion involving a paradox. In one sense it is the culmination of a natural tendency toward self-transcendence and, so, is a "basic fulfilment of our conscious intentionality"[102]. On the other hand, it has a supernatural quality that, "dismantles and abolishes the horizon in which our knowing and choosing went on"[103]. He explains how we find ourselves constituted by a new horizon "in which the love of God will transvalue our values and the eyes of

---

[98] Bernard Lonergan, *Method in Theology*, p. 112.
[99] Bernard Lonergan, *Method in Theology*, p. 112.
[100] Bernard Lonergan, *Method in Theology*, p. 101.
[101] Bernard Lonergan, *Method in Theology*, pp. 101-102.
[102] Bernard Lonergan, *Method in Theology*, p. 101.
[103] Bernard Lonergan, *Method in Theology*, p. 102.

that love will transform our knowing"[104]. With such an abundance of love in human consciousness, he speaks of how religious conversion normally leads to a moral conversion, stating, "moral conversion changes the criterion of one's decisions and choices from satisfactions to values"[105].

While *Method in Theology* presents a significantly expanded account of consciousness, it remains within the basic structure of the notion of emergent probability established in *Insight*. Now Lonergan offers a theologically enriched account of the culture of cosmopolis that is needed to reverse decline and promote progress in history. He states:

> There are needed, then, individuals and groups and, in the modern world, organizations that labour to persuade people to intellectual, moral, and religious conversion and that work systematically to undo the mischief brought about by alienation and ideology. Among such bodies should be the Christian Church[106].

The main focus of *Method in Theology*, is to outline the broad lines of a theological method that can help direct the church in performing this function as cosmopolis. Lonergan states, "a theology mediates between a cultural matrix and the significance and role of a religion in that matrix"[107]. He presents a notion of theological method based on two phases of theology, one which recovers a religious tradition and the other which mediates it to culture. His notion of four levels of consciousness helps him outline how there are four functional specialties in each of these phases. He explains how the finality of each functional specialty is analogous to the finality of one or other of the levels of consciousness. He lists the specialties of the first phase as: Research, Interpretation, History, and Dialectic. Those of the second phase he calls: Foundations, Doctrines, Systematics, and Communications[108].

2.3 Doran, Psychic Conversion, and the Dialectic of the Subject

Doran accepts all that Lonergan states about conversion in both *Insight* and *Method in Theology*. However, he suggests that Lonergan's thought is incomplete and sets out to expand it, starting with a notion of the subject and the kind of conversions we need. Ultimately, however, this produces and enriched

---

[104] Bernard Lonergan, *Method in Theology*, p. 102.
[105] Bernard Lonergan, *Method in Theology*, p. 225.
[106] Bernard Lonergan, *Method in Theology*, p. 333.
[107] Bernard Lonergan, *Method in Theology*, p. 3.
[108] See Bernard Lonergan, *Method in Theology*, Chapter 5, "Functional Specialties", pp. 121-138.

account of progress, decline, and redemption, as well as an enriched account of functional specialisation in theological method.

Doran begins his expansion of Lonergan's thought by paying more attention to the data of modern psychology. He does this in a major work, *Theology and the Dialectics of History*[109]. Doran expands what Lonergan has to say about our affective response to value, paying careful attention to the way our affectivity is related to our unconscious mind. Doran explores how emotions are released from the unconscious mind as we experience an affective response to value. He invokes the thought of Carl Jung to assert, "all psychic energy is distributed into complexes"[110]. At the same time, Doran goes beyond Jung by describing how complexes can assist the authentic exercise of conscious intentionality understood has having four levels:

> [Complexes] provide us with the images that we need for insight, or offer us memories that help us to discover ways of responding to new situations, or spontaneously to acquiesce in the progress of reflection that anticipates judgment, or to apprehend genuine values in an affectively charged way that will lead to action consistent with our affective response[111].

Doran also notes how complexes can also include distortions that result in "dramatic bias". Here Doran employs an account of the victimization of the psyche. He states that just as psychic energy can assist the self-transcendence of conscious intentionality, so also it prompts us to be "resistant to insight, true judgment, and responsible action"[112]. He explains that this happens because our psychic energy can find itself, "blocked, fixed in inflexible patterns, driven by compulsions, plagued by obsessions, weighed down by general anxiety or specific fears"[113].

Having begun his study of the role of the psyche in mediating between the unconscious mind and the conscious mind, Doran outlines a second role of

---

[109] Robert M. Doran, Theology and the Dialectics of History. Toronto: University of Toronto Press, 1990. Comments below are based broadly on a list offered in the Introduction of seven major ways in which Doran develops the thought of Lonergan: "The Question of the Situation"; "Psychic conversion", "Dialectic", "The Scale of Values", "The Dialectic of Culture", The Ontology of Meaning", "Praxis". A slightly altered order of presentation of themes is guided by a desire to anticipate links to the thought of Pope Francis.

[110] Robert M. Doran, *Theology and the Dialectics of History,* p. 229.

[111] Robert M. Doran, *Theology and the Dialectics of History,* p. 229.

[112] Robert M. Doran, *Theology and the Dialectics of History,* p. 229.

[113] Robert M. Doran, *Theology and the Dialectics of History,* p. 229.

the psyche. This role remains at the conscious level of mental operations: that of accompanying each level of intentional self-transcendence and setting up a duality within consciousness between "spirit" and "psyche". He states, there are "two distinct sets of data of consciousness, the data of intentionality and the data of the psyche". He then describes how consciousness "is immanently constituted as a taut equilibrium or, perhaps better, creative tension, between the organic and the spiritual", adding, "the sensitive psyche occupies, as it were, a middle ground between the organism and the spirit"[114]. He employs the analogy of the artist to explain how authentic living involves making decisions that find a *via media* between the dictates of spirit and the limiting claims of our bodily selves. He praises:

> The artist's capacities for abiding in a pattern of experience that excludes instrumentalization, that lets experience be for its own sake, that is a release, a letting be, a freedom for the emergence of the inevitability of form[115].

Doran then notes that the basic dramatic pattern of experience he describes as the basic, default, pattern of ordinary living must be able "to move into other patterns as these are required"[116]. Such patterns will often involve one's daily work, and will involve a switch to a more technical or theoretical pattern of experience. Consequently, his account of consciousness can extend to describing how the fundamental method of the human mind becomes differentiated into as a person engages in diverse jobs: the nurse, the plumber, the computer programmer, the historian, the philosopher etc. However, he stresses that the authentic person is capable of switching back and forward between these patterns of experience, always assuring that the dramatic pattern of experience, i.e. the basic life-direction that she or he is choosing for themselves, "governs these movements and controls them"[117].

In order to do justice to explaining the finalistic tension between psyche and spirit. Doran speaks of the "dialectic of the subject" and this requires that he expands on the following definition of dialectic that Lonergan had presented in *Insight*:

> There will be a dialectic if (1) there is an aggregate of events of a determinate character, (2) the events may be traced to either or both of two principles, (3)

[114] Robert M. Doran, *Theology and the Dialectics of History,* p. 55.
[115] Robert M. Doran, *Theology and the Dialectics of History,* p. 55.
[116] Robert M. Doran, *Theology and the Dialectics of History,* p. 54.
[117] Robert M. Doran, *Theology and the Dialectics of History,* p. 54.

> the principles are opposed yet bound together, and (4) they are modified by the changes that successively result from them[118].

Doran notes that Lonergan asserts that this notion of dialectic can be related to examples of healthy human functioning, what Lonergan calls, “an ideal line of pure progress resulting from the harmonious working of the opposed principles”, or unhealthy functioning, what Lonergan calls, “any degree of conflict, aberration, break-down, and disintegration”[119]. Doran notes that Lonergan has, in fact, been describing “two forms of the realization of a single but complex notion of dialectic”[120]. Speaking of the “two forms of opposition” implied by Lonergan, he calls the first “a dialectic of contraries”, and the second “a dialectic of contradictories”. He explains: “Contraries are reconcilable in a higher synthesis, while contradictories exclude one another”[121].

2.4 Perceiving the World Dialectically

Armed with this expanded description of subjectivity, Doran explores how the individual responds to most values a dialectical manner, one that appeals to both spirit and psyche. Ultimately, this will bring him to an expanded understanding of Lonergan’s theory of history, enriching a notion of the three heuristic categories of progress, decline, and redemption. However, in order to arrive at that point, he needs to deepen a reflection on the scale of values to which we respond at the fourth level of consciousness.

He states: “at least three of the levels of value—social, cultural, and personal—are understood to be structured in accord with the refinement on dialectic presented”[122]. At the level of personal values, he explains that the “conscious dialectic of the subject”, which he has already described, represents a dialectical response to personal value. He elaborates on this point by pointing out how dialectical consciousness needs to be assisted as much by symbols as by clear and distinct ideas. He explores how, just as ideas can be true or false, good or evil, so also symbols can express what is balanced or unbalanced, true or untrue, good or evil. Here he is able to quote Lonergan, who states in *Method in Theology* that symbols can stimulate us either to “development

---

118 Bernard Lonergan, *Insight*, p. 242.
119 Bernard Lonergan, *Insight*, p. 268.
120 Robert M. Doran, *Theology and the Dialectics of History,* p. 9.
121 Robert M. Doran, *Theology and the Dialectics of History,* p. 9.
122 Robert M. Doran, *Theology and the Dialectics of History,* p. 10.

or backsliding" and that, given the fallen nature of the human condition, the symbols that influence us usually need "transvaluation and transformation" as the subject grows in maturity and integrity[123].

After personal values, Doran speaks of how we respond dialectically to cultural values. First, he recalls what Lonergan, states that humans respond to cultural values because, "over and above mere living and operating, men have to find a meaning and value in their living and operating" and then adds, "it is the function of culture to discover, express, validate, criticize, correct, develop, improve such meaning and value"[124]. However, he does not find resources in Lonergan for explaining in what manner cultural values are constituted as a dialectic of contraries. On this matter, he turns to Eric Voegelin, who distinguishes between an "anthropological" mode of cultural expression and a "cosmological" mode. Concerning the pursuit of "anthropological truth", Doran suggests that Lonergan's epistemology and ethics is an excellent example, describing it as "reflection that the ultimate measure of human integrity is a reality beyond the cosmos[125]. By contrast, Doran finds in Voegelin's account of cosmological culture something that is lacking in Lonergan. He notes that, for Voegelin, the pursuit of cosmological truth involves recognizing that there should exist in human living "a harmony between human decisions and actions, on the one hand, and the rhythms and processes of nature, on the other hand". He adds that this involves recognizing "a synchronicity between culture and nature"[126].

Doran points out that, ever since the ancient Greeks spoke of the antimony between the philosophers and the poets, there has been a tendency in Western culture to emphasise the anthropological pole of culture at the expense of the cosmological pole and this has led to patterns of imperialism[127]. Consequently, he describes Western culture as often existing in a state of dialectic of contradictories. At the same time, he states that it is possible for a culture to remain in a cosmological framework and to refuse the development of anthropological pole of meaning, thus also falling into a state of dialectic of contradictories:

---

[123] Robert M. Doran, *Theology and the Dialectics of History,* p. 59 (quoting Method in Theology, p. 64).

[124] Bernard Lonergan, Method in Theology, p. 33.

[125] Robert M. Doran, *Theology and the Dialectics of History,* p. 216.

[126] Robert M. Doran, *Theology and the Dialectics of History,* p. 216.

[127] Robert M. Doran, *Theology and the Dialectics of History,* pp. 116-118.

> In the dialectic of culture, an exclusively cosmological set of constitutive meanings informs a culture that is not only dominated by myth and magic but also powerless against the oppressive forces of contemporary imperialistic or totalitarian ambition; and when anthropological constitutive meaning informs the way of life of the community at the expense of the cosmological attunement to the rhythms and process of the body and of nonhuman nature, the very intentionality of the anthropological breakthrough to a world-transcendent measure of human integrity is cancelled by human pride[128].

Regarding social values, Doran is able to look to Lonergan, who already described a the "dialectic of community". In *Insight*, Lonergan states: "Social events can be traced to the two principles of human intersubjectivity and practical common sense"[129]. Within this account, Doran highlights the fact that Lonergan does not only regard intersubjectivity as posing problems to the unfolding of practical intelligence but of having a value in its own right. He points out that, "the bond of mother and child, man and wife, father and son, reaches into a past of ancestors to give meaning and cohesion to the clan or tribe or nation"[130]. He agrees with Lonergan's position that social change, driven by the discoveries of practical intelligence, should be introduced with caution, and at times modified, so as to reduce the disruption caused to intersubjective community. He suggests that, at least in Western cultures, decline is often characterized by an emphasis on practical intelligence at the expense of intersubjective bonds.

On the basis of this analysis of the scale of values, Doran turns to speak of Lonergan's heuristic theory of history. A central theme of his book is that progress is to be understood as maintaining cultural values and social values in a state of dialectics of contraries. By contrast, decline is to be understood in terms of cultural and social values falling into a state of dialectics of contradictories. He states: "history is understood in accord with an analogy of dialectic"[131]. It remains to explain how Doran expands upon Lonergan's account of religious experience and a consequent account of the vector of redemption in history.

---

[128] Robert M. Doran, *Dialectics of History*, p. 146.
[129] Bernard Lonergan, Insight, p. 574.
[130] Bernard Lonergan, Insight, p. 237.
[131] Robert M. Doran, *Dialectics of History*, p. 479.

2.5 Grace and the Analogy of Dialectics

Up to this point, the analysis of Doran is primarily philosophical. He has expanded Lonergan's account of authentic human nature to include a conscious dialectic of the subject and has, correspondingly, spoken of both progress and decline in history as to be understood by analogy with this dialectic. His next challenge is to expand Lonergan's account both of the religious experience of the subject and, by analogy, of redemption in history. His account of the former involves the thought of Karl Jung on victimisation. His account of the latter involves a dialogue with the liberation theology of Latin America.

On the question of religious experience, Doran recalls what he has said about how the psyche can be "victimised" in various ways, but especially by being deprived of love in childhood. He recalls his description of the dramatic bias that occurs in individuals whose psyches include victimized complexes and who therefore experience a distorted ability to operate in each of the four levels of consciousness. Now he explores how God's grace, experienced in religious conversion, can heal this this problem. He begins with a discussion of how, even without grace, individuals can begin to recognize that they have problems with being authentic. He states that individuals who tend to live in a state of dialectics of contradictories can recognize that "something is wrong" and "can recognize that the complex is a victim"[132]. Next, he states:

> One can adopt an attitude of compassion in its regard. And finally, one can allow there to emerge from the recognition and the compassion a willingness to cooperate with whatever redemptive forces are available to heal the disorder of darkness and to transform the contorted energies[133].

Doran acknowledges that most psychotherapeutic techniques attempt to help individuals in this way, seeking to help individuals arrive at an act of insight into hurts from their past, resulting in an experience of catharsis and healing. However, he suggests that questions of moral impotence afflict efforts at finding healing by psychotherapeutic means[134]. He suggest that a vi-

---

[132] Robert M. Doran, *Dialectics of History*, p. 239.

[133] Robert M. Doran, *Dialectics of History*, p. 239.

[134] At this point, it is valuable to note that Doran conducts an extensive critique of modern psychoanalysis pointing out that almost no prominent practitioners in this field are intellectually converted and that they consequently fail to recognize that psyche possesses "a vertical finality toward participation in the life of the spirit", Robert M. Doran, Dialectics of History, p. 46. See also Robert M. Doran, *Dialectics of History*, Chapter 9, "Reorienting Depth Psychology".

cious circle easily emerges regarding our "willingness to negotiate emotional darkness", and points out that we need to employ resources that are being blocked by this emotional darkness if we are ever to gain insight into it. He concludes that, easily, "the power of the complex negates the possibility of self-transcendent behaviour"[135].

It is over-against this moral impotence that Doran speaks of the healing influence of religious conversion. Echoing Lonergan's explanations in *Insight* of the need for a supernatural solution to the problem of evil, Doran first explores the role that other individuals, or communities, can play in helping to heal victimised individuals. Here, however, he describes how people with victimized psyches put up any number of resistances to believing that the love of another is sincere. He explains that, often, what finally convinces such people is when one who loves them is prepared to enter into the "private hell"[136] of their suffering and to suffer with them: "Healing will frequently be mediated precisely in and through the suffering of the lover due to the darkness of the beloved"[137]. After first describing such love in terms of human interaction, he insists that:

> The love that can sustain the movement of the healing vector, however much it may be mediated through another person or a community, is itself beyond human capacity[138].

Here Doran expands upon on what Lonergan describes in terms of religious conversion. He explores the way that the energy of "being in love" extends toward the healing of the victimized psyche and the releasing, in consequence, of a capacity for authentic transcendence.

Next, while exploring the way individual subjectivity responds to God's grace, Doran begins to employ the insights of Latin American liberation theology, and to begin a shift to considering the consequences of religious and moral conversion for the vector of redemption in history. Continuing his exploration of the healing of victimisation, Doran suggests that the symbol par

---

[135] Robert M. Doran, *Dialectics of History*, p. 241.

[136] Doran states: "John Dunne has distinguished two types of human suffering. There is the hell of the night of private suffering, the suffering of isolation and victimization; and there is the night of the suffering of compassion and forgiveness . . . The hell of the night of private suffering is not redemptive. The night of the suffering of compassion and forgiveness is redemptive", Robert M. Doran, *Dialectics of History*, p. 114.

[137] Robert M. Doran, *Theology and the Dialectics of History,* pp. 243-4.

[138] Robert M. Doran, *Theology and the Dialectics of History,* p. 243.

excellence of a readiness to suffer for others is the cross of Jesus Christ[139]. He adds that a Christian will mostly find the ability to love selflessly if he or she employs the symbol of the cross as the constitutive meaning of his or her life. He asserts that, in this way, Christians seek a "participation in the specifically paschal dimension of 'what Jesus did'"[140]. He concludes that it is only because of the divine intervention in history, completed in the self-sacrificing act of Jesus Christ (and the resurrection that followed it), that our damaged psyches find healing. Here Doran recalls that Lonergan had drawn on a long tradition of Christian spirituality to assert that redemption in history requires that Christians behave in "fidelity to the just and mysterious law of the cross"[141]. Doran elaborates on this point employing some insights from psychoanalysis:

> Perhaps one knows oneself to be healed only when precisely the same material dynamics of victimization that once drove one into the hell of private suffering now can be responded to with compassion and forgiveness[142].

Next, Doran suggests that living in fidelity to the law of the cross expands naturally to a preferential option for the poor. Here he traces how, for the educated person, a spontaneous feeling of love of the poor can combine with an insight that, "global injustice at the most basic levels of vital, social, and everyday cultural values as the principle defining characteristic of the situation[143]. Consequently, according to Doran, at least for the one who enjoys both intellectual and psychic conversion, the insight dawns concerning the importance of "evoking a new set of cultural values"[144] that will be endowed with an effectiveness for "the transformation of polities, economies, technological structures, and intersubjective communities"[145].

Based on this analysis, Doran explains the task of theology. He speaks of how the study of the situation will hold a prominent place for the theologian and will pay special attention to how decline causes suffering for the poor. Next, such theologians will be prompt in considering questions that emerge from the situation as

---

[139] Because of the unique efficacy of the symbol of the cross, Doran asserts: "Christianity has some genuine universal significance" (Doran, p. 108), a theme he will develop in discussing a theology of religions.

[140] Robert M. Doran, *Theology and the Dialectics of History*, pp. 121-2.

[141] Robert M. Doran, *Theology and the Dialectics of History*, p. 113.

[142] Robert M. Doran, *Theology and the Dialectics of History*, p. 245.

[143] Robert M. Doran, *Dialectics of History*, pp. 422-3.

[144] Robert M. Doran, *Dialectics of History*, p. 423.

[145] Robert M. Doran, *Dialectics of History*, p. 12.

relevant to the functional specialty in which they operate. Returning to the overall aim of the collaborative exercise that is theology, Doran states, "The praxis of theology, I have concluded, is proximately the transformation of culture", more remotely, however, he speaks of how theology will be "endowed with the capacity to evoke a global horizon for economic justice"[146]. He describes such theological praxis to "the mentality, the development of consciousness, that Lonergan calls cosmopolis"[147] and concludes by describing the important role of a preferential option for the poor in the way it promotes a culture of cosmopolis:

> The church's participation in the specifically paschal dimension of "what Jesus did" is twofold. First . . . the Christian is invited to acknowledge Jesus' suffering and death as his or her own redemption. But, second, the Christian is than invited as well to have some share in the historical catalytic agency of that suffering and death as its power mediates the transition from a prevailing situation to an alternative one[148].

3. A Critical Realist Support for the Vision of Pope Francis

The task of this part is to make explicit the links between the thought of Doran and the teaching of Pope Francis, especially on integral ecology. The relevance of Doran is primarily to support and to offer explanatory depth to the vision of Pope Francis. And, indeed, various intuitions of Pope Francis can help direct the future of Doran studies. At the same time, an appeal to Doran can help bring coherence to the thought of the Pope, counterbalancing references sometimes made to philosophers and theologians who are not always either internally coherent in their own thought or compatible with each other.

3.1 Convergences between Doran and Pope Francis

There is a great deal in the thought of Doran that supports the teaching of Pope Francis, both on the notion of integral ecology and on other aspects of his teaching. In pursuing this question, it is relevant to study a text that Doran himself wrote about Pope Francis before he died in 2021. He states that: "Pope Francis is probably the first non-classicist Pope"[149]. He then speaks of similarities between Pope

[146] Robert M. Doran, *Dialectics of History,* p. 423.
[147] Robert M. Doran, *Dialectics of History,* pp. 356-7.
[148] Robert M. Doran, *Dialectics of History,* pp. 121-2.
[149] Robert M. Doran, "Forward", in Gerard Whelan, *A Discerning Church: Pope Francis, Lonergan, and a Theological Method for the Future.* Mahwah, NJ: Paulist Press, 2019, p. ix.

Francis and Lonergan both in terms of "common stances, horizons, basic options" as well as what he considers to be yet deeper "areas of actual convergence"[150].

Concerning general resemblances, Doran notes how both Pope Francis and Lonergan have tended to be criticised by representatives of both the "ecclesial right" and the "ecclesial left". In the case of conservative criticism of Pope Francis, Doran believes that this has much to do with the criticism offered of clericalism offered by the pontiff. He then points to Lonergan's criticism of "premature systematization" as an expression of bias and not the product of an appropriate accumulation of insights. Speaking of how biased ideas contribute to decline in history, he lists political and social philosophies, as well as theological attitudes, that have included such bias, "feudalism, liberalism, capitalism, socialism, nationalism, *clericalism*, and anticlericalism"[151]. Doran notes that, some members of the "ecclesial right" are "wary of the radical implication of [Lonergan's] work for ecclesial reform" and yet are not able to accuse him of heterodoxy. He then quips that they, "might perhaps wish that he were not orthodox, so that they could more effectively take issue with him"[152].

Turning to areas of deeper convergence between Pope Francis and Lonergan, Doran adverts to the differences that exist between the Argentinian theology of the people and other expressions of liberation theology that were prominent in the 1970s and 1980s. He recalls that, like the Argentinian theologians, Lonergan was alert to the dangers of employing Marxist categories in theology. He then points out that in *Theology and the Dialectics of History* he expands on such criticisms[153]. Here he notes that his views have been perceived by some as converging not only with that of Pope Francis but also with "Pope John Paul II's suspicions about both liberation theology and the Jesuits!" In explaining this point, he first makes a qualification, "my evolving work really did take seriously the liberation and Jesuit emphasis on a preferential option for the poor"[154]. Next, however, he acknowledges that it is, indeed, his opinion that, similar to the imbalances of the liberation theology, Jesuit "governance and pastoral planning" tended to speak of "the service of faith and the promotion of justice" in an unbalanced way[155].

---

[150] Robert M. Doran, "Forward", p. ix.
[151] Robert M. Doran, "Forward", p. ix. Emphasis in the original.
[152] Robert M. Doran, "Forward", p. x.
[153] See, Robert M. Doran, *Dialectics of History*, pp. 424-439.
[154] Robert M. Doran, "Forward", p. x.
[155] Robert M. Doran, "Forward", p. x.

Doran employs Lonergan's notion of the scale of values to explain this critique. Essentially, he sees the problem lying in a tendency to promote an activism for social change that neglects the need for transformation of personal and cultural values. He recalls that Lonergan describes how religious conversion should launch a process of healing from above; i.e. of one's ability to respond to personal values, cultural values, social values, and vital values in that order. He insists that this order must be preserved. Regarding the need to transform personal values—moral conversion—he cites the Daniel Berrigan, famous for anti-Vietnam-War protests, who, in spite of his leftist image, stated, "until the individual changes nothing changes"[156]. Concerning the need for cultural transformation, Doran invokes the vocabulary of Aquinas to state, "without the formal effects of habitual and actual grace moving in to the cultural domain and becoming social grace, personal conversion is shortchanged"[157]. He describes the task of cultural transformation as that of a cosmopolis that seeks "the establishment of the common meanings that would support and encourage the fuller transformation of the structures of polity and economy"[158]. He insists that it is only change in social values emerging from sincerely transformed cultural values "that would guarantee an equitable distribution of vital goods to the entire human community"[159]. Returning to relate Lonergan to Pope Francis, Doran states:

> I would not be surprised if this integration of faith and justice *through personal conversion and cultural transformation* will be recognized as defining the ministry of Jorge Mario Bergoglio/Pope Francis. And if this is the case, then have no doubt that the emphases of Bernard Lonergan and Pope Francis are already well on their way to convergence[160].

While Doran stresses the convergences between the thought of Lonergan and the teaching of Pope Francis, one can also explore the way in which such convergences are yet more evident in Doran's own thought. The originality of Doran's contribution lies, first of all, in his notion of psychic conversion. However, it also lies in a series of emphases in how to employ Lonergan's thought. One

[156] Robert M. Doran, "Forward", p. x.

[157] Robert M. Doran, "Forward", p. xi.

[158] Robert M. Doran, "Forward", pp. x-xi.

[159] Robert M. Doran, "Forward", p. xi. Doran concludes his criticism of the ecclesial left by commenting on the continuity of Lonergan's thinking with all that is best in Christian tradition, and adding the provocative statement, "resistance to Lonergan's work from the ecclesial left is due simply to the fact that, for all his creativity, which most acknowledge, he remains orthodox" (Ibid).

[160] Robert M. Doran, "Forward", p. xi. Emphasis in the original.

such emphasis is Doran's constructive engagement with themes from liberation theology. In order to appreciate how Doran contributes to reflection on integral ecology, it is important to reflect both on psychic conversion and on the intellectual, moral, and religious conversions, with which it is interwoven.

At the centre of the critical realist appeal for intellectual conversion is the hope that this can result in a community of cosmopolis being formed within culture that helps to promote ideas and values that reverse decline and promote progress in history. Relating this insight to those of *Laudato Si'*, one can add that promoting integral ecology should be central to such an evangelization of culture today. Next, an attitude of cosmopolis will need to be based on a cosmology of emergent probability. This metaphysical construct has an immense range of application. It helps to integrate a scientific understanding of the process of emergence in the universe with a notion of a hierarchy of being that places human beings at the apex of this process. Far from encouraging any kind arrogant, environmentally destructive, anthropocentrism, such an approach stresses the responsibility of humans to respect the integrity of the lower "schemes of recurrence" of inanimate and animate life, that, in fact, make the emergence and stable functioning of human life possible[161].

Two further characteristics of emergent probability are of importance. First, it can serve as a basis for interdisciplinary collaboration between science and philosophy, thus allowing an integration of ethics—and an acknowledgment of "objective values"—with rigorous thinking about technical solutions to human problems at the level of technology, economics etc. Secondly, this scientifically persuasive notion of the universe is open to theology. Lonergan's reflection on the "divine solution to the problem of evil" in *Insight* allows a heuristic theory of history to identify not only the vectors of progress and decline, but also that of redemption. Applied to questions of ecology, it supports the move made by Pope Francis to propose a sacramental vision of the universe. On this theme, Lonergan adds explanatory, metaphysical, depth to the relatively poetic expression of Pope Francis:

> The divine Persons are subsistent relations, and the world, created according to the divine model, is a web of relationships. Creatures tend towards God, and in turn it is proper to every living being to tend towards other things, so that throughout the universe we can find any number of constant and secretly interwoven relationships[162].

---

[161] Bernard Lonergan, *Insight*, pp. 57-92 and 126-162.
[162] Pope Francis, *Laudato Si'*, n. 240.

The adding of psychic conversion to intellectual conversion allows for yet further convergence with between critical realism and thinking about integral ecology. One might add that Doran's thought brings critical realism into closer convergence with romanticism than does that of Lonergan. In Part 1, it was noted that romanticism, especially as it was appropriated within the Catholic church, stresses certain themes, such as: the beauty of nature, spirituality, community, culture, and educating the young. These themes are prominent in *Laudato Si'* and one can reflect on how psychic conversion is a necessary addition to intellectual conversion in order to do justice to these dimensions of human living. An openness to beauty is something that occurs in the psyche. It is related to the capacity for imagination and symbolising that wells up from the unconscious and is evident, for example, in our dream life. Psychic conversion helps us appropriate this dimension of subjectivity, and not least the importance of symbol. Echoes of *Laudato Si'* are evident when Doran insists on a "two-way relation between the aesthetic and the theoretical"[163].

In the realm of spirituality, it is by an appeal to psychic conversion that he is able to explore how God's grace helps to heal victimized psyches, and how the process of our own healing encourages us to desire to participate in the healing of others. Consequently, Doran describes how we participate in the "law of the cross", whereby Christ returns good for evil, and how this helps us feel drawn to make a preferential option for the poor. Such themes find a direct echo in the theology of the people, so valued by Pope Francis, and are easily extended to a concern for the natural environment. Here one recalls the references to St. Francis of Assisi in *Laudato Si'* and the way that saint linked the "both the cry of the earth and the cry of the poor"[164].

On questions of culture, one of the characteristics of Doran's thought is to add a concern for cosmological culture to one for anthropological culture. Once again this echoes the respect for indigenous cultures evidenced by *Laudato Si'*. At a level of cultural critique, Doran's analysis of the imperialistic tendencies in Western culture is compatible with the critique of the technocratic paradigm that is central to *Laudato Si'*. Doran traces explains that that, in part, the technocratic paradigm is the result of the tendency of European culture—ever since the time of the animosity of the poets and the philoso-

---

[163] Robert M. Doran, "Lonergan and Von Balthasar", in *Theological Studies*, 58(1997), pp. 61-84, at 82.

[164] Pope Francis, *Laudato Si'*, n. 49.

phers in ancient Greece—to fall into a dialectic of contradictories where the anthropological pole of meaning is stressed to the exclusion of the cosmological pole. Once again, one finds a critical realist analysis bringing explanatory depth to a statement of the Pope:

> It is essential to show special care for indigenous communities and their cultural traditions. They are not merely one minority among others, but should be the principal dialogue partners, especially when large projects affecting their land are proposed. For them, land is not a commodity but rather a gift from God and from their ancestors who rest there, a sacred space with which they need to interact if they are to maintain their identity and values. When they remain on their land, they themselves care for it best. Nevertheless, in various parts of the world, pressure is being put on them to abandon their homelands to make room for agricultural or mining projects which are undertaken without regard for the degradation of nature and culture[165].

On a related issue, one finds echoes in Doran of when Pope Francis refuses to allow the excessive anthropocentrism of modern culture to be replaced by biocentrism, adding that this, "would entail adding yet another imbalance, failing to solve present problems and adding new ones"[166]. Doran would support this statement by explaining that Pope Francis is resisting allowing a dialectic of contradictories that exaggerates the anthropological pole to be replaced by one that exaggerates the cosmological pole.

3.2 Correcting a Certain "Looseness" of Philosophical Foundations?

Part 1 explored the philosophical foundations of Pope Francis as expressed in *Laudato Si'*. It noted reference by Pope Francis to a notion of polarity that is found in Romano Guardini and other Catholic authors influenced by romanticism. At the same time, it noted appeals to metaphysical realism that is clearly an attempt by Francis at "the broadening of the Guardinian framework"[167]. In fact, it is difficult to pin down exactly what philosophical foundations are operative in Pope Francis, as he seldom explicitly addresses this point. Commentators who express an uneasiness on this issue often find themselves referring more to general attitudes the Pope expresses in interviews with journalists, or impromptu talks, than with material that forms an explicit,

[165] Pope Francis, *Laudato Si'*, n. 146.
[166] Pope Francis, *Laudato Si'*, n. 118.
[167] Massimo Borghesi, "Polarity Model", p. 109.

written, part of his magisterium. Nevertheless, an honest presentation of a critical realist proposal to promote a notion of integral ecology should mention, also, how this might also serve a role of correcting certain tendencies in thinkers to whom Pope Francis likes to appeal. A first point of critique involves the notion of polarity in authors such as Guardini, as well as in certain interpretations of the *Spiritual Exercises of St. Ignatius*. First, however, one needs to identify convergences between the polarity thinking of such authors and the thought of Doran.

One recalls how Guardini speaks of the polarity of human existence that cannot be wished away but which form a structure of living and in which "concrete living persons" must make decisions, often finding a kind of *via media* between competing values[168]. Similarly one recalls how Pope Francis employs a similar notion of polarity to a reading of the *Spiritual Exercises*. This is evident when he states, "according to St. Ignatius, great principles must be embodied in the circumstances of place, time and people", and when he adds, "the Jesuit must be a person whose thought is incomplete, in the sense of open-ended thinking"[169]. Here there are clear overlaps with Doran, who speaks, as has just been noted, of each of the scale of values being characterised by dialectic of contraries, and how the "sensitive psyche occupies, as it were, a middle ground between the organism and the spirit"[170], and how authentic living involves making decisions that find a *via media* between the dictates of spirit and those of our bodily selves.

Similarly, one recognizes parallels with Guardini when Doran speaks of how moral decision-making should be conducted in an "aesthetic pattern of experience". Here, he agrees that, as in any approach to ethics, one should seek to do good and avoiding evil. However, he adds that one often has to choose between different goods and that this often involves making a decision about what kind of person one wishes to become, as a way of responding one's own particular vocation in the world. Here Doran appeals to an aesthetic criterion, praising an approach that, "lets experience be for its own sake, that is a release, a letting be, a freedom for the emergence of the inevitability of form"[171].

---

[168] Romano Guardini, *Der Gegensatz,* p. 16, cited in Krieg, *Romano Guardini,* p. 15.

[169] "Interview with Pope Francis", by Fr Antonio Spadaro SJ, 2013 September 2013, http://www.vatican.va/content/francesco/en/speeches/2013/september/documents/papa-francesco_20130921_intervista-spadaro.html.

[170] Robert M. Doran, *Dialectics of History,* p. 55.

[171] Robert M. Doran, *Dialectics of History,* p. 55.

If one can recognize similarities between polarity thinking and that of Doran, nevertheless, one has a lingering suspicion that there is something inadequate in the metaphysics of Guardini and other romantic-idealist thinkers. Here one recalls the question asked by one commentator of thinkers such as Johan Sabastian Drey, upon whom Guardini depends: "Did he, in fact, come down on one or the other side of the opposition? Or did he simply oscillate between the two, without being able to integrate them?"[172]. One wonders if the polarity thinking of authors such as these does not involve what might be called a kind of "blind praxis". Such authors seem to imply that initial action should be guided by nothing more than well-intentioned Christian-pastoral motivations. This contrasts with the approach of Lonergan who speaks of being orientated to "transcendental precepts" guided by intellectual and psychic conversion. Guardini seems to propose starting with relatively un-reflected-upon action and hoping that one will "intuit" the right thing to do as one engages with a reality in a "living concrete", that is characterized by polarities. As some commentators identify, in the case of Guardini, this allowed a number of unacknowledged prejudices to guide the kind of praxis in which he was engaging[173].

One must note that when Pope Francis employs polarity thinking to interpret the *Spiritual Exercises*, he is moving beyond what one can call a notion of blind praxis to speak of a praxis directed by an expectation to be guided by the movements of the Holy Spirit that Ignatius calls consolation. However, even here one has to recognize that a discerning attitude needs to be capable of being differentiated into an academic moment, where one recognizes that one is directed by transcendental notions and can be guided by the transcendental precepts[174].

Having offered this criticism of the polarity thinkers upon which Pope Francis relies, one must also consider the efforts made by Pope Francis at "the broadening of the Guardinian framework"[175]. Part 1 noted how the pastoral principle of Pope Francis, "reality is greater than ideas" is an expression of

---

[172] Wayne L. Fehr, *The Birth of the Catholic Tübingen School*, p. 259.

[173] Robert A. Krieg, *Romano Guardini*, p. 165, 20.

[174] See Gerard Whelan, "Robert Doran and the *Spiritual Exercises of St. Ignatius*: A Dialogue with Joseph Maréchal on the Meaning of the Application of the Senses", in *Ignaziana: Rivista di Ricerca Teologica*, Rivista web dal Centro di Spiritualità, Pontificia Università Gregoriana, ISSN 1828-2377, 26 (2018), pp. 268-281; "Ignatian Spirituality and the contemporary World: From Ignatius of Loyola to Bernard Lonergan", in *Hong Kong Journal of Catholic Studies*, (2020), N. 11, pp. 20-56.

[175] Massimo Borghesi, p. 109.

a desire to appeal to a metaphysical realism that provides a counterbalance to what he seems to judge as lacking in polarity thinking. Here, however, one has to ask if merely juxtapositioning such realist statements with what is otherwise polarity thinking conducts a successful broadening of the Guardinian framework. Is this internally coherent? Added to this, one suspects that, at times, the kind of realism that the Pope invokes is what Lonergan calls a "naïve realism", which expresses an "oversight of insight" that is found in the intuitionist thinking of thinkers such as Scotus. This approach is based on a cognitional theory that believes that reality can be known by an act of intuition, and not by a process of first gaining insight and then affirming this in an act of judgment. There can be an implicit anti-intellectualism in such realism.

Having noted the possible Scotist tendencies in certain appeals to reality of Pope Francis, one notices a shift in *Fratelli Tutti* toward a more critical realist position. Here the Pope explores how the different religions can promote dialogue in society that builds genuine social friendship. He notes that, in many places, culture has descended into relativism and that a religious voice can help cultural and social dialogue share "the conviction that it is possible to arrive at certain fundamental truths always to be upheld"[176]. In exploring how these truths are to be grasped, he comes close to articulating a process that passes though both insight and judgment. This is evident when he refers to "an enduring truth accessible to the intellect", how "inherent in the nature of human beings and society there exist certain basic structures to support our development and survival"[177].

In concluding this section, one can conclude that the notion of a heuristically oriented praxis, exercised by a person who is both intellectually and psychically converted is the best way of establishing the kind of "adequate anthropology" for which Pope Francis calls in *Laudato Si'*[178].

3.3 An Epoch-changing Papacy?

In conclusion, one can recall that Doran stated, "I have no doubt that the emphases of Bernard Lonergan and Pope Francis are already well on their way to convergence"[179]. Indeed, one can add that the voice if Pope Francis

[176] Pope Francis, *Fratelli Tutti*, n. 211.
[177] Pope Francis, *Fratelli Tutti*, n. 212.
[178] Pope Francis, *Laudato Si'*, n. 118.
179 Robert M. Doran, "Forward", p. xi.

is not entirely his own, but represents that of the Church of Latin America as articulated in the CELAM conference at Aparecida, Brazil, in 2007. This document represents the coming to maturity of a uniquely Latin American way of interpreting Vatican II, one which that sub-continent is ready to share with the universal church through the agency of this Argentinian pope. A great deal of this vision in encapsulated in what Pope Francis has to say in *Laudato Si'* about integral ecology. It is the intention of this chapter to provide philosophical and theological foundations that can ensure that this legacy can endure through future pontificates.

ECOLOGY AND LIFE

A reading of Laudato Si' from the African perspective

Adrien Lentiampa Shenge, SJ

1. Introduction

The problem addressed in the Encyclical *Laudato Si'* (LS) cannot leave anyone indifferent. The ecological question impacts everyone directly regardless of their culture: the global deterioration of the environment concerns us all collectively and each one personally. That is why, in this encyclical, Pope Francis addresses himself not so much to Catholics, believers or "people of good will" – as is usually the case for the texts of the magisterium –, but to "every person living on this planet" (LS, 3). The Pope speaks there, rightly, of "our common home"; for "nothing in this world is a matter of indifference to us" (LS, 3-6).

On that basis it is clear that the Pope's discourse in this Encyclical is intended to be a *dialogue* with *all*[1]. That is probably what justifies the style adopted since the document leans extensively on the utterances coming from many horizons. It thus has many references to the declarations of various episcopal conferences across the world; and numerous excerpts from the writings of authors coming from different cultural, religious and scientific horizons. This is what *Laudato Si'* intends as a discourse of dialogue, accessible to all because it draws on the experience of each one.

[1] The text is clear that the Pope intends to open a dialogue with his contemporaries. In fact, from the third section, we read: "In this Encyclical, I would like to enter into *dialogue* with all people about our common home" (Pope Francis, Encyclical Letter *Laudato Si' on Care for Our Common Home*, 24 May 2015, n. 3. Our emphasis). The Pope returns to this again in the fourteenth section in these words: "I urgently appeal, then, for a new dialogue about how we are shaping the future of our planet" (Pope Francis, *Laudato Si*, n. 14. Our emphasis).

DOI 10.32060/C.S.9.317-341

In truth, the nature of the problem posed by the ecological question requires such a dialogic approach. Indeed, "what relates to the Earth for which we are jointly responsible by our very existence [can only become] a subject of concern"[2] to all. Moreover, since it affects us existentially, the problem of the deterioration of the environment requires that each one of us, wherever we are, feels concerned and engaged in the protection of the environment. It requires each of us, with our own resources, to enrich this dialogue so that the concern for the degradation of nature is not limited simply to the level of speech, but also extends to our lifestyle, our way of inhabiting the world. Thus, it is important that Africa enters actively into this dialogue to enrich the discussion on the protection of "our common home".

It is with this dialogue that the present reflection would like to engage. To do this, it will seek to analyse the role of Africa in the debate on the deterioration of the environment in general, and in the text of the Encyclical, in particular. It will then extend the dialogue for which Pope Francis appeals, starting off from African rationality. From this dialogue, there could hatch a new understanding of our "being-in-the-world", and of our way of approaching and treating the ecological question.

Such a renewal of the approach to ecology is possible if one points to the cultural values which have always animated Africa and been the source of its life; values which have led African to inhabit the world in a way that respects and protects the environment. This set of values is crystallised in the "culture of life" as characteristic of African approach to the world. It is a culture which promotes life, not only on the individual level but life in its different aspects which also includes the world around us. Thus, the culture of life will give us a holistic vision integrating the question of the protection of the environment in a unified conception of the world.

2. *Laudato Si'* and Africa

Laudato Si' could hardly ignore African continent in its considerations. While it is possible to recognise several allusions to what is seen there, we count only two explicit references to Africa. They can be found in paragraphs twenty-eight and fifty-one. These two sections are in the first chapter of the Encyclical where the Pope describes "what is happening to our common

[2] Jean-Marc Ela, *Repenser la théologie africaine: Le Dieu qui libère*. Paris: Ed. Karthala, 2003, 118.

home" (LS 17-61), that is to say, the whole range of ecological problems to which we are confronted today. He pinpoints, one by one, the problems of pollution and climate change (LS 20-26); the question of water (LS 27-31); the loss of biodiversity (LS 32-42); the report of the deterioration of the quality of human life and social degradation (LS 43-47); global inequality (LS 48-52); as well as the weakness of reactions in the face of all these problems (LS 53-59). It is therefore a comprehensive diagnosis of the problem of environmental deterioration; and one can clearly observe that Africa is to be found in each of the aspects of this diagnosis.

2.1 Africa and the ecological tragedy bound up with the question of water

If we focus on the explicit reference to Africa in the Encyclical, we first come across it in the twenty-eighth paragraph, in the section dealing with the question of water. Here, it is emphasized that, whereas the problem of water affects a good part of the world, Africa remains the region where the question of the lack of running water is posed most acutely: "large sectors of the population have no access to safe drinking water", even though it is a commodity that is "indispensable for human life", not to mention its importance for food production and for the environment. Thus we can say that the question of water sums up, all by itself, the ecological tragedy denounced by the Encyclical: at one and the same time, water affects human life, health, food, security, not forgetting that it shares in the sustaining of land-based and aquatic ecosystems.

However, the question of water also has socio-political implications. In fact, its scarcity and subsequent drought are at the root of several migratory movements, with all their underlying problems. It is totally pertinent, from this point of view, that Halirou Abdouraman[3] invites us to look at the migration issue around the Lake Chad basin, with the cohort of conflicts that it has provoked for decades unceasingly. As such, "from the point of view of the neighbouring States, Lake Chad has long been perceived as a sensitive area, a spot for the potential expression of the expansionist intentions of the neighbours"[4].

---

[3] Cf. Halirou Abdouraman, "Le conflit frontalier Cameroun-Nigeria dans le lac Tchad: les enjeux de l'île de Darak, disputée et partagée" in *Cultures & Conflits* n° 72, hiver 2008, 57-76.

[4] Géraud Magrin & Lempalle, "Insécurités au lac Tchad: environnement et conflits", in Emmanuel Chauvin, Olivier Langlois, Christian Seignobos and Catherin Baroin (eds.), *Conflits et violences dans le bassin du lac Tchad*. Marseille: IRD Editions, 2020, 137.

This is proof that water can surely become a major political issue, and its control, as well as its distribution, a potential source of conflicts[5]. The United Nations has listed three hundred areas of potential conflicts and foresees a proliferation of "water wars"[6]. It is along these lines that Vandana Shiva[7], who has compiled a complete history on the subject, shows that a certain number of the most important conflicts of our time – and, above all, the Israeli-Palestinian conflict, but also the conflict between Syria and Iraq[8] – have often been aimed at controlling hydraulic resources, although generally disguised as ethnic or religious wars.

In Africa, it is not just the water of Lake Chad which is the source of conflicts that could degenerate into open warfare. Frédéric Lasserre has counted six of them in the last three decades[9]. Among these cases, we should cite that of the River Nile which is at the heart of a long and serious conflict over the sharing of its waters; a conflict involving at least four countries: Egypt, Ethiopia and the two Sudans. The 1959 treaty between Egypt and Sudan on the management of the waters of the Nile, signed shortly after a period of quasi armed conflict, did not put an end to tensions in the region, notably with its rejection by Ethiopia which was not a party to the treaty[10]. Such tensions are not to be minimised, especially when one recalls that, already in 1979, the then Egyptian president, Anouar el-Sadate, declared that "the sole factor that could trigger Egypt's entry into war is water". A few years later (in 1987), this was acknowledged by Boutros Boutros-Ghali, in a statement that has become famous: "the next war in the region [will be] over the waters of the Nile"[11].

If water has become a source of conflict, it is because of the strong disparity between the rich countries and the poor ones when it comes to its access. *Lau-*

---

[5] Cf. Frédéric Lasserre, "Conflits hydrauliques et guerres de l'eau : un essai de modélisation", in *Revue Internationale et Stratégique* n° 66-2007, 106. We recommend the reading of the whole dossier (with the evocative title "L'or bleu, nouvel enjeu géopolitique?") from which Frédéric Lasserre's article has been drawn. The dossier has been published under the editorship of Barah Mikaïl, (Cf. Barah Mikaïl [ed.], "L'or bleu, nouvel enjeu géopolitique?" in *Revue Internationale et Stratégique* n° 66-2007).

[6] Cf. Mohamed Larbi Bouguerra, *Les batailles de l'eau*. Paris: Éditions de l'Atelier, 2003. Cited by Philippe Hugon, "Vers une nouvelle gouvernance de l'eau en Afrique et en Amérique latine", in *Internationale et Stratégique* n° 66-2007, 66.

[7] Cf. Vandana Shiva, *La guerre de l'eau: privatisation, pollution et profit*. Lyon: Ed. Parangon, 2013.

[8] Cf. John K. Cooley, "The War over Water", in *Foreign Policy* n°82, 1991, 3.

[9] Cf. Frédéric Lasserre, "Le prochain siècle sera-t-il celui des guerres de l'eau?", in *Revue Internationale et Stratégique* n° 33-1999, 110. See also Frédéric Lasserre, "Conflits hydrauliques et guerres de l'eau : un essai de modélisation" [see note 5], 111.

[10] Cf. Frédéric Lasserre, "Le prochain siècle sera-t-il celui des guerres de l'eau?" [see note 9], 105.

[11] Cf. Peter Gleick, "Water and Conflict" in *International Security*, vol. 18, n°1, 1993, 85.

dato Si' rightly denounces the "growing tendency, despite its scarcity, to privatize this resource, turning it into a commodity subject to the laws of the market" (LS 30). For, in doing this, the law of the highest bidder comes into force over drinking water too, reducing this vital commodity to the level of a consumable good like any other. The Encyclical bids us to leave this logic of the marketplace to the extent that water is not only a consumable good, but the very condition of existence, since it is the essential element of (and for) life. In this sense, to deny someone access to drinking water is to deny his right to existence.

The position maintained here by *Laudato Si'* corresponds to what Philippe Hugon has called an "ethical approach"[12]. This is a interpretation that makes water a "primary good" in the Rawlsian sense, that is to say, a "useful good whatever the project of rational life"[13] and which, as such, cannot be replaced by another. That is why John Rawls claims that a just society is one which enables the possession of such primary social goods and which assures their equitable distribution among its members[14].

It is thus understandable that a country like South Africa has inscribed access to water on the list of rights guaranteed by its fundamental law, obliging the State to do everything possible for the achievement of this right[15]. The United Nations' Committee on Economic, Social and Cultural Rights has gone even further than the Constitution of the Republic of South Africa. Indeed, in 2002, it proclaimed that "the human right to water is (...) the preliminary condition for the realisation of all the others"[16]. This proclamation matches the Pope's statement that "access to safe drinkable water is a basic and universal human right" (LS 30).

---

[12] Philippe Hugon, "Vers une nouvelle forme de gouvernance de l'eau en Afrique et en Amérique latine" in *Revue Internationale et Stratégique* n° 66-2007, 68. We note that Philippe Hugon lists six possible approaches to the regulation of drinking water (see pages 67-69).

[13] John Rawls, *A Theory of Justice*, Cambridge: Harvard University Press, 1971, 62.

[14] Cf. John Rawls, *Théorie de la Justice. Op. Cit.* [note 13], 62. See also John Rawls, "Unité sociale et biens premiers" in *Raison Politique* n° 33-2009, 9-43.

[15] In fact, section 27 of the second chapter of the 1996 Constitution of the Republic of South Africa stipulates : "(1) Everyone has the right to have access to— (a) health care services, including reproductive health care; (b) sufficient food and *water*; and (c) social security, including, if they are unable to support themselves and their dependents, appropriate social assistance. (2) The state must take reasonable legislative and other measures, within its available resources, to achieve the progressive realisation of each of these rights".

[16] Comité des Droits économiques, sociaux et culturels de l'ONU, "droit à l'eau" in HCR - Conseil économique et social: Observation générale no 15 (2002) Le droit à l'eau (art. 11 et 12 du Pacte international relatif aux droits économiques, sociaux et culturels) (unhcr.org) (consulted on February 4th, 2021).

We should mark the breadth of the Pope's statement: access to water is not only a primordial or fundamental right; it is also a universal right, recognised as belonging to every human being wherever he or she lives. It is truly a right belonging to the human being as being. It follows that, to the extent that every right presupposes a corresponding duty, the right to drinkable water also implies a universal duty. That is why the Encyclical speaks of a universal obligation, that is to say an obligation which concerns each individual person with regard to those who do not have access to drinking water: "Our world has a serious social debt towards the poor who lack access to drinking water" (LS 30).

2.2 The ecological obligation towards Africa

The matter of debt is also taken up in the second paragraph where explicit reference to Africa is made, namely paragraph fifty-one. This speaks specifically of the "ecological debt" of the countries of the North towards those of the South. We should note that this paragraph is found in the section that deals with the planetary inequality which arises from the degradation of the environment. The assessment is made therein that: "the gravest effects of all attacks on the environment are suffered by the poor" (LS 48).

Paragraph fifty-one identifies three kinds of "ecological damage" suffered by Africa in the wake of the economic activities of the highly industrialised countries. This is a reference to the production by these industries of greenhouse gas with consequences on global warming which are also shared by Africa. Secondly, the Encyclical speaks of the ecological damage due to the exploitation of various African raw materials that enable those industries to function. Finally, there is the matter of the dumping toxic waste from those countries on the African continent.

On the issue of global warming due to industrialisation, we know that it is the result of the emission of greenhouse gases. Now, while it is true that the greenhouse effect is largely bound upon with natural phenomena, the fact remains that, since the eighteenth century, the proportion due to human activity – the human-made greenhouse effect – is in fact increasing to a disturbing level. It should be remembered that the eighteenth century ushered in the Industrial Revolution[17], with its huge need of fossil fuel. One can therefore

---

[17] To know more about the history of global warming in step with industrialisation, we refer to the studies published in 2002 by the Blaise Pascal University of Clermont-Ferrand (Chistoph Ber-

establish a causal link between the great development of industry – under its various designations – and the present tendency towards global warming[18].

While, for two centuries, industry developed more in the northern countries, global warming due to the gases emitted from the use of fossil fuel needed for the functioning of those industries has had a global and universal effect on the planet. Gas, of course, does not observe national boundaries: what is emitted at a given point on the planet has effects on global warming, not only on this part of the world but on the entire planet. Thus, global warming presents us with a real globalisation of the consequences of human activity on the environment, with no benefit from industrialisation directly profiting the entire world population.

Therefore, while industrialisation is concentrated in only one part of the world and profits a minority, its consequences on the environment affect the whole planet. It is in this context that Africa finds itself doubly the victim of human-made global warming. On the one hand, without the resources or technological means to remedy and possibly offset the consequences of climate change, its peoples increasingly suffer the ecological and economic repercussions of this globalisation of greenhouse gas without reaping any benefits from it; on the other hand, "a rise in temperature, together with drought, has proved devastating for farming" (LS 51). We find ourselves in a situation where Africa is increasingly impoverished because of the harmful consequences of the industrial opulence of the high-income countries.

As if that were not enough, these need of raw materials for those industries leads to further environmental damage on the African continent. The Encyclical cites explicitly "mercury pollution in gold mining or sulphur dioxide pollution in copper mining" (LS 51). However, it is also possible to pinpoint the devastating environmental consequences of the mining exploitation needed for the development of the electromechanical industry in advanced technology countries.

A country like the Democratic Republic of the Congo can serve as an illustration of this situation. Indeed, in a dossier published on February 4th, 2020,

nhardt and Geneviéve Massard-Guilbaud [ed.], *Le démon moderne. La pollution dans les sociétés urbaines et industrielles d'Europe*. Clermont-Ferrand: Presses Universitaires Blaise Pascal, 2002.

18 Cf. Franck Aggeri and Mélodie Cartel, "Le changement climatique et les entreprises : enjeux, espaces d'action, régulations internationals", in *Entreprises et Histoire*, n° 86 (2017), p. 6-7. Read also Franck Aggeri and Olivier Godard, "Les entreprises et le développement durable", in *Entreprises et Histoire*, n° 45 (2006), p. 6-19.

the Belgian newspaper *Le Soir*, revealed how the exploitation of cobalt in the region of Kolwezi, which ought to be a blessing for the local population, turned into a veritable ecological and public health disaster[19]. There is talk not only of the threat to biodiversity in the wake of the mining exploitation waste residues, but also of problems of congenital malformation due to inhaling the fine particles generated by the over-exploration of various mines. Indeed, high concentrations of cobalt in the urine of children were detected, as well as damage to the DNA of the newly born. All this without the standard of life of the neighbouring inhabitants of these mines showing any improvement. As the Pope says: these enterprises "operate in less developed countries in ways they could never do at home, in the countries in which they raise their capital" (LS 51).

It is the same thing with the dumping of waste on the black continent. Many reports coming from different structures and organisations show how Africa has become "the rubbish dump of the West", notably in the field of electronic waste, packed with components harmful to health and the environment[20]. We can also recall the scandal of the toxic waste from the oil industry dumped, in 2006, in a popular quarter of Côte d'Ivoire [Ivory Coast] by the oil company, Trafigura, based in the Netherlands[21]. It is as if, for the countries of the North, the life of Africans does not have the same value as that of their own people.

As if that were not enough, "Some strategies for lowering pollutant gas emissions", in fashion in the Northern countries, "call for the internationalization of environmental costs, which would risk imposing on countries with fewer resources burdensome commitments to reducing emissions comparable to those of the more industrialized countries" (LS 170). Thus, the poor, who are already the direct victims of the industrial needs of these countries, would be forced to forfeit their own need for development in order to bear the harmful consequences of industrialisation: a new injustice!

---

[19] Cf. Colette Braeckman, "Congo: le cobalt, désastre écologique et sanitaire", in *Journal Le Soir* n° 29, 4 February, 2020, 2-3.

[20] Cf. Cheikh Diop and Ramata Molo Thioune, (eds.), *Les déchets électroniques et informatiques en Afrique. Défis et opportunités pour un développement durable au Bénin, au Mali et au Sénégal*. Paris: Ed. Karthala, 2014.

[21] Cf. Amnesty International & Greenpeace, *Une vérité toxique. À propos du Trafigura, du Probo Koala et du déversement de déchets toxiques en Côte d'Ivoire*, September 2012 in https://reliefweb.int/sites/reliefweb.int/files/resources/afr310022012fr.pdf (consulted, on January 11th, 2021).

3. Thinking about ecology, starting from African rationality

3.1. Making Africa an ecological actor

The analysis of the references in *Laudato Si'* to Africa puts before us a dark picture, to say the least. This continent is therein portrayed as one of the great victims of the environmental degradation caused by human action. It is seen exclusively as the continent which is undergoing the ecological disaster, since this disaster is the result of what is decided, largely, elsewhere and does not depend on it.

It is true that in another (indirect) reference to Africa, the Encyclical refers to the continent in a way that is less dramatic. This occurs in paragraph thirty-eight where the Pope cites the Congo basin as one of the "lungs" of the planet, alongside with the Amazon. He reminds us "how important these are for the entire Earth and for the future of humanity" (LS 38), notably the role of their forests for absorption of greenhouse gases and for biodiversity.

To measure the essential contribution of the Congo basin in regulating the climate, it will not be unprofitable to take account, not only its forest whose impact on the "earth's breathing" needs no demonstration, but also its immense peat bogs. Indeed, these peat deposits, true natural carbon traps, are crucial for slowing down global warming. Now, scientists estimate that the world's largest global reserve of peat bogs is found in Africa, notably in this same Congo basin. If this world reserve has not yet revealed the whole extent of its secrets, a recent discovery, made around the locality of Lokoloma, in the Democratic Republic of Congo, has identified peat deposits extending over more than 145,500 Km^2 and storing some thirty billion tonnes of carbon, namely, the equivalent of three years of global emissions of greenhouse gas[22]. That only increases the urgency of preserving this area, with its forests and its rivers, or else risking depriving the planet of an irreplaceable protection or exposing it to unfortunate consequences[23]. Accordingly, one can say that the future of the ecological question is playing out and (also) will be played out in Africa[24].

---

[22] Cf. Greta C. Dargie, Ian T. Lawson, Tim J. Rayden, Lera Miles, Ed T. Mitchard, Susan Page, Yannick E. Bocko, Suspense A. Ifo and Simon L. Lewis, "Congo Basin peatlands : threats and conservation priorities", in *Mitigation and Adaptation Strategies for Global Change* n° 24 (April 2019), 669-686.

[23] Cf. Alain Huart, *Tourbières bassin fleuve Congo. Puits et bombe carbone*. Neufchâteau: Ed. Weyrich 2019.

[24] On Africa's contribution to the ecological question, see the article of Rigobert Minani Bihuzo, expert at the last Amazon Synod (Cf. Rigobert Minani Bihuzo, "L'intérêt de l'Église d'Afrique pour le synode spécial sur la pan-Amazonie" in *Congo-Afrique* n° 539 [November, 2019], 801-808).

That being so, one can only observe that this recognition of Africa's contribution to the planet's ecological balance is not sufficient to make of this continent a major actor in ecological reflection. In fact, when it is a question of ecology, Africa's position is essentially characterised by passivity: either one pities it, or else one recognises its natural resources as places of environmental well-being. At least, that is the impression one is left with at the end of the analysis of the allusions to Africa in *Laudato Si*. Besides, it is symptomatic that the three paragraphs referring (directly or indirectly) to the African continent are found in the chapter that treats the range of the ecological *problems* which face us today. No allusion to Africa in the systematic part of the Encyclical when, for example, it seeks remedies for the fight against the destructive culture of the environment. As much as to say, when it is a matter of the ecological question, Africa is to be placed in the column of the "passives" (it is to be sympathised with or protected), and not of the "actives"; it is on the side of the symptom, and not of the remedy.

This feeling is reinforced by the lack of references in the Encyclical to texts or positions coming from Africa. While *Laudato Si'* is distinctive in referring to the texts of different authors or episcopal conferences, there is only a single reference to a text coming from the Black Continent. It can be found in the fourteenth paragraph (introductory part of the Encyclical) and takes up the declaration of the South African episcopate which states that "Everyone's talents and involvement are needed to redress the damage caused by human abuse of God's creation"[25]. There is no other contribution from the episcopate, intellectuals or African tradition, whereas other parts of the world participate extensively in the discussion. Thus the voice of Africa seems non-existent or inaudible in discussions on ecology. This only serves to justify those who declare that "Africa has been largely missing from the thinking on environmental issues and political ecology"[26].

When looking at the environmental movements, we can have the same impression. In fact, it is impossible to deny that North America has always cut a pioneering figure in denouncing the destruction of natural environments caused by modernisation. In Europe, the environmentalist struggle has been nourished by the harmful effects of industrialisation in the past two centuries. Moreover,

---

[25] Southern African Catholic Bishops' Conference, *Pastoral Statement on the Environmental Crisis* (5 September, 1999), cited in Pope Francis, *Laudato Si'*, n. 14.

[26] Estienne Rodary, "Crises et résistants : les écologies politiques en Afrique" in Écologie et Politique n° 42 (2011), 19.

the detailed work of cataloguing ecological movements outside a European or North American context launched by Joan Martinez Alier and Ramachandra Guha[27] has enabled to draw attention to the rise of the environmentalist movements in the countries of Central and South America, as well as in the Asiatic continent, with India in the lead. But nothing about Africa! Hence, it is a short step to say that Africa, although well aware of the worst environmental situation recorded on Earth, makes no contribution to the global management of the environment[28]. Thus the bitterness of Estienne Rodary: "When the poor are brought into these reflections, Africa remains marginalised and does not appear"[29].

It would, perhaps, be fitting to nuance the opinion that insinuates that Africa is a "virgin land" where environmentalist sensitivity is concerned. Indeed, it is an established fact today, that "the privileged place of the first conservationist awareness was very often (...) the colonial soil, a heterogeneous mixture of scattered elements, indigenous, romantic or orientalist; the perceptible risk being generally that of an ecological and social disaster"[30]. Furthermore, Rodary himself, inspired by the works of the British historian of the environment at the origin of this discovery of colonial ecologism, Richard Hugh Grove,[31] acknowledged that the "explorations of the Europeans in the tropics, then, rapidly, the radical modifications made to the environments, brought about the emergence of an environmental sensitivity, prior to the American movements of the second half of the 19th century"[32]. It is true, however, that Rodary also points out that this sensitivity was more the result of the colonial administration in opposition to the practices of the natives. That is to say, "the conservation of "African" nature [proceeded] more from an exclusion than from a reappropriation, even though the asymmetries of power do not remove all capacity for action from the most excluded or the most defenceless"[33].

---

[27] Joan Martinez Alier and Ramachandra Guha, *Varieties of Environmentalism: Essays North and South*. London: Earthscan, 1997.

[28] Cf. Workineh Kelbessa, "La réhabilitation de l'éthique environnementale traditionnelle en Afrique" in *Diogène* n° 207, 2004, 25.

[29] Estienne Rodary, "Crises et résistants: les écologies politiques en Afrique" [see note 26], 20.

[30] Jean-Paul Deleage, "Aux origines de la science écologique: à propos de quelques ouvrages récents" *in Revue d'histoire des sciences*, tome 45, n° 4, 1992, 487.

[31] Cf. Richard H. Grove, "Science coloniale et naissance de l'écologie", Écologie Politique n° 2 (1992), 81-97. See also, Richard H. Grove, *Green Imperialism: Colonial Expansion, Tropical Island Edens and the Origins of Environmentalism, 1600-1860*: Cambridge, Cambridge Univ. Press, 1995).

[32] Estienne Rodary, "Crises et résistants : les écologies politiques en Afrique" [see note 26], 20.

[33] Estienne Rodary, "Crises et résistants: les écologies politiques en Afrique" [see note 26], 20.

Can we conclude that the crisis tied to the looting of the African continent is also augmented by a crisis in the ability of the social movements to react, and by a crisis in the ability to articulate and spread the environmental message? This would be to forget that the inventory of the forms of ecological movements made by Joan Alier and Tamachandra Guha dates prior to 1997, and that, since then, things have somewhat evolved. Rodary himself indicates that "discourse on the environment has spread in Africa over these last two decades just as elsewhere in the world"[34]. As for reflection on the environmental question in Africa, it would be enough to see the great literature actually developed there, in a strong pace[35], not forgetting those articles which do not date just from this decade.

Africa is not a simple black hole when it comes to the environment; and it cannot be considered as the sole victim of the environmental problem. Concern for ecology inhabits and must inhabit each son and daughter of this continent; this mobilisation must be at the same time political and social, nourished by a reflection calling forth the specific talents of African culture, in order to enrich the understanding of this crisis and to join the search for the relevant solutions. The appeal of the South African bishops, echoed by the Pope in the fourteenth paragraph of *Laudato Si*, finds there its complete meaning: Africa must contribute all the richness of its rationality and its culture, not only to repair the harm done to the environment, but also to construct a body of thought capable of sustaining the preservation of our common home. In addition to that, as the Encyclical refers to the document, *Pastoral Statement on the Environmental Crisis* of the Southern African Catholic Bishops' Conference, it indeed reminds us that "all of us can cooperate as instruments of God for the care of creation, *each according to his or her own culture*, experience, involvements and talents" (LS 14). Thus, the ecological question can be thought of starting from and by Africa, based on what constitutes its soul, its particular rationality and its culture.

---

[34] Estienne Rodary, Crises et résistants: les écologies politiques en Afrique" [see note 26], 20.

[35] We note, for example, the last two volumes of the collection "Climate and Culture" from Brill Academic Publishers (Cf. Ingo Haltermann and Julia Tischler (eds.), *Environmental Change and African Societies*, Leyde: Brill Academic Publishers, 2019 and Eckart Ehlers and Katajun Amirpur (eds.), *Middle East and North Africa. Climate, Culture and Conflicts*. Leyde: Brill Academic Publishers, 2021). Cf. also, Matthieu Teyomnou, *L'Afrique face aux défis de la dégradation de l'environnement et du réchauffement climatique de la planète. Tome I: Comprendre pour agir. Problèmes, analyses et diagnostic*. Paris: Éditions Campus ouvert, 2019.

3.2. The ecological crisis and the surrounding dominant culture

To understand the relationship between "culture" and the "ecological question", it is appropriate, first of all, to recall the long history of the word "culture". Its first traces go back to the eleventh century where the substantive, *cultura*, refers to a piece of land worked to produce plants, that is to say, the agriculture. In French, this significance remains to this day, as in the expression "culture vivrière", while the religious sense of the term – "*cultura*" also means the religious cult – dating from the same period, quickly fell into disuse and did not survive the eleventh century[36].

The *humanistic* sense of the term culture, which associates it with the things of the mind, must, therefore, be considered as a metaphorical and late meaning. In fact, it was only around the fourteenth century, with the humanists of the Renaissance, that the term culture began to be used in a figurative sense, applying it to acquired knowledge. The philosophers of the Enlightenment (such as Hobbes, Locke and Rousseau) bring this intellectualist sense to its acme. "Culture", therefore, refers to all the gains and developments of mental functions by means of intellectual process. It refers to all the knowledge accumulated by learned persons. This approach coincides with the rise and spread of education as the *medium* by which human mind is "cultivated", develops its knowledge.

It is with this meaning that the collective and identity sense of the word "culture" was gradually to be forged. Claude Rivière situates this new twist of the word during the nineteenth century under its Germanic form, *Kultur*[37]. In this case, "culture" refers to what is common to a group of individuals and welds them together, to distinctive values and features which enable a certain number of persons to be recognised as members of a group. It is in this sense that the Quebec sociologist, Guy Rocher, who is an authority on this subject, defined it as "a body bound together with ways of thinking, feeling and acting, more or less formalised, which, being learned and shared by a number of persons, serves, in a manner at once objective and symbolic, to group those persons into a particular and distinct community"[38]. Léopold Sédar Senghor, of the Académie française and from Senegal, is even more precise. For him,

---

[36] Cf. Claude Rivière, "Culture" in *Encyclopédie philosophique universelle.* Vol. II. *Les notions Philosophiques*, Paris 1990, p. 529. The following lines are also inspired by this article by Claude Rivière.

[37] Cf. Cristine Rivière, "Culture", p. 529.

[38] Guy Rocher, *Introduction à la sociologie générale.* Montréal: Ed. H.M.H, 1969, 88.

culture "is a certain way, proper to each people, of feeling and thinking, of expressing oneself and acting. And this "certain way", this character [...] is the symbiosis of the influences of geography and history, of race and ethnicity"[39].

We can say, then, that culture has a huge influence on behaviour, individual and collective. It determines the way of understanding the world and others. It is, as it were, our eye[40] on the world, our approach to what surrounds us. In a word, it corresponds to the view that a Human being has of himself and of the world which surrounds him. In this sense, culture has an effect on the ecological question and that it influences the way we perceive nature: our view of the world effectively conditions the management of our environment.

Consequently, we can understand that culture is to be considered as one of the transverse axes of the Encyclical *Laudato Si'*. Echoing the teaching of his predecessor, Pope François says that he is convinced that "the deterioration of nature is closely connected to the culture which shapes human coexistence[41]. He notes, for example, that the water problem – especially its waste– "is partly an educational and *cultural* issue, since there is little awareness of the seriousness of such behaviour within a context of great inequality" (LS 30). In general, *Laudato Si* traces an unswerving line between the present ecological crisis and what the Pope calls the "culture of waste", characteristic of the world around us accustomed to transforming everything into refuse (LS 21-22).

It is not enough to observe that the present ecological crisis is intimately linked to the ambient culture. We still must determine the aspects of this culture which account for this link. To do this, it would be helpful to identify the moment when the tendency toward the decline of the environment began.

Pascal Acot sets the rise of the awareness of the ecological problem in Modern Times[42]. It can be inferred that environmental degradation became flagrant and worrisome at that time. In fact, we know that this period was characterised by the signs of the first Industrial Revolution, and by the mental and material colonisation of the world by the West. These two events mark a radical break

---

[39] Léopold Sédar Senghor, *Les fondements de l'Africanité ou négritude et arabité*. Paris: Présence Africaine, 1967, 47.

[40] Not a static eye since culture can know evolutions and changes. It remains that these changes are imperceptible to the extent that they occur very slowly and over a long period. We also note "culture" is close to "civilization", although the latter term has always had comparative and highly discriminatory traits.

[41] Benedict XVI, *Encyclical letter Caritas in veritate on Integral Human Development in Charity and Truth*, 29 June 2009, n. 51, cited in Pope Francis, *Laudato Si*, n. 6.

[42] Cf. Pascal Acot, *Histoire de l'écologie*. Paris: PUF, 1988, 44.

in the type of relationship with nature (and their various representations) that all societies had encountered until then[43]. They inaugurated a new era, based on the claim to constant world dominion.

From there it is only a small step to take before considering nature as a "simple tool" at man's disposal. And that step was quickly taken by Francis Bacon, for instance. As proof, in his book, *New Atlantis*, he promotes a world where man is the centre, a society whose organizing principle is that of science, and whose sole aim is "the knowledge of causes, motions, and secret virtues enclosed in nature's bosom; and to give to the empire of man's mind the furthest possible extension"[44]. There we are confronted with a body of thought which assumes "the unavoidability of nature's domination by man's reasoning and economics, its desacralization and its mercantilization, all the way to the recent industrialization of life itself"[45]. Karl Marx would later speak of the will to "exploit the whole of nature, of the Earth in every sense"[46]. In a nutshell, man has been granted absolute authority over the biosphere.

From then on, this culture of domination took root in the mind of modern man who considers it as the only way to grasp reality. The cumulated influence of the Reformation and the Counter Reformation, then of the scientism in the following centuries, was going to consecrate this absolute anthropocentrism and install it in a lasting way. As Jean-Pierre Déléage has noted, "this Promethean vision of the submission of nature to the man [has ended up becoming] hegemonic in Western culture"[47], to the degree that anyone who dares to think otherwise is considered as a barbarian or a savage. The only relationship with the world around us is one of utility, with technology becoming the single interpretative paradigm of the universe.

The direct consequence of such an approach to the universe is to lead science gradually to abandon its mission to explain the phenomenal world and submit to the unbridled search for efficiency through the omnipresence of technology. It is a matter of giving a technical orientation to scientific activity.

---

[43] Cf. Jean-Paul Deleage, "Aux origines de la science écologique : à propos de quelques ouvrages récents" [see note 30], 477.

[44] Francis Bacon, *La Nouvelle Atlantide*. Gutenberg: The projet, 2008, 26.

[45] Donald Worster, *Les pionniers de l'écologie*. Paris: Ed. Le Sang de la Terre, 2009, 254.

[46] Karl Marx, *Fondements de la critique de l'économie politique*. Paris: Ed. Anthropos, 1967, 367.

[47] Jean-Paul Deleage, "Aux origines de la science écologique : à propos de quelques ouvrages récents" [voir note 30], 479.

This approach would later claim the pretty name of "instrumental reason", which is considered to be the natural fulfilment of modern science and its new expression.

True, Gilbert Hottois maintains that, *a priori*, the marriage between science and technology is not, in itself, a bad thing, to the extent, *a priori*, that they are not condemned to ignore each other, but called to adhere to each other very closely[48]. He observes, however, that, in practice, techno-science – the name he gives to this marriage – has quickly

> become a kind of absolute evil concentrating all the scourges of the time: technicism and technocracy, multinational capitalism, economic neo-liberalism, pollution, exhaustion of resources, greenhouse effect, American imperialism, globalisation, global injustice, loss of humanist values, etc[49].

Michel Freitag has also upheld the natural character of ambient technicism, arguing that man is naturally bound to technology. He thus affirms the ontological dimension of man's technicism, without making it however the only mode of human action. Indeed, for him, "ontologically, technology or technicity is an *essential dimension*, but nonetheless *partial or insufficient* for human action, and, more broadly, for every subjective activity in the world"[50]. In other words, while being one of its essential dimensions, technology alone cannot explain the dynamics of human action. Its "ontologization" rests on the substantial ontology of the human will. Hence human will is able to be true to itself; to remain the reflective master of the objectives of technology, faithful in that to its vocation of reaching towards self-awareness and awareness of the world. Unfortunately, it can also bow to technology, accepting, *ipso facto*, its own reification, conferring on technology and its dynamics an immediate ontological significance. *Everything would thus be a question of human freedom and responsibility*[51].

In his later works, Freitag would make the bitter observation that it was the second alternative that won out in contemporary culture. Indeed, wedded to

---

[48] Cf. Gilbert Hottois, *Philosophies des sciences, philosophies des techniques*. Paris: Odile Jacob, 2004, 14.

[49] Gilbert Hottois, "La technoscience: de l'origine du mot à son usage actuel" in Jean-Yves Goffi (ed.), *Regard sur les technosciences*. Paris: Vrin, 2006, 23.

[50] Michel Freitag, *L'oublie de la société. Pour une théorie critique de postmodernité*. Rennes: Presses Universitaires de Rennes, 2002, 245.

[51] Cf. Michel Freitag, *L'oublie de la société. Pour une théorie critique de postmodernité* [voir note 50], 246.

science, technology ended up attaining a certain degree of autonomy by virtue of the self-referential character of the techno-scientific logic. It is this logic which, in an autonomous way, channels the development of science (thus moving away from its aim of explaining man and the world), without directly considering human needs and aspirations[52]. Worse still, everything leads us to believe that it is technology that dictates to human beings (and to science) their needs and their aspirations, giving itself the status of "a power which knows and obeys no limit, internal or external, in its grip on reality"[53]. In this, Michel Freitag concurs with Jacques Ellul for whom, "when one looks for a "culture" or a humanism for our technical society, it always begins with the fact that the person in question is above all made for technology; the only great problem is that of adaptation"[54]. The drama in this situation is that this technologized person "has no point of reference, intellectual, moral or spiritual, as a starting point from which to be able to judge and criticise technology"[55].

It is indeed this "tendency, at times unconscious, to make the method and aims of science and technology an epistemological paradigm which shapes the lives of individuals and the workings of society" (LS 107) that the Pope identifies as the principal cause of our present ecological crisis. This approach, based on performance, does not hesitate to leave aside everything – person or object – which does not share in this performance or no longer fits it. The non-performing becomes rubbish, good for nothing, therefore destined to be chucked away. Today, this culture has taken a solid anchorage in the life of humanity, to the extent that the technocratic paradigm producing it "has become so dominant that it would be difficult to do without its resources and even more difficult to utilize them without being dominated by their internal logic" (LS 108).

Consequently, the resolution of the ecological crisis needs not only a factual commitment in favour of the environment and mankind, but, in a radical way, a change in the paradigm for understanding reality, a cultural renewal. For it is only such a renewal that will resolve the crisis, impelling from within an efficient commitment in favour of an integral ecology that binds together respect for the environment, justice and peace.

---

[52] Cf. Michel Freitag, "Totalitarismes: de la terreur au meilleur des mondes" in *Revue du Mauss* n° 25, 2005, 174.

[53] Michel Freitag, "Totalitarismes : de la terreur au meilleur des mondes" [voir note 52], 157.

[54] Jacques Ellul, *Le système technicien*. Paris: Calmann-Lévy, 1977, 349.

[55] Jacques Ellul, *Le système technicien* [see note 54], 352.

It remains that *Laudato Si* makes the bitter observation that "we still lack the culture needed to confront this [environmental and human] challenge" (LS 53) which assails us. Further, it recognizes that we cannot claim today to have a "sound ethics, a culture and spirituality genuinely capable of setting limits [to human action] and teaching clear-minded self-restraint" (LS 105). Hence the need "to show respect for the various cultural riches of different peoples" (LS 63) to leave the only key that interprets and transforms the reality imposed on us by the techno-scientific paradigm, and which has led us into the present ecological crisis.

Thus the change of paradigm consists not in the rejection of the advantages and benefits of the dominant culture of our time, but in its expansion and enrichment through other forms of wisdom. It is a matter of giving our contemporaries another point of reference that will allow them to critique their situation so as to remedy the flaws of "technologized man" signalled by Jacques Ellul. In that way, the monopoly and the omnipotence of the "techno-scientific" paradigm will be shattered in order to enable humanity to "recognize the need for changes of lifestyle" (LS 23), this cultural overture serving as the motivation and educational path necessary for that change.

3.3. African culture at the service of integral ecology

Africa's particular rationality can serve the environmental cause in offering modern people new points of reference with which to think their relationship with the world. In this way, the continent would get itself out of the passivity it is often kept in when it comes to the debate on ecology. Its peoples will drape themselves in the robes of power initiative and creative ambition for everything which concerns their destiny and that of the world.

It is quite possible to pose the question as to the African continent's unity of identity. We must then make clear that, when we speak of Africa, we are referring to the culture of sub-Saharan Africa, of so-called "Black Africa". Therefore, if, as Joseph Masson claims, "from the ethnological point of view, [the expression "Black Africa"] (…) in fact covers a number of differences"[56], and one must recognise that the scope of fundamental cultural elements common to the whole sub-Saharan Africa greatly exceeds those differences. Indeed, a careful observation of the African reveals a convergence in the conception of

[56] Joseph Masson, *Père de nos pères*. Rome: Ed. Pontificia Università Gregoriana, 1988, 10.

his being in the world[57], a factor that authorizes us to speak in general terms about the Black Continent. Our approach is thus opposed as much to the vague negationism "of those who think that there was no culture in Africa before its contact with the West", as to the narcissistic imperialism "of those who admit the existence of these cultures but yet demand their replacement by concepts imported from Europe; an attitude which consists in judging everything in the light of European criteria"[58].

Not that we are denying the permanent contact of the African culture with the outside world or its shift in the wake of this contact. We know well that no culture can claim to have remained self-sufficient and static throughout the centuries. Nonetheless the evolution of a culture can affect either its morphology, its institutions and its modes of expression, or, even the fundamental view of the world it harbours. If the morphology of a culture or its expression can evolve substantially in the short or long term, the view of the world that it harbours – its perspective of "sense bestowal", composed of inalienable certainties – will resist change for a longer time and more stronger way. This is what has appeared, moreover, in the analysis of Western culture, whose technical vein which rooted in the Renaissance has crossed the centuries so far as to guide our contemporary world to its form of "techno-science".

The perspective of sense bestowal, with its inalienable certainties – the way of one sees the world – thus constitutes the bedrock, the foundation which lasts over time and enables us the identification of a culture despite the changes of expressions it has undergone. It is this

> "bedrock" which supports all activity [...], ensuring that one cannot understand someone better than to start with "what he believes", that is to say, to start with his "bedrock", from the bedrock of his deepest certainties, a bedrock which is "invisible", more within himself and which harbours, simultaneously, his view of the visible and his conviction of the "invisible"[59].

---

[57] Cf. Charles Dundas, "Native Laws of some Bantu Tribes of East Africa", in *Journal of Royal Anthropological Institute* n° 51 (1921), 217.

[58] Adrien Lentiampa Shenge, "La culture de la vie et la promotion du pardon en Afrique", in *NRT* 133 (2011), 405.

[59] Ntima Nkanza, "La quête du divin en Afrique : autopsie d'une crise et grille possible de lecture" in *Telema* n° 128 (2007), 10.

Where Africa is concerned, we can say that life[60], as supreme value, constitutes the "bedrock of resistance" which has never yielded to the assaults of the cultural novelties and imports that have besieged it over the centuries. Indeed, a careful observation of African world reveals that reality has always been perceived and interpreted in terms of its dynamic aspects – such as fertility, the sense of the family, respect for the dead, as well as a certain harmony with nature –, all link to the burst and promotion of life. Life thus constitutes the filter through which African views reality.

Based on the supreme value of life, African's view of the world has been consecrated by the expression "life force" which Placide Tempels has highlighted to characterize what all the Bantus have in common: "there are in the mouth of the Blacks words which can be heard over and over again. They are those words express their supreme values, their highest aspirations. They are like variations on a leitmotiv rooted in their language, their thoughts and in all their acts and gestures. This supreme value is life, force, living strongly or life force"[61]. Life or "life force", therefore, appears as the substance, the bedrock of African view of the world. It is the characteristic feature of African rationality; a rationality that defines and fashions the frame in which interlocking concepts of person, community and nature are rooted and find their meaning[62].

Since this life force is the energy which affirms and reinforces life, determining the world and the extent of survival, it goes without saying that a

---

[60] This is the conclusion to which we arrived in our study of "the culture of life and the promotion of pardon in Africa" to which we have already alluded. This conclusion is based on the analysis of several texts from various horizons such as the documents concerning the two recent synods on Africa and the writings of various African philosophers of the rank of Placide Tempels, Fabien Eboussi Boulaga, Paulin Hountondji, Engelbert Mveng, etc. (See Adrien Lentiampa Shenge, "La culture de la vie et la promotion du pardon en Afrique" [see note 58], 407-411). We can go even further in the history of Africa and appeal to the "New Manden Charter", also known as the "Hunters' Oath", attributed to Soundjata Keita, founder of the Mali Empire. This Charter, which is more than seven centuries old, dating from 1222, is based on the primacy of life. Doesn't it open with this statement: "All (human) life is life. It is true that a life comes into existence before another life. But one life is not more ancient, nor more respectable than another life. Just as one life is not superior to another life"? It is this pre-eminence of life over all other things that founds the body of law, as we read in the second paragraph: "Every life being a life, any harm caused to a life requires reparation" (Read more in Youssouf Tata Cissé & Wâ Kamissoko, *Soundjata, la gloire du Mali*, Paris : Karthala, 2009).

[61] Placide Frans Tempels, *La philosophie bantoue*. Paris: Présence Africaine, 1965, 30.

[62] Cf. Innocent Chijindu Ngangah, "Vital Force, Personhood and Community in African Philosophy: An Ontological Study", in *The International Journal of Humanities & Social Studies* 7/10 (2019), 48.

significant life will be one which maximizes the life force in the aspects of a person's life[63]; and, among these multiple aspects of a person's life, there is the environment, there is the cosmos.

For an African, the expression "life" or "life force" does not refer only to people; it also concerns the plant and animal world; it refers to everything that exists or accompanies existence; and to the cosmos which houses them all and helps them to strengthen their life force. For, "beyond interpersonal relations, the African knows that the whole cosmos strives to make him grow so that, between him and the Creation, there is an indispensable interaction devoted to the mutual well-being of the human person and of the entire universe"[64]. In this way, there is, as it were, a link of mutual begetting, a mutual child-bearing, not only among the members of a community, but also between humans and the cosmos[65]. Put in another way, one cannot flourish as a person in African world unless one has a close relationship with the others and with the cosmos. The flourishing of the one leads to the well-being of the other, and *vice versa*.

There is no need, then, for African, to conjure up some kind of "heuristics of fear" to justify the responsibility which falls to him for the protection of nature. Such a responsibility deriving from the mutual begetting which binds him to the others and to the cosmos through the life which they all share as a gift received. The responsibility towards others and the environment stems from this life received. This is the basis of all ethics and all African ethics. The African is better placed to grasp what is at stake in the ecological crisis which we are living through, which is, ultimately, life[66]. That is as much as to say that one cannot understand the need for a commitment in favour of the environment, in favour of ecology, without referring to the value of life. Ecology and life march in step; and no one can separate the one from the other.

Moreover, in this connection, it is symptomatic that the various formulations of the categorical imperative on which Hans Jonas builds his complete theory of ecological responsibility in the age of techno-science revolve around the

[63] Cf. Ada Agada, "The African vital force theory of meaning in life" in *South African Journal of Philosophy* 39/2 (2020), 103.

[64] Bénézet Bujo, *Quelle Église pour un christianisme authentiquement africain? Universalité dans la diversité*. Fribourg: Schwabe Verlag, 2020, 44. See also 54-55 about mutual begetting.

[65] The principle of mutual begetting can also enable us to read with African eyes the Pope Francis's Encyclical Letter *Fratelli tutti on Fraternity and Social Friendship*, 3 October 2020.

[66] Cf. Jean-Marc Ela, *Repenser la théologie africaine : Le Dieu qui libère* [voir note 2], 124.

notion of "life", thus implying the centrality of the problem of life in every approach to the ecological question: "Act so that the effects of your action are compatible with the permanence of genuine human *life* on Earth", or "Act in such a way that the effects of your actions are not destructive of future possibility of such *life*", or "Do not compromize the conditions for the indefinite *survival* of humanity on Earth"[67]. Thus African rationality did not wait for the world to be threatened with destruction by man's technological power to make life the basis of its rationality.

The same is true of the jonasian archetype of the "newborn"[68], intensely decried by his opponents[69]. It would find strong support in the three-dimensional conception of family in Africa[70], which simultaneously encompasses the visible and invisible worlds, one of whose the three dimensions, the "not-yet-born" (alongside the two other dimensions, namely, the living and the dead) is an integral part of our world, thus already existent and not tied only to the future, to what is to come.

Emmanuel Lévinas could have also found in the African culture's principle of mutual begetting a metaphysical to found once and for all his rejection of the totalitarianism of the subject stemming from traditional ontology which consecrates the primacy of the ego. Indeed, in African rationality (as with Lévinas), the subject is not the measure of all things and does not constitute the constituent pole. For, in reality, he is constituted by others, by his community of life: it is that which gives him his identity and makes him what he is.

Ultimately, African rationality proposes an image of human being which is at the antipodes of the anthropology underlying the culture that led to the ecological crisis we have today. René Descartes is not totally foreign in the crystallization of such a culture. In exalting the human mind he made it, in a certain way, autonomous not only from the body but also from the world surround it, and which serves as its abode. In the end, only the mind, *res cogitans*, has the right to claim the status of truth and, therefore, of respectability.

---

[67] Hans Jonas, *The imperative of responsibility. In search of an ethics for the technological age*. Chicago & London: The University of Chicago Press, 1984, 11. Our emphasis.

[68] Cf. Hans Jonas, *The imperative of responsibility* [see note 67], 130.

[69] Cf. Bernard Sève, "Hans Jonas et l'éthique de la responsabilité", *Esprit* 165 (1990), 72-88. For a dense discussion on the question, we refer to our book on Ricœur (Cf. Adrien Lentiampa Shenge, *Paul Ricœur. La justice selon l'espérance*. Bruxelles: Ed. Lessius, 2009, 245-247.

[70] Cf. Bénézet Bujo, *Quelle Église pour un christianisme authentiquement africain ? Universalité dans la diversité* [see note 64], 43-54.

At most, the *res extensa* can benefit only from plausibility, with no proper guarantee. Consequently, it must be content itself with only a pragmatic function; with only being of service to what can assures the subject's well-being. Thus, behind the present ecological crisis, there is the problem of the identity of man as subject.

In an article on capitalism[71], we showed that the anthropological status of the subject promoted by Western culture is that of the individual. Now, the property of the "individual" is to understand himself essentially in the way that Paul Ricœur calls "*idem*-identity"[72], that is to say an identity which allows no room for alterity. Such an identity is always centered on distinctive identity, in that what it first postulates is what particularizes it. In that, "*idem*-identity" has always had a store of negativity to the extent to which what counts is its uniqueness. It is not for nothing that the notion of individual always goes hand in hand with that of autonomy; and the only conception of responsibility which it accepts is what derives from its own choice, its freedom. Unfortunately, as Jacques Ellul warned, "when one turns the individual into the supreme value, one ends up with a society that has disintegrated"[73]. The disintegration of the environment in a world based on the concept of man as "individual" was, therefore, unavoidable.

The African's idea of the subject is at the antipodes of the notion of the individual whose anthropological poverty could only lead to the abasement of man to the status of an object, thus opening the way to a culture of waste. If we had to choose a Western equivalent to African concept of human being, we would choose the notion of the person. Not only because, through its etymology, this notion refers to the face, thus highlighting the idea of relation and community as essential to the nature of man. But, above all, because, it is the opposite of the notion of the individual – who, insisting on his uniqueness, places himself in the category of things –, the notion of person introduces the idea of "singularity". This idea corresponds very well to what is fundamentally African: despite the common view about his supposed "communitarianism", his identity results more from the synthesis of the universal and the particular. His singularity means that he can never be abased to the level of

---

[71] Cf. Adrien Lentiampa Shenge, "De l'ère de l'individu à l'ère de la personne. Pour une économie à visage humain", in *Gregorianum* 94/4 (2013), 833-847.

[72] Cf. Paul Ricœur, *Oneself as another*. Chicago & London: The University of Chicago Press, 1992, 2-3.

[73] Jacques Ellul, *Propagandes*. Paris: Librairie Armand Colin, 1962, 106.

objects. Furthermore, and fittingly, Emmanuel Mounier, the great thinker of personalism, maintains that "the person is not an object, but truly what in each human being cannot be treated as an object"[74]. Therefore, let us agree with Paul Ricœur who thinks that "if the "person" returns, it means that it remains the best candidate for engaging in the [ecological] struggles [of our time]; I mean to say: a better candidate than all the other entities blown away by the upheavals [of the crisis]"[75].

Thus African rationality offers with regard to integral ecology the image of human being he must construct for himself. For, as Pope Francis notes in *Laudato Si'*, "There can be no ecology without an adequate anthropology" (LS 118). African rationality invites us to approach the environment with a vision which goes beyond the immediate and renounces the position of conqueror, to enter an attitude of wonder, conscious that we are intimately linked to everything that exists (LS 11 and 36).

4. Conclusion

While the present ecological crisis summons humanity "to recognize the need for changes of lifestyle" (LS 23), African rationality provides resources for reflection on cultural renewal, resources capable of sustaining an integral ecology. Instead of the culture of waste – therefore of death – which has a hold on man's current mind-set, this African rationality offers us a culture of life. It teaches us a concept of happiness which is not simply the sum-total of an accumulation of nature's bounty, but aspires to a balanced life where not only personal accomplishment but also communitarian achievement are met in a healthy environment.

An in-depth dialogue with African culture "will enable the rediscovery of the spirit of the Earth, the Air and the Water, or the tree violated, tortured and exploited because of human greed"[76]. Its own rationality will offer a means of reshaping our discourse on ecology. Its light will illuminate the endeavours of contemporary Western philosophy, as in the works of Hans Jonas or Emmanuel Lévinas, to nourish ecological reflection. It will give them new arms to react to the criticisms (and limits) habitually thrown at their proposals.

---

[74] Emmanuel Mounier, *Le personnalisme*, Que sais-je? 395, Paris 2001, 5.

[75] Paul Ricœur, "Meurt le personnalisme, revient la personne...", in *Lectures 2. La contrée des philosophes*, Paris 1992, 198.

[76] Jean-Marc Ela, *Repenser la théologie africaine: Le Dieu qui libère* [see note 2], 128.

In consequence, where the autonomy promoted by Western culture, at the root of the ecological crisis, stipulates that "each individual disposes of an absolute monopoly over his life", to the extent that "the life of each individual is his own"[77], African rationality, based on the principle of mutual begetting, reminds us that life comes to us first from our ancestors, sources of fertility, before the conjugal union of our parents who were invited, in turn, to strengthen it by transmitting it to their children[78]. Where imperialism of the subject, denounced by Levinas, sees the implement, enjoyment and consumption as the only possible relationship to establish with the world[79], African culture appeals to an "ethic of solidarity with our mother, the Earth"[80]. Where, for the technocratic culture, fulfilment consists of possession and consuming, the African culture will tell us that "one can only flourish when being in a close and harmonious relationship with others and with the universe"[81].

---

[77] Jeremie T.A. Rostan, *Le capitalisme et sa philosophie*, Édition électronique, 2009, 9, www.institutcoppet.org/wp-content/uploads/2012/06/Rostan-Le-capitalisme-et-sa-philosophie.pdf (consulted on April 14th, 2021).

[78] Cf. Marie Viviane Tsangu Makumba, *Pour une introduction à l'africanologie*. Fribourg: Ed. Universitaires de Fribourg, 1995, 135-136.

[79] Cf. Emmanuel Lévinas, *Totality and Infinity. An essay on exteriority*. Pittsburgh: Duquesne University Press, 2000, 131-133.

[80] Jean-Marc Ela, *Repenser la théologie africaine: Le Dieu qui libère* [see note 2], 128.

[81] Bénézet Bujo, *Quelle Église pour un christianisme authentiquement africain? Universalité dans la diversité*, [see note 64], 54.

HEIDEGGERIAN FOUNDATIONS IN POPE FRANCIS?

Ferenc Patsch, SJ

1. Introductory Hermeneutical Remarks

The title of this essay seems to be a provocation. Is it legitimate to search any traces of the Heideggerian thinking in the magisterium of Pope Francis? Can the German thinker offer some valid insights to the Church's official theology, particularly after his involvement, intensively discussed since the late 1980s in the literature, in the Nazi movement and especially after the more recent publication of his *Black Notebooks (1931-1941)*, which revealed some of his anti-Semitic ideas[1]? Is it not sacrilege to assume that there is (and there can be) such a thing than "Heideggerian foundations" in a theology *in sé* and *per sé*? Can it be anything other than part of a discrediting campaign, which tries to designate, eventually for sinister church-political purposes, the pontificate of Pope Francis proving his putative "unacceptable heterodoxy" and "heresy", as certain ecclesial contexts wish to see him?

(a.) *On the Project and Its Forerunners*. – In this essay I will study the delicate question whether or not it is possible to find connections between the Heideggerian philosophy and the Magisterial teaching of Pope Francis.

[1] Cf. The first series of Heidegger's so-called Black Notebooks, spanning the years from 1931 to 1941, were released in three volumes (*GA* 94; 95; 96) of the *Gesamtausgabe* in the spring of 2014. The debate is earlier, however. Heidegger's presumed Nazism was taken up in the 1960s in several books and articles, including important publications of Paul Hühnerfeld and Guido Schneeberger, but gained real intensity in the late 1980 with critical works by Victor Farias, Hugo Ott, and more recently, in the 2000s, by Emmanuel Faye. Cf. Ingo Farin and Jeff Malpas (ed.), *Reading Heidegger's Black Notebooks 1931-1941*. Cambridge (Massachusetts) – London: The MIT Press, 2016, ix; 3.

DOI 10.32060/C.S.9.343-384

The idea of retracing various sources (or "foundations") in papal teaching is neither new nor surprising. Pope Francis' theological vision has already been linked with many different thinkers (such as Romano Guardini, Thomas Berry, Leonard Boff, Jacques Maritain, Bernard Lonergan, Emmanuel Lèvinas[2], etc.) and more recently there have been some serious attempts to identify his spiritual heritage in various theologians (Hugo Rahner, Miguel A. Fiorito, Gaston Fessard, Pope John Paul II, just to mention some of them)[3]. A few authors have suspected/hypothesized a Heideggerian link too[4]. This line is mainly represented by those researchers who highlighted the influence of Romano Guardini on Pope Francis (it is well known that Guardini's view on technology has been influenced by Heidegger)[5]. We will confirm and contin-

[2] Kristof Oltvai, for instance, suggested a "phenomenologist" Bergoglio who has links with Emmanuel Lèvinas, Jean-Luc Marion, and Pierre Hadot; Cf. Kristof Oltvai, "Bergoglio among the Phenomenologists: Encounter, Otherness, and Church in *Evangelii Gaudium* and *Amoris Laetitia*", (https://www.degruyter.com/view/journals/opth/4/1/article-p316.xml) [last access: 28.06.2021]. Others, for instance, Gregory Rupik, presupposes the influence of Pierre Hadot; Cf. Gregory Rupik, "The Technocratic Paradigm as Promethean Vice: Pope Francis and Pierre Hadot on Nature and Humanity" (https://tspace.library.utoronto.ca/bitstream/1807/81414/4/Rupik_Gregory_201pdf711_MA_thesis) [last access: 28.06.2021]. The most common thesis, which is based on the explicit confession of Pope Francis, talks about the influence of Romano Guardini. See also our note 74 below.

[3] Cf. Santiago Madrigal Terrrazas, "Maestros espirituales del papa Francisco: Hugo Rahner, Migual A. Fiorito, Gaston Fessard", in *Manresas* 93-2021, pp. 5-18; Ryszard F. Sadowsky, "The Concept of Integral Ecology in the Encyclical *Laudato si'*", in *Divyadaan*, 27/1-2016, pp. 21-44. Others hypothesize the influence of Methol Ferré, Amelia Foretti, Hans Urs von Balthasar and even Georg Wilhelm Friedrich Hegel. Cf. Massimo Borghesi, *Jorge Mario Bergoglio: Una biografia* intellettuale. Milano: Jaka Books, 2017.

[4] Just for the sake of examples: Lisa Sideris notes laconically that "Francis [...] characterizes the technological mindset as one that enframes natures as lifeless, bloodless facts or raw material" and doing so he uses a "language reminiscent of Heidegger's key concept of nature viewed as a standing reserve". Cf. Lisa Sideris, "Techno-Science, Integral Thought, and the Reality of Limits in *Laudato Si'*", in *The Trumpeter*, 34/1-2018, pp. 14-35, at 18. Brian Gilchrist "perform[s] a media ecological analysis of *Laudato Si'* by comparing Pope Francis' plans to overturn the technocratic paradigm with Martin Heidegger's methods for rejecting *ge-stell*". Cf. Brian Gilchrist, "Papal Media Ecology: *Laudato Si'* as a Medium of technocratic Resistance", in *Journal of Communication and Religion*, 40/1-2017, pp. 56-75, on Heidegger: 63-66; 68-73, here at 66.

[5] Among the few worth to mention is, for instance, Thomas Kalary, professor of Philosophy at Suvidya College Bangalore. Cf. T. Kalary, "The Philosophical Foundations of the Encyclical *Laudato Si'* of Pope Francis", in *Jnanadeepa*, 22/1-2018, pp. 159-174. Kalary responds on the question, "who is the immediate philosophical mind behind *Laudato Si'*?", saying: a Heidegger filtered through Romano Guardini. (Beside Heidegger, Guardini was influenced mostly by Martin Buber and Max Scheler). See also some blogs (that is, publicist-journalistic level), for instance, Paul Collins, *Encyclical's Ground-Breaking Critique of Technology* (https://www.eurekastreet.com.au/article/encyclical-s-groundbreaking-critique-of-technology) [last access: 28.06.2021]; and a Dominican friar:

ue this trend putting however this issue in a broader context. We argue, it is not necessary to limit the possible Heideggerian effects to those that came through Guardini. How come? Nowhere can be found any explicit indication or evidence on the fact that Pope Francis had ever read Heidegger (whereas Bergoglio in 1986 begun a doctoral dissertation in Frankfurt on Guardini's philosophy).

(b.) *Contextuality, Interpretivity and Collaboration.* – An attempt to retrace the origins of Pope Francis' ideas beyond the explicitly referred authors should not be surprising. Magisterial documents signed by popes do not fall above from heaven like meteorites. Theological thoughts are contextual. The Spirit is at work in the Church and in her magisterium throughout human hands. The plural in this sentence is intentionally used: particularly in modern times, when Pontifices tend to promulgate a huge quantity of writings (encyclical letters, apostolic exhortations, pastoral messages, letters, homilies, etc.), it seems to be obvious that they write their texts with the help of certain theologians and/or theological circles. The end result, which they sign before promulgating the official text, should be considered as results/productions of a team-work. Given the incredible amount of scientific production within each and every theological sub-disciplines, the necessity of specialization (hardly comprehensible for non-specialists), and the interdisciplinary character of theological and pastoral activity of the pontiffs, it is inevitable that many theologians (as individuals and as groups) will be involved in the process of issuing a papal document. Collaboration is a necessary condition of remaining authoritative. However, these "helping hands", for their part, are also under the influence of a whole range of other thinkers (philosophers and theologians) and it is a highly complex task to retrace the origin of their own intuitions and terminology inbuilt, even involuntarily, into the official magisterial texts. In fact, an author and his or her sources are inseparable. Examples include the fact that neo-scholastic theologians (those who used to dominate magisterial teaching until the middle of the 20th century) primarily referred to Saint Thomas of Aquinas. Even here, however, Thomas read primarily through the lens of the 16th century theologian cardinal Cajetan, the founding father of the baroque scholasticism. By contrast, it is interesting to note that more recent magisterial theologians employ different authors, even modern ones, and when they quote

Bonaventure Chapman, in Bonaventure Chapman, "Technology and Vision", in *Dominicana* (https://www.dominicanajournal.org/technology-and-vision/) [last access: 28.06.2021], etc.

the Doctor Angelicus they tend to employ different interpretative keys than those provided by Cajetan.

In editing magisterial documents, one may wish to have a more transparent way of proceeding in the future. For instance, even collaborators could be named. After all, getting help from someone is not a shame – on the contrary! No pope is a polymath/polyhistor, nor does he have to pretend to be one! With regard to the overspecialized and highly differentiated modern intellectual landscape (in theological as in other disciplines), pontiffs have every right to be helped by experts. After all, this is an honest process: if popes give detailed instructions to their "ghost writers" (or the group of them), before they begin to work, explaining them in details their intentions, concepts and the basic theological direction they want to express, no one can question the authorship. In this way, the signed final document will reflect, indeed, the view of the one who commissioned the task: the "*mens auctoris*" is the one of the pope. However, it would be more convenient if the other authors were named and if posterity did not have to guess (or research using the difficult historical-critical methodology) to know who also wrote the text. A good example of this kind of collaboration was realised between two popes, when *Lumen fidei* (2013), for the first time in the history of papal encyclicals, was written by two pontiffs: started by Pope Benedict XVI (and, presumably, some of his collaborators), and finished by Pope Francis (and, most probably, some other experts, different than the previous ones) while at least two different styles can be distinguished in it.

(c.) *The Novelty.* – Papal encyclicals and their promulgation became a popular way of exercising papal power over the universal church roughly since the end of the 19th century. Presumably, as commonsense dictates, ecclesiastical authorities have had asked for assistance when they wrote such important texts. The novelty of in Pope Francis' magisterium is primarily that nowadays it is possible talk openly about this reality! Concerning the authorship of the encyclical *Laudato si'* (2015), for instance, Pope Francis himself has acknowledged the assistance of several people at the various stages of drafting his document. An authoritative statement about this fact came from Cardinal Peter Turkson, President of what was then called the Pontifical Council for Justice and Peace (since 2017, it has been called the Dicastery for the Promotion of Integral Human Development). At the press briefing announcing the encyclical, the drafting of which, the Cardinal said, exemplified the kind of dialogue for which the pope has called (and which he uses repeatedly in homilies and addresses). Cardinal Turkson said:

> This type of dialogue was also employed as the method of preparation that the Holy Father embraced in the writing of the Encyclical. He relied on a wide range of contributions. Some, in particular those from many Episcopal Conferences from all the continents, are mentioned in the footnotes. Others who participated in the various phases of this work all the way to the complex final phases of translation and publication, remain unnamed. The Lord knows well how to reward their generosity and dedication[6].

In his recent publications entitled *Let Us Dream* (2020), even Pope Francis has addressed openly the topic as he stated:

> After my election as Pope, I asked by experts on climate and environmental science to assemble the best available data on the state of our planet. Then I asked some theologians to reflect on that data, in dialogue with experts in the field from across the world. Theologians and scientists put their hands together until they reached a synthesis[7].

Kevin Irwin, an authoritative commentator of papal magisterium, suggests, on the one hand, that the reason for setting the record straight about authorship is that "when a papal document says things which a reader dislikes, the parlour game that ensues is to say that the pope himself did not write that section or those words, and so they are not authoritative or binding" while "[t]his pre-emptive strike by Cardinal Turkson [which] clarifies issues about authorship" which is "an invitation to dialogue and a reminder that, despite collaborators and the rather lengthy process, the [entire] document is part of the ordinary magisterium"[8]. On the other hand, however, the very fact that

---

[6] Vatican Press, "Press Conference for the Presentation of the encyclical letter Laudato Si' of the Holy Father Francis on the Care for Our Common Home] Bulletin (June 18, 2015), http://press.vatican.va/content/salastampa/en/bollettino/pubblico/2015/06/18/0480/01050.html#eng [last access: 28.06.2021]. Kevin W. Irwin, states: "This is what the Pope himself said on the plane flying to Manila (on January 15, 2015): 'Cardinal Turkson and his team prepared the first draft. Then, with some help, I took it and worked on it, then with a few theologians I made a third draft and sent a copy to the Congregation for the Doctrine of the Faith, to the second section of the Secretariat of State, and to the Theologian of the Papal Household [...]. Three weeks ago, I got their responses back [...] all of them constructive. Now I will take a week of March, an entire week, to complete it. I believe that by the end of March it will be finished and sent out for translation. I think that if the work of translation goes well [...] then it can come out in June or July". Cf. Kevin W. Irwin, *A Commentary on Laudato Si'. Examining the Background, Contributions, Implementation, and Future of Pope Francis's Encyclical*. Mahwah NJ: Paulist Press, 2016, 98, n. 2: 135.

[7] Pope Francis – Austen Ivereigh, *Let Us Dream: The Path to a Better Future*. New York: Simon & Schuster, 2020, p. 4.

[8] Cf. Kevin W. Irwin, *A Commentary on Laudato Si'*, pp. 98-99.

the pope has collaborators in drafting such a document, collaborators who come from a variety of fields (sciences, economics, social sciences and, of course, many different fields of theology) and locales (Europe, Asia, Africa, North and Latin America), is a clear sign of an "unofficial though very real collegiality" at work in drafting this document under Francis's authorship. The greater transparency of this process is a novelty of the pontificate of the Argentine Pope[9].

In this essay I will not offer a historical-critical reading of Pope Francis' writings[10], rather than presuppose the unity of his texts. I will not guess about the identity of the "ghost writers" either. My aim is simply to highlight some aspects of a possible "Heideggerian influence" (from any source) which has been, as a whole, rather underestimated in the literature, until now (even in the case of the *Laudato Si'* where it is the most evident). As I will show, there is a clear presence of some Heideggerian thoughts in Pope Francis (while, of course, the differences are also not negligible). We will study three aspects (which I name "Parallelisms") in this respect: the conditionedness of philosophical and theological thinking (1#); the criticism on technocracy (2#); and a contemplative spirituality as an alternative proposal for the future (3#). My research concludes with the affirmation (which counts as an answer to the question in the title) that, on the one hand, despite of all the remarkable similarities, there are not enough elements to label the magisterium of Pope Francis as "Heideggerian", however some parallelisms (at least the ones we study bellow) are most probably more than just *meta-literal* ones.

2. Parallelism 1#: Being Conditioned (History, Tradition, Context): A Hermeneutic Turn in Philosophy and Theology

We live in a hermeneutical epoch. Due to a number of circumstances – such as the cultural crisis of our epoch, the postcolonial political reorganization of the world order, the emergence of a world civilization, the expansion of technology (a process which we has just started and which is destined, as the so called third wave of the industrial revolution, to change radically our lives) and, intertwined with this, the loss of sense of a classical, humanistic tradition, etc. – the problem of interpretation have come to assume an ever more important role in recent ac-

[9] Cf. Kevin W. Irwin, *A Commentary on Laudato Si'*, pp. 98-99.

[10] Some references to textual disharmonies and irregularities in the text can be found, however, in the note 30 below.

ademic discourse. The so-called hermeneutic problem has emerged as a central topic first in philosophy and, after having been given autonomous elaboration, has begun to win acceptance even in the theological realm.

A first parallelism we study between Martin Heidegger and Pope Francis is related to the hermeneutical aspect of their philosophical and theological reasoning. In professional language people speak of a "hermeneutic turn" in humanities (*Geisteswissenschaften*) and also in the hard sciences[11]. What are the main characteristics of this epochal change in realm of thinking? How is it emerged in Heidegger's reflections (1.) and how is it manifested in Francis' magisterial documents (2.)? Let us now address these questions.

2.1. *Heidegger.* – It is commonplace to acknowledge Heidegger as one of the "founding fathers" of contemporary philosophical hermeneutics, maybe the most original one among all. To understand the meaning of this bold statement we need to begin with explaining in simple words what essentially "hermeneutics" means; then to take a brief look at the history of this new approach called philosophical hermeneutics; and finally to point out its main characteristics.

(a) *What Does Hermeneutics Mean?* – Every discourse which claims to be relevant, sensate, and understandable, needs to start with references to the ordinary world (that is to say, phenomenologically). So we do well if we explain the interpretative—hermeneutic—character of human understanding also with a reference to everyday life. Instinctively, we all know that educational background, personal insights, and life experiences matter in terms of our way of thinking and acting. These factors provide for us an intellectual framework to understand reality as such—with all its epistemological and relational aspects and other nuances[12]. Our prejudices— *Vorurteilen* in a

[11] To assume that the hermeneutic turn has been occurred only in humanities implies an unsustainable epistemological dualism between "explanation" (*Erklärung*) and "understanding" (*Verstehen*) and a latent ontological distinction between nature and spirit. This is a language which the post-Diltheyan hermeneutic philosophical tradition has overcome.

[12] To quote an authoritative hermeneut in this regard: Hans-Georg Gadamer in his major work *Truth and Method* writes, "In fact history does not belong to us; we belong to it. Long before we understand ourselves through the process of self-examination, we understand ourselves in a self-evident way in the family, society, and state in which we live. [...] That is why the prejudices of the individual, far more than his judgments, constitute the historical reality of his being". Hans-George Gadamer, *Truth ad Method*. London: Sheed & Ward, 1989, pp. 276-277 (ed. or. *Wahrheit und Methode*: *Grundzüge einer philosophischen Hermeneutik*. Tübingen: Mohr, 1969, p. 261).

Gadamerian sense—effectively influence the way we interpret everything that we read, hear, or view. On the one hand, they allow understanding in general, because without at least some previous experiences of things of the world we were not at all interested in them[13]. On the other hand, however, they prevent us seeing reality as it is. It is not only our *personal* "filter bubbles" (within which we constantly live in the context of a postmodern information society) that influence our political and social views. The news itself, presented to and consumed by a *community*, are also constructed: the meaning of the media content is determined by the editorial policies, the articulated or secret agendas, of newspapers and periodicals which we consult, either in printed or electronic form. It is not just the picking and choosing from the so called "objective facts", but also their presentation that is an interpretative issue. Editors give the news a decisive, and sometimes distorted, spin through shedding light on it in favour of a preconceived purpose.

There is also a third level: culture. Beyond the new social media and the printed media we belong, there is the *culture* we inhabit which, in an ever more subtle ways, provides us with "self-evident facts" that we take for granted as elements of our belief system and knowledge[14]. In this sense, there are no *brutum factum* (mere fact), since facts are *always interpreted*[15]. Information can be understood differently depending on the culture, personal history, the worldview of the information monopoly (editors of the TV station, radio show, or online news service), and the socio-cultural context of the communication itself. Brief: we belong inseparably to a hermeneutical situation. In the Western society, the postmodern life-world (*Lebenswelt*) reveals the essen-

---

[13] On this historically *interested* aspect of understanding (to understand something *significant*), Gadamer writes: "In any case, understanding in the human sciences shares one fundamental condition with the life of tradition: It lets itself be *addressed* by tradition. [...] However mediated this significance may be, and though it may proceed from a historical interest that appears to bear no relation to the present [...], the real fulfilment of the historical task is to determine anew the significance of what is examined. But the significance exists at the beginning of any such research as well as at the end: in choosing the theme to be investigated, awakening the desire to investigate, giving a new problematic". Hans-George Gadamer, *Truth and Method*, p. 282.

[14] Cf. Peter Ludwig Berger and Thomas Luckmann, *The Social Construction of Reality. A Treatise in the Sociology of Knowledge*. London – New York: Penguin Books, 1966.

[15] The radicalised form of this insight is formulated by Friedrich Nietzsche: "Against that positivism which stops before phenomena, saying 'there are only facts', I should say: No, it is precisely facts that do not exist, only interpretations". Friedrich. Nietzsche, *The Will to Power*. New York: Vintage Books, 1968, p. 267. Heideggerian and Gadamerian Philosophical Hermeneutics do not share Nietzsche's radically perspectivist perception, safeguarding the notion of truth claims present in positivism.

tially interpretative (or, more technically speaking, hermeneutic) character of our human existence.

This interpretative-hermeneutic aspect becomes particularly important, both from philosophical and theological point of view, when we are living a period of intense change in technology, society, and culture. Events, cultural phenomena, and Church documents can all be read throughout different (personal, social, historico-cultural) perspectives or "lenses": they can be interpreted according to different, but preconceived, prejudices, ideas or ideological points of view. In a Western context, passionate debates can occur based on a "liberal" or a "conservative" point of view, often using emotive and condemnatory language[16]. This reminds us that we need to proceed in a hermeneutically conscious way: i.e. reading church documents in full and on their own terms (and not just filtered by any strongly ideological preconceptions) and aware of the contextual (historical, cultural, linguistic) embeddedness of our own existence as readers. In fact, the conditionedness of every opinion, view, and conviction (including our own) are undeniable. Knowing reality as it is (*in se*), without the filtering effect of our personality, historical and cultural situation (*quoad nos*), is simply impossible – this is what the study of the phenomenology and hermeneutics reminds us. Ideas we employ did not descend suddenly into the world, but have had a long maturation process in the history of human thought. Let us now explore this point further[17].

(b) *The Hermeneutic Turn.* – Although in ancient Greece and elsewhere reflection on the laws of the interpretation in case of ambiguous texts began historical times ago, these primitive methodological considerations are not yet "hermeneutics" in contemporary sense. There are at least three phases or steps

---

[16] Kevin W. Irwin notes that one of the difficulties today for an American public is that often the prejudices of those interpreting church documents reflect the various sides of the American political debate, with numberless categories and labels used (liberal, conservative, progressive, neoconservative, etc.). Cf. Kevin W. Irwin, *A Commentary on Laudato Si'*, p. 103.

[17] In theology, this new emphasis on interpretation is generally used as an antidote to objectivistic conceptions and reductionist tendencies of the discipline's methods. None of the above-mentioned insights would have been imaginable without that dramatic change in the 20th century which has been brought about in the philosophy of Martin Heidegger, especially through his *Being and Time* (cf. Martin Heidegger, *Being and Time,* Harper & Row, New York 1962; ed. or. *Sein und Zeit.* Tübingen: Niemeyer, 1979[15]; hereafter *SZ* and *BT*). I have explained this issue in a more detailed way in a longer publication (cf. Ferenc Patsch, *Hermeneutika I. A filozófia jövője és a jövő teológiája.* Budapest: Jezsuita Kiadó, 2018). At this point it is enough to stress the simple fact that Heidegger makes interpretive understanding the central mode (*Art*) of human existence (*Dasein*).

of this long maturation. *Classical* hermeneutics started as a theological and legal methodology governing the interpretation of the Holy Scripture (*hermeneutica sacra*), and the application of written law (*hermeneutica iuridica*). Lawyers and Theologians laid down the foundations of Western civilisation analysing paragraphs of codices and the sacred text of the Bible. At the dawn of the *Modernity* (particularly under the effect of the Protestant Reformation) hermeneutics developed into a general theory of human understanding through the work of two German romantic historian, Friedrich Schleiermacher (1768-1834) and Wilhelm Dilthey (1833-1911). The most important development, however, was given to this renewed discipline by Martin Heidegger who developed hermeneutics into a fundamental *philosophical* perspective. Later on, under his influence, this discipline was worked out, more systematically, by Hans-Georg Gadamer (1900-2002), Paul Ricoeur (1913-2005), and some other postmodern scholars. In the early Heidegger hermeneutical theory still in a pre-systematic and embryonic state, whereas in the post-Heideggerian hermeneutic tradition this becomes articulated explicitly as theory. All these approaches attempt to shed light on the hidden presuppositions of our seemingly self-evident conceptions and taken-for-granted thought models. Each in their own way seek to make explicitly conscious the hermeneutical situation that is operative.

Hermeneutical philosophy recognizes the quest for understanding and meaning is a fundamental characteristic of our finite human existence. There are two consequences of this. First, that every bit of our knowledge hides an interpretative aspect (understanding equals interpreting) and, secondly, that our knowledge is inevitably finite and limited (conditioned from historical, linguistic, socio-cultural, etc. points of view). In this sense, contemporary hermeneutics is not focusing any more, as it did in the past, on understanding and interpretation exclusively of written texts. Rather, from the time of Schleiermacher and especially since Heidegger, its scope became more general: it includes all those elements in the world that somehow convey meaning and yet require interpretation. These include literary texts but extend to works of art, historical events, human actions, etc. Thus, philosophical hermeneutics (in the vein of Heidegger, Gadamer, Ricoeur, and their followers) is not merely a method, but rather a fundamental perspective on human existence and human self-understanding that has ontological relevance. As philosophical hermeneutics developed, the role of the interpreter in the process of interpretation became increasingly apparent. It becomes clear, that the thinking mind is never a "*tabula rasa*": our existence is rather imbedded into a socio-cultural, historical, and, not least, linguistic context which conditions permanently our understanding of being (*Seinsverständnis*). Reality unfolds as understanding gradually brings into play our hidden prejudices, confronting them with new experiences and modifying them in a never-ending circular movement (the hermeneutic circle). In this sense, instead of being an abstract "theory", philosophical hermeneutics is a general approach to reality as a whole (with evidently ontological claims).

These basic Heideggerian intuitions had enormous history of effects (*Wirkungsgeschichte*) on practically all disciplines (as we noted above, not just on the so called human sciences or *Geisteswissenschaften*). With the words of David Hoy: "To speak of a revolution in the history of thought is perhaps too grand, but certainly there has been a general movement that can be called the 'hermeneutic turn"[18].

(c) *The Main Characteristics of Hermeneutics: Consciousness of Contexts and Limitations*. – Now, let us turn concretely to a short analysis of a paradigmatic Heideggerian text to be able to define more precisely what is meant today under the term "hermeneutics". Despite of the fact that we find important references on the pages of *Being and Time* (1927) (*SZ/BT*) to Heidegger's intense previous engagement with the question[19], in his *opus magnum* there is not much more than a half a page left to explicitly explains his understanding of hermeneutics. The whole explanation, which is reduced to 28 lines in the first German edition (in the context of unfolding the phenomenology of Dasein), is inserted in the introductory methodological part of his philosophical project (cf. *SZ/BT* 7§C). Here, Heidegger differentiates three different levels of interpretation as he distinguishes three different levels of meaning of the word "hermeneutics". In the "primordial signification of the word (*in der ursprünglichen Bedeutung des Wortes*)", hermeneutic designates simply the "business of interpretation (*das Geschäft der Auslegung*)" (*SZ* 37/*BT* 62). This is the level of the jurists and priests engaged for millennia with ancient and

[18] David Couzens Hoy, "Heidegger and the hermeneutic turn", in Charles B. Guignon (ed.), *The Cambridge Companion to Heidegger*. Cambridge-New York-Victoria: Cambridge University Press, 1993, pp. 170-194, at 170.

[19] In a famous footnote Heidegger remarks the fact that since the winter semester of 1919/20 he dealt regularly with the question of the "Hermeneutics of facticity" (*Hermeneutik der Faktizität*) (cf. SZ 72, the translation of the footnote is missing from BT). The most important document in this regard is the course held in 1923 (*Wintersemester*) at the University of Freiburg, with the title *Ontology: Hermeneutics of Facticity* [*GA* 63], right before Heidegger left to Marburg to become a full professor there (between 1923-28).

complicated texts striving to understand them. At a second level, for Heidegger, that kind of philosophical activity is known as "hermeneutic" when it "works out the conditions (*Bedingungen der Möglichkeit*) of possibility of any ontological investigation" (ibidem). This is, however, still not the deepest level of the semantical realm, for, in this particular sense[20], the word "contains the roots of what can be called 'hermeneutic' only in a derivative sense (*nur abgeleiteterweise*)". This level of significance corresponds to the initiative of Schleiermacher and Dilthey – we can add – while these first semi-philosophical hermeneuts wanted to understand hermeneutics as "the methodology of those human sciences which are historiological in character" (*SZ* 38/*BT* 62).

Finally, there is a third, but "philosophically *primary* (*den, philosophisch verstanden,* primären *Sinn*)" sense of "hermeneutic" for Heidegger, where the term means the "analytic of the existentiality of [human] existence (*einer Analytik der Existenzalität der Existenz*)" (*SZ* 38/*BT* 62)[21]. How to make sense of this complex and slightly enigmatic expression? To understand it better, it is worth turning to the end of *Being and Time*, where Heidegger expounds Dilthey's and York's teaching on historicality. Here, he returns defining "hermeneutics" as "the way […] understanding enlightens itself (*Selbstaufkärung […] [des] Verstehens*)" (*SZ* 398/*BT* 450). To explain this latter (and, at the same time, primary) sense, Heidegger distances himself from the approach of Schleiermacher, Dilthey and York, for whom hermeneutics simply means "methodology of historiology". He considers this to be hermeneutics "only in a derivative form (*in abgeleiteter Form*)" (*SZ* 398/*BT* 450). By contrast, Heidegger makes it clear that his own initiative is radically different from the one of his romantic predecessors: while they were dealing with a methodological project, Heidegger wants to describe, initially and for most (*zunächst und zumeist*), how *things are* (phenomenologically) and his description claims to be an *ontology*, or, more precisely, the *fundamental* ontology of Dasein. In this third and very special sense, interpretation is the analytic of the human existence and the self-enlightening of understanding which is not reduced to dealing exclusively with a written text but which considers the whole world to be a text. In this letter sense, understanding is the Dasein's mode of being (*Seinsart*)[22].

---

[20] Heidegger specifies: "This hermeneutic works out *Dasein*'s historicality ontologically as the ontical condition for the possibility of historiology". (SZ 38/BT 62).

[21] "As an interpretation of *Dasein*'s Being" (SZ 38/BT 62).

[22] "As projecting, understanding is the kind of Being of *Dasein* [...]" (*Das Verstehen ist, als Entwerfen, die Seinsart des Daseins [...]*) (SZ 145/BT 185)

With this Heideggerian statement, hermeneutics become a universal philosophical discipline. To sum up its long history: in the first (*Classical*) period, hermeneuts dealt meticulously with methodological-exegetical questions (to understand normatively the Bible and juridical Codexes), remaining yet however completely blind to their own conditionedness in the process of understanding. (*Nota bene*: in the *Summa* of Saint Thomas of Aquinas, no matter how brilliantly it is written, had absolutely no *articulus* about historicality, linguisticity, etc.). In the period of Enlightenment (*modern*) hermeneuts (mostly German romantic historians: Ranke, Droysen, Dilthey) noticed for the first time that in different historical periods people thought differently: this is the moment of the discovery of the so-called historical consciousness (*historisches Bewusstsein*). But they, too, thought they could look at events objectively, from above, as they were free from influences/being influenced and saw things from an absolute (unhistorical) perspective. Only in its third period – as hermeneutics (in its *primary* sense) became the "analytic of [...] [human] existence" (*SZ* 37-38/*BT* 62) and the "way [...] understanding enlightens itself" (*SZ* 398/*BT* 450) – hermeneuts has been discovered, that our own point of view is not absolute either, but conditioned in many aspects. With Heidegger, hermeneutics reached the consciousness of *its own limitations* in noticing the conditionedness of his own point of view. Dasein become conscious of his own historicality, recognizing, with Heidegger's words, "the historicality of the historian's existence" (*SZ* 394/*BT* 446) itself. In this sense we can define Philosophical hermeneutics, laconically, as the *awareness of our own limitations and the positive recognition of our own conditionedness*. Are these elements present in some way also in the magisterium of Pope Francis?

2.2. *Pope Francis.* – We have mentioned above that the hermeneutic turn has been greatly effective among many different sciences. In fact, it has been taken various forms, including poststructuralist cultural studies, deconstructive literary studies, interpretive anthropology and social science, critical legal studies, and, last but not least, post-colonial theological studies[23]. In the following I will argue that we can consider the "style"[24] of the magisterium of

[23] "As projecting, understanding is the kind of Being of *Dasein* [...]" (*Das Verstehen ist, als Entwerfen, die Seinsart des Daseins [...]*) (SZ 145/BT 185)

[24] The word here is used as Christoph Theobald used in Christoph Theobald, *Christentum als Stil. Für ein zeitgemäßes Glaubensverständnis in Europa*. Freiburg: Herder, 2018.

Pope Francis as a consequence of the so called hermeneutic turn in theology and a greater awareness of our own limitations.

(a) *The Acknowledged Limits of a Magisterium.* – I have already argued elsewhere that the most surprising novelty of Pope Francis' magisterium consists of the courage with which he expresses an awareness of the historical-cultural situatedness of theology in general and of his own magisterial teaching in particular[25] (cf. *LS* 105). This can be clearly seen in some of the strong-worded passages of *Evangelii gaudium*. At the very beginning of his long text, the Pope is urging a "healthy 'decentralization'" in exercising papal power and he reflects also on the limits of his own initiative to write the document: "I have chosen not to explore [...] many questions which call for further reflection and study" (*EG* 16) while, as he writes, he does not believe that "papal magisterium should be expected to offer a definitive or complete word on every question which affects the Church and the world" (*EG* 16). Francis holds "not advisable for the Pope to take the place of local bishops in the discernment of every issue which arises in their territory" and, in this sense, he proves to be "conscious of the need to promote a sound decentralization" (*EG* 16). A similar idea returns in a later point of the document, when the Pope reiterates: "Neither the Pope nor the Church have a monopoly on the interpretation of social realities or the proposal of solutions to contemporary problems" (*EG* 184). Referring to Pope Paul VI, Pope Francis adds:

> [i]n the face of such widely varying situations [in which the church finds herself], it is difficult for us [popes] to utter a unified message and to put forward a solution which has universal validity. This is not our ambition, nor is it our mission. It is up to the Christian communities to analyze with objectivity the situation which is proper to their own country[26](*EG* 184).

The frequency with which the Pope address this issue of self-limitation may surprise us. At a later point of the same apostolic exhortation, Francis writes again: "In dialogue with the State and with society, the Church does not have solutions for all particular questions" (*EG* 241). In the already cited encyclical

---

[25] Ferenc Patsch, "Rivelazione, contesto, verità. Il Magistero di Papa Francesco in tempo di transizione", in Michelina Tenace (ed.), *Dal chiodo alla chiave. La teologia fondamentale di Papa Francesco*. Vaticano: Libreria Editrice Vaticana, 2017, pp. 45-72.

[26] Saint Paul VI, Apostolic Letter *Octogesima Adveniens,* 14 May 1971, n. 4; in Acta Apostolicae Sedis 63/1971, p. 403; Cf. Pope Francis, Apostolic Exhortation *Evangelii gaudium* on the Proclamation of the Gospel in Today's World, 24 November 2013 (hereafter *EG*), note 152.

letter *Laudato Si'* (2015), the Pope reproaches the attitude of forgetting one's own limits. In this ecological context we read: "Every age tends to develop a scarce self-awareness of its own limits" (*LS* 105). Of course, the Popes does not intend to weaken credibility of the one truth of revelation authentically interpreted by papal magisterium. On the contrary: to carry out the Pope's program and to build – on the footsteps of John Paul II and Paul VI – a "culture of life" and a "civilization of love" (*LS* 231) which are based on the Gospel, Francis considers it necessary that we observe "[the] limits that a healthy, mature and sovereign society must impose" (*LS* 177)[27] and, at the same time, also that the ecclesiastical magisterium develops a new (post-industrial) spirituality, sensitive to otherness and its context.

The same vision is present in the post-synodal exhortation *Amoris laetitia* (2016). The document places in the foreground the importance of discernment as an essential aspect of moral decision, focusing on the conditional and contextual aspect of revealed truth. The Pope underlines, once again, our human limits when he speaks of how the moral development of the individual "advances gradually with the progressive integration of the gifts of God" (*AL* 295) and that this will occur "by different stages of growth" (c.f. *AL* 300; 303; 312). In short, one of the main characteristics of the magisterium of Pope Francis is the desire to understand the consequences of the contextuality of theology taking into account its historical, cultural and linguistic conditions and recognizing these effects on the interpretation of ecclesiastical doctrine. Now, the question arises: what methodology does the Pope employ to achieve this particular purpose?

(b) *Hermeneutics, But Not Only: The Metaphysical Element.* – We have seen above how Pope Francis' writings reflect a hermeneutic approach in doing theology. It would be, however, one-sided and even misleading to suggest that his approach to the Christian Revelation is simply based on interpretative principles. There is also another side of the same coin. To prove this statement let us analyse another text which will nuance the picture and stress an (anti-Heideggerian) metaphysical aspect of the Pope's methodological principles. Interestingly enough, this reference is to find in an unexpected place, in the apostolic exhortation on the call to holiness in today's world, *Gaudete*

---

[27] The particular reason is that we are conditioned even in our religiosity: "Cultural limitations in different eras often affected the perception of [...] ethical and spiritual treasures [...]".Pope Francis, Encyclical Letter *Laudato Si' on Care for Our Common Home*, 24 May 2015 (hereafter *LS)*, n. 200.

et exsultate (2019). Here, Francis first legitimates the plurality of theological interpretations affirming:

> It is not easy to grasp the truth that we have received from the Lord. And it is even more difficult to express it. So we cannot claim that our way of understanding this truth authorizes us to exercise a strict supervision over others' lives. Here I would note that in the Church there legitimately coexist different ways of interpreting many aspects of doctrine and Christian life (*GE* 43)[28].

The underlying conviction of Pope Francis here is a metaphysical one: "the doctrine" and "the understanding and expression of it" (cf. *GE* 44) must be distinguishable. We can see here an element that goes beyond a mere phenomenological-hermeneutic method: for Francis, the "thing for us" and "thing as it is in itself" (with Kantian terminology, the *phenomenal* and the *noumenal* reality) cannot be all the same as radical hermeneuts and phenomenologist suggest. Although metaphysical (transcendental) elements have emerged, the predominance of the hermeneutic approach seems to be undeniable. At least until now. And it should not be surprising: ultimately, it is required by the very nature of theology (and its "principle of the incarnation"). The Pope writes:

> [The doctrine of the Church] is not a closed system, devoid of the dynamic capacity to pose questions, doubts, inquiries [...]. The questions of our people, their suffering, their struggles, their dreams, their trials and their worries, all possess an interpretational value [in the official Italian translation: *valore ermeneutico*] that we cannot ignore (*GE* 44).

With these and similar considerations, Pope Francis' text apparently wants to protect theologians from a nostalgic rigidity, fashionable in certain parts of the Church, and promote an opened and "attentive mentality" which includes a "readiness to listen" (*GE* 172) and an "obedience to the Gospel" (*GE* 173) – all considered to be hermeneutic "virtues" (or characteristics) of a post-colonial theological "stile". Indeed, for Francis

> [Doing theology] is not a matter of applying rules or repeating what was done in the past, since the same solutions are not valid in all circumstances and what was useful in one context may not prove so in another" (*GE* 173).

[28] Cf. in this respect the appreciation of doubt and a moderate perspectivism by *Laudato Si'*: "On many concrete questions, the Church has no reason to offer a definitive opinion; she knows that honest debate must be encouraged among experts, while respecting divergent views" (*LS* 61). And further, we read about a recognition of plurality in proceeding to find adequate solutions: "There are no uniform recipes, because each country or region has its own problems and limitations". (*LS* 180).

The universal claim of Francis' magisterium is that the magisterium itself is always hermeneutically circumscribed by the actual context of theology.

However, as we have noticed above, it would be certainly mistake, indeed a huge over-simplification, to affirm that Pope Francis has *tout court* a "Heideggerian methodology". Despite some undeniable philosophical-hermeneutical aspects of this theology, there is a main difference between his theological conception and that of Heidegger, and it lies in the role attributed to metaphysics. Heideggerian philosophy results to be clearly anti-foundationalist, whereas the theology of Pope Francis – despite of some of his anti-metaphysical tendencies[29] – maintains (both at a terminological and a content level) an undoubtable metaphysical foundation. This aspect is particularly present in the encyclical letter *Fratelli Tutti* (2020). At the cost of some stylistic and theological inconsistencies, chapter six of the encyclical employs some powerfully classical and foundationalist metaphysical language. Examples abound with reference to: "objective truth"(*FT* 206; 212), "universally valid principles" (*FT* 206), "adherence to the most basic of truths" (*FT* 207), "relativism" (*FT* 209), "fundamental truths", "enduring values" (*FT* 211), "enduring truth accessible to the intellect" (*FT* 212), "intrinsic worth" vs. "material objects" (*FT* 213), "universal moral demands" (*FT* 213)[30]. These departures from Heidegger, however, should not obscure further similarities and parallelisms.

---

[29] Cf., for example, one of Pope Francis' addresses with a clearly historical and salvation-economical (and in this particular sense "anti-metaphysical") conclusion: "The Christian doctrine is not a closed system, unable to generate questions, doubts, and interrogations, but it is rather a living thing which can cause restlessness, and is able to animate. It has a non-rigid face, has a body which moves and develops, has a sensible flash: the Christian doctrine is called Jesus Christ. (*La dottrina cristiana non è un sistema chiuso incapace di generare domande, dubbi, interrogativi, ma è viva, sa inquietare, sa animare. Ha volto non rigido, ha corpo che si muove e si sviluppa, ha carne tenera: la dottrina cristiana si chiama Gesù Cristo*)". Pope Francis, *Incontro con i rappresentanti del V Convegno nazionale della Chiesa italiana* (10 novembre 2015), https://www.vatican.va/content/francesco/it/speeches/2015/november/documents/papa-francesco_20151110_firenze-convegno-chiesa-italiana.html [last access: 28.06.2021].

[30] Just to quote one paradigmatic sentence: Francis' texts affirms (in a typically anti-Heideggerian vein): "As believers, we are convinced that human nature, as the source of ethical principles, was created by God, and that ultimately it is he who gives those principles their solid foundation". Pope Francis, Encyclical Letter *Fratelli tutti on Fraternity and Social Friendship*, 3 October 2020 (hereafter *FT*), 24. By contrast, in the previous chapter (chapter five) we find a much personalistic phraseology as we read about "initiate[ing] processes" (*FT* 196), "tenderness" (*FT* 194), "to practice love in [...] interpersonal relationships" (*FT* 193), "charity" (*FT* 183; 185), "social love" (*FT* 183), etc.

3. Parallelism 2# The Critical Analysis of the Present: Breaking With Technocratic Rationalism and Scientism

Perhaps the shortest and the most apt description of our epoch is that we are living in a technological age. We can see this affirmation to be proved in each and every aspect of daily life from the most profane parts of it to the most sacred ones. For instance, we fight against pandemic by counting infected and dead (and we believe in our "exact" numbers!) organizing vaccinations in industrial measures. Our pastoral ministry is no exception: we make pastoral plans rationalizing even church life, at times going to extremes with this. Western thought, under the influence of modern science, employs numerical quantification, algorithms, or formulas to bolster its findings and submit nature to human control—often with catastrophic ecological consequences.

This has not been always the case. The Aristotelian science, that reigned in the ancient and medieval periods, has been radically transformed by Bacon and Descartes into a scientifically and philosophically reductionist rationalism. Our life is still ruled by this modern cultural paradigm, but this in increasingly demonstrating an unsustainable character. In their respective philosophical and theological fields, both Heidegger and Pope Francis made efforts to change this view. Both offer a radical critique of what they call a technocratic paradigm.

3.1. *Heidegger.* – In one of his brilliant essays, the Canadian Catholic philosopher Charles Taylor notes that Heidegger's importance in the contemporary intellectual context lies partly in the fact that "he is perhaps the leading figure among that small list of twentieth-century philosophers who have helped us emerge […] from the grip of modern rationalism"[31]. Heidegger's contribution in this regard is indeed a breakthrough: it represents a kind of milestone in that painful and difficult process through which modern western culture shifted from its rationalistic worldview. It is rightly said that Heidegger's work represents an epochal change from that particular paradigm. His first merit, however, is to describe the situation and then to offer a diagnosis.

---

[31] Charles Taylor, "Engaged Agency and Background in Heidegger", in Charles B. Guignon (ed.), *The Cambridge Companion to Heidegger*. Cambridge – New York – Victoria: Cambridge University Press, 1993, pp. 317-336, at 317. In Taylor's list there are also Wittgenstein and Merleau-Ponty. He attributes, however, a pre-eminent role to Heidegger putting him in front of them as a realiser of a kind of "breakthrough".

(a) *The aspect of the Mathematical.* – From his very first philosophical lecture held in the University of Freiburg (*Wintersemester* 1919), the young Heidegger criticized all the major current philosophical schools of his time (neo-Kantianism of Marburg, Husserl's phenomenology, etc.)[32] blaming them that they narrow down (*beschraenken*), in a one-sided and absolutizing way, the notion of knowledge and the concept of reality reducing them to that of the mathematical and natural sciences (*mathematische Naturwisseschaft*) (*GA* 56/57,83) holding them as ideal (*Ideal*, *Vorbild*) of scientific rigor (*Strenge*) (*GA* 63,71). The topic would be recurrent in following years. As Heidegger progressively formulates his phenomenological-hermeneutical method (opposing it to an epistemologically oriented "worldviewish philosophy"/*Weltaschaugsphilosophie*)[33], he argues against the theoretical reification (*theoretische Vergegenstaendlichung*) of things (*GA* 59,62) and ubiquitous notion of scientific objectivity (*GA* 56/57,98-99; *GA* 58,52;57;77;91; *GA* 61,50; etc.). Nearly ten years later, in 1935-36, Heidegger returns to the topic at greater length, in an important lecture held at the University of Freiburg on *What Is a Thing?* [*Die Frage nach dem Ding*] (*GA* 41). Heidegger explains here how, in a broad sense, "the mathematical" (*das Mathematische*) (*GA* 41,69) mentality has always been present in the Western thinking since its origins. As it is well known, at the entrance of Plato's Academy a warning was addressed to the visitors: *Ageometretos medeis eisito* (Let no one who has not grasped the mathematical enter here!). For Heidegger, this motto, which proposes the centrality of *mathesis*, serves as an important aspect of the whole ancient Greek period of our culture which. He describes this mentality a "working-condition" (*Ar-*

---

[32] Heidegger exercised a devastating critique on each and every contemporary trend of philosophising: Spengler's biologism, Worldviewish/*Weltanschauung* philosophy, psychologism, historicism, existential philosophy, epistemologically oriented scientific philosophy, and even Dilthey's romantic hermeneutics and life-philosophy (i.e., the perspective centring around the conviction that philosophy's own most object is life or historical life). Heidegger's overall attempt at "re-appropriation" goes, schematically speaking, like this: epistemologically oriented scientific philosophy is criticised for not being scientific enough, life philosophy is accused of failing to grasp life itself, existential philosophy is charged with not seizing upon existence, historicism is called to account for losing sight of nothing less than history itself, and, last but not least, phenomenology is challenged for not being phenomenological enough (indeed, for being "un-phenomenological"). Cf. Fehér M. István, "Heidegger and the Hermeneutic Turn of Philosophy", in *Kellék*, 11/12-1998, pp. 7-22; and also David Couzens Hoy, "Heidegger and the hermeneutic turn", in Charles B. Guignon (ed.), *The Cambridge Companion to Heidegger*. Cambridge-New York-Victoria: Cambridge University Press, pp. 170-194, at 170.

[33] The polemic against *Weltanschauungsphilosophie* is a very frequent/common topic in Heidegger's early writings: cf. for instance *GA* 59, 9-12, etc.

beitsbedingung) and "clear work-delimitation" (*klare Arbeitsbeschränkung*), that not only characterized ancient Greek culture but which has remained effective until today in Western culture (cf. *GA* 41,76). Over the course of time the scientific revolution would occur. Figures like Newton and Galileo would assure that the scientific and technical method (*wissenschaftliche und technische Methode*) would no longer be a marginal aspect of culture, but would come to gain prominence in the Western mind. Philosophical consequences of the scientific revolution would first be expressed in the thought of Bacon, Leibniz, and Descartes and became a elevated to the position of rationalistic metaphysics (*rationale Metaphysik*) with Wolff and Baumgarten[34]. Heidegger tries to identify the deepest roots of this general cultural phenomenon.

(b) *The Underlying Metaphysics of Natural Sciences.* – One of the most emblematic representatives of this trend is René Descartes. In one of his most famous (unfinished) treatises, entitled *Regulae ad directionem ingenii* [*Rules for the Direction of the Mind*] (1701), this rationalist philosopher urges that "we [have to] reject all knowledge that is only probable, and [...] declare that only those things ought to be believed which are perfectly known and of which there can be no doubt"[35]. With these words are very well expressed the mathematical-metaphysical basic feature (*mathematisch-metaphysische Grundzug*) of the modern scientific paradigm as *Scientia universalis* "to which, as only one, which gives measure, everything is turned and directed (*auf die als die Eine, maßgebende alles hingewendet und eingerichtet ist*)" (*GA* 41,101). For Heidegger, these words note the end of the conception of the Aristotelian (ancient Greek and Medieval) principle, according to which natural bodies have their own guiding inner principles (through which they were both created and implanted by God) since they are *ens creatum*. The modern conception forgets this conception and re-writes the metaphysics of this matter, no longer holding the idea that different natural bodies have different natures. Now, all of nature comes to be seen as a coherent set of bodies in motion. Embrac-

---

[34] Bacon in his *New Organum*, a work written as an affront to Aristotle's *Organum*, attempts to demonstrate the illegitimacy of the old sciences and establishes a new scientific method. As Bacon puts it: "the sciences we now have are no more than elegant arrangements of things previously discovered, not methods of discovery or pointers to new results". Francis Bacon, *New Organon*, (ed. Lisa Jardine and Michael Silverthorne). Cambridge: Cambridge University Press, 2000, aphorism 8, p. 34.

[35] René Descartes, *Regulae ad directionem ingenii = Rules for the Direction of the Mind: a bilingual edition of the Cartesian treatise on method* (ed. and transl. by George Hefferman). Amsterdam-Atlanta (GA): Rodopi, 1998, rule 3.

ing the rationalist scientific methodology as a ubiquitous paradigm, modern scientists typically focus exclusively on the aspects of objects that can be quantitatively measured and, in doing so, stipulates a conception of nature in accordance with the interests of modern science. The result of this attitude is characteristically and eloquently expressed by Heidegger:

> We feel that what zoology and botany investigate concerns animals and plants and how they investigate it may be correct. But are they still animals and plants? Are they not machines duly prepared beforehand? (*GA* 41,40)

A whole new cultural-scientific paradigm has been created here which in the modernity is valid everywhere, and puts the emphasis exclusively on that which is calculable (that is, mathematical)[36].

(c) *A Paradigm Change?* – As we noted above, for Charles Taylor, Heidegger's work demonstrates a major "cultural turn" in the history of philosophy. This shift can be understood correctly as a "paradigm change" or "shift of a cultural paradigm". Since these notions as interpretative keys do not come from Heidegger himself, but are derived from secondary literature[37], it seems to be useful to explain the viability of this approach.

A paradigm shift is a period of time when the usual and generally accepted way of thinking about something and the praxis of doing things change completely. Such a fundamental change in approach or underlying assumptions can happen both at a communitarian and a cultural level. If a "cultural paradigm" changes, all the scattered, habitual and well established practices of a group loses its self-evident character. These terms became popular after the publication of the famous book of the American historian of science, Thomas Kuhn, entitled *The Structure of Scientific Revolutions* (1962)[38]. The funda-

---

[36] Cf. Casey Rentmeester, *Heidegger and the Environment*. London – New York. Rowman & Littlefield International, 2016, 39-56.

[37] The expression "paradigm shift" in the context of Heideggerian philosophy is used by Hubert Dreyfus. Cf. Hubert L. Dreyfus, "Heidegger on the connection between nihilism, art, technology, and politics", in Charles. B. Guignon (ed.), *The Cambridge Companion to Heidegger*, pp. 289-316, at 299-300.

[38] Cf. Thomas Kuhn, *The Structure of Scientific Revolutions*. Chicago: University of Chicago Press, 1970. The (probably ante meta-literal) Heideggerian insight of Kuhn is that there is a so-called paradigm that guides the scientists' practices and that this paradigm cannot be justified or falsified or described as a set of rules. It serves, rather, as a criterion of rules. As Kuhn further points out: "The shared paradigm [is] a fundamental unit for the student of scientific development, a unit that cannot be fully reduced to logically atomic components which might function in its stead". at p. 11. See also William Blattner, *Heidegger's Being and Time: A Reader's Guide*. London – New Delhi – New York – Sydney: Bloomsbury, 2006, p. 19.

mental idea of the volume is that in the history of science there occur such changing times – which Kuhn names "scientific revolutions" – when paradigms themselves become the focus of conflict. Each paradigm is an interpretation which seeks to be a dominant narrative and each tries to rationalize and justify itself as what constitutes "normal science". For nearly two millennia it was the Aristotelian paradigm that reigned over the European scientific mind. At the beginning of the scientific revolution things changed dramatically. From that time on, *the* paradigm *par excellence* for modern science was Newton's *Principia Mathematica*. It means that for more than two centuries virtually all agreed that Newton had seen exemplary problems—asked the "right" questions—given exemplary solutions after using the "right" method, and produced exemplary justifications for his claims. Consequently scientists after Newton sought to employ the paradigm of Newton, thus being assured that they were doing good science. Not even this, however, was the final destination. At the beginning of the 20th century, the Newtonian paradigm was replaced by the one of Albert Einstein. According to Kuhn, such paradigm shifts constitute the "scientific revolutions". To put it in a provocative way, after the theory of relativity as the most recent paradigmatic change, scientists not only do, but also *see* things differently: they live, so to say, in a different world and deal with different reality[39].

In sum, Heidegger's contribution to the 20th century world philosophy can be considered as a major turnaround (a real "cultural shift") in the history of Western (and global) thought because he was the one who addressed the current rationalistic and technocratic paradigm with the most radical criticism, thereby contributing to its overcoming. Can something similar be identified in the theological analysis of Pope Francis?

3.2. *Pope Francis.* – As he himself has repeatedly emphasized, Pope Francis stands firmly in the tradition of the Second Vatican Council. Before the Council, the whole edifice of theology seemed to have been completed: the best Neo-scholastic theologians, openly or covertly, talked about the necessity of the declaration of some failing dogmas to put even the last brick in place in this magnificent, logically coherent and unquestionable, spiritual cathedral: the dogma on Mary Co-Redemptrix (a Catholic theological concept concern-

---

[39] Cf. Hubert L. Dreyfus, "Heidegger on the connection between nihilism, art, technology, and politics", pp. 289-316, at 299-300 (on technology, more specifically: 301-311).

ing the Blessed Virgin Mary's role in the redemption of all peoples) and the elaboration of the failing Josephology (the doctrine about her Holy Spouse) seemed to also to be missing. Vatican II unexpectedly changed the horizon of theological reflection. It entered into dialogue with the modern world and its spiritual, political and scientific currents, and thereby opened up a radically new perspective. It does not seem to be an exaggeration to call this a paradigm shift, of which Francis is the heir and sequel. However, there are never sharp breaks in theology.

(a) *Continuity: Some Limits Of the Hermeneutic Of "Change"*. – Of course, in theological context, the rhetoric of sheer progress is not viable. At least for two reasons. Firstly, within humanities (and from this particular point of view theology is also considered a human science) novelty never comes without precedent and every thought belongs to the tradition where history is always at work (*Wirkungsgeschichte*). Secondly, theology has a special, distinct, and indestructible relationship to its own tradition, considering that it is intended to interpret a specific historical event, the revelation of God in Jesus Christ. Therefore, "change" can never be a radical novelty and it cannot mean a radical new beginning (*neuer Anfang*) in the Heideggerian sense of the word. A Catholic theological discourse on tradition in general, and on the Vatican II in particular, must always remain within the theological boundaries of the "hermeneutics of continuity" and must reflect what Pope Benedict XVI said in his 2005 Christmas address to the Roman Curia, in which he emphasized a "hermeneutic of reform, of renewal in the continuity of the one subject-Church". In this talk, the Pope stated that the true spirit of the Council is "in the impulses toward the new that are contained in the texts". Pope Benedict adds:

> On the one hand, there is an interpretation that I would call 'a hermeneutic of discontinuity and rupture'; it has frequently availed itself if the sympathies of the mass media, and also one trend of modern theology. On the other, there is the 'hermeneutic of reform', of renewal in the continuity of the one subject-Church which the Lord has given to us. She is a subject which increases in time and develops, yet always remaining the same, the one subject of the journeying People of God.

The hermeneutic of discontinuity risks ending in a split between the pre-conciliar Church and the post-conciliar Church. It asserts that the texts of the Council as such do not yet express the true spirit of the Council. It claims that they are the result of compromises in which, to reach unanimity, it was found necessary to keep and reconfirm many old things that are now pointless. However, the true spirit of the Council is not to be found in these compromises but instead in the impulses toward the new that are contained in the texts[40].

Heidegger's claim for a radical new beginning in the philosophical tradition, as we have seen above, is not comparable with the call for reform and for new theological impulses launched by Pope Francis. However, there is also room for innovation in theology. Especially when we consider how fast the world in which today's theological reflection is taking place is changing.

(b) *A "Change of Epoch" Instead of the "Epoch of Change"*. – In his apostolic exhortation *Evangelii Gaudium* (2013), Pope Francis speaks about our time as a "turning-point" in history of humanity (*EG* 52). This "epochal change", he explains, has been set in motion "by the enormous qualitative, quantitative, rapid and cumulative advances" occurring "in the science and in technology" and by their "instant application in different areas of nature and of life" (*EG* 52). Later on the Pope faces the "challenge of the urban cultures" (*EG* 73) and elsewhere, in his apostolic exhortation *Amoris laetitia* (2016), he puts before our eyes the "current situation of the family" (*AL* 32-59), all read in the hermeneutical key of this epochal cultural change or paradigm shift. This current social-cultural (and metaphysical) situation, however, is analysed more in depth in his first social encyclical (*nota bene*: not *just* an ecological encyclical!) entitled *Laudato Si'* (2015). Here, Pope Francis characterizes his task as getting "to the roots of the present situation, so as to consider not only its symptoms but also its deepest causes" (*LS* 15). In this sense, the Pope writes about a "New Era" (*LS* 102) into which "humanity has entered" in the last couple of centuries due to various technological inventions. With the words of the Pope:

> Humanity has entered a new era in which our technical prowess has brought us to a crossroads. We are the beneficiaries of two centuries of enormous waves of change: steam engines, railways, the telegraph, electricity, automobiles, aeroplanes, chemical industries, modern medicine, information technology and, more recently, the digital revolution, robotics, biotechnologies and nanotechnologies (*LS* 102).

---

[40] Benedict XVI, address to the Roman Curia, Christmas Greeting, December 22, 2005, http://w2.vtican.va/content/benedict-xvi/en/speeches/2005/december/documents/hf_ben_xvi_spe_20051222_roman-curia.html [last access: 28.06.2021].

Soon after (cf. *LS* 104), the Pope mentions the transforming power of nuclear energy, biotechnology, information technology, knowledge of our DNA, among other abilities which we as humankind have acquired. These tools and new sciences have given us knowledge and power (a "tremendous" and unprecedented power) to have "an impressive dominance over the whole of humanity and the entire world" (*LS* 104). The problem, and enormous danger, in this issue is that "nothing ensures" that these incredible possibilities will be used "wisely". The experience so far has augured well for the future (nuclear bombs dopped, technology used by Nazis, Communists, deadly arsenal of current weapons, etc.). The Pope expresses his preoccupation also for the "extremely risky" fact that the power to employ these technologies has been concentrated in the hands of a disproportionately small part of humanity (cf. *LS* 104)[41].

However, neither Heidegger, nor Pope Francis can be considered as an anti-technological, that is to say a premodern, thinker. In another paragraph Francis specifies: "It is right to rejoice in these advances and to be excited by the immense possibilities which they continue to open up before us", for – and here the Pope quotes John Paul II – "science and technology are wonderful products of a God-given human creativity"[42]. This kind of activity, the modification of nature for our own purposes and utility, has distinguished human nature from other animals from the beginning. Technology itself, and here Pope Francis cites his immediate predecessor, Benedict XVI, "expresses the inner tension that impels man gradually to overcome material limitations"[43]. These popes beyond a shadow of a doubt recognised that "technology has remedied countless evils which used to harm and limit human beings" (*LS* 102) for which "progress", especially in the field of medicine, engineering and communications[44], we can be grateful.

[41] Even in the *FT,* Pope Francis expresses his ambiguous opinion towards progress: "Technology is constantly advancing, yet how wonderful it would be if the growth of scientific and technological innovation could come with more equality and social inclusion" (*FT* 31).

[42] Saint John Paul II, "Address to Scientists and Representatives of the United Nations University", Hiroshima 25 February 1981, in AAS 73/1981, p. 422.

[43] Benedict XVI, *Encyclical letter Caritas in veritate on Integral Human Development in Charity and Truth*, 29 June 2009, n. 69 in AAS 101/2009, p. 702.

[44] In another context Pope Francis expresses his serious doubts about the benefices of digital communication: cf. *FT* 7;31;42-50 and stresses its dangers especially in the lives of the young generations. Cf. Pope Francis, Post-Synodal Apostolic Exhortation *Christus Vivit,* Loreto 25 March 2019, 86-90.

What is, however, the main problem with the so-called technocratic paradigm? Why does Pope Francis feels compelled to raise his word against at least some of the elements of this kind of progress? This he did in an important speech, on December 21, 2019 at the occasion of the eve-of-Christmas talk to the Roman Curia, saying: "what we are living is not simply an epoch of change, but a change of epoch"[45]. Following his inductive methodological approach (see-judge-act), in *Laudato Si'*, Pope Francis accuses the "Cultural paradigm" (*LS* 108) in force, and its "technological culture" (*LS* 112), to be the major cause of global inequality. The Pope quotes the Bolivian Bishops' Conference according to which "[b]oth everyday experience and scientific research show that the gravest effects of all attacks on the environment are suffered by the poorest" (*LS* 48)[46]. The symptom described and the problem is identified: its name is the "dominant technocratic paradigm" (*LS* 101, 112). For a possible change, the Pope tries to understand (and criticise) more profoundly this phenomenon.

(c) *Criticism of the Dominant Technocratic Paradigm*. – For Pope Francis, technology is an ambiguous phenomenon. The basic problem with it, in a deeper sense, is that "humanity has taken up technology and its development according to an undifferentiated and one-dimensional paradigm" (*LS* 106). This paradigm, continues the Pope, "exalts the concept of a subject who, using logical and rational procedures, progressively approaches and gains control over an external object" (*LS* 106). The method used by this subject is "the scientific and experimental method", which must be considered in itself already as a "technique of possession, mastery and transformation" and "manipulation". During human history "men and women have constantly intervened in nature, but for a long time this meant being in tune with and respecting the possibilities offered by the things themselves. [...] Now, by contrast, we are the ones to lay our hands on things, attempting to extract everything possible from them while frequently ignoring or forgetting the reality in front of us" (*LS* 106). A harmony has been broken between human beings and material objects; their relationship has become confrontational. The hopeless and unrealistic idea of "infinite or unlimited growth" has been embraced by economists, financiers and experts

[45] The famous frase goes: "*quella che stiamo vivendo non è semplicemente un'epoca di cambiamenti, ma un cambiamento d'epoca*" (Roman Curia on December 21, 2019): https://www.youtube.com/watch?v=mHm8QQymCps (11:30) [last access: 28.06.2021].

[46] Cf. Bolivian Bishops' Conference, Pastoral Letter on the Environment and Human Development in Bolivia El universe, don de Dios para la vida (23 March 2012), 17, cited by *LS* 48 (note 26).

in technology and on this illusory and wishful thinking many of them are presupposing "an infinite supply of the earth's goods" (cf. *LS* 106). These critical thoughts of the Pope are followed by a basically Heideggerian style affirmation, according to which "technological products are not neutral" (*LS* 107): we as humanity often have "the tendency", at times unconscious—from which many problems of today's world stem—to make "the method and aims of science and technology an epistemological paradigm which shapes the lives of individuals and [...] society", which ultimately proves to be "just one sign of a reductionism which affects every aspect of human and social life" (*LS* 107).

Francis' encyclical articulates and defends an "adequate anthropology" and a proper vision of "reality" that effectively grounds – as well as limits – all human endeavours. In particular, his quest to locate the root causes of our present environmental and social crises engages him in a critique of a totalizing "technocratic paradigm" (*LS* 101;109;111;112; 122) or even "tecno-economic paradigm" (*LS* 53) that stands in opposition to what Francis considers genuine reality in all its complexity. Rationalist scientism, in this sense, encourages delusional forms of instrumentalization of others and reification of nature[47]. This paradigm deepens global inequality, that is the difference among poor countries and rich countries: "The foreign debt of poor countries has become a way of controlling them, yet this is not the case where ecological debt is concerned" (*LS* 52) which causes "perverse structures" (*LS* 52) within the one political and economic organization of our single human family. Next, making explicit reference also on the well-known Heidegger scholar Romano Guardini (cf. *LS* 105), Pope Francis passionately criticizes the "technological mind", which "has paradoxically ended up prizing technical thought over reality"[48]. He also criticizes the "technoscience" (*LS* 103) that promotes also "myth of progress" (*LS* 60) and a sort of "reductionism" (*LS* 112, in the Italian translation, rightly, "*potere oggettivante*") which seeks to promote "the principle of the maximization of profits" (*LS* 67). This kind of "absolute domination" (*LS*

[47] Cf. also Lisa H. Sideris, "Techno-Science, Integral Thought, and the Reality of Limits in Laudato Si'", in *The Trumpeter*, 34/1-2018), pp. 14-35, at 14.

[48] Since this kind of technological mind-set "sees nature as an insensate order, as a cold body of facts, as a mere 'given', as an object of utility, as raw material to be hammered into useful shape; it views the cosmos similarly as a mere 'space' into which objects can be thrown with complete indifference". Romano Guardini, *Das Ende der Neuzeit. Ein Versuch zur Orientierung*. Würzburg: Werkbud-Verlag, 1965, p. 63 (English ed. *The End of the Modern World*. Wilmington: ISI Books, 1998, p. 55). See *LS* 115.

67)[49] from the part of an "instrumental way of reasoning" (*LS* 195)[50] strives to serve the "immediate results, posses, manipulation, egoism" (*LS* 203) by employing the "logic of violence, exploitation, and selfishness" (*LS* 230).

It is enough to refer to the ubiquity and danger of this trap to understand why is it so urgent to find an alternative. Rationalist scientism is currently one of the most widespread ideologies in Western culture and society (maybe together with materialistic consumerism). Being so obviously taken for granted in both the Western civilization, and, increasingly, elsewhere in our globalized world, it constitutes, at least "unconsciously", *the* mind-set *par excellence* of many modern persons, including a huge number of pious Christians. The widely accepted character of the phenomenon disguises the fact that, at the end of the day, scientism is a totalitarian ideology that generalizes the worldview derived from modern science to embrace the whole of reality and our knowledge of it. According to the main tenets of this dogma, anything that modern science cannot study is either unreal or irrelevant. It brushes aside in cavalier fashion the spiritual dimension of human person and considers the sacred reality is our life-world (*Lebenswelt*) to be devoid of reality or, which is basically the same, mere subjectivism[51]. In its rationalistic form, scientism is, therefore, one of the greatest obstacles, along with the current truncated image of human reality, itself related to this false ideology, that stands in the way of a restoration of an authentic view of nature and of man's relation to it. This endangers life on earth itself. As we have seen above, this preoccupation is undoubtedly one of the main points of convergence intellectual universe of Heidegger and that of Pope Francis. Is there an alternative? After taking account of the phenomenon from philosophical and theological point of view, now we should focus on the possible antidotes.

---

[49] "[I]t would [...] be mistaken to view other living beings as mere objects subjected to arbitrary human domination. When nature is viewed solely as a source of profit and gain, this has serious consequences for society" (*LS* 82).

[50] "An instrumental way of reasoning, [...] provides a purely static analysis of realities in the service of present needs, is at work whether resources are allocated by the market or by state central planning" (*LS* 195).

[51] One could certainly conceive of a world in which modern science exists as a legitimate but limited form of knowledge of the natural world and not invading other areas. The point is to let existing the scientific paradigm alongside other (human) sciences and their methodologies so that the non-mathematical dimensions can be integrated into a universal reign of knowledge. What has happened, at least not so far, is something rather different: many today consider modern science to be the only legitimate knowledge of reality. In this regard the above-mentioned big book of Hans-Georg Gadamer (a Heidegger-disciple) is crucial provided that it want to rehabilitate extra-scientific (artistic, historical, ethical, etc.) truth claims. Cf. Hans-Georg Gadamer, *Truth and Method.*

4. Parallelism 3#: An Ecological Spirituality as Alternative – Towards a Liveable Future

From an environmentalist point of view, modern life is clearly unsustainable and, what is more, self-destructive[52]. It becomes evident if we think about the fact that seemingly innocuous acts of contemporary life – one's daily routine in commuting to work by car, buying groceries and exotic fruits from the supermarket, and using the water closet 6-7 times every day, etc. – have detrimental consequences on a global scale. Is there any solution, any viable way out from this trap? Can we learn something useful from Heidegger and Pope Francis in this respect? And does their advice converge in some respects?

There are three presuppositions of this ending chapter. The first, following on the footsteps of Casey Rentmeester[53], is that if we heed Heidegger's call to dwell (*wohnen*) on things before we act, and we do this with the attitude of "letting beings be" (*Gelassenheit*), we would take into philosophical consideration how our actions are related to the environmental crisis – and we can find a way out of it. The second is that in his magisterial teaching (and particularly in his encyclical *Laudato Si'*), Pope Francis, theologically concretises this path giving practical inspirations. And finally, the third presupposition is that these two previous suggestions can have a relation among them. The task ahead of us is to prove these bold claims.

4.1. Heidegger – At first sight it might seem surprising to put Heidegger's name in connection to the ecological movement. However, there has in fact been some secondary literature on this matter and it is not unusual any to make this link anymore. As Casey Rentmeester puts is:

> Since one of Heidegger's main goals was to critique the reified traditions of Western philosophy and open up new ways of understanding reality, he is a fitting candidate for an eco-phenomenological approach to environmentalism[54].

---

[52] Even *LS* affirms: "the present world system is certainly unsustainable (cf. *LS* 61) and it has a "perverse logic" (*LS* 197).

[53] Cf. Casey Rentmeester, *Heidegger and the Environment*. London – New York: Rowman & Littlefield International, 2016.

[54] Casey Rentmeester, *Heidegger and the Environment*, p. xiv. Cf. also Michael E. Zimmerman, "Heidegger, Buddhism, and deep ecology", in Charles B. Guignon (ed.), *The Cambridge Companion to Heidegger*, pp. 240-269.

(a) *A Surprising Forerunner of Eco-Phenomenology*. – On a personal level Heidegger famously felt a deep connection to his work and the place in which it was carried out. This existential-ontological embeddedness of human existence has been discussed in detail in several of his writings. In the famous "existential analysis" of *Being and Time* (1927), Heidegger explains Dasein as being-in-the-world (*In-der-Welt-Sein*). "Thrown into its 'there' (*in sein 'Da' geworfen*), every [human] Dasein has been factically submitted to a definite 'world' – its 'world'" (*SZ* 297/*BT* 344).

> Resoluteness, as authentic Being-one's-Self (*eigentliches Selbstsei*), does not detach Dasein from its world, nor does it isolate it so that it becomes a free-floating 'I'. And how should it, when resoluteness as authentic discolsedness (*Entschlossenheit*), is authentically nothing else than Being-in-the-world? (*SZ* 298/TB 344).

This very existential structure of human Dasein – the Being-in-the world – will be expressed also "existentially" by Heidegger whose daily life was immersed in the changing of the seasons and nature's own course. Instead of viewing his environmental surrounding as an aesthetically pleasing landscape or a recreational opportunity and thereby merely subsuming it in terms of his own interests, Heidegger urged others to live, and lived himself, within the environment and through the environment in a way that is respectful of it[55].

If we pose the question, what an everyday existence would look like for someone who dwells, one potential answer to this question could be seen in looking at Heidegger's everyday existence. The philosopher was born and raised in a small town, Meßkirch, in the Swabian Alps and in his youth he spent many hours reading masterpieces of the western intellectual tradition on nature. Later on, as professor at the University of Freiburg he liked to retire for many days to send time in his hut in Todtnauberg in the mountains of Germany's Black Forest (about 15 kilometres of distance from the city). Despite of the fact that he did own a house also in the Freiburg, he preferred to live and work in the countryside where he could write more calmly in solitude. Heidegger learned to live with the land and felt a strong connection to his own roots: to the landscape and its people. He even liked to compare his own work with the work of the peasants (*die Arbeit der Bauern*) (cf. *GA* 13, 10). For instance, when Heidegger was offered a professorship at the University of

[55] Cf. Casey Rentmeester, *Heidegger and the Environment,* p. 78.

Berlin, which was – and still is – perhaps the most prestigious philosophy position in all of Germany, he simply rejected the attractive offer. As he explains his surprising decision in one of his meditative writings entitled "Creative View: Why do We Stay in the Provinces? (*Schöpferische Landschaft: Warum bleiben wir in der Provinz?*), he made reference to the "inner relationship (*Zugehörigkeit*) of [his] work to the Black Forest and its people" (*GA* 13,10-11): "My whole work [...] is hold and lead by the world of these hills and their peasants (*Meine ganze Arbeit [...] ist von der Welt dieser Berge und ihrer Bauern getragen und geführt*) (*GA* 13,11). For "the guiding-line for all philosophical inquiry [is] at the point where it *arises* and to which it *returns*" (*SZ* 38/*BT* 62), this Beginning remains also Heidegger's spiritual destiny.

(b) *A Spirituality In Heidegger's Late Philosophy: Dwelling as Letting Beings Be.* – Heidegger is famously not an easy author. While his best-known work, *Being and Time*, is challenging but discernible through careful study, his later philosophy – produced in the second and especially the third stage of his *oeuvre*[56] – has the reputation of being notoriously obscure and extremely difficult to grasp. It is clear, however, that there is a general trajectory within his thought of development of his thinking about the question of being (*Seinsfrage*). At the same time, it is also clear that there are some constants of each and every period of his life. Attentive readers rightly note that, over time, there is an increasing sensibility for spiritual topics. This aspect enables us to draw parallels with Pope Francis' theology. The philosophical project, however, remains demanding. The guideline of Heidegger's later thought is a sharp-eyed critique of the Western metaphysical tradition as a whole and an attempt to offer a glimpse at what a new way of revealing entities might look like[57].

---

[56] As is known, Martin Heidegger has divided his own thinking according to a so called three phase model, which has been taken over by some of his major interpreters, for instance Otto Pöggeler and Emil Kettering. This three-phase model – which replaced the long-established division of Heideggerian thinking into two parts (before and after the turn/*Kehre*), is as follows: 1) the question of the meaning of being (in the 1920s) in the time when *Being and time* has been written; 2) the question of the truth of being (first half of the 1930s and the 1940s), which interprets the truth as an event (*Ereignis*); 3) Being as clearing (*Lichtung*), determined by the topology of being (a question concerning the place of this *Lichtung*, with a special attantion of the role of the language) (from about 1947). Cf. Erzsébet Kerekes, Az idő kairologikus jellege a heideggeri hermeneutikai fenomenológiában (Kluj Napoca: Egyetemi Műhely, 2013), 9, note 4. Cf. also Frederick A. Olafson, "The unity of Heidegger's thought", in Charles B. Guignon (ed.), *The Cambridge Companion to Heidegger,* pp. 97-121

[57] Cf. Casey Rentmeester, *Heidegger and the Environment,* p. 61.

An appropriate question at this point might be how to accomplish a mindful dwelling (*wohnen*) in one's everyday life. How can we let a thing emerge? How do advance from a general awareness of being *as* things which capture our attention? In a short essay laconically entitled as *The Thing* (*Das Ding*), in 1950, Heidegger provides an answer:

> When and in what way do things appear as things? They do not appear *by* means of human making. But neither do they appear *without* the vigilance of mortals. The first step toward such vigilance is the step back from the thinking that merely represents – that is, explains – to the thinking that responds and recalls. (*Der erste Schritt zu solcher Wachsamkeit ist der Schritt zurück aus dem nur vorstellenden, d.h. erklärenden Denken in das andenkende Denken.*)" (*GA* 7, 183 – highlight in the original)[58].

From our part, it is clear that this sort of thinking requires a respectful responsiveness to things as they show themselves and a reflective capacity to appreciate things in their contextual relationships with other entities. Dwelling in the world, in this sense, is a spiritual "activity": it is something which is more receptive than active, more grateful than careless, more meditative than rationalistic; a mental temper or turn of mind which is light years away from the general Western attitude.

There are some further later events to quote in this regard. In 1949 Heidegger was invited to give a public lecture in his own hometown, Meßkirch, to commemorate the posthumous birthday (centenaries) of one of his fellow citizen, a German composer, Conradin Kreutzer. The interesting thing about the event was that the 60-year-old, now a famous philosopher, had to speak in front of a relatively simple audience. Heidegger, however, was able to express deep philosophical content even in very simple words and so can give us a glimpse as to what his most particular to his way of thinking – the thinking that responding and recalling (*andenkende Denken*) – looks like. After having lamented about the pervasive "thoughtlessness" (*Gedankenlosigkeit*) that occurs all too often and is one of the main characteristics of the today's world[59], Heidegger states:

---

[58] Martin Heidegger, "The Thing", in *Poetry, Language, Thought*. New York: Harper & Row, 1971, pp. 163-186, at 181.

[59] Heidegger states that today "one gathering follows on the heels of another", (*GA* 16, 518) but such "gatherings" are not gatherings at all since a respectful mindfulness of the occasion is lacking. In an attempt to truly commemorate the event, Heidegger states: "It is enough if we dwell on what lies close and meditate on what is closest; upon what which concerns us, each one of us, here and now; here, on this patch of homeground; now, in the present hour of history". (Martin Heidegger,

So, there are two sorts of thinking, which are legitimate and necessary in their own way: the calculative thinking and the meditative thinking (*So gibt es den zwei Arten von Denken, die beide jeweils auf ihre Weise berechtigt und nötig sind: das rechnende Denken und das besinnliche Nachdenken*)" (*GA* 16,520).

He goes on to reflect on the relationship between humans and their homeland (*Heimat*): a notable life, in this case the one of Conradin Kreutzer, can teach us how to live our own lives. An important aspect of dwelling that comes through in this passage is that dwelling involves dwelling on an event in its situational context. The commemoration, as a moment of gathering, attains the status of a significant event if there is a thoughtful reflection occurring wherein one respects its importance[60].

In a short text writing in 1951 with the title "Building, Dwelling, Thinking (*Bauen, Wohnen, Denken*)", Heidegger explains: the good gardener is the model of the way that human beings should live and work. A gardener is attentive to what his or her plants need to reach their fruition and to blossom in a healthy manner. Heidegger calls this sort of living, once again, "dwelling" (*wohnen*). He states: "To dwell […] means to remain at peace within the free, the preserve, the free sphere that safeguards each thing in its essential unfolding. *The fundamental character of dwelling is this preserving*"[61]. The Heideggerian notion of dwelling – we learn from another short essay with the title borrowed from the German poet Hölderlin, "To dwell poetically" ("*…dichterisch wohnet der Mensch…*") (1951) – is to live in a way that does not encroach on the bringing-forth of *phusis* but that takes part in this very bringing-forth (cf. *GA* 7, 189-208). Already the ancient Greeks understood *phusis* as a self-emerging. Though the pre-Socratics understood being as *phusis*, the Western metaphysical tradition eventually restricted its understanding of being to *oussia*, which eventually led to viewing beings as standing in isolation[62].

One last aspect to make this point clear can still add some new elements to the topic. It is important to highlight that Heidegger is not a pre-modern think-

Discourse on Thinking. New York – Hagerstown - San Francisco – London: Harper & Row, 1966, p. 47) (*GA* 16, 520).

[60] Cf. Casey Rentmeester, *Heidegger and the Environment,* p. 77.

[61] Martin Heidegger, "Building, Dwelling, Thinking", in David Farrell Krell, *Basic Writings: Revised and Expanded Edition*. New York: Harper Collins, 1993, pp. 343-363, at 351 (*GA* 7,151), NB: Translation modified by Casey Rentmeester, in *Heidegger and the Environment*, p. 72).

[62] Casey Rentmeester, *Heidegger and the Environment,* p. 74.

er who tries to suggest that we must completely refrain from utilizing nature for our own benefit altogether (for instance using natural entities as resources). Rather, he is simply arguing against viewing nature and its entities merely as resources (that is to say, object of calculation, use and exploitation). In an important essay in this respect, entitled "The Question Concerning Technology" [*Die Frage nach der Technik*] (1953) (*GA* 7, 5-36)[63], Heidegger offers the telling example of an old windmill. He acknowledges that this structure turns the wind into power, however, he points out that in doing so it does not use violence on nature: it still allows the wind to be wind (cf. *GA* 7,15). In this distinction Heidegger is trying to envision and conceptually elaborate a way of living in the world that respects the ways of nature and allows natural beings to run their own course rather than exercising mastery over them. In the same vein, says Heidegger, an old bridge allows for dwelling. Through this example, the Thinker of the Black Forest explains to the reader how this contrasts with the way in which humans live in the contemporary period. Concerning contemporary behaviour, he offers the example of the hydroelectric plant. He states,

> The hydroelectric plant is not built into the Rhine River as was the old wooden bridge that joined bank with bank for hundreds of years. Rather the river is dammed up into the power plant. (*Vielmehr ist der Stromin das Kraftwerk verbaut.*) What the river is now, namely a water power supplier, derives from out of the essence of the power station (*aus dem Wesen des Kraftwerks*)" (*GA* 7,16)[64].

Whereas the hydroelectric plant treats the river as merely a supply of energy to be exploited with maximum efficiency, the bridge respects the river as a river and allows it to flow on its own terms. The difference is well articulated: in the first case, the bridge allows for an appropriate relation to nature: in coming across an old bridge that joins two banks, one can still respect the banks as banks and the river as river. Whereas in the second case, in coming across a hydroelectric plant, one gets the impression that the river is simply there to supply energy to the plant. In being so, the hydroelectric plant does not let beings be the beings that they are in the way that the bridge does[65].

---

[63] Martin Heidegger, "The Question Concerning Technology", in *The Question Concerning Technology and Other Essays*. New York: Gardland Publishing, 1977, p. 14.

[64] Martin Heidegger, "The Question Concerning Technology", p. 16. Cf. Casey Rentmeester, *Heidegger and the Environment*, p. 73.

[65] Rentmeester comments the passage in a Heideggerian vene as follows: " [...] since the ener-

As we have see above, Heidegger does offer an alternative vision of a more appropriate way in which to understand the world around us and comport ourselves toward nature in his later thoughts on dwelling. The conception he elaborates can be summarise with the notion of letting beings be (*Gelassenheit*) which is also a title of an inspired writing from 1933 (*GA* 13; 16). Although the common association of letting something be is to adopt a passive and indifferent attitude, Heidegger stresses that letting be is not a matter of simply stepping back and passively allowing beings to unfold. In fact, letting be is not tantamount to some sort of withdrawal from interaction with beings, but rather, as Heidegger himself stresses, it lies "outside the distinction between activity and passivity [...] because it does not belong to the domain of the will" (*GA* 77, 109)[66]. In this quality, letting beings be (*Gelassenheit*) is a consequence (and, in a way, a summary) of authentic dwelling (*wohnen*), that is to say, a key expressions of the late Heideggerian philosophy's spirituality.

(c) *"Shepherds of Being" vs. Metaphysical "Gestell"*. In a famous essay, entitled *Letter on Humanism [Brief über den "Humanismus"]* from 1947—from which we count the beginning of the third and last period of Heidegger's oeuvre—the mature Heidegger turns back to the vexed question: what the essence of human being is and what it should be. Here, once again, the rationalism of the Western philosophical tradition is disqualified with a single large-scale gesture. Heidegger states that both the medieval-scholastic *animal rationale* and the Modern "subjectivity" (*Subjektivität*) miss the point, for humans are "ex-sisting counter-throw of being" (*ek-sistierende Gegenwurf des Seins*). And further we find the most well-known (semi-biblical) metaphor: "The human being is not the lord of beings. The human being is the shepherd of being. (*Der Mensch ist nicht der Herr des Seienden. Der Mensch ist der Hirt des Seins*)" (*GA* 9, 342)[67].

Like Pope Francis, Heidegger does not argue against technology as such or even specific technologies devices. What he argues against is, as he puts

---

gy that is supplied to the plant is ultimately utilized to quench human interests, buildings like the hydroelectric plant only reinforce the idea that nature is something to be mastered or conquered by humans and manipulated to our interests. The bridge, on the other hand, does not reinforce this attitude since it is built into the land and allows the landscape to be what it is [...]". Casey Rentmeester, *Heidegger and the Environment,* p. 73.

[66] Martin Heidegger, *Country Path Conversations*. Bloomington & Indianapolis: Indiana University Press, 2010, p. 70.

[67] Martin Heidegger, "Letter on Humanism", in William McNeill (ed.), *Pathmarks*. Cambridge: Cambridge University Press, 1998, pp. 239-276, at 260.

it, becoming "so firmly shackled to these technical devices that we fall into bondage to them" (*GA* 16, 518)[68]. For Heidegger, the point is that "enframing", or "framework" (*Gestell*), a designation for the technological world is, like "total mobilisation" in wartime, fundamentally a metaphysically dangerous program. Its danger consists in the fact that, on the one hand, it prevents us from thinking and, on the other, that it makes us violent, aggressive and disrespectful towards being[69].

In summary, we started from the premise that the Heideggerian call to "dwell" (*wohnen*) on things and to maintain an attitude of "letting beings be" (*Gelassenheit*) can help us address the current ecological crisis. The more we dwell on our actions and the relationships that are essentially implicated in them, the more we "dwell" among things in a liberated way, which implies treating things with a respectful deference to what they are in and of themselves. Let us now focus on some concrete suggestions from the part of Pope Francis, in the hope that some parallelism will rise among the two authors.

4.2. Pope Francis.– Similarly to the critical proposal of Heidegger, also the Magisterium of Pope Francis in general, and his thoughts on "integral ecology" in particular, can be read as a passionate (and even compassionate) call for "taking care" for our world (which is meant here as our "common home" in the "created" world).

(a) *A Bottom-up Method.* – The encyclical letter *Laudato Si'* points out, that "a blind confidence in technical solutions" (*LS* 14; cf. 20) are not able to helps us out of the present crisis. We require, instead, "a new and universal solidarity" (*LS* 14). The Pope insists that there is an "urgent challenge" to "protect our common home", in the hope that "things can change" (*LS* 13). The task in front of us includes a concern to bring the whole human family together to seek a sustainable and integral development, in strict collaboration with the worldwide ecological movement (cf. *LS* 14, cf. 166, 201, in a critical way 136) for the struggle for the sake of the environment (*LG* 91). In what way?

The inductive methodology of the encyclical, which imitates that of the well-known pastoral constitution of Vatican II, *Giaudium et spes*, follows the structure: to see, to judge, to act. In Chapter one of *Laudato Si'*, entitled "What is

[68] See the translation in Martin Heidegger, *Discourse on Thinking*. New York: Harper & Row, 1966, p. 45.

[69] Martin Heidegger, *The Question Concerning Technology and Other Essays*, p. 20.

happening to our common home?", the Pope urges a "desirable change" toward an integral and sustainable human development (cf. *LS* 20) calling humanity "to recognize the need for changes of lifestyle, production and consumption" (*LS* 23) in order to combat global calamities and preoccupations such as global warming, the rise in sea level, increased acidification of the oceans, carbon dioxide pollution, the loss of tropical forests, the extinction of part of the planet's biodiversity, and scarce availability of essential resources like drinking water, energy and agricultural production. Francis refers to "a very solid scientific consensus" (*LS* 23) which testifies to a disturbing and unprecedented destruction of ecosystems, "with serious consequences for all of us" (*LS* 24), including a decline in the quality of human life and the breakdown of society. After describing the empirical problems, he focuses to the solutions.

(b) *An Ecological Spirituality*. – In the third part of his document, Pope Francis outlines a "post-industrial" alternative (cf. *LS* 165). This includes a theological reflection that has two main characteristics: spirituality (cf. *LS* 75) and an ecological sensibility (cf. *LS* 202). In fact, the title of the Chapter Six of *Laudato Si'* is "Ecological Education and Spirituality". The ambiguous term "spirituality" is used in the encyclical prevalently in positive sense[70], asserting that the light of the faith has something essential to offer (cf. *LS* 93-94). Unlike Heidegger's path, Francis' approach is explicitly a theological one: the Mystery of Christ and his Incarnation serves as orienting light for the transition into a better future for humanity. Further key-concepts are "harmony" with the "fragile" creation (*LS* 78), a the return to a "simplicity" which "goes hand in hand with sobriety and humility" (*LS* 222), "ecological conversion" (*LS* 216-221, spec. 217), and "integral Christian spirituality" (*LS* 237). In the vision of the Pope, less is more: "integrity" and "harmony" (*LS* 10;66;82; IV) can be put to the service of the "common good" (*LS* 18;23;54;129;156-158).

Laudato Si' also invokes the ancient Christian concept of "universal fraternity" (*LS* 228 cf. also *LS* 11;70;82;93;201;221), term which five years later will play a central role in the second social encyclical letter of Pope Francis, *Fratelli Tutti* (2020). By invoking this notion, the Pope urges all to practice "care", both to the natural environment and to other human individuals, in-

[70] "Spirituality", as an ambiguous word, is surprisingly often used in the text of *LS*. The term comes up mostly in positive context (*LS* 63; 105; 111; 125; 202, 210, 216, 222, 225, 231, 236, 240) sometimes however with neutral (*LS* 111), or even definitively negative (*LS* 75) connotations. Cf. also "fraternal dimension of spirituality" (*FT* 86) and also "spirit of fraternity" (*FT* 165).

voking the Biblical exhortation to "'till and keep' the garden of the world" (*LS* 67,71,124). The Pope exhorts all to maintain this attitude to other individuals regardless of ethnicity, sex, culture, or religion. Indeed, he insists that self-giving love can also become the font of "intense spiritual experiences" (*LS* 232). He describes such experiences as including both profane deeds and activities, such as appreciating music, art, and maintaining contact with the nature; and explicitly religious experiences such as prayer or participating in the Eucharist. He describes all such experiences as aiming toward the realisation of a "spirituality of global solidarity" (*LS* 240). For Pope Francis, civic and political love are not separated form cosmic love (*LS* 236): a "spirituality" (rightly translated in Italian, "*mistica*") is "capable of inspiring us" whereas "doctrine alone" is not enough (*LS* 216). In this respect, he cites several Christian saints, including Saint John of the Cross, Saint Bonaventure and, above all, saint Francis of Assisi. However, he also referes to a non-Christian, Sufi Muslim, "spiritual writer" (in the Italian: "*maestro spirituale*"), Ali al-Khawas, (*LS* 233), although only in a footnote (cf. note 159). The essential message of universal fraternity in Pope Francis is this: through community-based good actions:

> The world, and the quality of life of the poorest, are cared for, with a sense of solidarity which is at the same time aware that we live in a common home which God has entrusted to us" (*LS* 232).

The Pope explains that, from a theological perspective, "everything is connected" (*LS* 91, 117, 138; 240) and "interrelated" (*LS* 120)[71]. This implies that such an interpretation of reality is not merely created by our free will, nor is it the product of a reductionist arbitrariness[72].

(c) *Beyond Reductionism*. – Pope Francis goes beyond a mere secular "Green' rhetoric" to speak of the inseparability of the "cry of the earth" and the "cry of the poor" (*LS* 49). This gives an unquestionable (post-anthropocentric) Christian tone to the analysis of *Laudato Si'*. The Pope's vision focuses on the good signs and positive elements, in order to overcome the "dominant technocratic paradigm" being attracted by the truth, goodness and beauty.

---

[71] *NB*: for Pope Francis, "interdependence" is related to "co-responsibility" (cf. *FT* 127).

[72] On ecological spirituality a useful paper is Ivan Platonvjak, "The Ecological Spirituality in the Light of *Laudato si*'", I *Nova prisutnost,* 17/1-2019, pp. 75-91, which treats extensively the Ignatian aspects of the encyclical (pp. 87-90), without any reference to a possible link to Heideggerian thought (although, in the text of the encyclical the latter is much more obvious than the former).

Liberation from the dominant technocratic paradigm does in fact happen sometimes, for example, when cooperatives of small producers adopt less polluting means of production, and opt for a non-consumerist model of life, recreation and community. Or when technology is directed primarily to resolving people's concrete problems, truly helping them live with more dignity and less suffering. Or indeed when the desire to create and contemplate beauty manages to overcome reductionism through a kind of salvation which occurs in beauty and in those who behold it (*LS* 112).Pope Francis stresses that an attraction to the aesthetic can counterbalance tendencies toward reductionism (*LS* 92;107 cf. also 54).

> The desire to create and contemplate beauty manages to overcome reductionism through a kind of salvation which occurs in beauty and in those who behold it. An authentic humanity, calling for a new synthesis, seems to dwell in the midst of our technological culture, almost unnoticed (*LS* 112)

He speaks of the value of "art and poetry" and of "interior life and spirituality" (*LS* 63) and describes how these qualities are found in a special way in indigenous cultures (*nota bene* such themes will be further explored in his apostolic exhortation *Querida Amazonia*), which preserve "various cultural riches" toward which "respect must [...] be shown" (*LS* 63; on "aesthetics" and "poetry" cf. also *LS* 199 and 233, nota 159). He then insists on "the urgent need for us to move forward in a bold cultural revolution", insisting, all along, that "science and technology are not neutral" (*LS* 114).

While Pope Francis criticises technocratic, reductionist, thinking, he does not indulge in blind anti-modernism, any more than Heidegger does. He states:

> Nobody is suggesting a return to the Stone Age, but we do need to slow down and look at reality in a different way, to appropriate the positive and sustainable progress which has been made, but also to recover the values and the great goals swept away by our unrestrained delusions of grandeur (*LS* 114).

Instead of premodern way of thinking, the Pope promotes "another type of progress", one which is "healthier, more human, more social, more integral". Indeed, reflecting on Christian theology, he states that, "liberation from the dominant technocratic paradigm" (*LS* 112) requires, at least partly, a revision of a certain classical and modern reading of Christian doctrine. Offering an example of this process of revision, he offers an exegesis of the *Book of Genesis*, speaking against any interpretation of it that permits "absolute domination" of nature by humans, but rather an attitude of, "tilling", "cultivating", "caring", and "protecting". In a terminology which is reminiscent of Heidegger's, the Pope states:

> Although it is true that we Christians have at times incorrectly interpreted the Scriptures, nowadays we must forcefully reject the notion that our being created in God's image and given dominion over the earth justifies absolute domination over other creatures. The biblical texts are to be read in their context, with an appropriate hermeneutic, recognizing that they tell us to 'till and keep' the garden of the world (cf. *Gen* 2:15). 'Tilling' refers to cultivating, ploughing or working, while 'keeping' means caring, protecting, overseeing and preserving. This implies a relationship of mutual responsibility between human beings and nature (*LS* 67).

5. Concluding Remarks

Much more could be said about possible parallelisms between Heideggerian thought and Pope Francis' theology. Further similarities include other aspects of their relations to the Enlightenment, their conservative attitude, their aversions against unchangeable systems, etc.) Differences include their relations to Christianity, their political attitudes, and their approach to metaphysics. Now, however, as a conclusion, let us be as concise and intellectually honest as possible.

Most of all, a correct hermeneutics is needed concerning the notion of "caring" at work in our two thinkers. The notion of "tilling" or "keeping" (*LS* 67) in Pope Francis is not *exactly* the same as the Heideggerian "care" (*Sorge*). Similarly, the Pope's understanding of "social friendship" (*FT* 2;5;6;94;99;142) is rather far from the "Being-with" (*Mitsein*) of Heidegger[73]. Even when vocabulary seems to indicate morphological convergences, the basic concepts of each thinker at times exhibit important divergence. For example, the Christian

[73] At this point Jin Xiping notes: "When I taught Heidegger's BT in China, I noticed that in his phenomenology of life, especially in the texts about *Mitsein,* Being-With-Others, one cannot find any serious discussion about friendship, love, or trust in others, all of which are very important concerns both in traditional Chinese culture and in Christianity. The same note was made by Scheler (Scheler, *GA* Vol. IX, 294), where Scheler criticises Heidegger for only using *Angst* and Death to characterise the life of human beings. He points out that 'this is not the case (*Aber so ist es nicht*)'; Scheler continues: the human being 'first turns to the immanent world, not because of repulsion, Angst and fleeing from oneself, but rather because of 'eros (Love)' [...] What opens the world for us, is love, not angst". Jin Xiping, "Heidegger's Conception of Being-With [Mitsein] and his Simple Designation of Social and Political Reality in the Black Notebooks", in *Front. Philos. China,* 11/3-2016, pp. 415-429, here 416-417.

conception of "Brother" is *not* the same as the fundamental ontological term Dasein. In this sense, it becomes obvious that Pope Francis is not a "Heideggerian"; the two personalities are completely different in character, profile, and shape. Even their goals are incompatible. To put it simply: in Heidegger's case we witness a revolutionary genius who reveals—and creates—an autonomous reading of reality, disclosing Being on its timely horizon. By contrast, in the case of Pope Francis, we witness a benevolent pastor of the universal world-church who interprets—and discloses—the divine revelation of the Christian faith, addressing his evangelical message to those who are entrusted to him, especially the poorest of the poor and the most vulnerable. However, at least from the three aspect treated above (embraced Conditionedness; rejected Technocratic Scientism; and possible Ecological Alternatives), *some of the parallelisms are undeniable*. And – as I tried to demonstrate it above – most probably *not* just in a *meta-literal* sense[74].

Abbreviations of Heidegger's Works Cited

SZ/BT Sein und Zeit. Fünfzehnte, an Hand der Gesamtausgabe durchgesehene Auflage. Tübingen: Niemeyer, 1979. Translated as: Being and Time (J. Macquarrie and E. Robinson) New York: Harper & Row, 1962.

GA 7 Vorträge und Aufsätze. Frankfurt am Main: Klostermann, 2000.

GA 9 Wegmarken. ed. by F.-W. von Hermann. Frankfurt am Main: Klostermann, 1996. Translated as: Pathmarks. (ed. W. McNeill). Cambridge: Cambridge University Press, 1998.

[74] Romano Guardini – who is explicitaly mentioned in *LS* 203 (and 5 times refered in the footnotes: 83;87;92;144;154) (cf. also *EG* 224; Papa Francesco, La mia porta è sempre aperta. Una conversazione con Antonio Spadaro [Milano: Rizzoli, 2013], 90, etc.) is the evident link between the Heideggerian thought and the magisterium of Pope Francis. He is the number one candidate for mediating between the two thinkers. However, as we argued above, most probably there are some more (anonymous) collaborators of Francis being responsible for the influence showed above. On the one hand, on the relationship between Francis and Guardini, see Massimo Borghesi, "The Polarity Model: The Influences of Geston Fessard and Romano Guardini on Jorge Mario Bergoglio", in Brian Y. Lee and Thomas L. Knoebel (ed.), *Discovering Pope Francis: The Roots of Jorge Mario Bergoglio's Thinking.* Collegeville Minnesota: Liturgical Press Academic, 2019, pp. 93-113, especially 104-113; and, on the other hand, on the relationship on Heidegger and Guardini, an informed contribution, with many details from the Black Notebooks: George Pattison, "Why Heidegger Didn't Like Catholic Theology: The Case of Romano Guardini", in Mårten Björk and Jayne Svenungsson (ed.), *Heidegger's Black Notebooks and the Future of Theology*. Cham Switzerland: Palgrave Macmillan, 2017, pp. 77-98.

GA 13 Aus der Erfahrung des Denkens. Frankfurt am Main: Klostermann, 1983.

GA 16 Reden und andere Zeugnisse eines Lebensweges, 1910-1976. Frankfurt am Main: Klostermann, 2000.

GA 41 Die Frage nach dem Ding: zu Kants Lehre von den transzendentalen Grudsätzen. Frankfurt am Main: Klostermann, 1984. Translated as: What Is a Thing? (trans. W. B. Barton Jr and V. Deutsch). Chicago: Henry Regnery Company, 1967.

GA 56/57 Zur Bestimmung der Philosophie. Frankfurt am Main: Klostermann, 1987. Translated as: Towards the Definition of Philosophy (trans. T. Sadler), London: Continuum, 2002.

GA 58 Grundprobleme der Phänomenologie. Frankfurt am Main: Klostermann, 1993.

GA 59 Phänomenologie der Anschauung und des Ausdrucks: Theorie der philosophischen Begriffsbildung. Frankfurt am Main: Klostermann, 1993.

GA 61 Phänomenologische Interpretationen zu Aristoteles/Einführung in die phänomenologische Forschung. Frankfurt am Main: Klostermann, 1985. Translated as: Phenomenological Interpretations of Aristotle (trans. R. Rojcewicz). Bloomington: Indiana University Press, 2001.

GA 63 Ontologie: Hermeneutik der Faktizität, Frankfurt am Main: Klostermann, 1988. Translated as: Ontology: The Hermeneutics of Facticity (trans. J. van Buren). Bloomington: Indiana University Press, 1999.

GA 77 Feldweg-Gespräche 1944/45. Frankfurt am Main: Klostermann, 1995.

"LEADERSHIP CAPABLE OF STRIKING OUT ON NEW PATHS"*

The Contribution of Female Intelligence to the Integral Ecological Paradigm Delineated in the Encyclicals Laudato Si' *and* Fratelli tutti

Elisabetta Casadei

1. Introduction: a new *forma mentis*

"It is easier to change governments than the mind of the people", so the popular saying goes, meaning that it is easier to change geopolitical assets than the paradigms of thought and the established practices of the masses.

This is the point.

The two most recent Encyclicals, *Laudato Si'* (LS) and *Fratelli tutti* (FT), make a point of saying that the care for our "common home" and for "universal fraternal love" are not questions of technical-legal adjustments (LS 60.106.109.111.190; FT 179), but of culture, paradigms of thought. "The problem – remarks the 'Pope of ecology' and of 'human fraternity', after having condemned "what is happening to our common home" (LS chap. 1) and the "dark clouds over a closed world" (FT chap. 1) – is that we do not yet possess the culture necessary for tackling this crisis and so there is a need to construct "leadership capable of striking out on new paths" (LS 53), as well as "of

* Pope Francis, Encyclical Letter *Laudato Si' on Care for Our Common Home*, 24 May 2015, n. 53. Pope Francis, Encyclical Letter *Fratelli tutti on Fraternity and Social Friendship*, 3 October 2020, n. 177.

DOI 10.32060/C.S.9.385-410

[healthy] politics which is far-sighted and capable of a new, integral and interdisciplinary approach to handling the different aspects of the crisis" (FT 177).

In this connection, the two papal documents carry out a comparison between two paradigms of thought and action: the Technocratic (or Techno-Economic) Paradigm and the hoped for Integral Ecological Paradigm.

In search of the foundations of integral Ecology, we shall ask ourselves in this paper what *mental structure* lies behind the two Paradigms, what type of *intelligence and sensitivity* they reflect (and not what philosophy or ideology) with particular attention to the Integral Ecological Paradigm.

We believe that Gender Studies can throw light on the (neuro-biological, psychical and cultural) foundations of the two Paradigms, to the degree in which they provide scientific bases for the differences of gender.

The thesis of our study is that the Integral Ecological Paradigm reflects the peculiarity of the intelligence of the female gender, while the Technocratic Paradigm, free from its corruptions, reflects better the intelligence of the male gender.

As a presupposition and to clear the field immediately of prejudicial presuppositions, by 'male gender' and 'female gender', we do not automatically understand individual males and females – and perhaps come to the hasty conclusion that the corruption of the Technocratic Paradigm is the consequence of male socio-cultural domination, as a good part of the feminist[1] and ecofeminist[2] movements would like to think – but modes of cognitive and emotional perception, of relations, of thought and action which can belong to both individual males and females, but in different measures and expressed

---

[1] Teresa Toldy, "Someone is Missing in the Common House: The Empty Place of Women in the Encyclical Letter 'Laudato si'", in *Journal of the European Society of Women in Theological Research* 25-2017, pp. 183-188.

[2] Extensive description of this phenomenon is found in the monographic number of the journal, *Deportate, esuli, profughe* (Rivista telematica di studi sulla memoria femminile dell'Università Cà Foscari di Venezia), 20-2012 (monographic number on Ecofeminism), in particular, Bruna Bianchi, "Ecofemminismo: il pensiero, i dibattiti, le prospettive", in *Deportate, esuli, profughe* 20-2012, pp. I-XXVII. Ecofeminist comments on *LS* can be found in Nontando M. Hadebe, "The Cry of the Earth is the Cry of Women: Ecofeminisms in Critical Dialogue with Laudato si'", in *Grace & Truth* 34/2-2017, pp. 50-65; Sue Rakoczy, "The Mission Spirituality of Laudato Si': Ecological Conversion and the World Church", in *Grace & Truth* 34-2017, pp. 66-78; Sue Rakoczy et al., "A Review Article on Laudato Si'", in *Journal of Theology for Southern Africa* 157-2017, pp. 135-171; David von Schlichten, "Strange as this weather has been: teaching Laudato Si' and Ecofeminism", in *Journal of Moral Theology* 6-2017, pp. 159-168; Anne M. Clifford, "Pope Francis' Laudato Si' on Care for Our Common Home: An Ecofeminist Response", in *Proceedings of the Catholic Theological Society of America* 72-2017, pp. 32-46.

in different ways and to different degrees. In fact, our focus will be on the *intelligence of the gender* and not on individual males and females (or other); such as could be males (from the point of view of biology, identity and sexual orientation) with ways of perceiving, thinking and acting closer to a different gender factor which is not typically male, and *vice versa*.

Moreover, we shall not be treating the two genders as bipolar, namely autonomous and exclusive (if more masculine, then less feminine), but as two articulations of the one human species in which gender factors simply distinguish *tendencies*.

Thus, our contribution will be arranged in four parts: 1) the formation and characteristics of the male and female genders and the possible different gender factors, that is to say, what can define the 'male' and the 'female'; 2) the description of the two Paradigms, as these emerge from the two Encyclicals (nature, method, objectives, effects and values); 3) the identification of the 'male' and the 'female' in the two Paradigms as their possible anthropological foundations; 4) in conclusion, the hypothesis of a possible integration of the two Paradigms, on the basis of their foundations in gender, and the indication of a working way forward consisting of an *integrated leadership*, aimed at tackling the global challenges outlined in the two Encyclicals.

2. Gender difference factors

2.1 Formation of gender: causes and levels of structuring of sexual gender

It seems useful to begin from the causes which structure sexual gender and from the levels on which these causes interact, to show that the boundaries between nature and culture cannot be easily identified. This is not the place for a detailed argument; we must limit ourselves to indicating that, in general and in the briefest of summaries, it is possible to identify two groups of causes (biological and cultural) of sexual gender and three levels[3]:

a) the level – let us say – of biology (genetic inheritance and uterine and extra-uterine environment) and culture (e.g. Latin or Chinese culture, social class, religion, etc); b) the level – let us call it – of roots, *viz.* biological (chromosomes, hormones, gonadic and genital sex) and cultural (family and education, socialisation, mass media, etc); c) the level – keeping the met-

[3] On this subject, we are following Richard Lippa, *Gender, Nature and Nurture*. New Jersey-London: Lawrence Erlbaum Associates, 2005², pp. 81-85.

aphor – of the trunk, the branches and the leaves which mirrors the actual condition of the subject in his or her identities, *viz.* biological (structure and functions of the central nervous system, level of hormones), psychical (cognitive, affective and relational abilities, character), cultural (social relations, position and social role, etc.) and spiritual (superior cognitive abilities, interests, ideals and values).

Personal traits, attitudes, abilities, identities and behaviour of the two genders are the product of the interaction of these three levels, which are not static or compartmentalised, but rather interconnected (each depends on and is influenced by the others) and interacting in different ways according to the different stages of the subject's life (prenatal, infancy, puberty, adulthood). A dynamism which the biological, human and social sciences are still not able to explain in detail, not least because of their polarisations over the subject, which they often see to be explained either from the side of biological theories or from that of socio-environmental theories[4].

2.2 Differences of gender

Focusing our interest on the origins of the subject, namely, on the level of the biological and cultural roots, studies on the prenatal life indicate that, after the chromosomes[5], the hormones are the most important factors for the development of sexual identity insofar as they are responsible for the gonadic and genital development as well as the development of the brain, and, so, the

[4] Cf. Richard Lippa, *Gender, Nature and Nurture*, pp. 85-118; 219-233; Richard Lippa, "Gender Differences in Personality and Interests: When, Where, and Why?", in *Social and Personality Psychology Compass* 4/11-2010, pp. 1098-1110; Richard Lippa, "The preferred traits of mates in a cross-national study of heterosexual and homosexual men and women: An examination of biological and cultural influences", in *Archives of Sexual Behavior* 36-2007, pp. 198-208; Richard Lippa, "Sex differences in sex drive, sociosexuality, and height across 53 nations: Testing evolutionary and social structural theories", in *Archives of Sexual Behavior* 38-2009, pp. 631-651.

[5] The sexual X chromosome, which constitutes about 2,5% of male DNA and double that of the female, possesses around 1200 genes; whereas the Y chromosome, present only in males, constitutes approximately 2% of their DNA and possesses only 90 genes, including the SRY (*Sex-determining Region of the Y chromosome*) gene, responsible for the primary determination of the male sex; unfortunately, we do not know enough yet of the major genetic contribution of the X chromosome. On the other hand, the autosomes, that is, the non-sexual chromosomes, have little influence on sexual identity, but only on the male and female traits (e.g. baldness), since their genes are activated or deactivated (genetic regulation) depending on various factors (hormones, intra- and extra-uterine environment, etc.). In this field, studies are just beginning and, from the methodological point of view, they have more success when they concentrate on chromosomal sexual disorders.

central nervous system[6]. Moreover, their influence extends to the whole life of the subject through the hypothalamus and the pituitary gland 'in dialogue' with the gonads, peaking at the time of puberty.

Although they still suffer from important methodological obstacles[7], the Hormonal Theories of sexual differences claim, in general, that in the prenatal stage, the androgens have a greater effect on sexual development than the oestrogens since they are activated first. The effects of the former on the behaviour are greater aggression (impulsivity), sexual desire, a capacity for risk, visual-spatial ability and concentration on objectives; while the effects of the

[6] For the subject of the entire section, we depend on: Judith Owen Blakemore, Sheri Berenbaum and Lynn Liben, *Gender Development*. New York-London: Psychology Press, 2009; Richard Lippa, *Gender, Nature and Nurture*, 2005[2].

On neurophysiological foundations, cf. Ivanka Savic, (ed.), *Sex differences in the human brain, their underpinnings and implications*. Amsterdam: Elsevier, 2010; Jill M. Goldstein et al., "Normal sexual dimorphism of the adult human brain assessed by in vivo magnetic resonance imaging", in *Cerebral Cortex* 11-2001, pp. 490-497; Susan Golombok and Stephan Hamann, "Sex differences in the responses of the human amygdala", in *Neuroscientist* 11-2005, pp. 288-293; Susan M. Resnick, "Sex differences in regional brain structure and function", in Peter W. Kaplan (ed.), *Neurologic disease in women*. New York: Demos Medical Publications, 2006, pp. 15-26; Rhoshel K. Lenroot et al., "Sexual dimorphism of brain developmental trajectories during childhood and adolescence", in *NeuroImage* 15-2007, pp. 1065-1073; Emilia Lentini et al., "Sex differences in the human brain and the impact of sex chromosomes and sex hormones", in *Cerebral Cortex* 23/10-2013, p. 2322-2336; Larry Cahill, "Equal ≠ the Same: Sex Differences in the Human Brain", in Brian Glovin (ed.), *Cerebrum 2014: Emerging Ideas in Brain Science*. Washington, DC: Dana Press, 2015, pp. 35-46; Melissa Hines, "Neuroscience and sex/gender: Looking back and forward", in *The Journal of Neuroscience* 40/1-2020, pp. 37-43.

Among the scientific studies on gender differences, cf. the following; for further in-depth treatments, we refer to the extensive bibliographies assembled in the studies cited above: Linda Mealey, *Sex Differences. Developmental and Evolutionary Strategies*. San Diego (USA): Accademic Press, 2000; Lee Ellis et al., *Sex differences. Summarizing more than a Century of scientific research*. New York: Psychology Press, 2008; Diane F. Halpern, *Sex differences in cognitive abilities*, (1992). New York: Psycology Press, 2012[4]; Linda Brannon, *Psychological Perspectives*. London: Routledge, 2017[7]; Louann Brizendine, *The Female Brain*. New York: The Doubleday Broadway Publishing Group, 2006; tr. it. *Il cervello delle donne*, Milan: RCS, 2007.

[7] One thinks, for example, of the difficulty of measuring the hormones in the pre-natal stage and, above all, in the first weeks of the foetus in which the gonadic (6th-7th) and genital (9th-10th week) developments begin, given that the amniocentesis procedure is dangerous and the measuring of the hormones in the mother's blood may not be sufficient; furthermore, the ethical difficulty of intervening in the foetuses and babies with doses of hormones (operations possible instead on animals); finally, seemingly still unresolved is the question whether it is the hormones that influence the behaviour or *vice versa*, or both (for example, aggressive parents alter the production of hormones in their babies). Therefore, the fields of research that can be undertaken more easily are those on animals and on subjects affected by sexual syndromes (DSD).

oestrogens are greater linguistic ability, sensitivity and emotional intelligence and an inclination for the care of others[8].

2.2.1 Cerebral-cognitive differences

Where cerebral-cognitive differences are concerned, it is possible to distinguish a double effect of the hormones on the body and so, also, on the brain: an effect of 'organisation', which is both structuring and permanent, in the pre- and post-natal stages; an effect of 'activation' in the adult period, that is to say, on the neuro-physiological functions and so is temporary.

Today's techniques of magnetic resonance neuro-imaging (MRI) enables the collection of the 'structural' and 'functional' differences of the brain of males and females, measuring the quantity of blood and of oxygen which flow in the tissues. This has made it possible to assert that the male brain is anatomically and functionally different from that of the female: if the male brain is bigger and possesses greater neuronal connections (white matter)[9], the female one has a greater density of neurons (grey matter); women use the right hemisphere more with a greater interconnection between the two hemispheres, whereas men use the left hemisphere more and have a greater intra-connection with the result that the male brain works with a greater lateralisation (specialisation and greater use of one hemisphere at a time).

In fact, the functional studies show that women use both hemispheres in language, whereas men chiefly use the left; the male brain works analytically and sequentially, and develops a greater visual-spatial capacity (linked to the androgens and to lateralisation), whereas the female one works in a way that is synthetic and simultaneous and develops mnemonic and emotional capacities.

In this field of enquiry too, the results are still at an early stage and it remains to be understood whether it is the brain which influences behaviour or the latter (and so the culture) which influences the cerebral functions, or else both, and in what way and to what degree[10].

---

[8] Cf. Judith Owen Blakemore, Sheri Berenbaum and Lynn Liben, *Gender Development*, pp. 39-68; 133-169; Richard Lippa, *Gender, Nature and Nurture*, pp. 81-154.

[9] The myelinic membrane which covers the neurons for the transmission of impulses.

[10] Cf. Judith Owen Blakemore, Sheri Berenbaum and Lynn Liben, *Gender Development*, pp. 169-176; Richard Lippa, *Gender, Nature and Nurture*, pp. 99-102; Grazia Attili, *Il cervello in amore. Le donne e gli uomini ai tempi delle neuroscienze*. Bologna: Il Mulino, 2017, pp. 175-195.

Operative alongside the biological factors in structuring sexual gender are the cultural and social ones through gender stereotypes, that is, directing the nature towards what 'male' and 'female' signify in a particular period and culture. However, it appears important to make clear, as has been clearly pointed out by R. Lippa – who coined the *Gender Diagnostic* (GD) method, focused on the interests of the persons – that these factors are to be rooted in the nature and can shape it but not replace it[11].

From the intellectual point of view, males and females are not distinguished by different IQ's, even if the dimensions of their brains are different, but rather by the different abilities which they manage to develop. As we have anticipated, males are distinguished by a greater capacity for analysis, a tendency to precision and detail, even in language; the breakdown of complex structures into their simple elements, sequential progression of thought, planning of actions, memory of the past. Women, on the other hand, display a greater fluency in oral and written communication; creative and artistic ability; simultaneous thought. If males tend to suffer more from mental illnesses (including those of personality and sexuality, mental and verbal handicaps, drug abuse), for females, the troubles are those of mood (anxiety, depression, panic attacks, anorexia nervosa)[12].

2.2.2 *Affective-emotional differences*

To debunk the stereotype that women are more emotional than men, today there is the possibility of measuring technically the physiological modifications relating to the feelings and emotions (e.g. through blood pressure, temperature, hormones and receptors in the blood). This shows that men too are highly emotional and that wide differences are revealed only in connection with certain emotions (e.g. aggression, sexual desire) linked to testosterone.

Measurements obtained from self-reporting (e.g. through questionnaires) and behavioural observations carried out in various cultures show instead that, generally, women perceive their emotions with greater intensity and duration, express them better (with verbal, para- and non-verbal language) and tend to hide the emotions that can upset others; men, on the other hand, are more reluctant to reveal their own emotions on account of social conditioning which presses them to maintain positions of control of the situation and inde-

[11] Cf. Richard Lippa, *Gender, Nature and Nurture*, pp. 155-172; 67.

[12] Judith Owen Blakemore, Sheri Berenbaum and Lynn Liben, *Gender Development*, pp. 78-94; Richard Lippa, *Gender, Nature and Nurture*, pp. 33-37.

pendence (for example, fear or vulnerability), which is why they show higher levels of antisocial behaviour.

There are medium-small differences over how men and women tackle difficulties emotionally (concentrating on the problem or managing emotion) here women stand out for their empathetic capacity including sympathy and mind reading (knowing how to guess the thoughts, desires and feelings of others).

Finally, the results concerning self-esteem and the concept of self are interesting: for the former, the differences are small to the advantage of men when they are adults; for the latter, women see themselves in relation to others (and that has an influence on their self-esteem), while men see themselves independently of others, instead identifying themselves by their social positions (e.g. employed, Italian etc.), even if these results vary according to cultures[13].

2.2.3 Relational differences

Where this subject is concerned, the analyses prefer to concentrate on babies since they are less influenced by their culture. For the purposes of our study, we shall limit ourselves to briefly pointing out two among many aspects of relationality: collaboration in groups and morality.

With regard to the former, men and women are equivalent in their capacity for negotiating and, ultimately, taking on risk. However, in the exercise of leadership, important differences appear: women are more focused on relations between the members of the group, more inclined to welcome other points of view, to share and to put their own interests in second place; men, on the other hand, concentrate more on the objectives to be reached, on the tasks to be carried out and on the procedures to realise them. In general, the former exercise a more participatory and transformational leadership (based on trust, esteem and innovation) and the latter one that is more autocratic (based on the objective and the order of the processes), but substantially equal in leaving freedom to members of the group[14].

As for morality, men manifest higher levels of antisocial and criminal behaviour while women appear more adaptable to rules and social practices; however, they seem to lie more so as not to hurt others and are more attentive

---

[13] Cf. Judith Owen Blakemore, Sheri Berenbaum and Lynn Liben, *Gender Development*, pp. 95-125; R. Lippa, *Gender, Nature and Nurture*, pp. 18-44.

[14] Cf. Judith Owen Blakemore, Sheri Berenbaum and Lynn Liben, *Gender Development*, pp. 103-106; Richard Lippa, *Gender, Nature and Nurture*, pp. 22-26.

to people and their needs rather than to rules (moral and legal), more abstract and universal. In other words, women seem to manifest more of a 'morality of care', rather than a 'morality of law'[15].

2.3 Psychological keys to the differences of gender

At this point, we ask if there can be any interpretative keys to these differences of gender, namely, if there could be a 'female logic' and a 'male logic' in perceiving, feeling, elaborating, thinking, deciding, communicating and acting. The question as to what is understood by 'female' and by 'male' is not new (on the basis of sexual orientation? Cognitive abilities? Social behaviour and roles?) and, from the historical point of view, we find ourselves in a rather favourable climate, both from the scientific point of view (thanks to new diagnostic techniques), and the cultural one (relieved from ideological pressures)[16].

In this paper, we shall briefly present three proposals (the first two rather recent) concerning factors which mark the differences of gender (gender difference factors), female and male, respectively: 'People-Things', of the United States psychologist, Richard Lippa[17]; 'Empathy-Systematising', of the English Sir Simon Baron-Cohen[18], 'Emotivism/ Agreeableness'-Emotive Stability/Domination', which emerge from the largest meta-analysis of gender

[15] Cf. Judith Owen Blakemore, Sheri Berenbaum and Lynn Liben, *Gender Development*, pp. 118-121; Richard Lippa, *Gender, Nature and Nurture*, pp. 21-22; 225-226.

[16] Indeed, now past are the *exclusive* positions of the two sexual genders (conceived as opposites) of the years 1920-1960 and the *independent* ones of the years 1970-1990, which conceived them as autonomous (*goal oriented* men and *people oriented* women, so as to distinguish the two genders with concepts of *Instrumentality* – the male – and *Expressivity* – the female); whereas still to this day, theories of *social construction* flourish, which sees genders as cultural products. Cf. Richard Lippa, "On Deconstructing and Reconstructing Masculinity–Femininity", pp. 168-191; Richard Lippa, *Gender, Nature and Nurture*, pp. 55-80.

[17] Cf. Richard Lippa, "Gender-Related Individual Differences and the Structure of Vocational Interests"; Richard Lippa, "On Deconstructing and Reconstructing Masculinity–Femininity", in *Journal of Research in Personality* 35-2001, pp. 168-207; Richard Lippa, *Gender, Nature and Nurture*, pp. 30-33; 67-80; Richard Lippa, "Gender Differences in Personality and Interests"; in particular, the very thorough meta-analysis of Rong Su, James Rounds and Patrick I. Armstrong, "Men and Things, Women and People: A Meta-Analysis of Sex Differences in Interests", in *Psychological Bulletin* 135/6-2009, pp. 859-884.

[18] Simon Baron-Cohen, *Questione di cervello. La differenza essenziale tra uomini e donne*. Milano: Mondadori, 2004 (orig. ed. *The essential Difference. The Truth about the Male and Female Brain*. New York: Basic Book, 2003).

differences, performed with measuring of the personality, considered in the study led by the Italian, Emiliano Lambiase[19].

2.3.1 People-Things (R. Lippa)

This distinguishes the female gender in its orientation towards 'People' from the male gender in its orientation towards 'Things'. This appears to be one of the most reliable interpretative keys on account of its methodology and its anthropological foundations. In fact, Richard Lippa coined a new methodology, the Gender Diagnostic (GD) which, instead of focusing on personality traits (often difficult to observe, especially psychological expressions), measures the actual interests and occupations of the subjects (e.g. hobbies, professions), indications – according to the Author – of the expression of the personality, identity and values with which they identify themselves[20]. The GD has the advantage of observing both the *natural* and the *cultural elements* of the sexual difference, by measuring the interests of a *specific group* in a *space* and *time* with a particular *cultural mark*. From the anthropological point of view, this signifies capturing the person in his or her striving towards what gives them satisfaction and joy, that is, in his or her intentional and voluntary dynamism towards the *ends which are peculiar to him or her* and which, therefore, constitute the realisation (goodness) of his or her being.

The Californian psychologist does nothing other than utilise the RIASEC model of occupational interests (*Realistic*, *Investigative*, *Artistic*, *Social*, *Enterprising*, *Conventional*)[21], which was elaborated by J.L. Holland at the end of the nineteen fifties[22], and take advantage of the deep structure which D.J.

---

[19] Emiliano Lambiase, Cristina Torraca and Bianca Lagioia, "Differenze sessuali nelle principali dimensioni della personalità", in *Modelli della mente* 1/2018-2/2017, pp. 7-54.

[20] Bibliographical references on the GD can be found in Richard Lippa, "Gender-Related Individual Differences", p. 998. On the meaning of "interest", cf. R. Su – J. Rounds - P.I. Armstrong, "Men and Things, Women and People", p. 860.

[21] *Realistic*: interest in using machines and tools, manipulating objects and instruments (e.g. mechanic, carpenter); *Investigative*: interest in the study of phenomena with the aim of controlling them (e.g. physicist, biologist, economist, sociologist); *Artistic*: interest in manipulating material, words (e.g. painter, actor, writer); *Social*: interest in the training, care and direction of people (e.g. teacher, social worker, priest); *Enterprising*: interest in managing things and people in order to realise objectives (e.g. politician, entrepreneur); *Conventional*: interest in handling data and processes (e.g. official, accountant, IT consultant).

[22] Cf. John L. Holland, "A personality inventory employing occupational titles", in *Journal of Applied Psychology* 42-1958, pp. 336-342, and the subsequent adjustments up to 1997.

Prediger[23] had identified there, according to whom these interests could be summarised in two fundamental directions: People/Ideas and Things/Data[24] (cf. Figure below).

According to Lippa, the People-Things dimensions are most correlated with the GD, and a series of meta-analyses (covering different cultures and wide temporal spaces) have effectively (80-90%!) indicated that men are considerably oriented towards Things, namely, to have interests and to follow activities which have to do with machines, tools, equipment and objects, whereas women are oriented towards People, that is, to have interests and follow activities which have to do with the care of and interaction with People[25]. This happens from the age of 2-4 years, it is found in homosexuals (gay and lesbian) and transsexuals, and, above all, it is revealed in free societies, namely, in cultural contexts where people can be most free from cultural conditioning with regard to gender stereotypes.

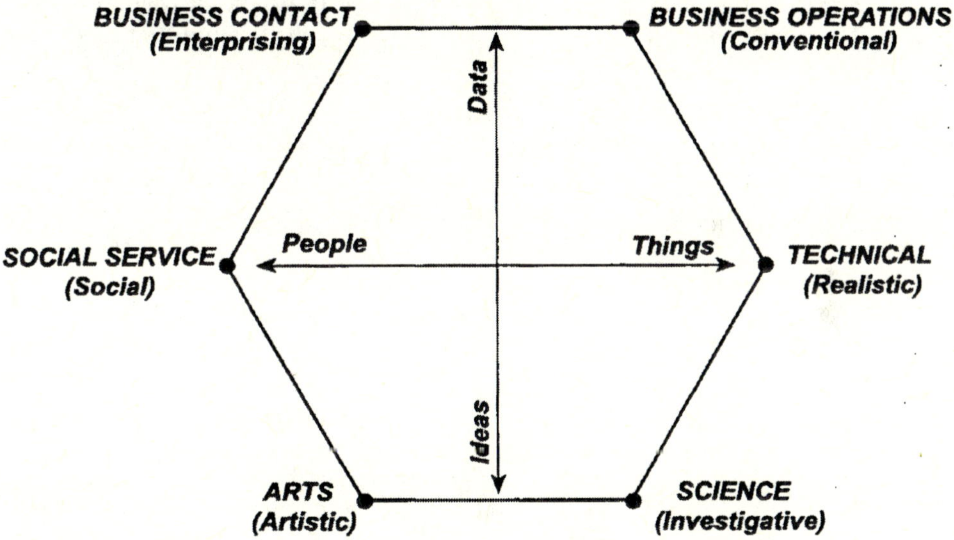

[23] Dale J. Prediger., "Dimensions underlying Holland's hexagon: Missing link between interests and occupations?", in *Journal of Vocational Behavior* 21-1982, pp. 259-287.

[24] *People*: orientation to being concerned about, taking care of and interacting with people; *Ideas*: orientation to the use of creative thought; *Things*: orientation to working with machines, tools, objects; *Data*: orientation to the use of calculating, executive and functional thought.

[25] The result has been corroborated by the meta-analytic study of Su, Rounds and Armstrong, cited above, which confirms, additionally that the Ideas-Data distinction does not constitute a Gender Difference Factor.

2.3.2 Empathy-Systematising (S. Baron-Cohen)

Our second proposal comes from one of the greatest researcher of autism. Its orientation could be called 'natural' in the sense that it finds sexual gender in the same male and female cerebral structure: the female capacity for 'empathy' and the male for 'systematising' – Baron-Cohen maintains – are rooted in precise regions of the brain, which coincide more or less with the right and left hemispheres respectively[26]. One could, therefore, claim that this proposal integrates with that of Lippa, demonstrating the biological factors behind the orientations to People-Things[27].

Baron-Cohen's thesis can be briefly described: the female brain is 'programmed', that is, structured to be 'empathetic' – capable, namely, of "recognising the thoughts and emotions of others and of reacting with suitable sentiments" – and so has its object is emotions[28]; the male brain, on the other hand, is 'programmed' to be a 'systematiser' – that is, to understand and elaborate natural, technical and abstract systems (e.g. logic, mathematics, linguistics) and social ones (e.g. politics, economics) – and so has systems for its object[29].

The author maintains that the cultural factors are explicable only on the basis of neuro-biological ones. This is shown, for example, by the studies on the preferred toys of babies of a few months, on the male and female abilities of animals and the theories of the hereditary nature of the difference between the sexes[30]. In particular, the male capacity for 'systematisation' depends on

---

[26] Simon Baron-Cohen, *Questione di cervello*, pp. 6-17. However, the author tends to make clear that he is dealing with average statistics, that is to say, his thesis holds for the "gender" and not for individuals, since there could be males with a female "brain" and *vice versa* (*ibid*, 11), as we ourselves have also made clear in the introduction.

[27] As the same Lippa recognises (cf. *Gender, Nature and Nurture*, pp. 32-33).

[28] Cf. Simon Baron-Cohen, *Questione di cervello*, pp. 3-6 (*ibid*, 4). Empathy, therefore, is the ability to come into harmony with the thoughts and feelings of others, that is, not just understanding them, but "feeling them" within oneself so that it has a cognitive and an affective component (cf. 28-36).

[29] That means understanding the functioning of something, namely, the laws which regulate it, and its elementary characteristics with the aim of foreseeing its movement and constructing another system. Cf. Simon Baron-Cohen, *Questione di cervello*, pp. 5-6; 74-82.

[30] Cf. Simon Baron-Cohen, *Questione di cervello*, 103-118; 142-160. In line with the proofs adduced by Maccoby and Jacklin in 1974, cited by Lippa, according to whom it is clear that biology has a role in the differences of gender on account of: a) the age in which sexual differences emerge; b) the consistency of sexual differences within cultures and throughout history; c) among the species; d) the relation between the physiological factors and the behaviours which manifest sexual differences, cf. Richard Lippa, *Gender, Nature and Nurture*, p. 137; Eleanor E. Maccoby and Caro N. Jacklin, *The psychology of sex differences*. Stanford (CA): Stanford University Press, 1974.

the high (but not very high) levels of testosterone in the pre-natal stage of the 'organisation of the brain' – mentioned above – and on the visual-spatial area of the right hemisphere[31], whereas the female capacity for 'empathy' has its neurobiological bases in the cerebral area assigned to language (frontal left hemisphere) and the emotions (amygdala, pre-frontal cortex and superior temporal sulcus)[32].

Thus, an 'empathetic brain' is more competent at the empathetic reading of faces and minds, at verbal language (fluency, vocabulary, memory) and non-verbal communication, at the establishment of cooperative and reciprocal relations and the welcoming of new and weak people, namely – we can conclude – in being centred on People[33]; whereas a 'systematising' brain is more competent at collecting details, the rules of functioning, orienting itself to reading maps, constructing spaces, classifying, reaching superior *status*, that is – we can conclude – being focused on Things[34].

2.3.3 *Emotivity/Agreeableness-Emotive Stability/Dominance (meta-analysis on personality)*

A third criterion for distinguishing female and male is provided by the extensive analyses of the personality carried out with the best-known diagnostic models: Big Five (5 great dimensions of the personality)[35], (male) Instrumentality and (female) Expressivity, and the 16 factors of Cattell which, unfortunately, it is not possible to expound here. Methods which manage to detect with difficulty the Gender Difference Factor, as Lippa's above-mentioned study of 2010 has shown and, finally the critical summary produced recently by E. Lambiase, C. Torraca and B. Lagioia, which we are about to set out in its essential lines.

This study analyses the principal meta-analyses carried out on the subject since the nineteen nineties, which look at samples which cover long periods, many cultural regions across five continents and different ages. The results show that the personality traits which differentiate masculinity from femininity – according to the Big Five factors – are principally those of *Neuroticism*

---

[31] Cf. Simon Baron-Cohen, *Questione di cervello*, pp. 119-129.

[32] Cf. Simon Baron-Cohen, *Questione di cervello*, pp. 127-136.

[33] Cf. Simon Baron-Cohen, *Questione di cervello*, pp. 37-73.

[34] Cf. Simon Baron-Cohen, *Questione di cervello*, pp. 83-102; 41.47-50; 56.

[35] 1. Extroversion; 2. Agreeableness; 3. Conscientiousness; 4. Neuroticism; 5. Openness.

(which, here, we have called *Emotivity* in the sense of vulnerability, impulsivity, anxiety, depression, embarrassment, rage) and *Agreeableness* (trust, altruism, sincerity, tenderness, cooperativity, empathy) for women, and those of *Emotive Stability* (opposed to Neuroticism, as a control of emotions and security) and *Dominance* (control of other people, antagonism, vigilance, coldness) for men; with differences observed among different cultural regions (wider in free societies where there is greater sexual equality), periods and ages, and with a superimposition of the profiles of about 10-24%. The study concludes that this convergence presses for the consideration of the effects of the underlying biology on the gender difference.

3. The 'Technocratic' and the 'Integral Ecological' Paradigms in LS and FT

The focus on and the comparison between the two Paradigms is explicitly a key point of the Encyclical of 2015, which applied them to the "common home". They are also presupposed and implied in FT which, this time, applies them to the possibility of a concrete realisation of "universal fraternity"[36].

By 'paradigm', *Laudato si'* understands, in general, "a way of understanding human life and activity" (LS 101). An ethico-cultural specification of its Platonic etymon of 'exemplar', 'explanatory model' of phenomenal reality[37].

In the first five years since *Laudato Si'*, an extensive literature has concentrated on the Integral Ecological Paradigm (and less on the much better-known Technocratic one) from various perspectives (theological, ethical, sociological, juridical)[38], but it seems that *a study of the profound anthropological*

---

[36] *FT* does not mention the expression "ecological or integral paradigm", preferring the expression "integral human development" (nn. 21.107.112.118.129.133.138.169.235.257.276), "integral growth" (n. 113), "integral fulfilment of persons" (120); "well-integrated view" (n. 167), "integral approach" (n. 177), "well-integrated view" (n. 167) and "integral knowledge of reality" (204). While it mentions the "technocratic paradigm" only once (n. 166).

[37] The *para-deigma* (from greek *deiknumi* = to show, present), indicates the "beyond" (*parà*) what is "shown" (*deiknuo*), that is, what lies behind and explains what is presented, constituting it as its cause and matrix (cf. Plato, *Euthiphro,* 6e; *Timaeus*, 28a-b). In Aristotle, it is a logical argument permitting inference from a known argument to one which is not yet known (*Analytics I*, 2,24). It enters contemporary scientific language with Thomas Kuhn, *The structure of scientific revolutions*. Chicago-London: The University of Chicago Press, 1968, with the double meaning of *disciplinary* and *exemplary* matrix.

[38] Cf. Michael Hanby, "The gospel of creation and the technocratic paradigm: reflections on a central teaching of Laudato si'", in *Communio* 42/4-2015, pp. 724-747; Mauro Cozzoli, "L'enciclica Laudato si' paradigma di una morale ecologica", in *Medicina e Morale: Rivista internazionale di Bioetica* 64/5-2015, pp. 737-780; Lorenzo Orioli, "Laudato si' and the New Paradigm of Catholic Environmental Ethics:

structure of the two paradigms, with its focus on gender difference, has not yet been tackled, at least in the present state of studies.

We shall give a brief description of the two Paradigms as they emerge from LS and are implied in FT.

3.1 The Technocratic Paradigm

3.1.1 Nature, Method and Objectives

The dominant Paradigm, also called the Techno-Economic Paradigm (LS 53.203), is defined as "a way of understanding human life and activity [which] has gone awry, to the serious detriment of the world around us" (LS 101) and indicated as cause of the global crisis[39].

Its deviation consists in the methodology which it adopts in relation to reality such as to render it "undifferentiated and one-dimensional": it is a purely scientific method, logical-rational and excluding other valid forms of knowledge (ethical, religious, aesthetic) (LS 106.199; FT 204.211.275). This method is understood as a "technique of possession, mastery and transformation" of reality, considered as "formless, completely open to manipulation" and such as to arouse the illusion of the absence of limit and of "unlimited growth" (LS 106).

Therefore – it is important to make clear (!) – the deviance is not in the technology or in the economy considered in and of themselves, but in their methodological use within a Techno-Economic Paradigm, which has its fulcrum in human frailty: the desire for dominion without limits (*kratos*), which the Christian tradition calls 'concupiscence':

> My criticism of the technocratic paradigm involves more than simply thinking that if we control its excesses everything will be fine. *The bigger risk does not come from specific objects*, material realities or institutions, *but from the way that*

Reflections on Environmentalist Movements in Italy", in *Journal of Agricultural and Environmental Ethics* 29/6-2016, pp. 931-943; Xabier Pikaza Ibarrondo, "Laudato si': paradigma tecnocrático y crisis ecológica", in *Estudios franciscanos: publicación periódica de Ciencias Eclesiásticas de las Provincias Capuchinas Ibéricas* 118/462-2017, pp. 15-43; Simone Hegel Bolson and Napoleão Miranda, "O novo paradigma jurídico-ambiental defendido na encíclica laudato si: a ecologia integral eo desenvolvimento sustentável", in *Revista Internacional de Direito Ambiental* 7/19-2018, pp. 311-332; Christophère Ngolele, "African Wisdom in Dialogue with Laudato si': An Environmental Ethics Based on the Paradigm of Recognition and Sacred Care", in *Journal of Catholic Social Thought* 16-2019, pp. 5-19.

[39] In fact, the technocratic paradigm is described mainly in Chapter III of *LS* (nn. 101-136), which is devoted to the causes of the ecological crisis (cf *LS*, chap. I, nn. 17-61), and also indicated as a cause of the social crisis (cf. *FT*, chap. I, nn. 9-55; *ibid,* 7.166.177.275).

> *they are used.* It has to do with human weakness, the proclivity to selfishness that is part of what the Christian tradition refers to as "concupiscence": the human inclination to be concerned only with myself, my group, my own petty interests. Concupiscence is not a flaw limited to our own day. It has been present from the beginning of humanity, and has simply changed and taken on different forms down the ages, using whatever means each moment of history can provide[40].

A method, moreover, is recognized as inadequate when it is intent on immediate and pragmatic results (LS 36; FT 17.21.109), aimed at contingent and fragmented interests and places priority on being realised in a short time (LS 54.122). Such a method which moves under the banner of the fragmentation of knowledge, of reality itself (LS 110.138.201) and, above all, of subjects incapable of facing reality in the perspective of 'us' (FT 7.16-17.30.116-117). A methodology, therefore, unprepared to consider and unable to tackle the problems at their ethico-philosophical and religious root, that is, in the light of the ultimate and penultimate horizons of the three relations between human beings, between human being and the creation and between human being and God. (LS 50.61.110-111.119; FT 271-276).

A "globalised and standardising" Paradigm which imposes its logic and its lifestyle even on politics and the economy, reduces freedom and creativity (LS 108-109; FT 12-14.177).

3.1.2 Vision of creation and of the person

The profound reason why reality is conceived as totally at human disposal and disconnected is the anthropocentric vision of the Technocratic Paradigm according to which the human being is autonomous from nature and above it like a god (LS 115). Failing to recognise his place in creation, the human being becomes deaf to the cry of his brother (FT 18-27; 108) and of nature (LS 122-123). He is schizophrenic between his own dignity and the exaltation of technology or, on the contrary, of 'biocentrism' (LS 115-118).

3.1.3 Effects

The effects of this Paradigm constitute the reason why the first 'ecological' Encyclical was written, addressed even to those who are not 'of good will' (LS 3) but who simply dwell on the planet, just like FT.

---

[40] Pope Francis, *Fratelli tutti*, n. 166 (our italics); cf. Pope Francis, *Laudato Si'*, n. 105.

In *Laudato Si'*, these effects are condensed in the chapter devoted to "what is happening to our common home" (chap. I, nn. 17-61), which goes from the deterioration of the world (pollution and rubbish, climate changes, destruction of ecosystems and loss of biodiversity, to take only some examples) to the deterioration of the quality of human life (poverty, diseases, death, forced migrations), coming finally to social degradation (of the cities, fragmentation and social exclusion, the loss of reflective thought skewed in favour of calculation) (LS 43-47) and politics (superiority of technology and finance over international politics) (LS 54).

FT concentrates precisely on these last effects, sketching "the dark clouds over a closed world": dreams of political and social integration shattered, precipitating in nationalistic chasms on account of the inability of politicians to think and to dialogue (FT 10-12; 15-17; 25-28); deconstruction of historico-cultural identity and of critical thought (FT 13-14; 51-53); production of 'human rejects' on the world level (the elderly, unemployed, sick, unborn, new slaves) (FT 18-24); globalisation and progress without a common direction (FT 29-31); the illusion of communication and digital information (FT 42-50); a vulnerability, which the present pandemic has finally helped to unmask (FT 32).

The most important datum highlighted, especially by the analysis of the first Encyclical and pervading the whole document (LS 15), is the profound link between environmental and human degradation (at all levels) (LS 48), together with strong and recurrent denunciations of the fact that paying the price for all this are the "poor", victims of the "culture of waste" (LS 22.49.29.45.51-52), and that disfiguring the creatures implies obscuring the reflections of the Creator (LS 33.69.77.84-85.87).

3.1.4 Values

If we had to identify the ideals and values which inspire, animate and sustain the way of thinking and acting according to this Paradigm, we could indicate at least three of them, strongly connected with one another as the three angles of an equilateral triangle: dominance, progress and acceleration. 'Dominating' with the aid of technology, that is, possessing and consuming without limits as practical and univocal signs of 'progress' (LS 105.109; FT 13), according to an altered rhythm of life and work, namely, one that is out of 'sync' compared with the slow rhythms of biological, psychical (perceptive and affective) and spiritual (ethical and cognitive) life (LS 18; FT 48-50).

3.2 The Integral Ecological Paradigm

3.2.1 Nature, method and objectives

The Integral Ecological way of "understanding human life and activity" does not follow the pattern of 'dominion' (*kratos*) between human being and nature, but that of 'friendship' – in the sense of reciprocal belonging – between human being and nature and between man and man human being until the attainment of 'universal fraternity':

> Men and women have constantly intervened in nature, but for a long time this meant *being in tune with* and respecting the possibilities offered by the things themselves. It was a matter of *receiving* what nature itself allowed, as if from its own hand. Now, by contrast, we are the ones to lay our hands on things, attempting to extract everything possible from them while frequently ignoring or forgetting the reality in front of us. *Human beings and material objects no longer extend a friendly hand to one another*; the relationship has become confrontational[41].

> No one can experience the true beauty of life without relating to others, without having real faces to love. This is part of the mystery of authentic human existence. "Life exists where there is bonding, communion, fraternity [...] Our love for others, for who they are, moves us to seek the best for their lives. Only by cultivating this way of relating to one another will we make possible a *social friendship* that excludes no one and a fraternity that is open to all. [...] A love capable of transcending borders is the basis of what in every city and country can be called "social friendship". Genuine social friendship within a society makes true universal openness possible. [...] [Therefore], *universal fraternity and social friendship are thus two inseparable and equally vital poles in every society*[42].

A friendship founded on a "common origin", divine and ever present (LS 11.75.80; FT 272-274), in which "everything is connected": environment and living things, living things among one another, time and space, physical elements and environmental and social phenomena (LS 42.70.86.89.92.138-139); these last integrated as the different sides of a polyhedron, rather than as points equidistant from the centre of a sphere (FT 215). Thus, the Integral

---

[41] Pope Francis, *Laudato Si'*, n. 106 (our italics); on reciprocity, cf. nn. 67-68. The Integral Ecological Paradigm is described mainly in its theoretical roots of a biblical nature (cf. cap. II: *The Gospel of Creation*) and detailed from the social point of view in Chapter IV (*An Integral Ecology*) and, above all, in *Fratelli tutti*. In both Encyclicals, there is no explicit citation either of the scientific sources or of the various philosophical concepts (except Christian ones).

[42] Pope Francis, *Fratelli tutti*, nn 87.94.99.142 (our italics); cf., also, 93.106.

Ecological Paradigm can be described as 'pluridimensional and relational', strongly distinct from the one that is "undifferentiated and unidimensional".

Consequently, reality is conceived as a loop of "open systems" (including societies which are "open" to integrating all, FT 87-117, *ibid* 97), open also to transcendence, and integrated (as a working hypothesis), by entering into communication with one another (LS 79); furthermore, always in search of an equilibrium, a synthesis (on the social level in the 'social dialogue' in a 'culture of encounter', FT 199-224), such that the Paradigm is always *in fieri*, flexible and dynamic (LS 121.144).

This vision has enormous consequences for methodology too: it cannot avoid being *multidisciplinary* in order to interpret and transform reality, that is, it is suitable for considering and integrating other forms of knowledge, beside the technico-scientific one, such as aesthetics and art, ethics, history and philosophy, the theology and spirituality of the different religions (LS 63.199-201; FT 204.211.275). In fact, the Integral Ecological Paradigm postulates the ability to make a 'broad use of reason', knowing how to move along different horizons (close, penultimate and ultimate) and how to respect and integrate different epistemologies. In sum, instead of the 'fragmentation' of knowledge and reality, it proposes the 'integration of the branches of knowledge' and the 'connection' of reality, including social reality in the longed for 'universal fraternity'.

Consequently, the solutions which are sought for the current global crises (cf. chap. 1 of LS and FT) cannot be only technical, partial and immediate, but 'integral'. That is, solutions which consider the interactions of the natural systems among themselves and with the social systems, and of the latter among themselves (LS 139.164.111.119.183.197; FT 179); and these solutions will take a long time.

The objectives of the Integral Ecological Paradigm are formulated within the horizon of the "goals of human activity" (LS 61), which can be identified and condensed in two concepts which refer to each other: "sustainable integral development" and "common good" (LS 13.18; FT 112.177-178). The latter includes respect for the person, social well-being and security, the development of intermediate groups according to the principle of subsidiarity, social peace and distributive justice[43]; elements which require, "logically and

[43] Elements that coincide with the Thomist vision of the common good, as we have shown in our study, E. Casadei, *Vita interiore, discernimento politico e impegno civile. L'influsso della contemplazione filosofica sul giudizio ci-vile e politico in Tommaso d'Aquino*, Soveria Mannelli (CZ): Rubbettino, 2013, 275-297.

inevitably", solidarity, the preferential option for the poor and justice between the generations, but also the reform of political institutions (global and local), both to protect the 'common home' as well as to defend the meaning and value of human life (LS 156-162; FT 106-127; 170-197).

3.2.2 Vision of the creation and of the person

The Integral Ecological Paradigm reflects the biblico-Christian vision of creation (LS chap. 2), which sees three fundamental, closely connected relations, but ones which are damaged and disfigured by man's refusal to accept his own status as a creature: the relation of *sonship* between human being and God, the *fraternal* relation between human being and human being, and the relation of "*responsible reciprocity*" between human being and the creation. These relations are called to be reconciled and integrated as true indicators of progress (LS 66-68.194). This *relational vision* constitutes the final horizon of the 'connection' of the whole of reality, as mentioned above.

With the male-female distinction, the person is seen to be inserted in a 'universal family', who live in a 'common home', and who share a future and a common project (LS 89.164.202; FT cap. 3); he nurtures this project through work, animated and supported also by a vision of the world which is not only technical but also contemplative, in which the person is superior to inanimate things (LS 124-128).

3.2.3 Effects

The consequences of this Paradigm are clearly highlighted by the reformulation of the principles of human life according to the 'principles of integral ecology', which were set out above: the economy, social and political life, culture and health (LS chap. 4; FT chaps. IV-V).

An ecological economics which puts at its centre the person and the common good (LS 141; FT 179); a social ecology in which the institutions (from the family to the international community) foster and protect human relations and those with the environment (LS 142; FT 114-117); a cultural ecology which is not limited to more ecological cities but knows how to integrate, initiating a dialogue between the environment (land, rivers, plants, animals) and culture (history, art religion) as well between cultures in such a way as to respect the different interpretations of the 'quality of life', not lapsing into homogeneous solutions which act as cultural levellers (LS 143-146; FT 133-

150.220); and a human ecology in which the quality of life is measured in its broadest sense (physical, psychical and spiritual) (LS 147-155).

3.2.4 Values

The redefinition of the concept of 'progress' (LS 112.194; FT 33), which originates from another way of understanding human life and the world, also implies a kind of 'transvaluation' of values in relation to the dominant paradigm. Ideals, goals and forms of human action which, in our opinion, could be listed around three cardinal and transcendental (in the classical sense of the term) values of thinking and acting: guarding, integrating and contemplating.

Guarding, in the sense of receiving and accompanying what has already been inscribed in the reality of nature (instead of dominating it, exploiting it and transforming it without limits) (LS 106), including the dignity and the sociality inscribed in human nature (FT 106-107.87); but also in the sense of nurturing it and bringing it to perfection through work, science and education (LS 67.78.124.105.202-232; FT 112). Guarding, therefore, as capacity for generativity, both in an intragenerational and an intergenerational sense (LS 160-162).

Integrating, in the sense of welcoming, including, connecting, dialoguing, cooperating, sharing as well as 'fraternising' locally and globally, as style of thought and practice (LS 13.137-162.185.197.219; FT chaps. III-IV.VI).

Contemplating, in the sense of developing the capacity for awe and wonder, for knowing how to enjoy the beauty of nature, of things and of relations in freedom; for knowing how to enjoy the expansion of time, of contemplative rest and of festivals (LS 11-12.18.85-86.220-227.233-237; FT 139).

4. The foundations of gender in the two Paradigms

To arrive at the result of our study, we are now going to put together the material covered so far (§§ 1-2), that is, to trace in the two Paradigms the Gender Difference Factors, namely, the elements which constitute the keys with which to interpret 'male' and 'female' intelligence and sensitivity'.

It is not difficult to observe that the Technocratic Paradigm can be best interpreted by the orientation or factor 'Things', 'Systematising' and 'Emotive Stability/Dominance', whereas the Integral Ecological Paradigm looks to the orientation or factor 'People', 'Empathising' and 'Emotivity/Agreeableness'. Even though it does not lack cross-contaminations, which are not possible to indicate here, it would be worth the effort to investigate the relationship further in a future study.

Before going into detail, *we must take account here of the premises indicated in the introduction, concerning the understanding of the male and female gender.*

4.1 The Technocratic Paradigm and the keys to gender: 'Things', 'Systematising', 'Emotive Stability/Dominance'

The affective-volitive orientation to concentrating on 'Things', on objects, and to transform them and employ them through the use of tools and machines, renders the male gender particularly suited to the Technocratic Paradigm, which sees reality as other than itself, objectified, reified, and appropriate to be transformed and utilised.

The intelligence which tends to 'systematise' objective reality, to identify and classify systems, to understand their characteristics and functioning and to formulate laws, with the aim of imagining and constructing other systems, is also suitable for understanding reality in its details (fragments), grasping the logic of each 'system' as distinct from the others. An intelligence that is calculating and executive, inclined to concentrate on procedures, to order processes, to find technical and functional solutions, sees in the scientific, logical-rational method, precisely that of the Technocratic Paradigm, its privileged way and, in the 'undifferentiated and unidimensional' nature of this Paradigm, its sphere of action.

Thus, the trait of Dominance, intertwined with vigilance, control of others and of the emotions, which characterises the personality of the male gender, finds its anthropological roots in its affective and intellectual inclination, and this is clearly in harmony with the values (and disvalues) and goals of the Technocratic Paradigm (possession, dominion, utility, consumption, progress, immediacy and pragmatism); as well as with its pronounced anthropocentrism; its tendency to identify and occupy physical, cognitive and institutional spaces; but to disregard the sense of limit, of time and generosity, and, finally, to downplay 'whatever is not functional'.

4.2 Integral Ecological Paradigm and the keys to gender: 'People', 'Empathising', 'Emotivity/ Agreeableness'

The orientation towards care, protection, growth, the development and fostering of the person and, generally, of life in all its forms renders the female gender particularly in harmony with the Integral Ecological Paradigm, which sees its values in the protection and fostering of what is inscribed in nature and its goals in the sustainable and integral development of people, as individuals and as members of communities (of all levels).

Particularly in tune with this Paradigm is also the female gender's orientation to interacting with people, weaving relationships (*Agreeableness*) and, therefore, to welcoming and inclusion, with attitudes of altruism, gratuity and trust important aptitudes for weaving solidarity between human beings and nature and between human beings.

Relations, moreover, of a cooperative and reciprocal type, aimed at developing capacities for dialogue, sensitivity and responsibility, namely, the sense of 'us' and of our 'common home', are observed in the female gender.

In order to do all this, the female gender makes use of an empathetic and creative intelligence, aimed at collecting unprecedented connections ('friendships'). These are founded not primarily on instrumental and calculating processes of thought, but precisely on empathetic affective-emotional thought. This form of intelligence, together with a pronounced linguistic and relational capacity, enables 'broad use of reason' and so of moving more easily between the different branches of knowledge (art, philosophy, religion, as well as the exact and social sciences) and thus of integrating the various systems with which the intelligence grasps reality. This enables this gender of intelligence to have an integral view of reality (polyhedric instead of spherical) and to better enter into the discovery of its possible and progressive integration, also thanks, indeed, to their propensity for dialogue. In other words, the female gender can welcome, include and integrate better as a style of thought and action.

Furthermore, this orientation, together with that of care for the person and for life in general, enables the female gender to better develop a contemplative attitude towards reality, gratuitous and pleasing, in which time, instead of space, is experienced in an expanded way.

Finally, their emotivity, synonymous with vulnerability and fragility, makes the female gender more aware of their own nature as creatures and particularly sensitive to the fragility of others and to every kind of poverty, which it manages better toassume empathetic and tender attitudes towards.

In conclusion, since they are affectively and intellectually disposed to taking care, gathering and welcoming differences and integrating them, and contemplating reality with *sym-pathy* and *em-pathy*, the female gender seems to be more at home in a 'pluridimensional and relational' Paradigm and more open to transcendence, rather than in a homogenous and undifferentiated one, flattened on anthropocentrism.

5. Conclusion and paths for action: Paradigms and Leadership integrated

At the conclusion of our study, the first datum which seems to emerge is that the two 'genders of intelligence' are not mutually exclusive, nor could they exist in separation. Consequently, even the two Paradigms, in which the two gender of intelligence are reflected, could coexist; *on condition*, obviously that the Technocratic one recovers from its methodological *vulnus*, which is the cause of its degeneration. The *magna quaestio*, however is the following: how can they coexist?

5.1 Integrability of the two Paradigms

In our opinion, the Technocratic Paradigm, even if freed from its methodological *vulnus*, cannot exist by itself since it does not possess a 'sense' in the more literal meaning of 'orientation and direction of its progress' – as denounced also by the two Encyclicals. Orientation to *pre*-ultimate goals, namely, to the dignity of the person, as an individual and as a member of the global community and of the 'common home". In fact, it is concentrated on 'instrumentality' instead of 'finality'; on fragments instead of integration; on the individual instead of the community of which the individual is a member; on work instead of rest, namely – in Aristotelian terms – on the *part* instead of the *whole*; on the *potency* instead of the *act*; on *what is for the substance* instead of on the *substance*.

It is precisely by referring to Aristotle's political philosophy that one is able to understand that the Integral Ecological Paradigm is not a simple paradigm but can be understood better as a *meta-paradigm*, 'directive and architectonic', that is, political and integrating. In fact, Aristotle conceived technique, economics, the means of communication (rhetoric) and military science as dimensions subordinate to politics, understood as "directive and architectonic", in the sense that it competed with them to give a *direction*, an *architecture* to the formers[44].

5.2 Integrated leadership

Thus, the papal appeal to "a better kind of politics", founded on social friendship and capable of an integral approach to the problems (FT chap. V; LS 189-198), and a "leadership capable of striking out on new paths" (LS

---

[44] Cf. Aristotle, *Nicomachean Ethics*, I, 1194a1-b11.

53), can also be read as an appeal for an *integrated leadership*, "capable of a new integral and interdisciplinary approach", of channelling progress in a new way (LS 197.191; FT 177).

This does not mean – putting it tritely – to clear more space for women in the 'corridors of power', but with the risk of remaining in the same dominant Paradigm (!), but means at least two other things: first, making more space for the intelligence that is more properly that of the 'female gender', which can belong – as we have said – to both men and women; second, forming an integrated or mixed leadership, not only in politics but in all the sectors of social life, capable of generating other models of economics, work, education, research, free time, health care and concern for the environment[45] and even the life of the Church and interreligious dialogue.

This model of leadership, which has evolved in recent years[46], as well as the emerging feminist leadership[47], supersedes the other paradigms of leadership (visionary, coaching, pacesetting, democratic, affiliative, commanding), in that it places its focus on abilities, capacities and values which emerge from the male and female genders: empathy, integration, creativity, tendency for caring, on the one hand; on the other, assertiveness, entrepreneurship, systematisation, concentration on objectives and processes. It is a matter of responsible leadership, sustainable and effective, capable of decoding cultural processes, developing modes of communication, combining technological

---

[45] "Filling the gap between the awareness of the ego-system and the reality centred on the eco-system is the principal challenge of leadership today", Otto Sharmer and Katrin Kaufer, *Leadership in un futuro che emerge. Da ego-sistema a eco-sistema: nuove economie e nuove società*. Milano: Angeli, 2014 p. 25, cf. further, pp.115-120 (orig. ed. *Leading from the Emerging Future. From Ego-System to Eco System Economies*. San Francisco CA, Berret-Kohler Publisher).

[46] Cf. Anna Guarnieri, *La leadership integrata. Una prospettiva culturale innovativa per le organizzazioni*, LibreriaUniversitaria, 2008/2009; Claudio Massa e Felice Strollo, "Apprendimenti di leadership tra generi", in *Rivista AIF: Periodico dell'Associazione italiana formatori* 96-97-2013, pp. 28-33; Thoma K. Kelemen et al., "When Does Gender Diversity Enhance Team Performance? The Dual Need for Visionary Leadership and Team Tenure", in *Journal of Applied Social Psychology* 50/9-2020, pp. 501-511; Michael Simmons, *New Leadership of women and man: building an inclusive organization*. Aldershot: Gover, 1996; Leonard Karakowsky and Jakob P. Siegel, "The Effects of Proportional Representation and Gender Orientation of the Task on Emergent Leadership Behavior in Mixed-Gender Work Groups", in *Journal of Applied Psychology* 84/4-1999, pp. 620-631; Suzanne Verhulp, *The effect of gender on the evaluation of a leader: and different leadership style*, LAP Lambert Academic Publishing, 2012.

[47] Object of study from the nineteen seventies, there still exists today an extensive literature on the subject which has seen an increase in recent years, especially in the fields of business, administration, health and scientific research.

and social innovation, fruit of relations, sharing of knowledge, concentration on complex and long-term objectives (e.g. social, environmental)[48].

However, we also hold that such integration is not to be understood in a symmetrical way as if the models of leadership were perfectly complementary and could be integrated with strategies of problem solving, but rather in an asymmetrical way, since integration does not constitute a starting but rather an arrival point, which can be reached from the capacity of the two genders, not only to recognise each other in their alterities, but also to accept each other and know how to walk in *syn-ergy* with values, objectives and processes; and that is due not to nature in the biological sense or to intelligence, but rather to freedom ('freed' by affective and spiritual maturity).

The tragedy of the pandemic has compelled us to slow down and rethink values, meanings and objectives (LS 113-114): "worse than this crisis is the tragedy of squandering it"[49]!

---

[48] Cf. Claudio Massa e Felice Strollo, "Apprendimenti di leadership tra generi", p. 29.

[49] Pope Francis, *Homily for Pentecost*, 31 May 2020.

ETHICAL-THEOLOGICAL FOUNDATIONS OF INTEGRAL ECOLOGY

Prem Xalxo, SJ

Integral Ecology, as the relational paradigm of interconnection and interdependence amongst human beings, non-human beings and everything that exists on the face of the earth, is at the heart of *Laudato Si'*[1]. The three pivotal axioms predominantly used in *Laudato Si'*, viz., "everything is interconnected" (*LS* 70, 92, 120), "everything is connected" (*LS* 117), "everything is closely interrelated" (*LS* 137, 142) substantiate the integral meaning of the word *ecology* as the study of "the relationship between living organisms and the environment in which they develop" (*LS* 138). Appearing nine times (*LS* 10, 11, 62, 124, 137, 156, 225, 230), integral ecology presents human reality in an interdisciplinary and holistic way, and defines all the dimensions of human life, dignity and existence, demonstrating "a more integral and integrating vision" of those realities (*LS* 141). As a three-dimensional relational paradigm, integral ecology contains various ethical and theological foundations for a renewed and vigorous commitment to "enter into dialogue with all people" (*LS* 3) who are "united by the same concern" (*LS* 7) to care for *Our Common Home*.

From faith perspective, an authentic and integral human development entails ethical and theological considerations, which also become the foundations of integral ecology, and invite us to approach the book of creation with an open heart in order to be able to face the ethical and theological challenges

[1] Pope Francis, Encyclical Letter *Laudato Si' on Care for Our Common Home*, 24 May 2015. Hereafter LS.

DOI 10.32060/C.S.9.411-431

of the contemporary world, which seems to be progressively individualistic, globalized and secularized. A thorough reading of the book of creation facilitates doing theology on one's knees and understanding the meaning of *fides quaerens intellectum*, and challenges us to respond to the call to ecological conversion and to promote the reconciliation of all creation "in Christ and for Christ" (Col 1:16).

Integral ecology highlights a cosmic vision of the concept of *imago Dei* according to which the human being is an integral part of creation. Being created in the *image and likeness* of God (cf. Gen 1:26-27), human beings are capable of a relationship with God, but not to "be like God" (cf. Gen 3:5). They are entrusted with the mission of accompanying and taking care of creation (cf. Gen 2:15), and not subjugating or dominating it so that they continue to "be fruitful and multiply" (Gen 1:28). Therefore, faith in God the Creator serves as an impetus to be open to an integral relationship with God, with others and with the entire creation.

The ethical foundation of integral ecology outlines the urgency of the formation of an ecological awareness that takes into consideration the whole creation to see, judge and act according to the principle *bonum est faciendum et prosequendum et malum vitandum*. In one way or another, integral ecology traces the desire of the human beings to repair and heal the rupture of the three-dimensional relationship, and to accept the resolution to restore order and harmony in creation. The Catechism of the Catholic Church (*CCC*) rightly affirms, "God wills the interdependence of creatures. The sun and the moon, the cedar and the little flower, the eagle and the sparrow: the spectacle of their countless diversities and inequalities tells us that no creature is self-sufficient. Creatures exist only in dependence on each other, to complete each other, in the service of each other" (*CCC* 340). With their way of treating the environment, the human beings become the authors, the protagonists, and also the victims of the damage caused by their anthropogenic activities. Following this ethical line, integral ecology also presupposes the awareness that the human beings have done things that they should not have done against creation, and they failed to do what they should have done to take for *Our Common Home*. An ethical-theological approach to integral ecology is an earnest invitation and a challenge to take responsibility for the damage done, and to an ecological education (*LS* 209-215) to reinforce the value of life, dignity and human existence in communion with God the Creator, with one another, and with all creation for the integral well-being of every creature on the face of the earth.

1. Theological Foundation

Right in the beginning of the Second Chapter of *Laudato Si'*, Pope Francis asks a pertinent question, "Why should this document, addressed to all people of good will, include a chapter dealing with the convictions of believers?" Faith and religions make rich contribution toward an integral ecology and the full development of humanity (cf. *LS* 62). As the relational paradigm, integral ecology defines human relationship with God the Creator, and reminds the Christian faithful that "their duty towards nature and the Creator, are an essential part of their faith"[2]. The theological foundation of integral ecology indicates the way to see creation from the eyes of Jesus, invites the faithful to ecological conversion and to promote reconciliation with the entire creation in Christ and for Christ (cf. Col 1:16).

1.1 Creation: Fruit of Divine Love

Benedict XVI writes in *Caritas in veritate*, "N*ature expresses a design of love and truth.* It is prior to us, and it has been given to us by God as the setting for our life. Nature speaks to us of the Creator (cf. Rom 1:20) and his love for humanity. It is destined to be 'recapitulated' in Christ at the end of time (cf. Eph 1:9-10; Col 1:19-20). Thus it too is a 'vocation'"[3]. The theme of love, a key word that reinforces the entire scenario of interconnected relationships, recognizes not only God's love for the human being, created in his *image and likeness* (cf. Gen 1:26-27), but also affirms that "by the word of the Lord the heavens were made" (Ps 33:6), and the earth was filled with "the steadfast love of God" (Ps 33:5). The love of God for his creation is expressed in the goodness of creation, which is evident from God's proclamation after each day's creative work, "God saw that what he did was good" (Gn 1:4; 10; 12; 18; 21; 25; 31). The goodness of creation affirms "God's loving plan in which every creature has its own value and significance" (*LS* 76).

Before creating everything on the face of the earth, God wills and decides to be the creator. His creating words resonate his love for every created being, "Let us make humankind in our image, according to our likeness" (Gen 1:26). Because of God's will and creative work, the entire creation came out

[2] Saint John Paul II, Message for the World Day of Peace *Peace with God the Creator. Peace with All of Creation*, 1 January 1990, n. 15.

[3] Benedict XVI, *Encyclical letter Caritas in veritate on Integral Human Development in Charity and Truth*. Rome: LEV, 29 June 2009, n. 48.

of nothing in the order of love. Love is the vital force and the guiding principle for the interconnected human relationship with God, with fellow human beings, and with all the creatures. The order of God's love in creation, as "the fundamental moving force in all created things" (*LS* 77), lays the theological orientation to integral ecology, and opens up the new ways to show the most profound meaning of the interconnected and interdependent relationship. The Most Holy Trinity is the model par excellence of the communion of love and the most intimate interconnected relationship. Creation itself is a Trinitarian process in which God the Father creates everything through the Son in the Holy Spirit. The interconnected joint act of creation in perfect communion of the Father, the Son, and the Holy Spirit is the theological foundation of integral ecology. The design of love, unity and communion amongst the Three Persons of the Trinity – the Father as "the ultimate source of everything, the loving and self-communicating foundation of all that exists", the Son "through whom all things were created", and the Holy Spirit who is "intimately present at the very heart of the universe" continues to inspire and bring new pathways to maintain and nurture the interdependent relationships in creation (cf. *LS* 238). The intimate communion of the Most Holy Trinity indicates that "everything in the universe is connected with everything, at every point and at every moment. Ecological spirituality is that which allows our gaze to recognize in all existing things - side by side and each connected to the other – a sign of the presence of God"[4].

As the design of love, the entire creation presents "a glimpse of his infinite beauty and goodness" of God the Creator (*LS* 12). The order of love is reflected in the goodness of creation and in "the greatness and the beauty of creatures" (Wis 13:5). By sharing His "goodness and love with creation and humanity God unconditionally declared all creation as «indeed very good» and affirmed the «integrity of creation». The story of creation is, indeed, the story of God's purely gratuitous love"[5]. Indeed, as a magnificent book, the entire creation offers a glimpse of God's beauty, goodness and grandeur. The goodness and grandeur of creation reflect the glory and the perfection of God. Belonging "to the order of love", everything in creation is the fruit of love.

[4] Anselm Grün and Leonardo Boff, *Un nuovo modo di pensare: diventare uno*. Padova: Edizione Messaggero, 2017, p. 112. Translation is mine.

[5] Federation of Asian Bishops' Conference, *For All the Peoples of Asia. Federation of Asian Bishops' Conference Documents* (2007-2012) Vol. 5, Quezon City 2014, p. 69.

By creating the human beings in His own *image and likeness*, God inscribes in them the Trinitarian imprint and enjoins them to participate in His creative activities through the Holy Spirit who "possesses an infinite inventiveness, proper to the divine mind, which knows how to provide to untie the knots of even the most complex and impenetrable human events" (*LS* 80). The Trinitarian vision of creation, founded on communion and inter-relationship between the three Divine Persons of the Most Holy Trinity, is a source of inspiration for making a serious commitment to take care of *Our Common Home*, to promote the concept of integral ecology and to collaborate with God the Creator in the creative Divine activities *to till it and keep it* (Gen 2:15). In the order of love, "God created the world, writing into it and order and dynamism that human beings have no right to ignore" (*LS* 221), but to become the responsible stewards of the entire creation.

1.2 Emancipation of Creation in Christ

Jesus Christ, "through whom all things were created" (*LS* 238), is the beginning and the end, the *Alpha* and *Omega* of the entire creation, which is inscribed emphatically by St. Paul in his Letter to the Colossians, "He is the image of the invisible God, the firstborn of all creation; for in him all things in heaven and on earth were created, things visible and invisible, whether thrones or dominions or rulers or powers – all things have been created through him and for him. He himself is before all things, and in him all things hold together" (Col 1:15-17). St. John the Evangelist sums up the continuity between the mystery of creation and the Incarnation, "In the beginning was the Word, and the Word was with God, and the Word was God. He was in the beginning with God. All things came into being through him, and without him not one thing came into being" (Jn 1:1-3). Both creation and the Incarnation are two inseparable aspects of one great mystery, that is, the recapitulation of all things in Jesus Christ through an eternal divine plan "for the fullness of time, to gather up all things in him, things in heaven and things on earth" (Eph 1:10). The intrinsic and intimate relationship between creation and the Incarnation means that "the God who creates is the God who saves; indeed, only by living in Christ does the new creation unfold"[6]. By taking the human form, Jesus Christ became one with the entire humanity in all its realities except the

[6] Ilia Delio and *al.*, *Care for Creation: A Franciscan Spirituality of the Earth*. Cincinnati: St. Anthony Messenger Press, 2008, 202-203.

reality of sin (cf. Heb 4:15), and continues to create and recreate everything, and manifests his glory and presence alive and active through the Holy Spirit.

The life, words and deeds of Jesus Christ demonstrate that he "lived in full harmony with creation" (*LS* 98), not as an ascetic separated from the world, but as a person lived among the people, worked with his hands, and preached the Good News of the Kingdom with the parables related to the ordinary human life. With his exemplary life of love, mercy, and compassion, he exhibited the primacy of "harmony, justice, fraternity and peace" (*LS* 82). The mystery of the Incarnation in which the Divine Word becomes flesh and dwells amongst peoples affirms the importance of creation and the entire humanity in Jesus Christ. His life lived among the human and earthly realities of that time, his words and deeds, and in the end, his death on the cross and the resurrection affirm that he was present in the beginning of creation and will be present "to the end of the world" (Mt 28:20). The ultimate destination of the human beings and of all creation is emancipation in Jesus Christ, when God will be "all in all" (1Cor 15:26), and in the end all will meet "face to face" with the infinite beauty of God (cf. 1Cor 13:12) in the eternal bliss. The emancipation of creation in Jesus Christ signifies the fullness of human life and existence when "we will find ourselves face to face with the infinite beauty of God (cf. *1 Cor* 13:12), and be able to read with admiration and happiness the mystery of the universe, which with us will share in unending plenitude" (*LS* 243).

Human beings are "endowed with intelligence and love, and drawn by the fullness of Christ", and therefore, the emancipation of all created order in Jesus Christ reminds them of their responsibility to take care of *Our Common Home* by participating in the divine design of love "to lead all creatures back to their Creator" (*LS* 83), because "each creature reflects something of God, [and the Risen Lord] is intimately present to each being, surrounding it with his affection and penetrating it with his light" (*LS* 221). Jesus Christ is before all things and all subsist in him (cf. Col 1:15-19), he sustains and maintains the entire creation and thus, he is the foundation on which all creation rests (cf. Col 1:17). The fact of Jesus Christ being the *Alpha* and the *Omega*, the beginning and the end of the entire creation demonstrates the perennial continuity between creation and Incarnation (cf. Jn 1:1-18). The mystery of Incarnation is the key to reveal the secret of creation (cf. 1Cor 8:6), which is moving towards its source, that is, Jesus Christ, and it needs to be recapitulated in him (cf. Eph 1:10) in order to be freed "from its slavery to decadence, to enjoy the same freedom and glory as the children of God" (Rom 8:21).

1.3 Seeds of God's Presence in Creation: Sacramentality of Creation

The entire creation manifests the presence and glory of God, as the Book of Psalms rightly proclaims, "The heavens are telling the glory of God; and the firmament proclaims his handiwork. Day to day pours forth speech, and night to night declares knowledge. There is no speech, nor are there words; their voice is not heard; yet their voice goes out through all the earth, and their words to the end of the world" (Ps 19:1-4). God created the entire universe out of his own divine will and on every aspect of creation, he left his imprint by being "intimately preset to each creature", which "ensures the subsistence and growth of each being", and out of His goodness, He "continues the work of creation" (*LS* 80). By creating human beings in His *image and likeness*, He made his imprint more evident. His commandment to "be fruitful and multiply" (Gn 1:22; 28) and "to till and keep" (Gen 2:15) the created order, marks the divine presence in creation always alive and active. The mystery of the Incarnation is the unfolding of the perennial divine presence in creation, in which the Word becomes flesh and makes His dwelling on earth. *Laudato si'* describes the unfolding of the universe in creation as God's splendid and glorious handiwork "who fills it completely" as an ideal, which is "not only to pass from the exterior to the interior to discover the action of God in the soul, but also to discover God in all things" (*LS* 233).

Taking into consideration the interconnectedness of all created living beings and the perennial presence of God in His created order, the entire creation could be characterized as a sacrament in an analogous and limited sense, which "discloses the Creator's presence by visible and tangible signs"[7]. Creation, as the gift of self of God, the Creator, is an efficacious sign and effective means of God's divine revelation. He speaks to human beings in and through creation, and "this «God-talk» renders all creation holy; all [becomes] sacrament"[8]. Human beings experience the greatness of God in the beauty and the grandeur of creation; and in and through creation, they praise and revere Him, because His "incorruptible spirit dwells in all things" (Wis 12:1), and He is "truly discoverable and discovered on earth"[9]. Thus, creation becomes the

---

[7] United States Conference of Catholic Bishops, *Renewing the Earth: An Invitation to Reflection and Action on Environment in Light of Catholic Social Teaching* (4 November 1991) in https://www.usccb.org/resources/renewing-earth (20.10.2021).

[8] Ilia Delio and *al.*, ed., *Care for Creation*, 202.

[9] Kevin W. Irwin, "Religious Teaching on the Protection of the Earth's Resources", in Kevin W.

expression, symbol and sacrament to reach God. He is "profoundly present to all things and God is expressed in all things, so that each creature is a symbol and a sacrament of God's presence and Trinitarian life"[10]. God's presence in creation invites human beings to treat everything that is on the face of the earth "not just as means to human fulfillment but also as God's creatures, possessing an independent value, worthy of our respect and care"[11].

Laudato si' presents the teachings of Saint Bonaventure that "contemplation deepens the more we feel the working of God's grace within our hearts, and the better we learn to encounter God in creatures outside ourselves" (*LS* 233). Commenting on the sacramental vision of Saint Bonaventure, the proponents of the Franciscan spirituality of creation accentuate that "God's glory is reflected in the sacred order of creation. Because the world expresses the Word through whom all things are made (Jn 1:1), every creature is in itself a 'little word' of God. In this respect, the whole creation is both sacramental and incarnational, that is, every aspect of creation is a 'little incarnation' of the divine Word"[12].

Sacramentality is in no way an affirmation of the sacredness of creation; rather it provides us with an efficacious and visible instrument to express goodness, to praise God the Creator and to bless Him.

> For Christians the theological resources for an ecological renewal of faith can be found not only in biblical texts and doctrinal tradition, but also, and no less fundamentally, in the 'sacramental' character of nature itself. To say that nature is sacramental simply means that, even apart from biblical revelations, nature in all its beauty and diversity reveals the divine mystery – not just to Christians, of course, but to people of many traditions. A Sacrament is anything through which we are gifted with a sense of the sacred, and it is especially nature's beauty and vitality that have communicated to human an impression of the divine[13].

Irwin and Edmundo D. Pellegrino, *Preserving the Creation: Environmental Theology and Ethics*. Washington D.C.: Georgetown University Press, 1994, p. 88.

[10] Denis Edwards, "The Discovery of Chaos and the Retrieval of the Trinity", in Robert J. Russell, and *al.*, ed., *Chaos and Complexity: Scientific Perspectives on Divine Action*. Vatican City: Libreria Editrice Vaticana, 1996, p. 162.

[11] United States Conference of Catholic Bishops, *Renewing the Earth: An Invitation to Reflection and Action on Environment in Light of Catholic Social Teaching* (4 November 1991).

[12] Ilia Delio and *al.*, ed., *Care for Creation*, 45.

[13] David M. Lodge and Christopher C. Hamlin, ed., *Religion and the New Ecology*. Notre Dame (Indiana): University of Notre Dame, 2006, p. 233.

The sacramental approach is "typically allied with a sense of the interrelationship of all forms of life"[14], where human beings and all other life forms are seen as "a unified, single community with an overarching purpose: the exaltation and joy of existence, praise of the divine, and participation in the great liturgy of the universe"[15]. In fact, the entire sacramental system depend on God's free gifts found abound in creation; for examples, the fundamental elements of the Eucharistic celebrations, that is, the bread and the wine are "the fruits of the earth and the work of human hands". Those fruits of the earth become the efficacious means to reach God, praise and revere him. Saint Bonaventure invites all to open one's eyes and ears to see and hear God's marvellous deeds in creation and emphasizes that "one person who is not illumined by such great splendor in created things is blind. Anyone who is not awakened by such great outcries is deaf. Anyone who is not led by such effects to give praise to God is mute"[16].

The concept of sacramentality of creation, a cursor to the epiphany of the Divine, is brought out emphatically by St. Ignatius of Loyola in his *Spiritual Exercises* how God dwells in creatures, in the elements, giving them being, in the plants vegetating, in the animals feeling in them, in human beings, animating them, giving them sensation, making them to understand, and likewise; making a temple of them, being created to the likeness and image of His Divine Majesty[17]. We can easily conclude from the Ignatian spirituality that "an experience of the presence of God in and through creation, and a sense of being in communion with fellow human beings and other life forms provide with an impetus to be at home with oneself and with others"[18]. Such understanding is a call to have deep respect and care for creation.

1.4 Sin: Rupture of Relationships

At the heart of integral ecology dwells the intimately interconnected human relationships not only with the fellow human beings and every aspect of human life, but also with God and with every created living beings and

---

[14] *Ibid.,* 235.

[15] Thomas Berry, *The Christian Future and the Fate of the Earth*. New York: Orbis Books, Maryknoll 2009, p. 66.

[16] Bonaventure, *Itinerarium Mentis in Deum* 1.15, cit. in Ilia Delio and *al.*, ed., *Care for Creation*, 204.

[17] Elder Mullan. New York: *The Spiritual Exercises of St. Ignatius of Loyola*, 1914, n. 235.

[18] Prem Xalxo, *Current Ecological Crisis and Its Moral Dimensions*. Ranchi: Xavier Publications, 2008, p. 14.

non-living objects on the face of the earth. An interconnected and interdependent relationship founded and nurtured on reciprocal harmony and order ensures a state of peace, prosperity, and, theologically speaking, a state of grace. The Book of Genesis depicts a vivid picture of harmonious interrelated human relationship with "everything that [God] had made", which "was very good" (Gen 1:31). The walking of the Lord God "in the garden in the time of evening breeze" (Gen 1:8) demonstrated the state of order and harmonious interconnected relationship, which is destroyed by the sin of disobedience. In the context of interconnected and interdependent relationship, sin could be defined as the rupture of three-dimensional human relationship with God, with fellow human beings and with the entire creation. In terms of relationality, sin "describes the personal history of one who was created for communion and refuses it. It is the rejection of our identity as part of an interdependent world in which God's power as creative source expresses itself through the shared power of other creatures"[19]. Pope Francis cites Patriarch Bartholomew to substantiate the concept of "ecological sin",

> "For human beings [..] to destroy the biological diversity of God's creation; for human beings to degrade the integrity of the earth by causing changes in its climate, by stripping the earth of its natural forests or destroying its wetlands; for human beings to contaminate the earth's waters, its land, its air, and its life – these are sins". For "to commit a crime against the natural world is a sin against ourselves and a sin against God[20].

Excessive and unbridled anthropogenic activities disturb and destroy the order and harmony of creation as willed by God, who has written into it "an order and a dynamism that human beings have no right to ignore" (*LS* 221). The destruction of the order and the dynamism of creation is going against the divine will and mandate to "till and keep the earth" (Gen 1:15). It alienates humans from God, the Creator, from fellow human beings and from creation itself, which is a form of sin. In the light of the deadly sins, would it be possible to define some "crimes against ourselves and sins against God" (cf. *LS* 8)? Reflecting on the seven Capital Sins, we can put pride at par with the human denial of God the creator, greed with the illusion of reshaping the earth, envy with the unquenchable thirst for more and better, gluttony with indifference toward poverty and hunger, sloth with ignoring the reality of the ecologi-

---

[19] Ilia Delio – *al.*, ed., *Care for Creation*, 50.

[20] Pope Francis, *Laudato Si'*, n. 8.

cal degradation, lust with the excessive exploitation of natural resources, and wrath with the arrogance and violence against creation.

"Ecological sin" in action is the breaking of harmony and order in creation, "The brutal consumption of Creation begins where God is not, where matter is henceforth only material for us, where we ourselves are the ultimate demand, where the whole is merely our property and we consume it for ourselves alone. And the wasting of creation begins when we no longer recognize any need superior to our own, but see only ourselves"[21]. The negation of God results into the replacement of God with new "golden calf" (Ex 32:1-20; 1Kgs 12:25-33) in the form of wealth, consumerism, hedonism and narcissism, which causes the unrestrained exploitation of natural resources and the destabilization of the order in creation with the sole purpose of achieving maximum profit. Indeed, such unbridled exploitation of natural resources and the replacement of God with new earthly gods affirms that "without the Creator the creature would vanish"[22]. These ecological sins are the forms of human alienation not only from God, not only from one another and from the true self, but also from the whole creation. They represent the abusive use of God's gift to all humanity and the destruction of order and harmony willed by God.

The "ecological sin" of omission is visible in the globalization of indifference and irresponsibility to world hunger and poverty. It could be compared with the act of putting one's head in the sand like the ostrich and pretending not to see the misery and suffering abound around; it is "the refusal to accept responsibility for those to whom we are connected; thus, it is the refusal to accept the 'other' of the relationship (the 'Thou') as the one who addresses us, discloses our responsibility and calls us into question"[23]. The "ecological sin" of omission is evident in the attitude and life of those, who strive to have even *more and better* despite having an abundance of riches, and thus, deprive others of having fundamental human needs. Because of their indifference, those, who do not have the most basic needs to survive, struggle to stay alive every day with the sole possibility of surviving "from the crumbs that fall from the table" (Mt 15:27).

---

[21] Benedict XVI, *Meeting with the Clergy of the Diocese of Bolzano-Bressanone,* 6 August 2008.

[22] *Gaudium et spes*, n. 36.

[23] Ilia Delio – *al.*, ed., *Care for Creation*, 50.

2. Ethical Foundations

The axioms of interrelatedness and interdependence of everything in creation are the defining pointers to the ethical foundation of integral ecology. Being created in the image and likeness of God, human beings have the capacity to establish and nurture relationship, which is the nucleus of integral ecology and the pivotal guiding principle to define human ethical behavior. The three-dimensional human relationship with God, with fellow human beings and other life forms in creation define an integral parameter to see, judge and act ethically for the good of the human person and the entire humanity. An orderly and harmonious relationship with oneself and others help one to listen to the voice of God, which always summons him or her "to love good and avoid evil [...] do this, shun that"[24].

One of the famous Marian hymns says, "No one can live as an island, journeying through life alone", which reflects a Seventh Century scintillating poem written by John Donne, "No man is an island entire of itself; every man is a piece of the continent, a part of the main [...]; any man's death diminishes me, because I am involved in mankind. And therefore never send to know for whom the bell tolls; it tolls for thee"[25]. Definitely, we cannot live without relating to our fellow human beings and surrounding ambiance. *Our Common Home* provides us with "a fundamental conduit for relationship" (*LS* 67), which is built on interpersonal contacts, mutual rapport, and interactions with a firm conviction that "human life is grounded in three fundamental and closely intertwined relationships: with God, with our neighbour and with the earth itself" (*LS* 66).

The contemporary rapid advances in science and technology, particularly the spectacular development of the communication technology, has turned the world into a global village, and enhanced enormous increase in interconnectedness, which points to the relational dimension of every human person. No doubt, "it is through relationships that we move forward toward full human growth and development. Without relationships we cannot 'become'. To be totally devoid of relationships, to be completely shut off, to be totally isolated runs contrary to human need and health – it is to frustrate

[24] *Gaudium et spes*, n. 16.

[25] John Donne, *Meditation XVII, Devotions Upon Emergent Occasions*. Wheaton College, Illinois: Christian Classics Ethereal Library, 1995, p. 1; John Donne, *Donne's Devotions*. Cambridge: Cambridge University Press, 1923, p. 98.

the fundamental truth of being human"[26]. Indeed, human life flourishes and reaches its fullness in communion with others in a harmonious relationship with other forms of life, and that is why "all God's creatures are worthy of moral consideration, as a sign of the worthiness imparted by God"[27]. As social and dialogical beings, our life does not belong only to each one of us, but also "to family members, friends, the community. [...] Every time someone dies, an entire world disappears and is lost forever"[28]. Such deeper sense of being related and interconnected with fellow human beings proves to be the soul of universal fraternity.

The ethical foundation of integral ecology stems from the interconnectedness of human beings with all other forms of life. The fact of being interrelated forces human beings to introspect their way of treating fellow human beings and other life forms. After John Paul II termed the ecological crisis as a moral crisis[29], Benedict XVI made it clear that "t*he way humanity treats the environment influences the way it treats itself, and vice versa*. This invites contemporary society to a serious review of its life-style, which, in many parts of the world, is prone to hedonism and consumerism, regardless of their harmful consequences"[30]. From the ethical point of view, it is not enough to intervene with economic incentives or deterrents, not even an ecological education to care for *Our Common Home*. What is needed most is "the overall moral tenor of society"[31]. Therefore, a possible reversal of the environmental crisis "will only come about when we recover a deeper sense for the relationality of human to particular ecosystems and part of the biosphere"[32].

---

[26] Richard J. Devine, *Good Care, Painful Choices: Medical Ethics for Ordinary People*. New York: Paulist Press, 2000², p. 17.

[27] James A. Nash, *Loving Nature: Ecological Integrity and Christian Responsibility*. Nashville: Abingdon Press, 1991, p. 96.

[28] Remo Bodei, "L'epoca dell'antidestino" in Daniela Monti, ed., *Che cosa vuol dire morire*. Torino: Giulio Einaudi Editore, 2010, p. 57. Translation is mine.

[29] Cf. Saint John Paul II, Message for the World Day of Peace *Peace with God the Creator. Peace with All of Creation*, 1 January 1990, nn. 5-6.

[30] Benedict XVI, *Caritas in veritate*, n. 51.

[31] *Ibid.*, 51.

[32] Michael S. Northcott, *The Environment and Christian Ethics*. Cambridge: Cambridge University Press, 2009, p. 123.

2.1 Essence of Human Behavior: Interconnectedness

Relationality or interconnectedness is one of the focal points integral ecology, which is defined and underlined in all-embracing axioms of "everything is interconnected" (*LS* 70, 92, 120), "everything is connected" (*LS* 117), "everything is closely interrelated" (*LS* 137, 142), which embraces the entire creation as *Laudato si'* substantiates, "when our hearts are authentically open to universal communion, this sense of fraternity excludes nothing and no one. It follows that our indifference or cruelty towards fellow creatures of this world sooner or later affects the treatment we mete out to other human beings" (*LS* 92). Without the interconnected relationships, human beings turn into an isolated desert, which is metaphorically used by Benedict XVI,

> And there are so many kinds of desert. There is the desert of poverty, the desert of hunger and thirst, the desert of abandonment, of loneliness, of destroyed love. There is the desert of God's darkness, the emptiness of souls no longer aware of their dignity or the goal of human life. The external deserts in the world are growing, because the internal deserts have become so vast. Therefore the earth's treasures no longer serve to build God's garden for all to live in, but they have been made to serve the powers of exploitation and destruction[33].

Neglecting a just and harmonious relationship with the fellow human beings and with the other life forms destroys the much-needed interpersonal relationship of care and concern. The concept of integral ecology based on interdependent three-dimensional relationship suggests that "all entities in the natural world are constituted by different and varying degrees of relationality. For example, atoms, cells, molecules, eco-systems, the solar system and the universe are all configured by a vast network of complex relationship"[34]. It has been clearly underlined that human beings are the authors, the protagonists, and the victims of the ecological degradation. They are part of the whole, and therefore, if they are "part of the problem within the ecological crisis, then [they] must become also part of the solution. It is increasing clear that the self-understanding of the human must change, but the relationship, or more accurately, the lack of a recognition of the deep relationship between the

[33] Benedict XVI, *Homily for the Beginning of the Petrine Ministry of the Bishop of Rome*, 24 April 2015.

[34] Dermot A. Lane, *Theology and Ecology in Dialogue: The Wisdom of Laudato si'*. Dublin: Messenger Publication, 2020, p. 43.

human and the world of nature must also be remedied"[35]. *Laudato Si'* emphasizes the need of the renewal of humanity itself, "there can be no renewal of our relationship with nature without a renewal of humanity" (*LS* 118).

The interconnectedness, as the essence of human behavior, results into the formation of harmonious social relations, communities of mutual love and care, which ultimately leads to the integral human development. The spirit of individualism, narcissism, and indifference to the suffering and pain of the fellow human beings destroys the thread of interconnectedness, and leads to a superfluous sense of self-sufficiency and isolation, which is the channel of death. In fact, "no one can face life in isolation [...]. We need a community that supports and helps us, in which we can help one another to keep looking ahead"[36]. Integral ecology with its foundation on relationality and human interconnectedness with the entire creation is perfect reminder of our duties towards the environment, which "are linked to our duties towards the human person, considered in himself and in relation to others"[37]. Neglecting a correct relationship with the fellow human beings, towards whom we have the duty of care and custody, destroys our relationship also with ourselves, with creation and with God.

2.2 *Imago Dei*: A Call to Ethical Living

Integral ecology brings to light a new meaning to the concept of *Imago Dei*. The fact of being created in the *image and likeness* of God makes human beings not only a little lower than the angels crowned them with glory and honor (cf. Ps 8:5), but also blesses them with the capacity to establish relationship with God the Creator and the entire creation. In the context of integral ecology, a cosmic vision of the *Imago Dei* inscribes human beings as the integral part of creation. A comprehensive ecological reading of the Creation narrative depicted by the Book of Genesis helps discover that the fact of being created at the end of God's creative activity makes human beings the dependent members of the terrestrial community and conditions them[38] with an august responsibility to take care of *Our Common Home*. Undoubtedly, various substantial characteristics, such as

[35] *Ibid.*, 33.

[36] Pope Francis, Encyclical Letter *Fratelli tutti on Fraternity and Social Friendship*, 3 October 2020, n. 7.

[37] Benedict XVI, *Caritas in veritate*, n. 51.

[38] Jürgen Moltman, Piero Stefani and *al.*, *La terra come casa comune: crisi ecologica ed etica ambientale*. Bologna: Edizioni Dehoniane, 2017, p. 20. Translation is mine.

conscience, the ability to reason and discern, the aesthetic sense, freedom, etc., distinguish human beings from other forms of life in creation; but they remain part and parcel of the created order, and find their true place in creation by maintaining the three-dimensional harmonious relationship with God, with fellow human beings and the entire creation.

> When human beings fail to find their true place in this world, they misunderstand themselves and end up acting against themselves: "Not only has God given the earth to man, who must use it with respect for the original good purpose for which it was given, but, man too is God's gift to man. He must therefore respect the natural and moral structure with which he has been endowed"[39].

Being created in the *image and likeness* of God, human beings are capable of establishing a relationship with God, but not to fall into temptation of becoming "like God" (cf. Gen 3: 5). They are entrusted with a mission "to till and keep" (Gen 2:15), to accompany, and to take care of creation, which is a fundamental dimension of the Covenant created by God with human beings, "be fruitful and multiply; populate the earth abundantly and multiply in it" (Gen 9:7). Therefore, human beings must strive to live *Imago Dei* by establishing just and harmonious relationship with God, with others and with creation, and thus attain the fullness of their life and existence.

2.3 An Integral and Holistic Vision of Human Life and Dignity

The concept of integral ecology described in *Laudato Si'* presents an integral and holistic vision of human life and dignity by defining the full-fledged flourishing of human life in its harmonious relation with all other forms of life on the face of the earth. To understand and approve the value of human life, dignity and existence, human begins must "retain an understanding of the whole human person in which the values of intellect, will, conscience and fraternity are preeminent. These values are all rooted in God the Creator and have been wonderfully restored and elevated in Christ"[40]. Integral ecology, as a paradigm of interconnectedness and interdependence, holds together the reality of human life and existence with the environmental issues and problems, which is "necessary for the creation of a more dignified environment" (*LS* 155). Benedict XVI emphatically describes that "*the way humanity treats*

---

[39] Pope Francis, *Laudato Si'*, n. 115.

[40] *Gaudium et spes*, n. 61.

the environment influences the way it treats itself, and vice versa. [...] Every violation of solidarity and civic friendship harms the environment, just as environmental deterioration in turn upsets relations in society"[41].

The notion of the quality of life is no longer connected only to the material well-being, riches and comforts in one's life; but to the socio-cultural and familial relationships that enrich us and make us what we ought to be, that is, the social beings. The environmental exploitation and degradation for the sake of the material well-being " Pope Francis speaks of the *transcendent human dignity*, which means "appealing to human nature, to our innate capacity to distinguish good from evil, to that 'compass' deep within our hearts, which God has impressed upon all creation. Above all, it means regarding human beings not as absolutes, but as *beings in relation*"[42].

In terms of interrelatedness and interdependence, the concept of environment receives an important hermeneutical stance as "a relationship existing between nature and the society which lives in it" (*LS* 139). The creative rational and moral capacities enable human beings "to serve as responsible representatives of God's interests and values, to function as protectors of the ecosphere and deliberately constrained consumers of the world's good"[43]. They are the integral "part of nature, included in it and thus in constant interaction with it" (*LS* 139) and care for *Our Common Home* should be an integral part of their lifestyle "which includes the capacity for living together and communion" (*LS* 228). And therefore, it is apt to conclude that human life flourishes and bears fruit in communion with others in a society and in a harmonious relationship with creation.

2.4 Ecological Conscience: Parameter to See-Judge-Act

The concept of integral ecology, which includes every aspect of human life and its connectedness with every form of life, accentuates the intrinsic value of all living beings and the need to respect their rights without over-emphasizing "an ethics in which natural rhythms become the determinant for ethical behavior"[44]. It demands prudent human responsibility toward the biotic world by providing a framework and motivation for ethically-inspired initiatives to care

---

[41] Benedict XVI, *Caritas in veritate*, n. 51.

[42] Pope Francis, *Address to the European Parliament*, Strasbourg, 25 November 2014.

[43] James Nash, *Loving Nature: Ecological Integrity and Christian Responsibility*, p. 105.

[44] John T. Pawlikowski, "Theological Dimensions of an Ecological Ethic" in Richard N. Fragomeni and John T. Pawlikowski, ed., *The Ecological Challenge: Ethical, Liturgical, and Spiritual Responses*. Collegesville: The Liturgical Press, 1994, p. 42.

for *Our Common Home* because all of creation and each living being has value in itself (cf. *LS* 33) and is "good and admirable in itself" (*LS* 140). Human being adorned with "unique capacity of knowledge, will, freedom, and responsibility" (*LS* 118), have the ethical mandate to care for *Our Common Home*. The fact of everything being interconnected and closely interrelated also opened up new vision of conscience to discern, decide and to act good and to avoid evil.

> We therefore need to cultivate a new set of values and virtues – including conservation of the environment, compassion for the excluded, courage to take bold decisions, and a commitment to work together in common purpose for the global common good. We need a full conversion of hearts and minds, habits and lifestyles, structures and institutions. Ultimately, it is about the habituation of virtuous practices, stemming from an intrinsic desire to do what is right. And here, the world needs good role models. So let religious leaders step up to the plate! Let us lead by example![45]

The invitation to see, judge and act in the interconnected web of all forms of life has a powerful and deep-rooted potential to add an all-inclusive reference point to the famous moral axiom, "*bonum est faciendum et prosequendum et malum vitandum*" (the good has to be done and pursued, and evil avoided). Such all-inclusive moral vision defines an action as right "when it tends to preserve the integrity, stability and beauty of the biotic community; otherwise wrong"[46]. The all-inclusive moral vision forces us to introspect ourselves with some of the intriguing relevant questions,

> When we ask ourselves what kind of world we want to leave behind, we think in the first place of its general direction, its meaning and its values. [..] What is the purpose of our life in this world? Why are we here? What is the goal of our work and all our efforts? What need does the earth have of us? It is no longer enough, then, simply to state that we should be concerned for future generations. We need to see that what is at stake is our own dignity[47].

---

[45] Peter K.A. Turkson, "Protect the Earth, Dignify Humanity: The Moral Dimensions of Climate Change and Sustainable Development", Workshop on "Protect the Earth, Dignify Humanity": The Moral Dimensions of Climate Change and Sustainable Development (28 April 2015), Casino Pio IV, Vatican City 2015, n. 5.

[46] John Rodman, "Quattro forme di coscienza ecologica" in Mariachiara Tallacchini, ed., *Etiche della terra: antologia di filosofia dell'ambiente*. Milano: Vita e Pensiero, 1998, 328: "Una cosa è giusta quando tende a preservare l'integrità, stabilità e bellezza della comunità biotica. È sbagliata quando tende altrimenti". Translation is mine.

[47] Pope Francis, *Laudato Si'*, n. 160.

Such questions challenge the overemphasis on the anthropocentric values and exclusion of the biocentric values encouraging thus human arrogance and commodification of earth resources[48], which ultimately are the common heritage "the fruits of which are for the benefits of all"[49]. The inability to see, judge or discern, and act from an ethical-theological perspective reinforces the "tyrannical anthropocentrism" (*LS* 68) and promotes a worldview which sees creation as a mere "*motore morta*" devoid of any spiritual or transcendental value. *Laudato si'* draws a clear line that "we are not God" (*LS* 67) to underline human limits, and exhorts the responsibility of "tilling and keeping" the earth for the full flowering of human life and dignity in harmony with all forms of life. Indeed, when we lose the vision of creation as the fruit of God's creation for the integral wellbeing of all forms of life, "we end up either considering nature an untouchable taboo or, on the contrary, abusing it"[50]. The ecological conscience enlightened, sustained and nurtured by the three-steps dynamics of see, judge and act widens the horizon to see the beauty and grandeur of creation, stimulates to discover the meaning, value and importance of each creature, and finally, paves the way to accept the entire creation as God's free gift for the integral human wellbeing in consonance with all other creatures.

3. Conclusion

The ethical and theological foundation of integral ecology enhances a vison of harmonious interconnected relationship amongst all life forms in creation. An awareness of being interconnected with all in the web of life helps to discover "the creative reason and the love from which the world was born and of which the universe speaks to us, if we pay attention, if our inner senses awaken and acquire perception of the deepest dimensions of reality"[51], and offers an ideal ethical and theological basis for an ecologically viable lifestyle, bringing about an appropriate change in "our way of thinking, feeling and living" (*LS* 216). It also accentuates that the call to care for *Our Common Home* is not a suggestion to "return to the Stone Age", but is an indication of the urgent need "to slow down and look at reality in a different way, to appropriate the

---

[48] Cf., James A. Nash, *Loving Nature: Ecological Integrity and Christian Responsibility*, p. 151.

[49] Saint John Paul II, *Peace with God the Creator. Peace with All of Creation*, 1 January 1990, n. 8.

[50] Benedict XVI, *Caritas in veritate*, 48.

[51] Benedict XVI, *Address to the Roman Curia Offering Them His Christmas Greetings*, 22 December 2005.

positive and sustainable progress which has been made, but also to recover the values and the great goals swept away by our unrestrained delusions of grandeur" (*LS* 114). The question arises, "what human interests and benefits must be sacrificed in this age of ecological crisis in order to serve the needs of other creatures and to enhance the health of this biotic community of which we and they are interdependent part"?[52]. The response to such intriguing questions is an all-out effort to reinstate the three-dimensional interconnected relationships by putting the ideas into practice because realities are greater than ideas, and "realities simply are, whereas ideas are worked out"[53].

The interconnected and interdependent vision of *Our Common Home* from the ethical and theological perspective keeps reminding human beings that "all that exists, coexists"[54]. In an interconnected web of life, all those voices, whether human or non-human, which are "often ignored, exploited, mistreated, and abused"[55] need to be heard for the integral human development for reestablishing *Shalom*, that is, the peaceful and harmonious living of all life forms in communion with one another and with God. Such vision can enlighten and encourage to look at creation with "fondness and wonder" (*LS* 97) with the gaze of Jesus. Nurtured by faith, his gaze must help us to see "things as Jesus himself sees them, with his own eyes: it is a participation in his way of seeing"[56] because "they have an intrinsic value independent of their usefulness" (*LS* 140).

As a concrete conclusion, the reflections on the ethical and theological foundations of integral ecology delineate two moral paradigms, which are essential to maintain a harmonious human interconnected relationship with all that exists in *Our Common Home*,

> first, an ecological consciousness which recognizes the integrity of creation and the interconnectedness of human life and all other life forms in nature should be part and parcel of one's way of life. Second, a morality which takes care of both human and all other life forms must be the moral point of reference for human behavior towards fellow human beings and other life forms

[52] James A. Nash, *Loving Nature: Ecological Integrity and Christian Responsibility*, p. 150.

[53] Pope Francis, Apostolic Exhortation Apostolic Exhortation *Evangelii gaudium* on the Proclamation of the Gospel in Today's World, 24 November 2013, n. 231.

[54] Lary L. Rasmussen, *Earth Community Earth Ethics,* Orbis Books, Maryknoll 1996, p. 324.

[55] Anne Marie Mealey, "Engaging with *Laudato si'* – Methods in Morals", *The Furrow* 66/11 (2015), 563.

[56] Francis, Encyclical Letter *Lumen fidei*, 29 June 2013, n. 18.

in nature. These two moral paradigms have strong potential for inculcating a sense of appreciation and gratitude to God for the gift of life, a respect for all life forms, and a harmonious relationship with the entire creation[57].

In the final analysis, the ethical and theological foundations of integral ecology pave the path to bring about some small but tangible differences in human life woven around interconnectedness and interdependence for the integral wellbeing of the entire humanity.

[57] Prem Xalxo, *Current Ecological Crisis and Its Moral Dimensions*, 257.

THE METHOD INFORMING *LAUDATO SI'* AND ITS CRITICAL DIALOGUE WITH THE EMPIRICAL SCIENCES

René Micallef, SJ

1. Introduction[1]

The term "integral ecology" could easily sound as fascinating as it may be mind-boggling. The idea of a science of the "ecosphere", human and non-human, embracing past, present and future, integrating the causal approach to reality typical of the natural sciences and the teleological approach typical of ethics and some of the humanities, could be seen as a wonderous and timely call to a scientific holism that is unprecedented in modernity. Yet many are tempted to dismiss such a call, by interpreting such invitation to holism as an invitation to return to a pre-modern, methodologically undifferentiated "science of everything", or to see in this a naïve or possibly dangerous project, a product of Catholic nostalgia, a new façade and language to resurrect contentious medieval claims about the morality inscribed "in nature"[2] and our ability to condemn "acts against nature". I believe that, even though some voices in the Catholic academia world might be tempted to interpret "integral ecology" in such a manner, this is not the direction in which we are being oriented by the

[1] Some parts of this text have been translated and adapted from an article I have previously published in Italian: René M. Micallef, "Laudato si' e la sua metodologia: un critico ascolto delle scienze empiriche?", in Humberto M. Yáñez, ed., *Laudato si'. Linee di lettura interdisciplinari per la cura della casa comune*. Rome: Gregorian and Biblical Press, 2017, pp. 13-48.

[2] Cfr. Edouard Herr, "Loi naturelle et Écologie humaine", in Pierre Coulange, Jean-François Lefebvre and Waltraud Linnig, (ed.), *La loi naturelle : Lieu de rencontre ou pierre d'achoppement?*. Paris : Éditions Silence et Parole, 2018, pp. 49-68.

DOI 10.32060/C.S.9.433-476

present Church Magisterium. Rather, the holism sought by "integral ecology" is one based on a serious dialogue between sciences — natural sciences, social sciences, ecclesiastical and religious sciences — which conforms to a method that respects the just autonomy and competence of each academic discipline.

To support my claim, I will focus on the major magisterial text on the topic, the encyclical *Laudato Si'* (LS), and study the way it engages other sciences in critical dialogue. Furthermore, to show that LS is not an exception, but crowns a search for an appropriate inter- and trans-disciplinary methodology spanning many decades, I will place its methodology in historical context. I believe that LS is a remarkable point of arrival and a great model for "integral ecology", yet it should also be seen as a point of departure for a project that is still taking shape. This is why I will also comment on its methodological limitations and shortcomings in view of the future of "integral ecology".

A quick glance at the table of contents of LS shows that its six chapters refer to the classical "see-judge-act" (S-J-A, or better, "observe & analyse – appraise – act") paradigm which has shaped Catholic Social Thought (CST), during and after Vatican II. Nevertheless, Pope Francis adapts this paradigm and uses it in his own manner, introducing some elements of newness: notably, as part of the process of "seeing" or "observing" reality, he includes a dialogue not only with the human sciences, but also with the natural sciences:

CHAPTER	TITLE[3]	METHOD
1	What is Happening to our Common Home	*Observing and analysing* - with the eyes of **natural sciences** and **economics** (S)
2	The Good News of Creation	*Contemplating* – with the eyes of **faith** (S/J)
3	The Human Roots of the Ecological Crisis	*Discerning* – with the understanding of **philosophical anthropology** (J)
4	An Ecology that is Integral	*Appraising* – with the prudence of **social ethics** (J)
5	Some Guidelines for Orientation and Action	*Advocating and supporting* – at the local and international political level - through **dialogic diplomacy and policymaking** (A)
6	Ecological Education and Spirituality	*Working and celebrating* - on a personal and community level - through **spirituality** (A)

[3] In the official English translation, Ch. 2 bears the title "The Gospel of Creation"; Ch. 4 "Integral Ecology" and Ch. 5 "Lines of Approach and Action"; my translations seek to reflect better the Italian and Spanish versions and the content of the chapters themselves.

CST, like pre-Vatican II Catholic Moral Theology, has a history of appraising human realities (and even "judging" them, in the stronger "moralistic" sense) and of suggesting courses of action. As we shall see below, what has more often been critiqued as weak, inconsistent, superficial, selective, or simply non-existent, in certain periods or documents, is the "observation" and "social analysis" part. Pope Francis, however, has been quite vocal in inviting the Church to listen and observe carefully before "appraising" or "judging"; he seems intent in promoting the S-J-A paradigm in its entirety, assuming the inductive elements of such a method.

My reflection below is divided into two major and one minor sections. Firstly, I will trace this debate on methodology in CST since Vatican II, including the insights which can be gleaned from the forward of the Apostolic Constitution *Veritatis gaudium* (VG, Part 1). Then I will focus on the first chapter of LS, using a critical approach, to see how the "ecological" encyclical "observes" and "listens" to the data of the natural and social sciences, and to the data analyses which such disciplines provide (Part 2). In this way I will try to learn, in the light of LS and CST, how scientists from various disciplines could contribute to make a fruitful reflection on "integral ecology", in dialogue with today's world (Part 3).

2. The semi-inductive methodological approach of the post-Vatican II Catholic Social Thought

2.1 Dialogue and Interdisciplinarity in *Veritatis Gaudium* (2017)

In paragraph 4 of the "Foreword" of *Veritatis Gaudium* (VG, approved on the 27th December 2017; promulgated on the 29th January 2018), Pope Francis reflects on the contribution of ecclesiastical studies to a Church of missionary outreach. He claims that his reflection is based upon the Church's "attentive listening to the Holy Spirit and to the deepest needs and most pressing questions of the human family", and identifies four criteria that emerge from the Second Vatican Council's teaching and recent Church experience:

a) First, the most urgent and enduring criterion is that of contemplation and the presentation of a spiritual, intellectual and existential introduction to the heart of the kerygma, namely the ever fresh and attractive good news of the Gospel of Jesus Christ […].

b) A second guiding criterion, closely linked to and flowing from the first, is that of wide-ranging dialogue, not as a mere tactical approach, but as an intrinsic requirement for experiencing in community the joy of the Truth and appreciating more fully its meaning and practical implications. [...]

c) From this follows the third fundamental criterion that I would propose: inter-disciplinary and cross-disciplinary approaches carried out with wisdom and creativity in the light of Revelation. What distinguishes the academic, formative and research approach of the system of ecclesiastical studies, on the level of both content and method, is the vital intellectual principle of the unity in difference of knowledge and respect for its multiple, correlated and convergent expressions. [...]

d) A fourth and final criterion concerns the urgent need for "networking" between those institutions worldwide that cultivate and promote ecclesiastical studies, in order to set up suitable channels of cooperation also with academic institutions in the different countries and with those inspired by different cultural and religious traditions. At the same time, specialized centres of research need to be established in order to study the epochal issues affecting humanity today and to offer appropriate and realistic paths for their resolution[4].

In the context, the discourse on inter- and trans-disciplinarity (3rd criterion) might be limited by some readers as one to be applied solely to "classical" ecclesiastical studies (e.g., interdisciplinarity between biblical exegesis, pastoral research, liturgy, theological ethics, dogmatic theology, "Thomistic" philosophy, "ecclesiastical" history and historiography, "ecclesiastical" social sciences, and so forth). Yet this reduction in scope easily justified: the 2nd and 4th criteria point to the need of interdisciplinarity between the ecclesiastical disciplines and other disciplines. In fact, Pope Francis envisions this dialogue between ecclesiastical disciplines as a template for healthy pluralism in modern university studies: "It must also show its effectiveness in relation to the fragmented and often disintegrated panorama of contemporary university studies and to the pluralism — uncertain, conflicting and relativistic — of current beliefs and cultural options". He sees a weak form of interdisciplinarity (looking at an issue from a series of yet-disjointed academic perspectives) as a positive first step, but encourages "trans-" or "cross-" disciplinarity, that is, the "strong form" of interdisciplinarity wherein the different disciplines dialogue critically among themselves, set against a common backdrop, which is the search for truth.

---

[4] Pope Francis, Apostolic Constitution *Veritatis Gaudium* on Ecclesiastical Universities and Faculties, 27 December 2017, n. 4; highlighting in bold added by the present author.

2.2 The Importance of Dialogue with the Natural and Social Sciences

The call to dialogue in VG is not surprising since the magisterium of the current pope had already shown the way before in many contexts. "Dialogue" is a term we find 24 times in LS (excluding the footnotes), a text which clearly seeks to dialogue not only with several academic disciplines, but also with many different audiences and cultures, different institutions, as well as religions and non-religious ideologies. For many of us theological ethicists, today, it may seem obvious that inter- and trans-disciplinary dialogue is a necessary thing — especially in the circumstances of today's world — and that it must take place at all levels of theological reflection, even touching the documents of the Church Magisterium. This epistemological vision, in turn, leads us to hope and assume that today in the 21st century, such documents do not appear *ex machina*: that they be themselves the fruit of dialogue between pastors, theologians and the lived faith of the faithful (analysed with social-scientific tools as is necessary), and furthermore that some of this dialogic background might become increasingly more explicit and visible to the final reader.

As we shall argue in this paper, this entails moving away from outdated editorial traditions based on a feudal architecture of scientific knowledge, that seek to create the impression of an "omniscient pontiff" able to speak with unquestionable authority on all scientific matters, given his authority regarding "theological" matters. This idea dates to the time when, in a dispute with modernity, theologians held onto medieval notions of their discipline as an undifferentiated, all-encompassing "queen of the sciences", a super-science containing within itself the most eminent expertise in everything else, from astrophysics to econometrics to palaeozoology. They refused to recognize that regarding certain concrete matters, a more reliable source of truth might be found through other "human" scientific authorities or via other methods, which theology needs to humbly (though critically) accept. VG's vision, in challenging this perspective, entails that this cross-disciplinary dialogue should be an exercise where the various disciplines listen to each other, mindful of their own strengths and limitations, and those of others.

Even so, not everyone within the Church, and among those we hope to engage in a fruitful dialogue, is willing to embrace such an epistemological framework. Some Catholics balk at the idea of dialogue: "Why should we dialogue with others if we already 'possess' the truth?" "Why should we dialogue with people who are hostile towards us, or have a worldview which is incompatible with ours?" I shall not attempt to answer these objections here — I believe the documents of the Second Vatican Council have already provided ample answers, more than fifty years ago. Rather I shall seek to deal with a question that seems more innocent but is just as pernicious: "Should we really take such dialogue, inter- and cross-disciplinarity seriously? Is it not safer to pay lip service to such an epistemological vision while keeping it at arm's length and making sure it does not challenge any of our long-held beliefs and traditions?"

My preliminary answer to that question is that the epistemological framework promoted in the Prologue of *Veritatis gaudium* allows us to avoid two dangers. On the one hand, it helps us avoid the "apologetic" or "rhetorical" use of social or natural sciences merely to "prove" or support social ethics claims. Intellectual honesty demands of the theological ethicist, no matter their level within Church hierarchy, to avoid the instrumental and selective use of the data of the empirical sciences, citing them only when he or she "agrees" with them. Instead, honesty demands of us that we adopt a critical but humble listening attitude which recognizes the competence of the other in their fields of expertise. This can be costly for the ethicist, who constantly and implicitly uses a sort of practical syllogism to apply values to concrete realities: "n is a case of x, x is bad/good → we should/shouldn't do n". If other sciences clearly concur in demanding that a certain reality be understood in a new or more complex way ("n is not a case of x" or "n is a case of x only when y"), then she or he may have to accept that that the "minor" clause of the syllogism (that which subsumes reality into the equation) may have been displaced, and hence, the ethicist's "traditional" conclusions may need to change[5].

Evoking, *a posteriori*, some "metaphysical" or "anthropological" differences between his/her worldview and that of the other scientist, in a desperate last-minute attempt not to reach such a conclusion, often hurts the credibility of Christian ethics much more than the veneer of constancy afforded by the pretence of "having always said the exact same thing over the centuries" can ever fend off critical voices. In other words, because there is one ultimate Truth in which all sciences are rooted, the ideal is to move from "weak" to "strong" interdisciplinarity wherein data and arguments from other disciplines are able to truly challenge some of the outdated or oversimplistic content of

[5] Such claims are not new or controversial; Aquinas, for instance, discusses the nature of practical knowledge in ST I-II, q. 94, a.4, *resp*.

ecclesiastical disciplines and thus push theological ethics and other practical fields of theology closer to that one ineffable light coming from God's creation and revelation.

On the other hand, dialogue is not a one-way street where we simply listen to and welcome what the other sciences provide; this is the other side of the "strong" interdisciplinary where ecclesiastic disciplines can become more proactive and energetic in the dialogue with other disciplines. While social and natural sciences seek objectivity and a "view from nowhere", research is conducted by humans, who seek human aims, and are funded by human money. Twentieth-century philosophers of science have shown us how not only data *analysis* is "theory laden" and implicitly "value laden", but even the mere *observation* of reality and the collection of "raw data". Truly honest, profound, critical and rigorous dialogue among academics should help us all unmask and challenge any lingering modern myths that society, markets, animal behaviour, desertification or other human and natural realities can be studied from a perfectly neutral position. Therefore, the idea that the "facts" that are presented to us by the social and natural sciences are completely devoid of any underlying anthropological and moral visions of the human being and of human societies is false. Such underlying visions are not "scientific" and they may be in contrast with ethical visions such as those held by Catholics.

Certainly, many features of the scientific method allow scientists (especially in the so-called "hard sciences") to overcome excessive bias and can provide us ethicists with dependable results for our moral reflection. Sometimes, however, we as external yet engaged observers of the scientific community could be more aware of the bias and ideologies lurking behind certain types of data analyses than scientists themselves, and may need to be proactive in demanding scientists to be faithful to their methods and best practices. Furthermore, after welcoming inputs from other disciplines, we theological ethicists are then required by our discipline to place these inputs against a background of meaning that the natural-/social-scientific method cannot provide them.

All of this points to a continual two-way exchange between ethicists and other scientists. True inter- and trans-disciplinarity and dialogue means a respect of one another's competence coupled with an ability to provide a constant critique which is not completely extrinsic to our colleague's discipline. Real dialogue implies the ability of dwelling the borderlands between disciplines, and learning to humbly (and clumsily, perhaps) use the language and resources of the other's discipline to raise questions that a particular colleague may have never dared consider. Scientific communities are human communities, marked with inevitable social dynamics and inclusion/exclusion mechanisms, which establish the "orthodoxy" and "mainstreaming" present in all scientific research, and which may routinely silence certain lines of inquiry even in the most rigorous and "disruptive" of traditions. Note that these considerations do not warrant facile skepticism regarding well-established scientific claims, as that promoted by anti-vaxxers or climate-change deniers, but helps scientists to enter into meaningful dialogue with serious critics.

This constant dialogue on the underlying anthropological and metaphysical assumptions is extremely challenging since it is *a priori* with respect to queries on particular issues in view of sustaining or reframing concrete moral norms of action. Real dialogue implies years and years of sharing and academic community-building, often gratuitous and open-ended. It is a wholly different attitude with respect to the cherry-picking attitude previously mentioned where the ethicist accepts, *a posteriori*, all scientific data that seems to confirm his/her previous claims (or that of his/her tradition) on reality and avoids him/her the challenge of dealing with change or more complex distinctions, while invoking 11th hour anthropological reasons to hastily dismiss any scientific data or analysis that might imply a different conclusion to their practical syllogism.

2.3 The Turning Point of the Council

Given the framework we have just sketched, the task of this text is to situate LS and the integral ecology movement it has engendered within this framework. We shall start with a historical overview of this framework and how it came to be. I think it is fair to say that the structure of the Encyclical *Laudato Si'* is the result of a long and fruitful dialogue between the social sciences, philosophical ethics and moral theology that has taken hold in the last 50 years. This dialogue has shaped CST in the conciliar and post-conciliar era, as elaborated by the official organs of the Church. However, such dialogue in turn was also the fruit of CST itself, which made great strides to engage other disciplines asking them to react to its analysis of social situations, and to take heed of those responses.

In the pre-conciliar era, the importance of such a dialogue was by no means obvious to Catholic thought. For centuries, morality was conceived mainly as a deductive science. This implies a discipline which by definition cannot learn anything substantial from the empirical world, and therefore from the natural or social sciences. Pre-conciliar theological ethicists had become accustomed to a "geometric" method, which establishes the great principles of morality as axioms (on the basis of their being logically deduced from self-evident claims, or on account of the weightiness of the opinions of past experts in favour or against such claims), and logically deduces concrete normative conclusions from such axioms. At some point in this syllogistic exercise, concrete data was fed in, mostly in an implicit and a-historical manner. It was taken for granted that reality was, in essence, simply what certain great ethicists or thinkers believed and argued it to be, while sitting in their university chairs or lecturing to disciples, at some point in modernity, medieval times, or antiquity. Few ethicists brought in their pastoral experience or dared to use it to challenge the claims of the great masters of the past. Rather, they felt that if the "mundane" reality were actually any different from what Aristotle, Aquinas or Augustine assumed it to be many centuries before the modern era, it was the task of pastors in sacristies and confessionals to use hermeneutical tools like *epikeia* and apply the clear and neat deductions based on the thought of these masters benignly, so to assuage the unease of the scrupulous or perplexed faithful, unable to live up to the lofty ideals that tradition promoted and enshrined in moral law. In any case, few if any felt the need to consult the lived experience of the faithful or read and reflect on the results of empirical research coming from "modern sciences", in an age where modernity itself was seen as an enemy of the Church and its moral authority.

To be sure, the fact that many famous scientists in Europe the 19th and 20th century were ideologically anti-clerical and anti-Catholic and used their fame to make anthropological and metaphysical claims far exceeding the bounds of their disciplines did not help to create an atmosphere of mutual trust. In spite of all this, Catholic Social Thought itself, starting with *Rerum Novarum* in 1891, was the result of a historical grassroots movement seeking to overcome this impasse, and listen and engage with the "new things" that modernity offers to ethics as food for thought[6]. It set the stage for a dialogue with modernity which blossomed in the late 1950s after a 50-year hiatus in some areas due to

[6] The major historical claims in this paragraph are generally taken for granted by CST scholars; the reader lacking familiarity with the history of CST may find adequate warrant and further information in academic CST handbooks, such as Kenneth R. Himes, (ed.), *Modern Catholic Social Teaching: Commentaries and Interpretations*. Washington D.C.: Georgetown University Press, 2005; Donan Dorr, *Option for the Poor: A Hundred Years of Vatican Social Teaching*. Dublin: Gill and MacMillan / Orbis Books, Maryknoll, N.Y. 1983; Departamento de Pensamiento Social Cristiano de la U. P. Comillas, *Una nueva voz para nuestra época*. Madrid: Universidad Pontificia Comillas, 2006.

the "modernist crisis". Some social encyclicals such as *Quadragesimo anno*, written to commemorate and develop *Rerum Novarum*'s teaching during that hiatus and prior to Vatican II, were important, beautiful and rich documents written with a similar desire to engage modern realities, yet methodologically, these documents were written from the neo-Thomistic perspective described above, strongly suspicious of any ethical reflexion "contaminated" by a "modern" inductive epistemology. The pre-conciliar CST, which we might call "CST version 1.0", did not completely discard the natural sciences and the human sciences, but they were seen mostly as apologetic tools to be used selectively to corroborate *a posteriori* the deductive conclusions of scholastic ethics. They were certainly not to be used to appraise or rethink social and natural realities, or to adjust traditional ethical conclusions regarding how such realities are to be conceived and morally appraised.

In the 1960s, Josef Ratzinger (later elevated to the papacy as Pope Benedict XVI) was one of the first to criticize this purely deductive way of proceeding of neo-scholastic morality, in truth closer to Spinoza's modern ethical thinking than to that of Aristotle[7] and Aquinas[8]. Ratzinger argued that "what is 'right according to nature', that is, what is truly human — if we want to call it that — is impossible to establish from a purely speculative or '*more geometrico*' point of view for the human being as he [or she] is"[9].

---

[7] "Nor is practical wisdom concerned with universals only – it must also recognize the particulars; for it is practical, and practice is concerned with particulars. This is why some who do not know, and especially those who have experience, are more practical than others who know; for if a man knew that light meats are digestible and wholesome, but did not know which sorts of meat are light, he would not produce health, but the man who knows that chicken is wholesome is more likely to produce health. Now practical wisdom is concerned with action; therefore one should have both forms of it, or the latter in preference to the former." (*Nicomachean Ethics* 1141b 15ss). Hence, for Aristotele, ethics is primarily an inductive science, and the "good" is apprehended via a certain kind of "perception" (1143a36-1143b5; 1142a25-3). However, the knowledge of universals (the ultimate and penultimate goods to seek in life, and the general values that direct us towards these goods) allows us to live virtuously (1140a26-28), by connecting concrete goods to the common good and to the ultimate end in life. Knowledge of universals is useful to attain a sense of the moral life as a whole, rather than as axioms from which concrete decisions can be derived deductively. Cfr. Bruno Niederbacher, "The Epistemology of Moral Belief in Aristotle's Nicomachean Ethics", in *Universitas. Monthly Review of Philosophy and Culture (Taiwan)* 396 (2007), pp. 37-58.

[8] In his commentary on the *Nicomachean Ethics*, Aquinas is in complete agreement with Aristotle on these matters (*In Ethicorum*, liber VI, lectio VII).

[9] Joseph Ratzinger, "Naturrecht, Evangelium und Ideologie in der katholischen Soziallehre. Katholische Erwägungen zum Thema", in Klaus von Bismarck and Walter Dirks, (ed.), *Christlicher Glaube und Ideologie*. Stuttgart: Kreuz-Verlag, 1964, p. 27.

Ratzinger and many other scholars consider *Mater et Magistra* (1961) as the first pontifical social ethics text to take the mentality of modern men and women seriously, as well as the input from modern science, which is what shapes such a mentality:

> We may suggest that the truly problematic tendencies, within the concrete formation of Catholic social thought — as it has existed until now — consist in the lack of attention to history, in order to rather favour speculation, to which we may add a strong bond to the world of the Middle Ages, characterized by hierarchical and static components as an ideological basis. Both features imply a strong option in favour of conservative postures. [...] All this probably has the consequence that the Social teaching of the popes — except for "Mater et Magistra", where a significant change of principle seems to emerge — has always remained as something extraneous to the mentality of the technical and industrial world. Despite the many good elements of guidance it contains, the Social teaching of the Magisterium has failed to touch the mental and linguistic structure of the human being who lives in a technical world[10].

On this basis of Ratzinger's claims, we may say that the transition to "CST version 2.0" thus begins with *Mater et Magistra*, and in the pontificates of John XXIII and Paul VI, we slowly start to see documents which are more inductive, more in dialogue with the lived reality of the faithful and of "people of good will", and more informed by the empirical sciences. One should note that, in various western countries, there was a certain dialogue between Catholic ethics and the scholarly analysis of new social realities in some areas of society and the Catholic community, long before *Mater et Magistra*, mainly sponsored by Catholic social movements. Such dialogue, in its more formal and cultivated embodiments (such as the "Social Weeks" in France) engendered a culture of listening to the modern sciences[11]. However, this spirit of exchange, wherein ethicists listen to rigorously analysed data based on empirical research, before pronouncing themselves on a social issue, failed to penetrate the Roman Curia — except in a form very filtered and diluted by Neotomism — until John XXIII decided to ask two Venetian diocesan priests, Pietro Pavan and Agostino Ferrari Toniolo, to start writing his social encycli-

[10] J. Ratzinger, "Naturrecht, Evangelium und Ideologie", p. 29.

[11] Edward Cahill, "The Catholic Social Movement: Historical Aspects", in Charles E. Curran and Richard A. McCormick (ed.), *Official Catholic social teaching*, Readings in moral theology no. 5. New York: Paulist Press, 1986, pp. 3-31.

cals, instead of the team of very "classical" Jesuit ethicists who usually wrote such texts for his predecessors[12].

This seemingly simple and "parochial" decision of bringing local theologian friends over to Rome "from back home" was a significant one. John XXIII, while not known as a great "intellectual", was from his pastoral experience keenly aware of the epistemological issues in Neothomist ethics, and clever enough to realize how deep the resistance in the Roman Curia would be to any change of approach. For instance, when writing the inaugural speech of the Council, *Gaudet mater Ecclesia,* Pope John sought to raise these issues. The (original) Italian version, which John XXIII himself cites literally in the *Discourse to the Sacred College and the Roman Prelature*, pronounced on the 23rd December 1962[13] (seemingly oblivious of the considerable "corrections" made by his translators in the official Latin version), affirms the need for doctrine to be "studied and exposed through the forms of inquiry and literary formulation of modern thought [...], measuring all things according to the forms and proportions of a predominantly pastoral magisterium". The Latin version of *Gaudet mater ecclesia*, on the other hand, omits the reference to the adoption of the research methods and forms of expression coming from modern thought, limiting itself to the distinction between the deposit of faith (the truth contained in Christian Revelation) and the forms in which it is enunciated[14].

In the course of Vatican II, Pope John decided to bring together the Neothomist Jesuit ethicists and the Venetian priests as *periti* (experts) in the drafting of the first versions of what eventually became *Gaudium et Spes*, the Council document intended to deal with the relation between the Church and the modern world, and to provide an ethical reflection on a number of concrete social, economic and political realities. Eventually, many other *periti* became involved and took over the drafting process, but in the process, Vatican Council II welcomed a new meth-

---

[12] The Venetian priests Pietro Pavan and Agostino Ferrari Toniolo (who came from the sphere of the "Social Weeks" in Italy), together with Jesuits like Georges Jarlot (who came from the world of social sciences) clashed especially with the Jesuits with a more neotomist approach, who were influential under Pius XI and Pius XII - such as Gustav Gundlach, Sebastiaan Tromp and Franz Hürth - especially during the preparation of the preliminary conciliar *schemas* that sought to deal with social issues, that is, the *De ordine sociali* and the *De communitate gentium* of 1960. Cfr. Giovanni Turbanti, *Un concilio per il mondo moderno. La redazione della costituzione pastorale «Gaudium et spes» del Vaticano II*. Bologna: Il Mulino, 2000, pp. 70-71.

[13] *Acta Apostolicae Sedis* (1963) pp. 44-45.

[14] Cfr. Christoph Theobald, *L'avvenire del Concilio: nuovi approcci al Vaticano II*. Bologna: Edizioni Dehoniane, 2016, pp. 86-87.

od for drafting CST documents, a method based on the work of mixed groups of theologians and ecclesial experts (coming from different cultural and pastoral backgrounds), and also of lay experts (women and un-ordained men who were experts in a number of non-ecclesiastical disciplines)[15]. Though some may claim that during the pontificates of John Paul II and Benedict XVI some of the social encyclicals do not give the impression of having been informed by a deep dialogue with the social sciences, with the experience of the faithful and with a wide range of positions within the Church, it is clear that in *Laudato Si'*, Pope Francis seeks to return to the more participatory methodology and the dialogical style inaugurated by the encyclicals of John XXIII and *Gaudium et spes*.

Let us however take a step back and look more closely at *Mater et Magistra* and the methodology which Pavan and Ferrari Toniolo start to introduce in this and other texts. This methodology, in fact, allows the Magisterium to grasp and integrate important ferments in Catholic thought. The drafters of *Mater et Magistra* were inspired by the reflection made within the Catholic Action movement, which for many years had emphasized the importance of an inductive epistemology when grasping and appraising social realities in a rapidly changing world. For this reason, *Mater et Magistra* adopted in 1961 the see-judge-act (SJA) method, which had already been used for decades by Joseph-Léon Cardijn and the *Young Christian Workers* (known in French-speaking countries as "JOC", and part of "Catholic Action") in their social reflection. Cardijn (created cardinal in 1965) already reflects on this method in a 1935 speech on the "Three Truths"[16]. For Cardijn, we cannot seriously engage in social ethics without observing reality. This invitation to adopt an inductive method is prudently and gently welcomed by *Mater et Magistra*:

> There are three stages which [are usually] followed [when translating the general principles of a social teaching] into practice. First, one [takes measure of] the [true state of affairs]; secondly, one forms a [diligent] judgment on it in the light of these same principles; thirdly, one [determines] what can [or] should be done to implement these principles [in such a way that the norms handed down might be applied according to customs suited to the relevant times and places]. These are the three stages that are usually expressed in the three terms: [behold], judge, act. (Mater et Magistra, 236)[17].

[15] Cfr. Giovanni Turbanti, *Un concilio per il mondo moderno*, pp. 223-224, 233-238, 246-248, 632-635.

[16] Joseph Cardijn, *Challenge to Action. Address of Monsignor Joseph Cardijn (Founder and Chaplain General of the Young Christian Workers)*. Chicago: Il Fides Publishers, 1955.

[17] This is mostly my translation of the "official" Latin text (the official translation into English is quite

We note, however, that the text avoids saying that the principles may be discovered or reformulated on the basis of the results of the observations, and this is probably why Ratzinger, in the text quoted above, is cautious in his evaluation of this document: "[…] 'Mater et Magistra', where a significant change of principle *seems* to emerge"[18]. The more fullhearted shift occurs during the Council:

> To carry out such a task, the Church has always had the duty of scrutinizing the signs of the times and of interpreting them in the light of the Gospel. Thus, in language intelligible to each generation, she can respond to the perennial questions which men [and women] ask about this life and the life to come, and about the relationship of the one to the other. We must therefore recognize and understand the world in which we live, its explanations, its longings, and its often-dramatic characteristics[19].

The Council does not speak of "perennial principles [and directives]", but of perennial questions that human persons ask about life and relationships. Ethical norms and principles are some of *responses* to those questions, responses which — to some extent — may need to be adapted in space and time, to different cultures and generations. They sometimes need to be reformulated in accordance with lived reality and read "in the light of the Good News" (which, properly understood, is the second "perennial feature" of this anthropological landscape, anchoring Christian ethics). In paragraph 46, the document returns to the question of methodology, speaking of the "urgent needs characterizing the present age", and stating that the Council has chosen to consider these issues "in the light of the Gospel and of human experience" (rather than "in the light of the principles of a perennial moral law"[20], as expected from CST methodology prior to the Council), and implying that we should do the same. Should the nuanced paragraphs of *Gaudium et Spes* leave

---

flawed or oversimplified, requiring many corrections that have been added in square brackets). Here, on the other hand, is the translation from the official Italian version, is probably the original text (rather than a re-translation from Latin): "When translating social principles and directives into concrete terms, one usually goes through three moments: surveyal of situations; evaluation of these in the light of those principles and directives; research and determination of what can and must be done to apply those principles and directives to the situations, according to ways and degrees that the same situations allow or demand. These are the three moments that are usually expressed in the three terms: see, judge, act".

[18] The italics have been added by the present author.

[19] Vatican Council II, Pastoral Constitution *Gaudium et spes* on the Church in the Modern World, 7 December 1965, n. 4.

[20] This was rather the perspective adopted in *De Ordine Morale Christiano*, a draft text on moral and social issues that the Council Fathers rejected, which we will mention below.

room for doubt, Paul VI takes up this issue in 1971 in *Octogesima adveniens* 4, probably having in mind those who after the *Humanae vitae* crisis wanted to get rid of the methodology adopted by *Mater et Magistra* and the Council:

> In the face of such widely varying situations it is difficult for us to utter a unified message and to put forward a solution which has universal validity. Such is not our ambition, nor is it our mission. It is up to the Christian communities to analyse with objectivity the situation which is proper to their own country, to shed on it the light of the Gospel's unalterable words and to [tap into the social teaching of the Church so to] draw principles of reflection, norms of judgment and directives for action.

Therefore, for Paul VI, what is "perennial" or "immutable" is the Good News. The method starts with "seeing" and "analysing" concrete, local situations "with objectivity" (that implies the use of appropriate scientific methods), while at the same time shedding on this reality the "light of the Gospel" which gives us eyes to see reality more profoundly (this concomitant kerygmatic operation not being at the same level, and hence in competition with, the required scientific observation and analysis). CST is a wellspring from which we draw out ("haurio", in Latin), those resources for our discernment: principles to reflect on this reality, norms to appraise it, and directives to act upon that judgement. What is implicit, but clear from the opening statement, is that such norms and directives must not come purely deductively from a "natural law" considered as a universal and immutable set of concrete precepts or "axioms". The "seeing" part of the methodology, as we note from these documents, is quite layered. It includes:

1. Observation and analysis *using objective methods* (natural and human sciences)
2. *Contextualized* observation of that same reality in *the light of the Gospel* (ongoing personal and communal nourishment by Scripture, enriched by professional exegesis)
3. Reflection on both levels of observation, nourished by the principles provided by the CST (as part of a longstanding tradition, and as presented in more recent texts by the Magisterium).

All this points towards a method that incorporates empirical elements, and which we can therefore describe as "inductive" or "semi-inductive"[21]. Such a

[21] Few philosophers of science today still believe in a project like that of the Vienna Circle, which

method implies acknowledging, with respect and prudence, the great values and norms handed down by tradition (those inherited from the deductivist and speculative theology of the manuals, and especially those that seem far from the "common sense" of our contemporaries) and, through a dialectical process, *placing them in critical dialogue* with the good practices and the correct ways of proceeding that we can derive from *today's contextualised inductive generalizations*. The latter include common sense, people's experiences and pastoral experience, but also the results of the empirical sciences which — in a more rigorous manner — unveil to us the experienced reality of our fellow believers and fellow citizens. Though most of this tradition refers more explicitly to social sciences when speaking of all this, when dealing with issues such as climate change, natural sciences must also have an important place at the table of dialogue, as LS teaches us by setting an example.

2.4 Classicist and Historically-Minded Approaches in Theology and Christian Ethics

In the previous paragraphs, I may have given the reader the impression that most issues regarding the most adequate CST methodology and its dialogue with other sciences were resolved during the Council and in its immediate aftermath. This may seem hard to reconcile with our initial comments on VG 4, and the appraisal of LS chapter 1 which we shall offer in the second part of this paper. In fact, both imply that 50 years after such a debate, we are still "hoping" that this vision might be put into practice more fully and admitting that we are not quite there yet. The reason for this is that the resistance towards this methodology and these new practices are deeply ingrained in some theological schools and in some parts of the Catholic world. The position of Paul VI and the Council reflects the strong desire of the council fathers to overcome the "classicist"

wanted to reduce all human knowledge to theories based on neutral empirical observations and logical operations, and therefore inductively "verified". The reflection, within the Popperian school, on the theory-ladenness of observation, and on the importance of falsification in order to approach scientific "truth" via corroboration, has made us discover that even in the natural sciences, the "inductive" method is never "perfectly inductive" in the ideological sense promoted by the Vienna Circle. I hereby make use of the strange terminology of "semi-inductive" for those readers who still hold this naive view of the inductive method (often unconsciously promoted by science communicators) and who, with reason, find it hard to imagine the ethicist as someone who discovers ethical truth through the stereotypical "empirical method", by first observing reality in a purely "neutral" and "objective" way and then elaborates "empirically verified" moral principles and norms.

vision of the world, typical of the *De Ordine Morali Christiana.* This draft or *schema* was presented to the Fathers at the beginning of the Council and immediately rejected (together with all the moral *schemas* written by the authors chosen by the Holy Office)[22] mainly due to their non-dialogical style and their arrogant presentation of human realities. Such documents however continue to fascinate seminarians and church leaders today due to the "clarity" and "authority" which they exude and which seem to have been lost in more recent texts espousing humbler and more nuanced approaches.

It may be said that in the conciliar documents, and especially in *Gaudium et Spes*, which is the main moral theology document of the Council, the council fathers adopted many elements of a *modern* vision of reality, avoiding however relativism, subjectivism and historicism which, in fact, are more characteristic of some *postmodern* approaches. Following Bernard Lonergan's terminology, some authors describe the moral theology of the Council — and the epistemology and moral teaching that derives from it — as "historically-conscious" or "historically-minded" that is, "aware of the historicity of human reality"[23]. The previous and opposing paradigm (pre-Vatican II, Neo-Thomist, manualistic, casuist) is labelled "classicist" by Lonergan; the issues inherent to this paradigm are present in "classical" authors, but probably the more acute problem lies within the *modern* mindset with which *classical* authors were read before the Council, reducing their thought to a static museum artefact. Thus, a somewhat understandable *classical* aversion to change is morphed into a *classicist* fixation with mummifying Antiquity using the rigid embalming fluids derived from Newtonian physics[24]. Richard Gula helps us understand Lo-

[22] Cfr. Philippe Delhaye, "The Contribution of Vatican II to Moral Theology", in *Concilium (English Edition)* 75 (1972) pp. 58-67; Sabatino Maiorano, "Coscienza e verità morale nel Vaticano II", in Marian Nalepa and Terence Kennedy, (ed.), *La coscienza morale oggi. Omaggio al Prof. Domenico Capone*. Roma: Edizioni Accademia Alfonsiana, 1987, pp. 259-278.

[23] Bernard J. F. Lonergan, "The Transition from a Classicist World-view to Historical-Mindedness", in Willimam F. J. Ryan and Bernard J. Tyrrell, (ed.), *A Second Collection: Papers by Bernard J. F. Lonergan S.J.* Toronto: University of Toronto Press, 1996, pp. 1-9; Bernard J. F. Lonergan, "Theology in its New Context", in Willimam F. J. Ryan and Bernard J. Tyrrell, ed., *A Second Collection: Papers by Bernard J. F. Lonergan S.J.* Philadelphia: Westminster Press, 1974, pp. 55-67.

[24] For those unfamiliar with Lonergan's thinking, his terminology may seem strange. The main point, as I understand it, is that modern readers do not think in a "classical" manner: they may have sufficient training to understand the texts of St. Thomas, Boethius or Aristotle well enough, but they do not read them as the intended audience would have, with its premodern or medieval mindset. They may however possess a "classicist" mindset, pretending that in some areas of their life they can easily bracket modernity, with its epistemological revolutions in the natural sciences

nergan's distinction between a "classicist" and a more "historically-minded" (conciliar and post-Vatican II) approach by constructing a useful table, which we will summarize and adapt in the points below[25]:

- **Metaphysical grounding**: A *classicist* vision of reality conceives the world as a complete and harmonious fixed reality, which can be "objectively" described in an abstract and universal way as a series of self-contained essences. A *historically-conscious vision*, on the other hand, sees the world a something dynamic that grows, changes and evolves in history. Reality is best described using concrete and historical concepts, that can tolerate and fit unique and exceptional cases.
- **Epistemological grounding**: A *classicist* approach is mainly deductive: it starts with abstract essences and self-evident truths regarding those essences and the nature of reasoning itself, and then deduces general and particular principles from these foundations: the results are assumed to be immutable and certain, as long as the syllogism used to derive them is logically flawless. A *historically-conscious* approach, on the other hand, is mainly inductive, since it starts with experience and observation, establishes generalizations on the basis of accumulated observations, and then applies these generalizations to new cases and situations fully aware that the generalizations are non-final and incomplete, and there

---

and in the understanding of history, its critiques of ancient metaphysics, its masters of suspicion, its economic and social organization, and so forth, and return to an enchanted pre-modern world, built from entities moved by an internal entelechy. "Classicist" thinking, therefore, is not exactly "classical" thinking, but rather a modern idealization and reconstruction of the thinking of people of another era, often done by modern people to "solve" the modern problems of a modern audience. Sometimes, the reconstruction is done along the lines of modern thought, by merely mirroring it and upending it.

For instance, one of the hallmarks of "classicist" thinking is the exaggeration of the importance of the deductivist notion of the sciences present in "Posterior Analytics" of Aristotle's *Organon*, (Bernard J. F. Lonergan, "Questionnaire on Philosophy: Response", in Robert C. Croken and Robert M. Doran, (ed.), *Philosophical and Theological Papers, 1965-1980*, Collected Works of Bernard Lonergan 17. Buffalo/Toronto: University of Toronto Press, 2004, pp. 355-357). Aristotle, however, was keenly aware that practical science cannot be rigidly axiomatized and deprived of an empirical contact with human experience and with the facticity of biophysical and socio-physical reality (*Nich. Eth.* 1141b 15ss). Ethical classicism thus presents to us as "premodern" something that, in large part, was a yet another fruit of modernity: a fusion (consolidated starting from 1600) of metaphysical essentialism, juridical legalism, and the search for particular norms that present themselves as "universal and immutable", mimicking those of Newtonian physics or Euclidean geometry.

[25] Richard M. Gula, *Reason informed by faith: foundations of Catholic morality*. New York: Paulist Press, 1989, pp. 32-33.

might be outliers which need to be handled individually, as well as novel cases that spur us to rethink and refine our generalizations. (I would add here that, of course, reality is a bit more complex: there is an important component of deductive reasoning within inductive approaches, as modern philosophy of science teaches us. We use deductive logic and previously-known theoretical claims when we design empirical studies. When we generalise, e.g., by fitting a curve onto data points, the resulting "observation" is often "theory-laden" and our generalizations and theoretical claims are often underdetermined by the data. Hence, a historically-conscious approach is by no means "purely empirical", but is rather both inductive and deductive.)

- **Ethical and moral implications**: Classicist ethics focusses on what is abstract and universal, often ignoring concrete issues. It insists on obedience to authority and to pre-established norms, and the need to submit to a fixed order of things seen as God-given and unquestionable. It tends to derive certain principles using a particular pre-modern way of observing (and metaphysically construing) physical and biological nature. Historically-minded ethics, on the other hand, seeks to deal with moral issues as they emerge in history, mindful of the historical reality of the persons involved and particular circumstances in which they emerge. It sees most moral norms as culturally and historically conditioned rules, which engage the conscience as imperative *prima facie* claims, and demand that it act responsibly and offer concrete parameters to guide its judgement.

This brings us to the advantages and disadvantages of each of these approaches, as outlined by Gula. The classicist vision promises clarity, simplicity and certainty, but it deems itself already in possession of the truth, hence it is disinterested in dialogue, but rather seeks to impose its claims on others considered to be in error to the extent that their beliefs do not coincide with its truth claims. In extreme cases, this approach risks becoming authoritarian, overly judgmental, arrogant and doctrinaire. The historically-minded approach promises to respect the uniqueness of each person in his or her historical circumstances, and teaches us the patience and humility to follow and accompany long-term processes of conversion within individuals and societies. Its main limitation is that some people may feel that its complex and nuanced approach, coupled with its tendency to relativize and historicize norms, does not offer enough clarity to enlighten their moral decision-making, and may too easily be mistaken for relativism and historicism, thus eroding the Church's moral authority.

To Gula and Lonergan's distinctions, I feel we should today add some further observations. The classicist approach seems to stem from a more pessimistic and Augustinian vision of human nature, freedom and conscience, where any deviation from a clear and monolithic order is associated with moral "deviance" and hence evil (e.g. *De Libero Arbitrio*, Book 2). The idea that such a deviation might fit into, or be a part of, a complex, multifaced "order" created by God, is inconceivable. God's rationality here is reduced to that of an architect fascinated by efficiency and simplicity or a physicist who applies Ockham's razor wherever possible; like the old Pythagoreans and Spinoza, this view adopts an approach inspired by mathematics and geometry to reduce ethics to a *certain, universal, and immutable* set of facts and rules which can presumably be formulated as *simple* axioms. This approach, 50 years after Vatican II, is still very attractive to people who are unduly suspicious of moral sophistication, but also justly wary of the games of rationalization and intellectualization that are sometimes used to laxly shroud narcissistic and self-centred behaviours contrary to the common good. The historically-conscious approach, on the other hand, seems to stem from a more optimistic and Thomistic vision of human nature, freedom and conscience[26], and seeks to offer guidance for decision-making that is sufficiently but not excessively "clear" (leaving real space for conscientious and prudential moral discernment, shaped by virtue, not pre-established general rules). It is ready to take outliers seriously and to assume deviation from the norm as something that can occur in good faith.

Lonergan's distinctions between the two paradigms are very enlightening, but the question remains: why are we still struggling, 50 years after Vatican II, to complete the paradigm shift to historical consciousness? I believe it has to do with persons and institutions and their resistance to change. We have always known that ethics debates have a way of capturing the attention of brilliant but rigid minds, of persons with little tolerance for complexity, such as Blaise Pascal and some of his Jansenist friends. If the reader will allow me to indulge in a non-expert psychological reading, I believe modern psychology can help us see how dangerous and inhumane certain persons with narcissistic or obsessive-compulsive tendencies can become when they embark on solving moral

[26] We have previously cited Aquinas' discussion of the practical nature of ethics in the *Summa Theologiae* and in his *Commentary on the Nicomachean Ethics* (cfr. footnotes 5 and 7).

issues, aiming primarily at the aesthetical pleasure of finding "perfect" and "elegantly simple" normative solutions, rather than seeking the better good possible in a given finite and historical situation. I believe it is exactly these kinds of persons who find a historically-conscious approach distasteful and unattractive, and who continue to champion the classicist model; yet, their endless quest for an unrealistic simplicity and clarity in the field of ethics continues to garner the support of other persons, typically ones with a low self-esteem who want to be constantly told by others or by "the rules" what to decide, and are uneasy with the notion of "personal discernment".

I mention this reading since I believe our ecclesiology has found ways, over the centuries, to resist such tendencies, even though it was not informed by modern psychological notions and theories. The Petrine leadership model has helped us place into positions of authority in the Church persons who are confident, humane and humble, knowing they are "flawed yet chosen" who have promoted healthy and benign approaches to ethics, overall: we are usually led by wise pastors, not by brilliant theoreticians. Yet, it is well-known that popes and pontifical dicasteries depend on various teams to draft documents and implement policies, and the transparency and accountability of the ghost-writers and ghost-deciders remains an issue in the 21st century. This is particularly concerning when *some parts* of a document are written by authors who espouse a classicist mindset, and who deep down believe that if taken seriously, the epistemology of Vatican II and of VG is dangerous since any changes in traditional ethical norms and principles undermine the honour and credibility of the Church. Since most documents are signed by the pope or a prefect of a dicastery, and there is no possibility of public feedback on the draft versions of the texts, it is hard for the critical reader to intervene respectfully before the final text becomes official and critique the parts written using a classicist mindset.

Pope Francis has often stirred controversy by speaking critically of such scholars reflecting on reality from their ivory towers, with no pastoral-mindedness, with no experience of close contact with the reality of their brothers and sisters (no "odour of the flock"), with a mercy-deficient and hence flawed conception of justice, with an inability to grasp the multifaceted or "polyhedric" nature of reality, and so forth[27]. Some authors, in fact, have re-

[27] Cfr. Pope Francis, *Evangelii Gaudium*, n. 24; *Fratelli tutti*, nn.145, 215; *Meeting with Polish Jesuits* (30th July 2016) https://www.laciviltacattolica.com/today-the-church-needs-to-grow-in-discernment-pope-francis-meets-with-polish-jesuits/ [accessible on 12/04/2021].

cently studied the similarities between the thought of Pope Francis and the historically-minded mind-frame promoted by Lonergan[28]. A well-known contemporary debate that clearly pits these two approaches is that concerning Chapter 8 of *Amoris Laetitia*, with the "dubia cardinals" adopting a classicist approach, and the defenders of Pope Francis' magisterium opting for historical consciousness[29]. Yet the some of the "dubia cardinals" had important roles in the Curia and continued to vet official texts even while writing public declarations patently hostile to a "historically-minded" approach to ethics. To be sure, Pope Francis has gradually made serious changes within the Roman Curia, seeking more pastorally-minded persons, yet I believe we still need to overcome certain traditions of "secrecy" and start moving towards a culture where all sources are cited, all teams can sign and own up to what they write, all conflicts of interest are declared openly.

2.5 Objections and Nuances

The reader may have a number of critical questions to raise at this point. Firstly: was all pre-Vatican II theological ethics "classicist"? Secondly: is the paradigmatic change in Vatican II truly as clear-cut as we have just presented it?

As regards the first question, I believe that the Council itself clearly notes the problems of pre-Vatican II moral theology in the document *Optatam totius* 16: "Special care must be given to the perfecting of moral theology. Its scientific exposition, nourished more on the teaching of the Bible, should shed light on the loftiness of the calling of the faithful in Christ and the obligation that is theirs of bearing fruit in charity for the life of the world." When situated in an organizational culture skilled at diplomatic parlance and not used to self-critique, these words resound as a harsh rebuke of the "un-evangelical, un-scientific, moralistic and charity-deficient" pre-Vatican II model, often denoted with the term "casuistic". Yet, within Catholic theology, I believe there are already many elements of a "history-conscious" approach in the era of High Casuistry, which according to Albert Jonsen and Stephen Toulmin goes from 1556 to 1656[30].

---

[28] Gerard Whelan, *A Discerning Church: Pope Francis, Lonergan, and a Theological Method for the Future*. New York: Paulist Press, 2019.

[29] Jean-Miguel Garrigues and Alain Thomasset, *Une morale souple mais non sans boussole*. Paris: Le Cerf, 2017; Maurizio Chiodi, "Coscienza e discernimento: quale rapporto con la norma? Sul capitolo VIII di *Amoris laetitia*", in *Teologia* 43 (2018), pp. 18-47.

[30] Albert R. Jonsen and Stephen Toulmin, *The Abuse of Casuistry: A History of Moral Reasoning*. Berkeley, CA: University of California Press, 1990, pp. 152-251. The Low (or "bad")

Such an approach, which matches casuistry training with scholastic ethics, both cognizant of the complexity of reality, was progressively lost as Catholic ethicists came under the scathing critiques of the Jansenists and sought safety and certainty to avoid being labelled as "lax". In 1879, Leo XIII tried to reintroduce the medieval moral tradition, and Aquinas in particular, with the encyclical *Aeterni Patris*; he probably realized how sterile the casuist-manualist approach had become (due to what Lonergan would call its "classicist" mindset). However, the "new" scholasticism (Neo-Scholasticism) that developed from this call to return to the writings of Aquinas, while wary of the dangers of Jansenism and ideologies inspired by modern natural sciences, was unable to shake off the fascination with "certainty" that these cultural adversaries cherished and promoted among the faithful and within academic circles.

As regards the second question, as we know from the research of Imre Lakatos on the resistance to paradigmatic shifts in the natural sciences[31], such shifts are never as "clear-cut" as Thomas Kuhn and his popularisers would have us believe. I guess this is even more the case in a discipline like theology which can easily resist falsification challenges, and in an institution imbued with respect for tradition, such as the Catholic Church. To be sure, if we were to dwell on the precise wording of the final texts that were approved at Vatican II — which are the product of complex negotiations among editors, amenders, and voting blocs — we will not find a *single* epistemological paradigm throughout. This does not mean that we can deny that a paradigm shift occurred on that occasion but rather that we should recognize that the paradigm shift from classicism to historical consciousness is not ubiquitous or perfectly consistent: it is still ambiguous, tentative, or embryonic in some

---

casuistry that arose after this period, however, is the result of a split between these two elements of post-Tridentine moral training of seminarians, giving rise, on one hand, to safe "casuistry exercises" performed in convents and seminary residence halls, and centered on trivial or surreal cases (no longer unrelated to people's lived experience), and, on the other hand, of a "geometric" (deductive and legalistic) moral theology presented in the manuals, and taught in the classrooms of Catholic seminaries and universities. It was most probably the scathing sarcasm of the mathematician Blaise Pascal and his Jansenist friends that bullied moral theology into a fear of complex and nuanced thinking, and an obsession with the search for "certain" answers, as many moral theologians gave up probabilist approaches and moved towards rigorism, so not to be accused of "laxity". This is how, in the 1800s, we end up with a Catholic Moral theology based on manuals that disdained serious reflection on lived experience and the empirical approaches that were being adopted in many other scientific disciplines.

[31] Imre Lakatos, *The Methodology of Scientific Research Programme (Philosophical Papers, I)*. Cambridge: Cambridge University Press, 1978, I.

paragraphs and documents, yet well in place in others. As noted above, the major documents (the "Constitutions" in general, and *Gaudium et spes* in particular) move carefully but surely towards a "modern" way of grasping reality, or at least away from the classicist paradigm.

It is important to note here that intellectual honesty requires that we steer clear from certain extreme views which naively or ideologically claim that Vatican II marks a definite rupture from past Church teaching[32]. Yet it is equally dishonest to claim that the epistemology of the Council was nothing but a continuation of that of previous Councils, and that its changes were mainly cosmetic[33]. There is today a copious amount of scientific literature which analyses the texts, the drafts and diaries and personal annotations of the main drafters of the documents approved by Vatican II, which I cannot engage here; what I believe we can conclude from all this research is that the Council *tends towards* such a more historically-conscious approach in many documents (and this is a central part of the reform of the Church adopted by the Council). Yet, it is well known that in the areas of sexual and family morality, and in bioethics, attempts to move away from the classicist paradigm — and thus to open a fruitful dialogue with the modern vision of the world, through the conciliar paradigm of "historical awareness" — have encountered considerable resistance in the post-conciliar Catholic Church. This resistance can still be seen in recent synods on the theme of the family and in the attitudes of some "conservative" Catholics towards the teaching of Pope Francis. However, in matters of social morality, the new paradigm is well accepted, and *Laudato si'* is a very clear example of the fruits that this "new" approach can bring[34].

Even so, today, some Catholic theologians and philosophers would like to return to Neo-Scholasticism and the "classicist" paradigm, stressing to the bitter end the (admittedly real) risk that readers of their texts might misapply the conciliar paradigm and succumb to the temptation of relativism or subjectivism[35]. The end (at times naively paternalistic) of not scandalizing or confusing

---

[32] Benedict XVI, "Address of His Holiness Benedict XVI to the Roman Curia Offering Them His Christmas Greetings", 22 December 2005.

[33] Cfr. John W. O'Malley, "'The Hermeneutic of Reform': A Historical Analysis", in *Theological Studies* 73 (2012), pp. 517-546.

[34] Richard M. Gula, *Reason informed by faith*, pp. 239-240.

[35] For instance, this is a conclusion of an article on *LS*, signed by R. R. Reno, editor of *First Things* (August/September 2015 issue of the journal): "The old scholastic systems were superseded by a wide variety of experimental theologies. I don't gainsay the need for and value of some of those experiments. But we can't deny the debilitating consequences. [...] So if we, as Catholics,

the faithful and less educated seminarians may be well intended and good, but the means — that of questioning the value of an Ecumenical Council, rejecting its methodological reform in the field of morality, and presenting Moral Theology as a purely deductive and speculative science (against the claims of this science's tradition and epistemology) — is often blameworthy. If to fight relativism we are willing to relativize the great tradition and the value of a Council, then we have already lost the battle.

2.6 A Return to the Semi-Inductive Method during the Francis Papacy

It can be said that the most emblematic thing that Pope Francis has done so far to promote a return to the semi-inductive method of the Council is the use of questionnaires to prepare the synods on the family in 2014 and 2015. However, the apostolic exhortation *Evangelii gaudium* (EG) makes numerous references to this method, taking up once again the "People of God" terminology of the Council, underlining the importance of the *sensus fidei* of the faithful (EG 119, 198)[36], and proposing a moral epistemology that takes its cue from experience:

> Goodness always tends to spread. Every authentic experience of truth and goodness seeks by its very nature to grow within us, and any person who has experienced a profound liberation becomes more sensitive to the needs of others. As it expands, goodness takes root and develops. If we wish to lead a dignified and fulfilling life, we have to reach out to others and seek their good[37].

In EG 77, Francis cites a little-known text by John Paul II, which invites us to seek spaces "where the most profound questions and daily concerns are shared, where deeper discernment about our experiences and life itself is undertaken in the light of the Gospel, for the purpose of directing individual and social deci-

---

are to be honest with ourselves, we must allow that we face a difficult season, at least as far as theological cogency is concerned. The men trained in the coherent old theological systems of the pre–Vatican II era have passed from the scene. The Church is now led by men who came of age during the Great Disruption. This will have an effect on Church teaching, I'm afraid, and it won't be in the direction of consistency and clarity" (http://www.firstthings.com/web-exclusives/2015/07/the-weakness-of-laudato-si).

[36] Dario Vitali, "Una chiesa di popolo: il 'sensus fidei' come principio dell'evangelizzazione", in Humberto M. Yáñez, (ed.), *Evangelii gaudium: il testo ci interroga*. Roma: Gregorian & Biblical Press, 2014, pp. 53-66.

[37] Pope Francis, Apostolic Exhortation *Evangelii gaudium* on the Proclamation of the Gospel in Today's World, 24 November 2013, n. 9.

sions towards the good and beautiful". Elsewhere, Francis also includes a subtle critique of those who claim that they can deductively derive, once and for all, a complete moral doctrine, full of universal and immutable norms:

> The Church has rules or precepts which may have been quite effective in their time, but no longer have the same usefulness for directing and shaping people's lives. Saint Thomas Aquinas pointed out that the precepts which Christ and the apostles gave to the people of God "are very few". Citing Saint Augustine, he noted that the precepts subsequently enjoined by the Church should be insisted upon with moderation "so as not to burden the lives of the faithful" and make our religion a form of servitude, whereas "God's mercy has willed that we should be free"[38].

Later, in EG 132-134, Francis makes a whole reflection on dialogue with the sciences. LS, however, seeks to go further: the encyclical not only includes a new dialogue with the natural sciences, but reinterprets Cardijn's linear logic of seeing-judging-acting in the light of the most recent debates on the "hermeneutic circle"[39]. We may summarize such debates as follows: there is no purely neutral view of reality: when we observe reality, we are always situated within it, in relationship with it, immersed in an action that is constantly transforming it. Our observation is conditioned by pre-comprehensions; our measurements are theory-laden. This realization leads us to conceptualize our knowing as a helix that is strengthened with every turn, where every new understanding of the world depends on old preconceptions, yet at the same time is consistent enough to enable us to question them. This helical logic of such academic epistemological debates, while not explained in detail, is hinted at in a very important passage of the Encyclical:

> Although each chapter will have its own subject and specific approach, it will also take up and re-examine important questions previously dealt with. This is particularly the case with a number of themes which will reappear as the Encyclical unfolds. [...]. These [themes] will not be dealt with once and for all, but reframed and enriched again and again[40].

The omitted text in the above citation (the "square brackets") is also important, but it first needs to be bracketed to fully appreciate the helical logic. Let us

---

[38] Pope Francis, *Evangelii gaudium*, n. 43.

[39] Joe Holland, and Peter J. Henriot, *Social analysis: linking faith and justice*. Maryknoll, N.Y: Orbis Books, 1983.

[40] Pope Francis, *Laudato Si'*, n. 16.

get to it now: LS 16 lists 7 "supporting axes" that function as "precomprehensions" in the Gadamerian and Ricoeurian sense. Here they are, reformulated in the form of working hypotheses: (a) there is an intimate relationship between the poor and the fragility of the planet (and therefore "ecological discourse" has also a social and humanistic dimension); (b) everything in the world is intimately connected; (c) the development paradigm and the forms of power that derive from the exaltation of technology are problematic and must be criticized; (d) we need to look for new ways of understanding the economy and progress that allow us to overcome the "throwaway culture" and propose a new lifestyle; (e) each creature has its proper value; (f) in our societies, there is a need for sincere and honest debates on ecological issues; (g) international and local politics must assume its grave responsibilities regarding the above matters.

Let us therefore proceed to the second part of our reflection, so as to study in greater detail how LS uses the "semi-inductive" method typical of Vatican II-style CST, while also engaging in dialogue with the natural sciences. For the purposes of this chapter, I will analyse solely on the first chapter of the encyclical letter[41].

3. The methodology used in LS, Chapter 1: Observing reality through the eyes of the natural sciences and of economics

Here is the structure of LS chapter 1, following the subtitles chosen the by the official editors:

<table>
<tr><th>Paragraphs</th><th>Topic</th><th>Comment</th></tr>
<tr><td>17-19</td><td>[introduction]</td><td></td></tr>
<tr><td>20-26</td><td>Pollution and Climate Change</td><td rowspan="3">Observation of realities: Dialogue with the natural sciences</td></tr>
<tr><td>27-31</td><td>The Issue of Water</td></tr>
<tr><td>32-42</td><td>Loss of Biodiversity</td></tr>
<tr><td>43-47</td><td>Decline in the Quality of Human life and the Breakdown of Society</td><td rowspan="2">Observation of realities: Dialogue with the human and social sciences</td></tr>
<tr><td>48-52</td><td>Global Inequality</td></tr>
<tr><td>53-59</td><td>Weak Responses</td><td>Observation of responses (cross-disciplinary)</td></tr>
<tr><td>60-61</td><td>A Variety of Opinions</td><td>Epistemological comments, paths to action (preliminary)</td></tr>
</table>

[41] I offer an appraisal and critique of the other chapters elsewhere: René M. Micallef, "*Laudato si'* e la sua metodologia".

This structure makes it quite clear that in Chapter 1, Pope Francis embarks on his hermeneutic journey by observing, on one hand, the state of the planet through the eyes of ecology and conservation biology, and, on the other hand, the situation of the poor in the world, using economic data. The structure also brings to the fore the three preliminary hypotheses:

a) that our "common home" is suffering from environmental degradation and the effects of climate change,
b) that the most negative effects fall on the poorest and most vulnerable people, and
c) that the current model of never-ending economic and technocratic expansion, fed by consumerism, hinders, rather than helps, the search for a way out of this crisis.

Implicitly and explicitly, this chapter (and to a lesser extent some paragraphs of the later chapters) engages in dialogue with various disciplines to weigh the veracity of these hypotheses.

The end goal of this dialogue is ultimately ethical and practical, as is usually the case with this literary genre, that of a CST encyclical. As can be seen from the concluding paragraphs (60-61), Francis adopts a humble approach that is aware of the epistemological limits of engaging the results of recent empirical research. Even so, he requires of his reader that they take seriously the scientific literature which speaks of a looming ecological crisis: "On many concrete questions, the Church has no reason to offer a definitive opinion; [she understands she needs to listen to and encourage honest debate][42] among experts, while respecting divergent views. But we need only take a frank look at the facts to see that our common home is falling into serious disrepair" (LS 61).

3.1 The Debate with the Natural Sciences

Let us focus on hypothesis (a) above: LS logically links climate change to pollution and environmental degradation, since the release of greenhouse gases, from a scientific point of view, is the most pernicious form of pollution

[42] The section in square brackets is my retranslation of the official English version, which omits Pape Francis' claim that the Church "needs to listen" to that debate. This claim is extremely clear in the Italian and Spanish versions (deemed closest to the "original" final draft), as well as in the Latin: "Ecclesia [...] intellegit meditatam disputationem esse auscultandam promovendamque inter scientiae cultores". This lapsus is probably indicative of a deeply ingrained mentality that the Church has nothing to learn from the empirical sciences.

and the most serious cause of present and future environmental degradation. Even though the encyclical restricts its reflection on climate change to just four paragraphs (23-26), most of the other tell-tale signs of the malaise of our "common home" — water scarcity and loss of biodiversity — are closely linked to climate change. The reflection on climate change is particularly important given the timing of the encyclical (published during the negotiations leading to the Paris Climate Accord in 2015, known as COP21): such a reflection was, at the time, probably the most novel, politically sensitive, and ecclesially polarizing section of the Encyclical (particularly in places such as the USA or Poland). In such a context, it is especially interesting to our study, since it is a section of LS wherein Pope Francis exhibits great courage and consistency, by remaining faithful to the SJA method, listening unflinchingly to mainstream science and deeming it stable enough to be used as a basis for moral judgement. Pope Francis offers, in this chapter, a clear presentation of the theses supported by the great majority of climate scientists and biologists to support that hypothesis. Today there is practically no independent and serious voice in the active scientific community that continues to argue that there has not been a rapid and constant increase in average temperatures (on a world scale) in the last decades, that such climate change is not caused — in large part — by human activity, or that such change does not pose a threat to the future of humanity[43].

Obviously, it is normal in the field of climatology that there are small divergences on the exact estimates of the average temperatures we will have in the future, and on which tools, methodologies and statistical models are the most reliable. It is also true that there are also some "natural" factors that can cause important climatic changes in the history of the planet, such as volcanic activity, changes in the Earth's orbit and axis, the solar cycle, and the periodic concentration of sunspots. It is a hallmark of intellectual honesty to *mention* these

[43] Sometimes the figure of 3% of "scientific sceptics" is cited, compared to 97% of scientists who would support the international consensus on climate change. This figure, however (taken from an article published in 2013 in the journal *Environmental Research Letters* by Cook et al.), was obtained using a very questionable method, which puts on the side of sceptics those articles on climate issues that do not explicitly say they are in favour of the consensus, even when the article is not directly concerned with the corroboration of such a consensus, or takes it for granted, or is written by an author well known for his position in favour of consensus. According to the latest studies, less than 0.1 of peer-reviewed natural-science articles sceptical of the consensus, meaning that more than 99.9% are explicitly in favour of the consensus or take it for granted. Cfr. James Lawrence Powell, "The Consensus on Anthropogenic Global Warming", in *Skeptical Inquirer* 39 (2015), p. 42.

technical and minor divergences from the main theory (and LS does mention them, in a parenthesis in n.23), but it would be equally dishonest to *dwell* on them in the context of an encyclical, since practically all experts agree they do not contribute significantly to the current crisis[44]. Doing so would seem as psychologically and ethically unhealthy as discussing the health risks of exposure to obscure rare metal carcinogens with a heavy smoker who has just been diagnosed with lung cancer and is itching to get out his cigarette box as you speak. Offering someone a seemingly credible pathway to denial and avoidance of necessary but painful change may be "emotionally intelligent", but "morally blameworthy".

When writing this chapter, Pope Francis and his team of experts were keenly aware of the economic forces behind climate change scepticism and denial, and their use of pseudo-scientific think-tanks and dirty money to convince the general public that scientific claims regarding human-induced climate change are not "certain" and that politicians should keep on stalling and avoiding the hard but necessary decisions[45]. Yet, in this chapter, instead of delving deeply into the complex scientific arguments to unmask climate sceptics and their political manipulation games, a strategy which might confuse or antagonize the readers who require the most convincing, LS sticks to familiar CST territory and moves on to the critique of the short-sighted and exploitative economic models that the financial backers of eco-sceptics are so desperate to preserve, no matter how many people may have to die or face misery due to climate-induced disasters or environmental degradation.

Even so, what could be useful in this context, but is missing in LS, is a common-sense argument to the effect that most climate scepticism makes little sense when analysed with a clear mind. Let us, just for the sake of argument, consider credible, relevant or still valid some of those debatable publications which call into question the fact that the cause of rising global temperatures is

---

[44] Cfr. International Astronomical Union, "Corrected Sunspot History Suggests Climate Change Since the Industrial Revolution Not Due to Natural Solar Trends" [accessed: 7.2.2016], http://www.astronomy.com/news/2015/08/corrected-sunspot-history-suggests-climate-change-since-the-industrial-revolution-not-due-to-natural-solar-trends; Chelsea Harvey, "No, Earth is not heading toward a 'mini ice age'", in *The Washington Post* (07/14/2015); Terry Gerlach, "Volcanic Versus Anthropogenic Carbon Dioxide", in *Eos* 92 (2011), pp. 201-202.

[45] Cfr. René M. Micallef, "Laudato si' si basa su una vera scienza? Trasmettere un messaggio etico nell'era della manipolazione e della polarizzazione commerciale dell'opinione pubblica", in Humberto M. Yáñez, (ed.), *Laudato si'. Linee di lettura interdisciplinari per la cura della casa comune*. Rome: Gregorian and Biblical Press, 2017, pp. 141-170.

mainly anthropogenic. Even if we were to accept that climate extremes were being caused, in part, *also* by sunspot concentrations or volcano eruptions or the like, we still would have to admit that such claims boil down to yet another admission that we *have a serious problem with climate change*. And if we have a problem, we humans need to deal with it, no matter whether we humans or the sunspots are the bigger "culprit". The only lever we *humans* have to limit global warming is, by definition, the attenuation of the *anthropogenic* causes (mainly, the production of greenhouse gases). Certainly, if we do not want low-lying coastal regions and islands (and the livelihoods of millions of people who live there) to be engulfed by rising sea levels, this is what we can and have to act upon, rather than expect some demigod to appear and start plugging volcanoes or dissipating sunspots. The lack of an argument of the sort is regrettable, since it would seem that the authors are avoiding here the more demanding form of cross-disciplinary recommended in VG 4.

Our *first conclusion* is that Chapter 1 of LS takes the current scientific consensus for granted, without bothering to really engage and wrestle with the sceptics. Some right-wing climate-change deniers might find the chapter's scientific pedigree mediocre at best, but for all other readers — people with enough common sense who do not allow themselves to be manipulated by the "merchants of doubt" — the natural-science part of the Pope's text is obviously well founded, not so much on the basis of the chapter's internal arguments, but rather on the basis of dependable extrinsic scientific knowledge which is no longer truly "debatable" and which the chapter subtly "hints at" while avoiding any clear citations of the "real" sources (see section 3.3 below).

3.2 The Debate with the Social Sciences, and with Economics in Particular

Let us now move to the economic side of the chapter; here Pope Francis needs to argue for the hypotheses (b) and (c) above, concerning the harmful effects of degradation and climate-change on the most vulnerable persons, and the issues with the underlying economic model. Francis' claims are not new; they fit within a longstanding debate in certain economics circles dating back to the 1960s.

John Sniegocki distinguishes four major trends among economists concerned by climate change and environmental degradation[46]. (a) *Market Lib-*

[46] John Sniegocki, "The Political Economy of Sustainability", in Christiana Z. Peppard and Andrea Vicini, (ed.), *Just sustainability: technology, ecology, and resource extraction*. New York: Orbis Books, Maryknoll, 2015, pp. 57-68.

erals believe that only free markets (with some incentives but little or no regulation) are able to guarantee the flexibility and creativity needed to find the necessary human and technological solutions to overcome the ecological crisis. (b) *Institutionalists* are also optimistic that the solution will come from market economies, but only if international markets are oriented towards the good of humanity with serious and cleverly designed out regulation. (c) *Radical Political Economists* believe that ecological sustainability is not compatible with capitalism (at least in its current incarnations), since this economic model depends on continuous economic and consumptive growth, and hence it is the very notion of a "market economy" that needs to be questioned. Finally, (d) *Grassroots Critics of Development and Neoliberal Globalization* go even further in their critiques of economic globalization and consumerism, claiming that the standard of living in Western countries is morally unacceptable or that it is neither achievable by all nor desirable for all. This latter position is often associated with the "degrowth (décroissance)" movement, which believes that global consumption and production may need to shrink, together with the GDP of certain regions and countries, so as to allow others to develop without causing further harm to the environment and to future generations.

Traditionally, CST has been seen to be closest to the position of the "institutionalists", promoting regulation of domestic and international markets, and seeking multilateral international solutions to the world's most pressing economic problems. In some passages of the encyclical, however, Pope Francis seems closer to the latter positions, those of the "radical political economists" and of the "grassroots critics". For example, in n. 46, he claims that "these are signs that the growth of the past two centuries has not always led to an integral development and an improvement in the quality of life"; in n. 54 he laments the weakness of international political responses as if suggesting that we should be more radical and proactive, since "the alliance between the economy and technology ends up sidelining anything unrelated to its immediate interests" we cannot expect much from summits and international institutions. In n. 59, he speaks of "bold decisions" and takes issue with those seeking "a licence to [carry] on with our present lifestyles and models of production and consumption". The encyclical returns to the issue many times in the other chapters[47]; an important text can be found in n.193: "the time has come to accept decreased growth in some parts of the world, in order to pro-

[47] Other interesting passages can be found in LS, nn. 106, 109, 141, 172, 178, 193*,

vide resources for other places to experience healthy growth", followed by a citation of Benedict XVI that points in the same direction.

From a methodological point of view, this approach to dialogue with economics poses two main problems. Firstly, one can understand that Pope Francis is mostly intent on dismissing (a), but at the same time does not want to choose between (b), (c) or (d) and so adopt one option as the single "official economics perspective" for CST or for his papacy. However, the fact of mixing and interweaving elements of the three in the text and not facing the claims of their proponents and the critiques of their opponents systematically gives the impression of a confused or opportunistic approach to interdisciplinary dialogue with economics.

Secondly, by adopting some of the claims of the "degrowth" movement without pausing to argue solidly in favour of this choice and allay the fears of those who see this position as overly radical and potentially catastrophic, he risks marginalising many Catholic economists in the English-speaking world who reject this view as a "fringe" position. (This view has become more "fringe" even in Continental Europe, in the years since LS was published.) From an epistemological point of view, this may allow certain right-wing critics to claim that the Pope's (or CST's) acceptance and espousal of data and analysis coming from non-ecclesiastical sciences is not based on an a balanced, humble and attentive listening to the consensus among scholars in those sciences (and hence to what is "mainstream" rather than "fringe"), but rather on some irrational ideological conviction of the Pope and his editorial team, or simply on a cherry-picking of "scientific" claims in view of what is most instrumentally useful to reach the moral conclusions that they decided *a priori* that they wanted to reach. But if it were so, another pope, in the first part of this chapter, might have just as well bracketed the consensus in climate science and conservation biology and espoused a "fringe" position, such as that of the eco-sceptics. In other words, does this approach to economics imply that CST and "integral ecology" are constantly tempted to dialogue with "mainstream" science only when this is convenient, and that their interdisciplinarity and transdisciplinarity are not to be taken very seriously?

I do not believe we should reach this conclusion, for two reasons. Firstly, the notion of a "scientific consensus" in economics is something quite different from that in natural sciences, and it is actually the "scientific consensus" in the natural sciences that has made us every more aware, in recent years, that the basic tenets and methodology of "mainstream" economics need to be radically questioned and challenged by serious attempts at falsification[48]. Today, we might in fact be facing a moment of paradigm shift for economics (and the recent COVID-19 pandemic has made that more evident). In recent years, many publications have argued, quite convincingly, that the models on which many economists base their beliefs tend to be quite dependent on the political ideology of the researcher[49]. Furthermore, most of the economic research that is published in leading journals focuses on the markets of rich countries and tends to project US market trends onto the rest of the world, without seriously studying developing country markets or taking their "anomalies" seriously (indeed, such "anomalies" are often not true anomalies, but indicators of parochialism and the weakness of the economic discourse that presents itself as "global")[50].

Secondly, the tendency to engage positions (b), (c) and (d), albeit in a seemingly confused and opportunistic manner, comes not so much from a "Latin American fascination with Marxism", as some seem to think given the geographic origins of Bergoglio, but from a serious interdisciplinary dialogue with the claims of renowned economists such as Jeffrey Sachs, Nicholas Stern, and the economists on the IPCC[51], leading to a growing belief within Catholic institutions that the true economics "mainstream" is to be found in a combination of these positions, rather than in Chicago-school economics. A movement of older and younger economics scholars known as the "Economy of Francesco"[52], is taking shape to rigorously flesh out the intuitions in recent

---

[48] Kate Raworth, *Doughnut Economics: Seven Ways to Think Like a 21st-Century Economist*. Vermont: Chelsea Green Publishing, White River Junction, March 29, 2018.

[49] An interesting example of the current tensions and politicization of economics is the intractable debate on whether or not immigrant workers compete directly with poor autochthonous workers, and therefore whether migration "make poor citizens poorer". Stephen Castles and Mark J. Miller, "Additional Case Studies: 10.2 The Borjas versus Card Debate: Is Labour Immigration Bad for U.S. Workers?" [accessed: 2.10.2012], http://www.age-of-migration.com/na/resources/casestudies.html; Giovanni Peri and Vasil Yasenov, Report: "The Labor Market Effects of a Refugee Wave: Applying the Synthetic Control Method to the Mariel Boatlift". Cambridge, MA; National Bureau of Economic Research, 2015, available at http://www.nber.org/papers/w21801.pdf.

[50] The Economist, "The useful science? Economics is highly parochial" [accessed: 1.11.2015], http://www.economist.com/news/finance-and-economics/21592652-economics-highly-parochial-useful-science; Diego Alonso Lasheras, "Evangelizzazione ed economia: denuncia e proposta", in Humberto M. Yáñez (ed.), *Evangelii gaudium: il testo ci interroga. Chiavi di lettura, testimonianze e prospettive*. Roma: Gregorian and Biblical Press, 2014, pp. 221-234

[51] Paolo Conversi, Fernando De la Iglesia Viguiristi and René M. Micallef, "Quello che sta accadendo alla nostra casa", in Emilia Palladino (ed.), *Approfondimenti e riflessioni sulla Laudato si'*, Gregorian and Biblical Press, Roma February 3, 2017, pp. 11-29.

[52] https://francescoeconomy.org/ [accessible on: 02/06/2021]

CST (including *Fratelli Tutti*). This brings us back to the reflection in the first part of our text, where we insisted that true cross-disiplinarity is not a one-way street, and theological ethics should not simply listen to the empirical disciplines but also challenge them and help them become more aware and critical of their arbitrary or ideological anthropological assumptions (while respecting their just autonomy and competence).

Our *second conclusion* is that while CST should avoid engaging its scientific interlocutors in ways which may seem like "cherry picking", what is considered as consensus or "mainstream" science is not as clear in all disciplines. Intelligent engagement with more "radical" positions can in some cases be warranted and fruitful, as long as it remains rigorous and intellectually honest. In some cases, critiques from CST and other ethics voices may provide a stimulus for certain disciplines to rethink certain assumptions and develop, and such a "two-way" cross-disciplinarity is the most desirable, if done in a manner respectful of each discipline. An important criterion through which CST often challenges the anthropological assumptions which shape experimental design and data analysis in many empirical sciences, and hence also their conclusions, is that of seeing economic (and other) realities through the eyes of the poor (cf. LS 49): each scientist from every discipline is also invited to do so, but it is obvious that they must bring with them the statistical tools and rigorous methods of their disciplines.

3.3 Recognizing the Real Sources and Interlocutors in a Debate

What we have said till now indicates that the first chapter of LS strives to be faithful to the interdisciplinary and dialogic method we have discussed above. However, there are two critical issues that I would like to raise which indicate, in my opinion, that there is still work to be done if we truly believe that such methodology should be adopted wholeheartedly in a text like this.

The first issue, hinted above, concerns the scientific sources and the interlocutors in the dialogue: often, these are present but carefully hidden from the reader. In this chapter, the drafters choose not to provide any data, statistics or scientific citations to support their claims. This is probably due to a classical (and "classicist") strategy in official Catholic texts, which like to appear independent of any particular academic school, and to make truth claims that are seemingly not reliant, in their grounding assumptions, on any natural-science or social-science data coming from secular and falsifiable (inductive) sources.

Firstly, there is a concern regarding accusations of impartiality if one school is cited but not its competing school of thought. Secondly, there is a fear that inductively-grounded non-ecclesiastical authors and publications, being inherently prone to falsification and the possibility of being discredited in the future are shaky ground to build on. It is precisely this falsifiability that makes them "scientific" (according to the Popper and many other philosophers of science), yet in the religious world there is a lingering fear of any modification of past claims, since anything but complete "inerrancy", even on practical and mutable matters, might cause scandal among the undiscerning faithful. Finally, the Church Magisterium, which claims to have the assistance of the Holy Spirit in making truth claims on matters of faith and morals, does not like to appear dependent on and subject to secular authorities in particular scientific fields while arguing claims that finally concern "matters of morals".

Thus, LS's manner of citing sources and dialoguing with the natural and social sciences continues to replicate old editorial practices which struggle with historical consciousness and with the idea that most ethical debates depend on knowledge from disciplines which the Church cannot control but needs to listen to and respect. Hence, we continue to witness these semantic games whereby the drafters consult the other sciences and then conceal the source, as though humbly admitting any dependence of the Magisterium's claims on knowledge provided by the Church's interdisciplinary partners were a shameful thing. The text, however, does not simply make a list of claims without referencing anything. There is a clever sleight of hand at play in the footnotes: LS often cites texts from the Episcopal Conferences on matters of natural and social sciences. Of themselves, the ecclesial documents cited by LS cannot perform the epistemological task of corroborating the claims coming from non-ecclesiastical sciences, given that pastoral texts written by bishops and their commissions are not intended or able to provide the social- or natural-scientific warrant needed to support such claims. (Catholic bishops receive no particular assistance from the Holy Spirit, nor have any particular competence, *qua* bishops, in matters pertaining to the natural and social sciences.) Yet, when we go and look at the ecclesiastical sources cited by LS, we often find that in some of those sources, the bishops are more forthcoming with their own sources and that they actually ground their claims in scientific sources (popular or academic) which they cite in their footnotes.

For instance, LS's footnote 26 cites a 2012 document of the Bolivian Bishops' Conference, *El universo, don de Dios para la vida*, which in turn cites scientific sources (e.g. the IPCC reports in paragraph 24, footnote 25[53]). LS's footnote 27 cites a document produced by two commissions of the German Bishops' Conference which likewise cites many reputable natural science and economics sources in its footnotes, including the IPCC reports and the Stern Review[54]. Hence, it is obvious that the ultimate "authorities" on which the claims of this chapter are based are not religious texts or church bodies, but rather the reports of the IPCC (Intergovernmental Panel on Climate Change) and the economic reflections of economists such as Nicholas Stern and Jeffrey Sachs.

This brings us to our *third conclusion*. I believe it makes little sense, at this point in history, for CST not to directly quote its natural-science and social-science sources, depriving the sceptical reader with an epistemologically-suitable source to verify what is being said. The excessive fear that some scientific studies might be questioned or falsified in the future reveals itself here as counterproductive and detrimental to the robustness of the argumentation and methodology adopted, in the light of the discourse made earlier on the nature of moral knowledge and its dialogue with the empirical sciences.

3.4 Engaging Detractors, Critical Voices, and Sensitive Issues

The second critique I would like to raise concerns the tendency, ubiquitous in LS, to "play it safe", and avoid engaging detractors, critical voices and sensitive issues. We have already mentioned the issue of the anthropogenic causes of climate change where the text treads too "carefully"; from the argumentative point of view, the text seems reluctant to enter into dialogue with any serious contrary positions, and this does not help to convince the sceptical reader. I am not sure of the effectiveness of such a strategy, which remains hostage to the fear of antagonizing or alienating "right-wing" readers, those most in need of taking this text seriously. Furthermore, it probably backfires in the case of moderate sceptics who might be aided by the encyclical to debunk myths and unmask fake news, and to eventually accept the claims of mainstream science, since one might argue that not even the Pope really takes the claims of the IPCC so seriously as to dare cite that "authority" among his sources. The existence of alternative claims by scientists who question that the main cause of global

---

[53] Carta-Pastoral-sobre-Medio-Ambiente-Conf-Episcopal-Boliviana.pdf (fmclimaticas.org.br) [accessible on: 02/04/2021]

[54] Climate Change: A Focal Point of Global, Intergenerational and Ecological Justice (ordosocialis.de) [accessible on: 02/04/2021]

warming is human activity are simply acknowledged in a parenthesis (LS 23), instead of being faced and discarded with a convincing argument.

There is however a more serious issue, in my opinion. The themes dealt with in numbers 20-41 (pollution, climate change, water resources, biodiversity, etc.) remind us of a classic manual of conservation biology (including references to juridical and socio-economic aspects as regards rights to use and management of resources)[55]. Discourse on conservation is concrete, tangible and focuses on a particular area or species with which people can identify; it can be the basis of campaigns which elicit strong emotional responses, and of films depicting the battles fought by conservation heroes. It is a good first step in creating an ecological consciousness, but often remains too limited in scope (concerned only with "this scenic lake" or "that cute mammal"). The second step, which LS seems hesitant to take, entails engaging "hard" ecology, i.e., the scientific discipline that studies the growth, survival and decline of the populations of living beings[56]. Scientific ecologists are wary of the abuse and misuse of the word "ecology" by people who like the idea of conservation but are unwilling to think in universal and abstract terms, deal with the harsh realities of survival in a world of competition and limited resources, do the math, and face the realities of population growth and collapse at the level of biomes and the biosphere. As long as we avoid debates on populations, including human populations, we are avoiding "ecology" in the scientific sense. To be sure, the issue of "loss of biodiversity" boils down to certain populations declining and others (fewer in number or diversity) replacing them. Yet, if "everything is connected", we need to connect deforestation in the Amazon and elsewhere also to increasing livestock populations, cattle belching to the release of methane (a very powerful greenhouse gas) and to climate change, and all this to meat production and consumption by an increasing human population. This does not mean that we all need to become vegetarians to "do the right thing", but rather that we should remember that meat ought be considered as a luxury and eaten in moderation, as Catholic Lenten traditions have always reminded us[57]. It is indeed surprising that LS avoids this issue.

---

[55] For example: Navjot S. Sodhi and Paul R. Ehrlich, (ed.), *Conservation Biology for All*. Oxford/New York: Oxford University Press, March 12, 2010; Fred Van Dyke, *Conservation Biology: Foundations, Concepts, Applications*. New York & Philadelphia: Springer, 2008.

[56] For example: Nicholas J. Gotelli, *A Primer of Ecology*. Sunderland: Sinauer Associates, Mass 2008[4th ed]; Robert E. Ricklefs, *The economy of nature*. New York: W. H. Freeman, 2008.

[57] An intelligent and balanced analysis of the issue is provided by authors such as Simon Fairlie, *Meat: A Benign Extravagance*, Permanent Publications, East Meon: White River Junction, VT, 2013.

Talking frankly about human population surges and the effects of overpopulation is not in itself "Malthusianism" and does not immediately imply the promotion of abortion or contraception. Many Popes have spoken about the ills of overpopulation in the past[58] and GS 50-51 talks about "responsible transmission of life" and "circumstances where at least temporarily the size of [...] families should not be increased". It is not the task of natural sciences to decide how great or small the human population on planet Earth should be, but once we as ethicists sketch what we mean by "dignified living conditions for humans" (conditions which can ensure the respect of their basic human rights and the possibility of their flourishing as humans during their earthly life), it is, yes, the task of natural scientists to estimate the volumes of resource consumption and waste generation that correspond to those living conditions for a given human population, and the impact on other creatures of the externalities. And theological ethics, as well as CST, should carefully listen to those numbers and reflect on them seriously. Yet, LS avoids doing that; the fear of population modelling and any hint of a concession to Neo-Malthusianism is so great that, for example, in n. 50, when the theme of the demographic explosion of the *Homo sapiens* species in recent decades is touched upon, the discourse becomes apologetic and defensive.

Indeed, how can we deal with the delicate issue of human population growth in a world of limited resources, resources that we cannot continue to exploit *ad infinitum* (LS 106)? I believe there are many intelligent ways to face the issue instead of skirting it while remain faithful to the Catholic tradition. Before projecting the polarizing debates surrounding *Casti connubii* (1930) and *Humanae vitae* (1968) onto the issue, it is good to remember that by the late 1700s and well before the invention of modern means of contraception, the population of France, an overwhelmingly Catholic country and one of the most densely populated in the world, had realized its land resources were limited, and started to effectively control its demographic expansion by limiting family size, mainly by delaying or reducing nuptiality[59]. Today, many agree

[58] E.g. Pius XII, *Exsul familia* ("invitabamus Nationes, quae territorii vastitate praestant et pari incolarum numero sunt expertes, ut homines vellent excipere, qui in frequentissimis commorarentur regionibus"); John XXIII, *Mater et Magistra*, nn. 153-154; 185-192; Paul VI *Populorum Progressio,* n. 37; Cfr. *Gaudium et spes,* n. 42.

[59] Edward Anthony Wrigley, "The Fall of Marital Fertility in Nineteenth-Century France: Exemplar or Exception? (Part I)", in *European Journal of Population / Revue Européenne de Démographie* 1 (1985), pp. 31-60.

that accelerating and increasing the access of girls and women to education is one of the most effective ways of reaching and reducing the human population peak in the next few decades; supporting education is of course not morally problematic for Catholics[60].

Many experts estimate that the world population peak will be reached in 2100[61], or maybe even earlier, possibly in 2040. At this peak, the world population is estimated to reach around 10 billion, compared with the current population of around 7 billion. The drafters of LS could have presented the issue and argued that current agriculture and aquaculture technologies and resources are already capable of producing food for 10 billion people; a population of 10 billion could be sustainable if we deal effectively with the problem of "food waste", especially by increasing the effectiveness of distribution networks (that is, of markets and transportation systems), and control the amount of cereals used in the production of meat[62], while seeking a gradual and smart transition to less polluting, destructive and water-intense forms of farming and fishing, which does not necessarily entail a decrease in the overall volumes of food produced.

Such a response would have been criticized as apologetic and overly optimistic, maybe, but it would have had the merit of not avoiding the issue. I believe that in the future "integral ecology" should go further and adopt an even more candid approach. For instance, it could recognize that, Catholic moral theology concerning "integral ecology" is wider that what can be found in official documents and that, in the current CST, that there are certain constraints on official Catholic discourse on the good of the planet given the nature, the audience and the history of such discourse. At present, such discourse, it seems, needs to submit to three previous considerations: (a) it must not openly promote the reduction of human population growth (in order not to give the impression of favouring contraception or abortion), (b) it must not question the principle that every human person has the right to be fed, clothed and have a roof under which to take refuge (which is implied in Catholic claims on equal human dignity of persons), and (c) it must not deny the right of every human person to work (cf. LS 124-129). Thesis (b) implies

---

[60] Samir KC and Wolfgang Lutz, "Demographic scenarios by age, sex and education corresponding to the SSP narratives", in *Popul Environ* 35 (2014), pp. 243-260.

[61] United Nations, *2015 Revision of World Population Prospects* (http://esa.un.org/wpp/Documentation/publications.htm, accessible on; 05/03/2016).

[62] Simon Fairlie, *Meat*.

that, even though "organic" agriculture may be very desirable,[63] a massive and rapid replacement of industrial agriculture with novel and eco-friendly forms of agriculture is not ethically acceptable until we can guarantee, at the very least, that in this new system of agricultural production, the poor will have no less food security than they have today; the change would have to be gradual and ensure people do not go hungry. Thesis (c) implies that even though, in theory, jobs in the service sector (especially those that do not promote environmentally harmful consumerism) might be "ecologically" preferable to jobs in manufacturing (especially those that produce goods designed according to the logic of planned obsolescence), we need to acknowledge the fact that the service sector typically requires skilled, educated workers, and the provision of quality education is a daunting task that takes time to bear fruit. This implies that it is often not possible to quickly create hundreds of millions of jobs in this sector so to absorb the mass of people who can no longer find a job in agriculture, fisheries or aquaculture in emerging countries. In the meanwhile, then, to guarantee a right to employment to these people, we may have to create "unnecessary" jobs for them in the industrial sector, and that implies tolerating a certain amount of consumerism and waste.

This kind of discourse is surely not palatable to those who would like a rapid ecological transition, and desire consumerism and its harmful social and environmental effects to end immediately; it is also the kind of discourse that is conveniently evoked, again and again, by many vested interests who wish to block or delay the transition. Nonetheless, at some point, we have to face the complex and dauting decisions that "integral ecology" entails and reflect on what jobs and what products and what behaviours should be phased out, and what others phased in, and at what point in the transition. All this requires discernment, a term which Pope Francis likes to use, but which irks people who prefer an "all-or-nothing" approach that pretends that complexity does not exist.

Therefore, the critical reader will immediately understand that, despite all the passion that Pope Francis has for the environment and the wellbeing of the biosphere — which will lead him later on in the text to make harsh criticisms of industrial agriculture (nn. 20, 131-134) and consumerism (nn. 34, 184, 203, 209-210, 215, 219) — he is not actually proposing simplistic or "ideological"

[63] Cfr. LS 129; Mark Graham, "The Unsavory Gamble of Industrial Agriculture", in Christiana Z. Peppard and Andrea Vicini, (ed.), *Just sustainability: technology, ecology, and resource extraction.* New York: Orbis Books, Maryknoll, 2015, pp. 105-116.

solutions, but rather, in the light of the CST tradition, that we discern the way forward. His speech is bounded by a series of ethical commitments made previously by the CST tradition (and which Francis accepts unreservedly) which imply that certain "evils" (such as industrial agriculture and consumerism) must be largely tolerated until real alternatives are found, for fear of unleashing even greater evils. This level of candidness probably is not to be expected from a papal text; it is rather the task of academic theological ethics to reflect on the implications of the various concrete options available to human action and develop further the field of "integral ecology".

This leads us to our *fourth conclusion*. The parenetic style of a social encyclical, laden with moral idealism that seeks to prod the reader to reflect and to act, implies that it can only do so much to "engage detractors, critical voices and sensitive issues" when facing a world of limited resources where not everything is "win-win" and certain goals can only be achieved by sacrificing others. This in turn means that such a text can inspire us, but cannot provide many concrete solutions. It is indeed natural for the proposals regarding action in the final chapters of a CST document built on the See-Judge-Act paradigm to remain sketchy and vague, and to avoid proposing simplistic solutions and one-size-fits-all recipes to deal with a moral problem such as the ecological crisis. Yet, what is important is that the "sketchiness" of the solutions phase does not undermine the seriousness, completeness, and honesty of the observation phase. The concern that we may find it hard to offer a complete ethical appraisal and concrete directives for action which correspond to a certain observation (such as that concerning the ecological implications of human demographic growth) should not result in a decision to avoid "seeing" this matter and to cherry pick the data and the debates we would like to have with the natural and social sciences. Furthermore, it is the task of "integral ecology" to flesh out some of the more concrete solutions in texts which are less "official" and more adequate for this purpose.

4. Some Lessons for the Future of Integral Ecology

In this very brief third section, I would like to draw some conclusions from the above debate to propose a set of foundations of "integral ecology" which is the purpose set by all the papers in this book.

On the basis of our first (3.1) and third (3.3) conclusion, above, I propose that integral ecology insofar as it seeks to be a serious approach to understanding reality, as well as a movement within the Church and within society, should build on universal ethics claims and official CST documents by (a) distinguishing clearly between mainstream and more radical or "fringe" positions among its natural- and social-science interlocutors, (b) being explicit in citing the sources, and (c) making use of mainstream positions intelligently and critically.

Furthermore, on the basis of our second (3.2) and fourth (3.4) conclusion, I propose that integral ecology should (d) have the courage to systematically engage the critics and sceptics of the positions its considers "mainstream" or it choses as main dialogue partners, (e) justify its use of dialogue partners when they are not considered "mainstream" in some countries or academic circles, (f) dare to respectfully challenge the acritical anthropological assumptions of certain disciplines when it deems them problematic, (g) be forthcoming as regards the logical implications of certain options it tends to favour, if they tend to be mutually exclusive in a world of limited resources and possibilities, and (h) dare to propose concrete solutions while fully aware of the above constraints.

5. Conclusion

Our journey through the history of CST's search for a more historically-conscious and cross-disciplinary methodology and through the pages of the first chapter of LS has allowed us to see, critically, how the encyclical strives to dialogue with the natural and human sciences, trying to listen carefully to human reality. LS manages quite well to apply the new methodological paradigm, addressing human persons in their current historical condition, promoting an ethics that humbly faces a particular theme, constructing an argument on the basis of perfectible and fallible empirical knowledge, remaining close to the experience of the faithful and the "people of good will" who are ready to allow their hearts and minds to be moved to ecological conversion. For this reason, it provides a useful methodological paradigm for "integral ecology".

We have also noted some shortcomings: LS does not always pause to recognize the complexity of the problems, and sometimes it avoids the confrontation with positions that evoke uncomfortable themes, or issues which may require the updating of prior CST claims to be fully addressed in the "judge" and "act" phase of the process. While this is understandable given the literary genre, context, and practical purpose of this particular text, the critical reader may feel short-changed in the "observe" phase of the process. This means that there is still room for improvement, both in official texts and in the reflections of Catholic scientists, ethicists and theologians, published by individuals or academic bodies: we have made some suggestions to orient the future of such "integral ecology" reflections.

I do not believe that these shortcomings and possible areas for improvement overshadow the quality and authenticity of the dialogue and interdisciplinarity present in the text, given the historical background we have sketched. It is unrealistic to hope that Pope Francis, who perceives his ministry above all as that of a pastor rather than an academic theologian, will resolve all the methodological issues that have plagued CST by signing a single encyclical on the specific theme of ecology. The enduring success of "integral ecology", I believe, lies in the progressive rediscovery and deepening of the appropriate, semi-inductive and interdisciplinary methodology that LS amply showcases, which, we hope, will continue to inform CST texts in the future, as well as the pastoral letters issued by episcopal conferences, and mainstream academic texts in theological ethics.

WHAT KIND OF CHURCH FOR AN INTEGRAL ECOLOGY?

A Synodal Approach to Care for our Common Home

Giacomo Costa, SJ

Synodality and *integral ecology* are two key words in the magisterium of Pope Francis. The latter is more common in social sciences and in the public arena, but it is often perceived as foreign to the Church's ordinary discourse and concerns. On the contrary, *synodality* has recently become a very fashionable word within the Chuch, especially since the Synod 2021-2023 was launched in October 2021, under the heading "For a Synodal Church: Communion, Participation, and Mission". In any case, they are technical terms belonging to semantic and disciplinary universes that are normally considered distant and mutually unrelated; this runs the risk of overshadowing a rich and complementary understanding of them, and especially of disconnecting us from the experience of faith in which they are rooted.

This contribution intends to show how they are articulations of the same basic vision: the notions of "journeying together" and "caring for the common home" complement and illuminate each other, and both draw inspiration from the Magna Carta of Pope Francis' teaching, the Apostolic Exhortation *Evangelii gaudium*[1]. Both guide the discernment of the Church of the 21st century in carrying out its task of proclaiming the crucified and risen Christ; for both, the joy of the Gospel is the main point of reference. Finally, both refer to the need for a conversion, both ecological and synodal, that deeply touches the Church, its hierarchical structures, and its approach to fulfilling its diakonia in service to the world.

[1] Pope Francis, Apostolic Exhortation *Evangelii gaudium* on the Proclamation of the Gospel in Today's World, 24 November 2013; henceforth EG.

DOI 10.32060/C.S.9.477-498

1. At the root of the synodal approach[2]: "journeying together"

The very identity of the Church is at stake in the synodal journey, as it concerns its life and mission. As the Preparatory Document of the Synod 2021-2023 recalls, "Our 'journeying together' is, in fact, what most effectively enacts and manifests the nature of the Church as the pilgrim and missionary People of God"[3]. This is why Pope Francis could state: "It is precisely this path of synodality which God expects of the Church of the third millennium"[4]. This synodal journey is set in a specific historical context, which needs to be considered in its specific traits.

1.1 An era of fragmentation

The world we live in offers us mixed signals. The ancient and ever new dream of fraternity and social friendship continues to fascinate our contemporaries. We realise this as we observe the growing awareness of the need to care for the common home as the ecological crisis deepens. Even a global tragedy such as the COVID-19 pandemic, "has indeed for a time raised awareness of being a world community sailing in the same boat, where the evil of one is to the detriment of all. We remembered that no one is saved alone, that we can only be saved together"[5]. At the same time, it has exploded the inequalities and inequities that divide mankind, manifesting their brutality, as it is evident in the case of access to vaccines.

On the one hand, the globalization process has transformed the world into one worldwide market, reducing human beings to cogs in the system; on the other, populism and fundamentalism are radicalising the defence of identities, whether cultural, social or religious, by building walls and turning the other into an enemy. The outbreak of war in Ukraine has confirmed this once more. But for decades the tragic plight of migrants all over the world has been a living testimony to how high and strong the barriers that divide the one human

---

[2] The expression "synodal approach" is our way of translating the Italian "*stile sinodale*", literally synodal style, synodal way of proceeding.

[3] Synod of Bishops, *Synod 2021-2023. For a Synodal Church: Communion, Participation, and Mission. Preparatory Document*, n. 1; henceforth PD.

[4] Pope Francis, *Address for the ceremony commemorating the 50th anniversary of the institution of the Synod of Bishops*, 17 October 2015.

[5] Pope Francis, Encyclical Letter *Fratelli tutti on Fraternity and Social Friendship*, 3 October 2020, n. 3; henceforth FT.

family still are. Even the new digital media, which can be considered the most characteristic trait of our time, are marked by a strong ambiguity. On the one hand, the new communication technologies are an extraordinary opportunity for dialogue, encounter, access to information and knowledge, socio-political participation and active citizenship; on the other hand, they are also a place of exclusion (starting from the gaps in accessibility), violence and manipulation (*dark web*, *fake news*, *hate speech*).

The aspiration to fraternity runs throughout human history, as accounts of the origins of the main religious traditions testify, but reality belies it: "In today's world, the sense of belonging to a single human family is fading, and the dream of working together for justice and peace seems an outdated utopia" (FT, 30). From this contrast arises an analysis that recognizes the paradox of our age: as globalization increases, so does fragmentation and isolation: "For all our hyper-connectivity, we witnessed a fragmentation that made it more difficult to resolve problems that affect us all" (FT, 7). It is a dynamic that cuts across all dimensions of social life:

> Local conflicts and disregard for the common good are exploited by the global economy in order to impose a single cultural model. This culture unifies the world, but divides persons and nations, for "as society becomes ever more globalized, it makes us neighbours, but does not make us brothers". We are more alone than ever in an increasingly massified world that promotes individual interests and weakens the communitarian dimension of life. Indeed, there are markets where individuals become mere consumers or bystanders. As a rule, the advance of this kind of globalism strengthens the identity of the more powerful, who can protect themselves, but it tends to diminish the identity of the weaker and poorer regions, making them more vulnerable and dependent. In this way, political life becomes increasingly fragile in the face of transnational economic powers that operate with the principle of "divide and conquer"[6].

Defensive reactions result in a new explosion of claims and counterclaims: "Ancient conflicts thought long buried are breaking out anew, while instances of a myopic, extremist, resentful and aggressive nationalism are on the rise. In some countries, a concept of popular and national unity influenced by various ideologies is creating new forms of selfishness and a loss of the social sense under the guise of defending national interests" (FT, 11). Both dynamics — that of a faceless globalism, and that of a desperate particularism — breathe

[6] Pope Francis, *Fratelli tutti*, n. 12.

new life into the "throw-away" culture, because they lead to some human beings being considered as second-class, who "can be readily sacrificed for the sake of others considered worthy of a carefree existence" (FT, 18).

Pope Francis' analysis deepens and has anthropological implications. The mentality underlying the dynamic just identified is a symptom of a radically individualistic conception of social bonds, understood as no more than a tool for the pursuit of individual interests, not as a shared commitment to the realization of a common project:

> There, even the possibility of acting as a neighbour is excluded; one is a neighbour only to those who serve their purpose. The word "neighbour" loses all meaning; there can only be "associates", partners in the pursuit of particular interests[7].

This is true in interpersonal relationships as well as among social groups and nations: at all levels, consideration of the common good has become increasingly marginalized and even incomprehensible. This mentality appears in the ecclesial community as well, as evidenced by its internal divisions and conflicts.

FT's analysis focuses in particular on the link that unites individuals and peoples. Already in 2015 the encyclical *Laudato Si'*[8] had identified a similar dynamic: "Everything is interconnected and everything is related" is the motto that best summarizes the paradigm of integral ecology it proposes:

> Ecology studies the relationship between living organisms and the environment in which they develop. This necessarily entails reflection and debate about the conditions required for the life and survival of society, and the honesty needed to question certain models of development, production and consumption. It cannot be emphasized enough how everything is interconnected. Time and space are not independent of one another, and not even atoms or subatomic particles can be considered in isolation. Just as the different aspects of the planet - physical, chemical and biological - are interrelated, so too living species are part of a network which we will never fully explore and understand. A good part of our genetic code is shared by many living beings. It follows that the fragmentation of knowledge and the isolation of bits of information can actually become a form of ignorance, unless they are integrated into a broader vision of reality[9].

---

[7] Pope Francis, *Fratelli tutti*, n. 102.

[8] Pope Francis, Encyclical Letter *Laudato Si' on Care for Our Common Home*, 24 May 2015; henceforth LS.

[9] Pope Francis, *Laudato Si'*, n. 138.

As the ending of this passage indicates, this awareness clashes with the reductionism of the technocratic paradigm that dominates contemporary culture, and that leads to an inability to perceive connections and relationships:

> The specialization which belongs to technology makes it difficult to see the larger picture. The fragmentation of knowledge proves helpful for concrete applications, and yet it often leads to a loss of appreciation for the whole, for the relationships between things, and for the broader horizon, which then becomes irrelevant. This very fact makes it hard to find adequate ways of solving the more complex problems of today's world, particularly those regarding the environment and the poor; these problems cannot be dealt with from a single perspective or from a single set of interests[10].

The price of this reductionism, which serves dominant economic interests, is the inability to find truly effective solutions: "Technology, which, linked to business interests, is presented as the only way of solving these problems, in fact proves incapable of seeing the mysterious network of relations between things and so sometimes solves one problem only to create others" (LS, 20). Ecological initiatives also run the risk of remaining prisoners of this reductionism: "To seek only a technical remedy to each environmental problem which comes up is to separate what is in reality interconnected and to mask the true and deepest problems of the global system" (LS, 111).

1.2 Creating a "we" dwelling together in this common home

In the vision of Pope Francis, a stark analysis of reality, of its contradictions and distortions, does not lead to discouragement. Again, from this point of view as well, the parallelism between LS and FT is compelling: in both, the same invitation to hope resounds, a hope which has profound theological origins. The first chapter of FT concludes by stating: "Despite these dark clouds, which may not be ignored, I would like in the following pages to take up and discuss many new paths of hope. For God continues to sow abundant seeds of goodness in our human family" (FT, 54). We find similar words in LS: "Yet all is not lost. Human beings, while capable of the worst, are also capable of rising above themselves, choosing again what is good, and making a new start, despite their mental and social conditioning. [...] No system can completely

[10] Pope Francis, *Laudato Si'*, n. 110.

suppress our openness to what is good, true and beautiful, or our God-given ability to respond to his grace at work deep in our hearts" (LS, 205).

This hope continues to nourish a commitment to changing course (cf. LS, 61), elaborating and pursuing a plan that responds to the contradictions of this age of fragmentation and remedies the grave consequences weighing on the shoulders of the poorest. In the lexicon of FT, "the human family's innate vocation [is] to fraternity" (FT, 26), while in LS, *integral ecology* names the alternative to the technocratic paradigm (cf. LS, 137). Frequently this project is expressed in terms of building a collective subject, which in this case means developing an ever wider and more responsible "we": if LS begins with a "concern to bring the whole human family together to seek a sustainable and integral development" (LS, 13), FT begins from the observation that "we need to think of ourselves more and more as a single family dwelling in a common home"[11](FT, 17; cf. also 217).

This is a key theme on which Pope Francis never ceases to insist, a steady and defining guiding value of his magisterium. It is rooted in his experience as Archbishop of Buenos Aires and in the life of the Argentine and Latin American Church, in particular in the elaboration of the so-called *teología del pueblo*[12]. The foundation for understanding this insistence on the need to "build a people" is to be found in chapter IV of EG, in a section tellingly entitled "The common good and peace in society", as if to confirm that the first apostolic exhortation of Pope Francis is a defining one for his pontificate.

According to EG, it is the people, as a collective subject that is authentically inclusive – i.e. capable of recognising and valuing each of its members –, who can take their destiny into their own hands and decide the direction of their development, thus representing the alternative to the anonymous "mass" that can only be dragged along by the forces of globalisation and consumerism. In fact, we read in n. 220: "People in every nation enhance the social dimension of their lives by acting as committed and responsible citizens, not as a mob swayed by the powers that be. [...] Yet becoming a people demands something more. It is an ongoing process in which every new generation must take part: a slow and arduous effort calling for a desire for integration and a willingness

[11] The original Spanish is more incisive: "*necesitamos constituirnos en un 'nosotros' que habita la casa común*" (literally: "we need to constitute ourselves into a 'we' that inhabits the common home").

[12] For a concise but accurate presentation, cf. Juan Carlos Scannone, "El papa Francisco y la teología del pueblo", in *Razón y Fe*, n. 1395-2015, pp. 31-50; more generally cf. Id., *La teología del pueblo. Raíces teológicas del Papa Francisco*, Maliaño (Cantabria): Sal Terrae, 2017.

to achieve this through the growth of a peaceful and multifaceted culture of encounter".

Building a people is thus a matter of processes, as FT, n. 217, also repeats: "What is important is to create *processes* of encounter, processes that build a people that can accept differences". In his predilection for concrete terms, Pope Francis frequently translates this imperative into an exhortation to set out on a journey, and to do so together: "Let us continue, then, to advance[13] along the paths of hope", he urges in n. 55 of FT, and in n. 113 he continues: "Let us return to promoting the good, for ourselves and for the whole human family, and thus advance together towards an authentic and integral growth". Similarly, at the end of LS he recalls that it is the Spirit, intimately present at the heart of the cosmos, who animates and inspires new paths (cf. LS, 238), and he repeats the invitation: "In union with all creatures, we journey through this land seeking God, [...] Let us sing as we go!" (LS, 244).

2. "Church and Synod are synonymous": from synod to synodality

"Synod means precisely 'to walk together'", Pope Francis recalled when he met with the Popular Movements on 28 October 2014[14]. In his approach, the technical, ecclesiological and even canonical meanings of the term "synod" can never be separated from the original meaning of traveling a stretch of road together; indeed, a kind of mutuality is created among them. On the one hand, the experience (even the physical experience) of walking together is a key to understanding what a synodal dynamic entails for a Church that conceives of itself as a People journeying together. On the other hand, the concrete practice of synodality gives the Christian community an experience that enables it to contribute more competently to the processes of building a people, of which it itself is a part, capable of committing "together to seek a sustainable and integral development" (LS, n. 13), i.e., to move from the fragmentation of the technocratic paradigm to the horizon of integral ecology and fraternity.

As already noted, integral ecology and synodality do not, therefore, belong to mutually exclusive realms: at the deepest level, they are closely linked,

[13] Again, the original Spanish seems more clear and concise: "*Caminemos en esperanza*", that is, "Let us walk in hope".

[14] Pope Francis, *Address to the participants in the World Meeting of Popular Movements*, 28 October 2014.

since both refer to the work of "building a people in peace, justice and fraternity" (EG, 221). For this reason, it makes sense to explore them in the light of one another. In order to do so, it is necessary first of all to retrace the development of the notion of synodality since the Second Vatican Council[15].

2.1 Establishment and evolution of the Synod

From a theological point of view, the post-conciliar understanding of synodality is rooted in the vision of the Church as sacrament and as People of God proposed by the dogmatic constitution *Lumen Gentium* (LG), with its recovery of the doctrine of "the common priesthood of the faithful" (LG, 10) and of the necessity of the active participation of all members in the life and mission of the Church. Bestowed with the dignity of children of God and with the gift and responsibility of proclaiming the Gospel to all, the laity are called to participate in the government of the Church according to their own tasks, roles, and means. The Spirit, in fact, gives them special charisms and graces, "makes them fit and ready to undertake the various tasks and offices which contribute toward the renewal and building up of the Church" (LG, 12). They are specifically "permitted and sometimes even obliged to express their opinion on those things which concern the good of the Church" (LG, 37). If the Holy Spirit is the principle of unity, making into one dynamic subject all the members of the Church, diverse in ministry, vocation and mission, the Eucharist is the source and summit of the believing community (cf. LG,

---

[15] The bibliography on this subject is rather voluminous and it is not possible here to review it all. An indispensable reference is undoubtedly the document of the International Theological Commission, *Synodality in the Life and Mission of the Church*, 2 March 2018, (http://www.vatican.va/roman_curia/congregations/cfaith/cti_documents/rc_cti_20180302_sinodalita_en.html, consulted on 21/03/2021), which will be referred to with the abbreviation SYN. Among the most recent contributions, we highlight two, significant on account of the profiles of the respective authors: Nathalie Becquart, "La sinodalidad, un camino de conversión comunitaria", in *Razón y Fe*, n. 1450-2021, pp. 175-183, appointed Undersecretary of the Synod of Bishops on 8 February 2021, and previously consultor of the Secretariat of the Synod; Michael Czerny, "Toward a Synodal Church", in *La Civiltà Cattolica,* n. 4093 (January 2, 2021), pp. 3-15, formerly Special Secretary of the Special Assembly of the Synod of Bishops for the Pan-Amazonian Region (2019) and currently interim Prefect of the Dicastery for Promoting Integral Human Development. For a more extensive reflection, see: Rafael Luciani and María del Pilar Silveira (eds.), *La sinodalidad en la vida de la Iglesia. Reflexiones para contribuir a la reforma eclesial.* Madrid: San Pablo, 2020; Rafael Luciani, Serena Noceti and Carlos Schickendantz (eds.), *Sinodalità e riforma. Una sfida ecclesiale*, Brescia: Queriniana, 2022; Jos Moons, "A Comprehensive Introduction to Synodality: Reconfiguring Ecclesiology and Ecclesial Practice", in *Roczniki Teologiczne*, 2022/2, pp. 73-93.

11; constitution *Sacrosanctum Concilium,* 10), in which the many grains of wheat become one bread. The Council thus points to the action of the Spirit who enlivens the Church through sacramental grace, and in a particular way through the Eucharistic celebration, as the source from which the ecclesial "we" springs. The Synod of Bishops, instituted by Paul VI on 15 September 1965 with the Motu Proprio *Apostolica Sollicitudo* and conceived as "a permanent Council of bishops for the universal Church" (*ibid.*), represents one of the greatest (but also controversial) fruits of the Council, inasmuch as it is a body with the task of supporting and advising the Pope in the governance of the universal Church.

This vision of synodality places the Church in a continuous process of rebirth and reform "from the grassroots", which we can connect to the concept of ecclesiogenesis or ecclesiogenetics found in the conciliar decree *Ad gentes*. It builds a dynamic and relational vision of the identity of the Church: a relational identity of communion and mission rooted in the Trinitarian and Eucharistic mysteries. The image of an inverted pyramid offers an understanding of hierarchical ministry intertwined with this vision of synodality. Pope Francis also emphasized this on a particularly solemn occasion, at the commemoration of the 50th anniversary of the institution of the Synod of Bishops: "Synodality, as a constitutive element of the Church, offers us the most appropriate interpretive framework for understanding the hierarchical ministry itself"[16].

A crucial step in the Church's awareness of the potential and significance of the Synod as an instrument was the apostolic constitution *Episcopalis communio*, promulgated by Pope Francis on 15 September 2018, which renews the guidelines for Synods, integrating the synodal event as such (the Synodal Assembly) into an ecclesial process in three phases – preparation, celebration, and implementation – which involves the people of God without setting aside the "all-some-one" dynamic[17]. In this way, Pope Francis "has profoundly re-

---

[16] Pope Francis, *Address at the Ceremony commemorating the 50th anniversary of the institution of the Synod of Bishops*, 17 October 2015. The reflection on the meaning and the exercise of hierarchical roles within a synodal Church is constantly growing; among many, cf. Alphonse Borras, *Communion ecclésiale et synodalité*, Paris: Editions CLD, 2018; Id. "Primauté et synodalité : Réflexions éclectiques d'un canoniste latin à propos d'un ouvrage récent", in *Nouvelle revue théologique*, 2/2021, pp. 290–304; Myriam Wijlens, "Reforming the Church by Hitting the Reset Button: Reconfiguring Collegiality within Synodality because of sensus fidei fidelium", in *The Canonist* 2017/8, pp. 235–261; Dario Vitali, *Verso la sinodalità*, Magnano (BI): Qiqajon, 2014.

[17] Cfr SYN, n. 64: "This ecclesiological vision invites us to articulate synodal communion in terms of 'all', 'some' and 'one'. On different levels and in different forms, as local Churches, regional

newed the Synod of Bishops, inserting it within the framework of synodality as a constitutive dimensionof the Church, at all levels of her existence"[18].

2.2 Synodality as a constitutive dimension of the Church

The lived experience of the Church as sketched above mirrors the evolution of its understanding, as the International Theological Commission itself notes:

> In the theological, canonical and pastoral literature of recent decades, a neologism has appeared, the noun "synodality", a correlate of the adjective "synodal", with both of these deriving from the word "synod". Thus, people speak of synodality as a "constitutive dimension" of the Church or tout court of the "synodal Church". This linguistic novelty, which needs careful theological clarification, is a sign of something new that has been maturing in the ecclesial consciousness starting from the Magisterium of Vatican II, and from the lived experience of local Churches and the universal Church since the last Council until today[19].

The word "synodality" thus takes on three distinct but closely related meanings. We express them in the words of the International Theological Commission:

> a. First and foremost, synodality denotes the particular *style* that qualifies the life and mission of the Church, expressing her nature as the People of God journeying together and gathering in assembly, summoned by the Lord Jesus in the power of the Holy Spirit to proclaim the Gospel. Synodality ought to be expressed in the Church's ordinary way of living and working. This *modus vivendi et operandi* works through the community listening to the Word and celebrating the Eucharist, the brotherhood of communion and the co-responsibility and participation of the whole People of God in its life and mission, on all levels and distinguishing between various ministries and roles.
>
> b. In a more specific sense, which is determined from a theological and canonical point of view, synodality denotes those *structures* and *ecclesial processes* in which the synodal nature of the Church is expressed at an institutional level, but analogously on various levels: local, regional and universal. These structures and processes are officially at the service of the Church, which must discover the way to move forward by listening to the Holy Spirit.

---

groupings of local Churches and the universal Church, synodality involves the exercise of the sensus fidei of the universitas fidelium (all), the ministry of leadership of the college of Bishops, each one with his presbyterium (some), and the ministry of unity of the Bishop of Rome (one)".

[18] N. Becquart, cit., 17. Cf. also Annemarie C. Mayer, "'For a Synodal Church': Equipping the Catholic Church on Her Way into the Third Millennium", in *Louvain Studies*, 2020/3, pp. 205-214.

[19] International Theological Commission, *Synodality in the Life and Mission of the Church*, n. 5.

> c. Finally, synodality designates the programme of those *synodal events* in which the Church is called together by the competent authority in accordance with the specific procedures laid down by ecclesiastical discipline, involving the whole People of God in various ways on local, regional and universal levels, presided over by the Bishops in collegial communion with the Bishop of Rome, to discern the way forward and other particular questions, and to take particular decisions and directions with the aim of fulfilling its evangelising mission[20].

In short, therefore, synodality as a constitutive and constituent form of the Church certainly cannot be reduced to some particular event or specific structure or procedure, but pervades them all. Pope Francis recalled in the aforementioned 50th anniversary commemoration of the institution of the Synod of Bishops: "as Saint John Chrysostom says, [...] 'Church and Synod are synonymous' [*Explicatio in Ps.* 149: PG 55, 493], inasmuch as the Church is nothing other than the 'journeying together' of God's flock along the paths of history towards the encounter with Christ the Lord"[21].

3. Characteristics of the synodal approach

For purposes of our research, the meaning of synodality as an approach, which is expressed in the ordinary manner of the Church's day-to-day life and in a particular way in the conduct of synodal processes, is most relevant. The effort to translate the synodal approach into concrete practices, procedures, instruments and structures requires a more complete development of its characteristics. We can turn to the PD for the Synod 2021-2023, which indicates "ten thematic nuclei that articulate different facets of 'lived synodality'" (PD, n. 30). They are: i) The journeying companions; ii) Listening; iii) Speaking out; iv) Celebrating; v) Co-responsible in the mission; vi) Dialogue in church and society; vii) With the other christian denominations; viii) Authority and participation; ix) Discerning and deciding; x) Forming ourselves in synodality. These ten nuclei are offered as guidelines to articulate the answer to the fundamental question guiding the whole synodal process: "A synodal Church, in announcing the Gospel, 'journeys together:' How is this 'journeying together' happening today in your particular Church? What steps does the Spirit invite us to take in order to grow in our 'journeying together'?" (PD, n. 26).

---

[20] International Theological Commission, *Synodality in the Life and Mission of the Church*, n. 70.

[21] Pope Francis, *Address at the Ceremony commemorating the 50th anniversary of the institution of the Synod of Bishops*, 17 October 2015.

Given the link that has already been highlighted between integral ecology and synodality, it is apropos to explore these characteristics of lived synodal processes further, even within the limited space available. This will allow us to sketch by analogy an approach characteristic of integral ecology, and above all to focus on what is required to promote what we could call "processes of integral ecology". After all, the need for such processes is now more imperative than ever, as witnessed by the denunciation of ecological, social and ethical degradation that we find in LS, beginning with attention to the cry of the earth and of the poor.

3.1 Listening and Dialogue

Listening is a key starting point: "A synodal Church is a Church which listens, which realizes that listening 'is more than simply hearing'. It is a mutual listening in which everyone has something to learn. The faithful people, the college of bishops, the Bishop of Rome: all listening to each other, and all listening to the Holy Spirit, the 'Spirit of truth' (*Jn* 14:17), in order to know what he 'says to the Churches' (*Rev* 2:7)"[22]. This listening requires a fundamental disposition of recognizing the partiality of one's own position and the need to take into account other points of view in order to truly walk a stretch of road together. The synodal approach requires the Church to allow itself to be challenged by the needs of those it encounters, without thinking that it always has a ready-made answer.

As Pope Francis warned while opening the 2018 Assembly of the Synod of Bishops, "It is this listening that creates space for dialogue. The Synod must be an exercise in dialogue"[23]. Dialogue is therefore the second key word that characterizes a synodal approach. As the International Theological Commission points out,

> Synodal dialogue depends on courage both in speaking and in listening. It is not about engaging in a debate where one speaker tries to get the better of the others or counters their positions with brusque arguments, but about expressing whatever seems to have been suggested by the Holy Spirit as useful for communal discernment, at the same time being open to accepting whatever has been suggested by the same Spirit in other people's positions, "for the general good" (*1 Corinthians* 12,7)[24].

---

[22] *Ibid.* These words, which refer to n. 171 of EG, are later taken up in SYN, n. 110.

[23] Pope Francis, *Address at the opening of the Synod of Bishops on "Young People, the Faith and Vocational Discernment"*, 3 October 2018.

[24] International Theological Commission, *Synodality in the Life and Mission of the Church*, 2 March 2018, n. 111.

Dialogue is also a cornerstone of Pope Francis' teaching. EG considers it an indispensable component of evangelization: within the family (EG, 70), in pastoral activities (EG, 31), and in preaching (EG, 137), but also among states, within society, between culture and the sciences, and with believers of other confessions and religions (EG, 238-259). In LS it is the backbone of chapter V, while FT is punctuated by frequent calls for dialogue, to which chapter VI is explicitly dedicated. For Pope Francis, dialogue is not a conversation for its own sake, nor a strategy to "conquer" the other, but an encounter among different people, united by a real desire to travel a stretch of road together. For this reason, it does not conceal or deny differences, tensions, or conflicts, but knows how to face them squarely in order "to make [them] a link in the chain of a new process" (EG, 227), and to make it "possible to 'build communion amid disagreement' " (SYN, 111, with reference to EG, 228).

Of great interest to us is a passage from FT that describes social dialogue in terms that converge singificantly with those we have encountered in examining the synodal approach, referring among other things to n. 108 of the Apostolic Exhortation *Querida Amazonia*[25].

> Authentic social dialogue involves the ability to respect the other's point of view and to admit that it may include legitimate convictions and concerns. Based on their identity and experience, others have a contribution to make, and it is desirable that they should articulate their positions for the sake of a more fruitful public debate… "Indeed, in a true spirit of dialogue, we grow in our ability to grasp the significance of what others say and do, even if we cannot accept it as our own conviction. In this way, it becomes possible to be frank and open about our beliefs, while continuing to discuss, to seek points of contact, and above all, to work and struggle together"[26].

It thus seems legitimate, and particularly fruitful, to link integral ecology and synodality again, in this case drawing inspiration from an exploration of dialogue as a component of the synodal approach. This will allow us to better sketch the major paths of dialogue that, according to LS, n. 163, are indispensable to "escape the spiral of self-destruction which currently engulfs us" and to which, as we mentioned, the encyclical dedicates all of chapter V. In light of what has been said here, it is now much easier to identify what Pope

---

[25] Pope Francis, Post-synodal Apostolic Exhortation *Querida Amazonía*, 2 February 2020; henceforth QA.

[26] Pope Francis, *Fratelli tutti*, n. 203.

Francis means when he speaks of "frank dialogue" (LS, 189), intending something more significant than the ordinary exercise of diplomacy or negotiation.

3.2 The method of discernment

A second distinctive element of a synodal approach is the use of discernment: "Exercising discernment is at the heart of synodal processes and events" (SYN, 113). Pope Francis insisted on this most vigorously when opening the 2018 Synodal Assembly:

> *The Synod is an ecclesial exercise in discernment.* To speak frankly and listen openly are fundamental if the Synod is to be a process of discernment. Discernment is not an advertising slogan, it is not an organizational technique, or a fad of this pontificate, but an *interior attitude* rooted in *an act of faith.* Discernment is the method and at the same time the goal we set ourselves: it is based on the conviction that God is at work in world history, in life's events, in the people I meet and who speak to me. For this reason, we are called to listen to what the Spirit suggests to us, with methods and in paths that are often unpredictable. Discernment needs space and time[27].

The method of communal discernment[28], which *Episcopalis communio* establishes as the form of every Synod, is typical of the synodal approach. The phases of the process (preparation, assembly, implementation) recapitulate the three major steps of the discernment process summarized in n. 51 of EG as recognizing, interpreting, and choosing. The close relationship between discernment and synodality, whereby the former is the method most proper to the latter, merits a much more in-depth discussion[29]; here we limit ourselves to pointing out some suggestions relevant for our present purpose.

In particular, adopting communal discernment allows us to focus on the purpose of synodal processes – to seek and find together the will of God – discarding the attitude of opposition that seeks to win over to one's own view

---

[27] Pope Francis, *Address at the opening of the Synod of Bishops on "Young People, the Faith and Vocational Discernment"* 3 October 2018.

[28] On the relationship between the synodal process and discernment in common, cf. Jacques Haers, "A Synodal Process on Synodality Synodal Missionary Journeying and Common Apostolic Discernment", in *Louvain Studies* 2020/3, pp. 215-238.

[29] In this regard, consider two previous works dedicated to the topic: Giacomo Costa, *Il discernimento.* Cinisello Balsamo (MI): San Paolo, 2018, with particular regard to par. 2 of ch. IV, entitled "Il cammino della sinodalità", pp. 99-110; Giacomo Costa, "Discernement et synodalité", in *Lumen Vitae* 2020/1, pp. 31-37.

those who have the task of making the final decision. The outcome of communal discernment, when the process is correctly carried out, does not envision winners and losers, but a synthesis ordered to the good in which differences can be recognized without being dissolved. While respecting obvious differences and the need to find appropriate "translations", this synthesis is a goal that can be of great interest for renewing processes of social dialogue, and more specifically of decision-making processes that require the involvement of many actors, even those in entrenched conflict. Indeed, this is the usual situation of decision-making processes that relate to projects with significant environmental impacts, to which nn. 182-188 of LS refer: to establish these processes in a spirit of integral ecology, methods must be found to get beyond the opposition between stakeholders – an opposition that, in subsequent negotiation, inevitably ends up rewarding those with greater resources, influence and power. It is here that it seems especially promising to work on ways to implement the insights of communal discernment.

3.3 Participation and culture of encounter

A final trait of the synodal approach that deserves to be highlighted here is the importance assigned to the involvement and participation of all members of the Church, which galvanizes a dynamic of co-responsibility. This is stated with unequivocal clarity in the *Final Document* of the Synodal Assembly of 2018.

> A characteristic feature of this style of Church is the valuing of the charisms that the Spirit bestows in accordance with the vocation and role of each of the Church's members, through a process of co-responsibility. [...] Motivated by this spirit, we can move towards a participatory and co-responsible Church, one capable of appreciating its own rich variety, gratefully accepting the contributions of the lay faithful, including young people and women, consecrated persons, as well as groups, associations and movements. No one should be excluded or exclude themselves[30].

Participation as a characterstic trait of synodality is further stressed by Pope Francis: "I would say that celebrating a Synod is always a good and important thing, but it proves truly beneficial if it becomes a living expression of 'being Church', of a way of acting marked by true participation. This is not a matter

[30] Synod of Bishops on Young People, Faith and Vocational Discernment, *Final Document*, 27 October 2018, n. 123, later taken verbatim in n. 206 of the Apostolic Exhortation *Christus vivit* (25 March 2019). Similar expressions can be found in SYN, n. 67.

of form, but of faith. Participation is a requirement of the faith received in baptism"[31]. The PD offers some hints on how to put this emphasis on participation into practice by pointing at the need to listen to the laity[32] (especially women and youth), to integrate the contribution of consecrated men and women, and to secure a space "for the voice of minorities, the discarded, and the excluded" (PD, 30). A genuinely synodal process needs to be inclusive[33]: "it will be of fundamental importance that the voice of the poor and excluded also find a place, not only that of those who have some role or responsibility within the particular Churches" (PD, 31).

This insistence on the richness represented by variety and on the importance of valuing each person's contribution immediately calls to mind another fundamental passage of Pope Francis' magisterium:

> Here our model is not the sphere, which is no greater than its parts, where every point is equidistant from the centre, and there are no differences between them. Instead, it is the polyhedron, which reflects the convergence of all its parts, each of which preserves its distinctiveness. Pastoral and political activity alike seek to gather in this polyhedron the best of each. [...] Even people who can be considered dubious on account of their errors have something to offer which must not be overlooked[34].

This is the first occurrence of the term *polyhedron*, which has become a key word of the pontificate of Pope Francis, capable of condensing into one image the idea of "a peaceful and multifaceted culture of encounter" (EG, 220). FT states clearly: "The image of a polyhedron can represent a society where differences coexist, complementing, enriching and reciprocally illuminating one another, even amid disagreements and reservations" (FT, 215). In this perspective we can understand synodality as the declination in the ecclesial sphere of the culture of encounter, which designates how collective subjects are constructed in the socio-political sphere. This illustrates the profound organicity of the magisterium of Pope Francis and why it is so fruitful to read its texts in parallel.

---

[31] Pope Francis, *Address for the opening of the Synod*, 9 October 2021.

[32] For a reflection on the role of laity and the status of their voice, cf. Alphonse Borras, "'Soltanto consultivo'? Sul valore del consigliare nella Chiesa", in *La Rivista del Clero Italiano*, 2016/5, pp. 377-393

[33] For a deeper reflection on the actual meaning of this call to listen to the voice of the poor, cf. Giacomo Costa and Paolo Foglizzo, "Sinodo: come la Chiesa ascolta la voce degli esclusi", in *Aggiornamenti Sociali*, 2022/3, pp. 165-172.

[34] Pope Francis, *Evangelii gaudium*, n. 236.

The term polyhedron does not appear in LS, but this does not mean that it is alien to the notion of integral ecology. On the contrary, in many ways it constitutes the backbone of the entire encyclical[35] which emerges at many points: for example, when LS underlines the need for dialogue among the various sciences and forms of wisdom, including art, poetry, and the wisdom that comes from religions: "If we are truly concerned to develop an ecology capable of remedying the damage we have done, no branch of the sciences and no form of wisdom can be left out, and that includes religion and the language particular to it" (LS, 63; cf. also. 110 and 201). The development of integral ecology requires the practice of a culture of encounter.

3.4. The exercise of leadership

The affirmation of each person's particular contribution that is implied by the image of the polyhedron also helps us to integrate the specific roles of those in authority and leadership positions. This is a particularly sensitive issue in the case of synodal processes, given the distinctive foundation and exercise of authority in the Church. As SYN makes clear, synodal processes do not take place in an anarchical space, but are hierarchically structured and very specific about decision-making and decision-taking:

> The synodal process must take place at the heart of a hierarchically structured community. In a diocese, for example, it is necessary to distinguish between the process of *decision-making* through a joint exercise of discernment, consultation and co-operation, and *decision-taking,* which is within the competence of the Bishop, the guarantor of apostolicity and Catholicity. Working things out is a synodal task; decision is a ministerial responsibility. A correct exercise of synodality must contribute to a better articulation of the ministry of the personal and collegial exercise of apostolic authority with the synodal exercise of discernment on the part of the community[36].

The relationship between the participation of all in the process and the decision-making responsibility of those in authority can be described in terms of the concept of *conspiratio fidelium et pastorum* developed in the 19th century by Cardinal Newman.

---

[35] Cf. Giacomo Costa and Paolo Foglizzo, "«Laudato si'»: un'enciclica poliedro", in *La Rivista del Clero Italiano*, 2015/7-8, pp. 485-496.

[36] International Theological Commission, *Synodality in the Life and Mission of the Church*, 2 March 2018, n. 69.

> If there is no synodality without primacy, the minister who is the leader is making the final decision from this whole spiritual process of listening and discernment that is embodied in the important notion of *conspiratio*. We might consider an image first developed by Cardinal John Henry Newman in the nineteenth century: the *conspiratio fidelium et pastorum*, literally the "breathing together of the faithful and the pastors". [...] The vision and practice of a synodal church as opposed to a clerical church is articulated in a relational anthropology and theology of ministry[37].

Within synodal processes, just as in the processes of communal discernment, the role of authority as a guarantor of unity stands out; nor could it be otherwise, given that discernment in common represents the proper method of synodality. According to Pope Francis, this is exactly the role of the pontiff within a Church that proceeds synodally[38]. In order to exercise this responsibility, it is essential to be aware that the institution within which power is exercised is not the leader's private property, but a common good to be safeguarded so that each and every one may find their place in it.

Notwithstanding the unique characteristics of the exercise of authority within the Church, these reflections are also relevant to understanding the kind of leadership that is needed to promote integral ecology and to manage the complex processes of consultation and decision-making that it requires. It is not by chance that LS addresses some stern words in this regard to the world of politics about responsibility for the situation of global degradation (cf., for example, LS, 16, 26, 54, 166), but at the same time proclaims great esteem and trust for the potential impact of a leadership capable of renewing its own inspiration (cf., for example, LS, 178, 181, 197).

4. Missionary Synodality and Integral Ecology

In a Church whose sole raison d'être is to fulfill the mission received from her Founder, the constitutive dimension of synodality certainly cannot be limited to the intra-ecclesial sphere, but is meant to become the manner in which the community of believers walks along the paths of history together with all men and women, with all the members of the human family and all

---

[37] Nathalie Becquart, "The pandemic, a kairos to foster synodality as a way to implement *Laudato si'* ", *UISG Bullettin*, n. 173-2020, pp. 15-20.

[38] Cf. Pope Francis, *Address at the Ceremony commemorating the 50th anniversary of the institution of the Synod of Bishops*, 17 October 2015.

the beings of the universe which "together form a kind of universal family" (LS, 89).

> Synodality manifests the 'pilgrim' character of the Church. The image of the People of God, gathered from among the nations (*Acts* 2,1-9; 15,14), expresses its social, historical and missionary character, which corresponds to the condition and vocation of each human person as *homo viator.* The path is the image that clarifies our understanding of the mystery of Christ as the Way that leads to the Father. Jesus is the way from God to man and from man to God. The grace-filled event whereby He made Himself a pilgrim by pitching His tent among us (*John* 1,14), goes on in the synodal path of the Church[39].

Therefore, the dynamics of synodality as well cannot fail to be ordered to the mission, in a process of constant extroversion to which the *Final Document* of the Synod on Youth refers "by speaking of synodality for mission, or missionary synodality: 'Making a synodal Church a reality is an indispensable precondition for a new missionary energy that will involve the entire People of God'"[40].

LS marks the moment when the Church becomes fully aware that, in the present circumstances, care for our common home is fully part of the mission of Christians:

> [F]aith convictions can offer Christians, and some other believers as well, ample motivation to care for nature and for the most vulnerable of their brothers and sisters. If the simple fact of being human moves people to care for the environment of which they are a part, Christians in their turn "realize that their responsibility within creation, and their duty towards nature and the Creator, are an essential part of their faith". It is good for humanity and the world at large when we believers better recognize the ecological commitments which stem from our convictions[41].

We thus discover a further link between synodality and integral ecology: the latter, as part of the Church's mission in the contemporary world, is one of the elements of that horizon toward which the urgency of missionary synodality

---

[39] International Theological Commission, *Synodality in the Life and Mission of the Church*, 2 March 2018, n. 49.

[40] Synod of Bishops on Young People, Faith and Vocational Discernment, *Final Document*, 27 October 2018, n. 118, with a cross-reference to SYN, n. 9.

[41] Pope Francis, *Laudato Si'*, n. 64, referring to Saint John Paul II, *Message for the 1990 World Day of Peace*, 8 December 1989, n. 15.

impels the ecclesial community. SYN expresses this clearly, in a passage in which the lexicon of LS clearly resounds:

> In fact, today, when growing awareness of the interdependence between peoples forces us to think of the world as our common home, the Church is called to demonstrate that her Catholicity and the synodal way in which she lives and works are a catalyst of unity in diversity and of communion in freedom. This is a significant contribution that the life and synodal conversion of the People of God can make to the promotion of a culture of encounter and solidarity, respect and dialogue, inclusion and integration, gratitude and gift[42].
>
> The Church's synodal life presents itself, in particular, as *diakonia* in the promotion of a social, economical and political life of all peoples under the banner of justice, solidarity and peace. [...] In this context, it is an important obligation and a criterion of all social action of the People of God to hear the cry of the poor and the cry of the earth, and to draw attention urgently, in determining society's choices and plans, to the place and the privileged role of the poor, the universal destination of goods, the primacy of solidarity and care for our common home[43].

In a similar vein, the *Final Document* of the 2018 Synod on youth emphasizes how the synodal approach is a clear prerequisite for the action of believers to be credible. It constitutes a commitment based on a multifaceted vision of the world and the importance of connections, and seeking to build a unity that values differences without homogenizing them; in short, the characteristics proper to the paradigm of integral ecology:

> The effective realization of a community with many faces also has a bearing on its relationship with the local area, on its openness to the social fabric and on its dealings with civil authorities. Only a united and diversified community is able to present itself in an open way and to shine the light of the Gospel upon the concerns of society that challenge us today: ecological questions, work, support for the family, marginalization, renewal of politics, cultural and religious pluralism, the quest for justice and peace, the digital environment[44].

---

[42] International Theological Commission, *Synodality in the Life and Mission of the Church*, n. 118.

[43] International Theological Commission, *Synodality in the Life and Mission of the Church*, n. 119.

[44] International Theological Commission, *Synodality in the Life and Mission of the Church*, n. 132.

In these words, a final element of fundamental importance emerges. As a factor in fulfilling its mission, the Church chooses synodality not for its functional character, as an organizational form that promises efficiency or as a method to achieve results, in the manner of internal organization decisions of a large company. When the mission of the Church is at stake – "to shine the light of the Gospel upon the concerns of society that challenge us today", in the words of the Synod – there must be a full correspondence between form and content. As pastoral theologian Enzo Biemmi wrote in a commentary on EG: "the Church evangelizes not only with words, but with the form she gives herself within history. Her organization expresses her mission [which ...] thus becomes the key to rethinking the face of Christianity, of the Church, of its pastoral care"[45].

It is on this most fundamental level that the connection between synodality and integral ecology is rooted. This is shown by the path taken by the most recent Synodal Assembly, the one dedicated in 2019 to the Pan-Amazonian region, which in its very title intertwines the perspective of the new paths of the Church with that of integral ecology, with a clear reference to the terminology of synodality. Its *Final Document*[46] invites the Church to undertake a fourfold journey of conversion, aiming at authentic reform, that is, to assume the form that is up to the task of fulfilling its mission in today's world. It is certainly not by chance that among the "interconnected dimensions" (n. 19) of this conversion are precisely the ecological one ("Towards an integral ecology based on the encyclical *Laudato Si'* ", as indicated by the title of the paragraph that includes nos. 66-70) and the synodal one: "To walk together the Church requires a synodal conversion, synodality of the People of God under the guidance of the Spirit in the Amazon" (86)[47].

Next, QA takes up the four dimensions of conversion and rethinks them with respect to the Church and to all humanity, in the form of four dreams (a category already known to be particularly dear to Pope Francis). The last one directly concerns the ecclesial community; and it is framed in terms proper to missionary synodality, even without naming it directly:

---

[45] Enzo Biemmi, *La nuova evangelizzazione alla luce* di *Evangelii gaudium*, speech in the Diocese of Bergamo, 4 June 2014, http://ftp.bergamo.chiesacattolica.it/PASTORALE_16-17/Biemmi.pdf, consulted 25/03/2021.

[46] Synod of Bishops - Special Assembly for the Pan-Amazonian Region, *Final Document*, 26 October 2019.

[47] On the categories of integral ecology, interculturality, and synodality in the "official" documents of the Special Synod for the Amazon, cf. Pasquale Bua, "Inculturation, Interculturality, Synodality", in *Urbaniana University Journal* 2020/2, pp. 91-115.

> The Church is called to journey alongside the people of the Amazon region. [...] The journey continues, and missionary efforts, if they are to develop a Church with an Amazonian face, need to grow in a culture of encounter towards "a multifaceted harmony". But for this incarnation of the Church and the Gospel to be possible, the great missionary proclamation must continue to resound[48].

A particular adjective distinguishes each of the other three dreams: social, cultural, and ecological (to be understood here in relation to both environmental ecology and human ecology). If we consider them together, in the light of LS, we have no difficulty in tracing out the fundamental dimensions of integral ecology, to which chapter IV of the encyclical is dedicated: indeed, the first two paragraphs are entitled "Environmental, economic and social ecology" and "Cultural ecology". With an expression that would be difficult to use in a pastoral context, we could say that their intertwining defines the dream of integral ecology. The Church is moved by the desire to contribute to the realization of this dream, starting from "her broad spiritual experience, her renewed appreciation of the value of creation, her concern for justice, her option for the poor, her educational tradition and her history of becoming incarnate in so many different cultures throughout the world" (QA, n. 60). In order to equip itself to do this, it cannot help but set out along the path of synodal conversion: once again, integral ecology and synodality go hand in hand.

---

[48] Pope Francis, *Querida Amazzonia*, n. 61.

'AN INTEGRALLY HIGHER QUALITY OF LIFE'*

Ecological Conversion

Bruno Bignami

1. What conversion is not

There is no year zero for humankind. There's always something that came before and has something to teach us. This also applies to conversion. It is a question of opening new paths, but there have also been attempts and temptations in the past. The former deserve the utmost attention: it's not a bad idea to praise the courage of those who have tried. Temptations, on the other hand, are precious so that we don't go backwards, either naively or voluntarily. It is possible to review the most widespread and deviant of these.

A first temptation is "pre-packaged conversion". It follows a simple pattern: others ask the questions, while the Church already has the answers ready on the table. A few examples: is there a problem of hunger in the world? You respond to it with "sharing"! Is there the drama of war? You invite "forgiveness": sacrosanct words, but if they remain as abstract principles, it is difficult for them to be credible. An abundance of obvious answers which leads to a festival of banalities is not enough for humankind. The issues at stake for theology and faith are twofold. First, the incarnation is denied, as if it were enough just to reaffirm principles or values to get them accepted on the ground. The encyclical letter *Fratelli Tutti* warns that "Great changes are not produced behind desks or in offices"[1]. Everything looks useless,

* Pope Francis, Encyclical Letter *Laudato Si' on Care for Our Common Home*, 24 May 2015, n. 194. Hereafter LS.

[1] Pope Francis, Encyclical Letter *Fratelli tutti on Fraternity and Social Friendship*, 3 October 2020, n. 231. Hereafter FT.

DOI 10.32060/C.S.9.499-520

in fact, if you don't point out paths that you intend to walk together. Moreover, we put ourselves on a step above, as if the Christian community is not involved in the problems of the world. Much more energetic seems the request of LS 220, according to which ecological conversion leads humankind to develop its creativity and enthusiasm, to put its hand to the dramas of the world. Personal and community involvement is required that enables us to feel part of it, putting ourselves at risk with the willingness to sacrifice oneself, making oneself a gift.

A second temptation is thinking that only others need conversion. Ecological sin could be pinned as a label onto whoever is not a believer. From time to time the recipients of the conversion would take on different team colours: atheists, non-Christians, non-Catholics. Or, within the Church itself, mutual accusations may arise between conservatives and progressives, priests and laity, members of the north or south of the world, associations and movements… It would launch the logic of the scapegoat, who is always outside and never inside ourselves[2]. Theatrics already seen and recognized in many fields: the fault lies with others! LS requires of us decisive steering: "Our goal is not to amass information or to satisfy curiosity, but rather to become painfully aware, to dare to turn what is happening to the world into our own personal suffering and thus to discover what each of us can do about it" (LS 19). Here too, involvement, as a result of awareness, is required, inviting us to see ourselves within a humanity on a journey and involved in a search. The theological question sees an incarnational flaw, but also a lack of reference to the Paschal Mystery. The other is not hell, but the one with whom redemption is possible. Salvation is in the cross and not in condemnation.

A third temptation is the 'middle way'. It's the *tout court* denial of conversion. It embodies the logic of private interest at all costs, true to the *business is business* line. The logic could be summed up in this criterion: if you used to do business by polluting, now you do it by embracing the *green*, decarbonizing or pushing the ecological agenda. Since sustainability pays off economically, it is implemented out of pure self-interest. In fact, 'eco and organic' suffixes are wasted on products and companies. The encyclical warns against such positions:

> It is not enough to balance, in the medium term, the protection of nature with financial gain, or the preservation of the environment with progress. Halfway measures simply delay the inevitable disaster. Put simply, it is a matter of re-

---

[2] René Girard, *Il capro espiatorio*. Milano: Adelphi, 2008[5].

> defining our notion of progress. A technological and economic development which does not leave the world a better place and with an integrally superior quality of life, cannot be considered as progress[3].

The middle way appears in all its traditionalist inconsistency. You change everything so as not to change anything. It's pure fiction. It does not call into question the relational and social mechanisms, which remain intact in their unjust logic. In fact, real change calls for "rethinking processes in their entirety" (LS 197). The downward compromises simply delay a meltdown that is already inscribed in the perverse logic of those who do not know how to rethink themselves in terms of the centrality of the human person within the production mechanisms that tend to homogenize and flatten. The goal is an integrally higher quality of life, where ecology and economy know how to safeguard humankind in its constitutive relational dimensions.

With the fourth temptation one falls victim to cosmic pessimism: everything is useless. The planet is so compromised that there is nothing left to do. Human attempts are futile, because humankind is not able to really change. This negative view is supported by the ecological crisis that in all its severity and struggle seems to attract sufficient awareness. It might be enough to analyze some recurring arguments to reinforce the idea that we will never manage it. There are, in fact, state leaders who argue from the point of view of being a sovereign nation without realizing that pollution or impoverishment of the ecosystem or the loss of biodiversity has no borders. Everything is connected. The Amazon is a World Heritage Site. Environmental disasters are suffered by many populations in the world. Alongside this, there are countries saying they need the possibility of polluting, now that they have entered the world economy and double-digit growth, while until yesterday they were relegated to the 2nd division of the world economic championship. If any country savours the goal of well-being, why should it deprive itself? Faced with such arguments, it is difficult to find a solution. Self-referentiality seems to reign, putting a big strain on builders of hope. Yet Pope Francis says he is convinced that humanity has 'the ability to work together in building our common home' (LS 13). Pessimism would be driven back in the name of realism: it depends on what variant of humankind we educate and call upon. Humankind has been capable of beauty in the past. It knows how to do good. It is a guardian of hope. On the basis of this fact, efforts should be concentrated on the formation of consciences.

---

[3] Pope Francis, *Laudato Si'*, n. 194.

The fifth temptation is called "technological illusion". It is invoked by many: technology, it is presumed, will be able to solve all our problems. Just rely on its extraordinary discoveries. Where there are no solutions yet, it's only a matter of time. Scientific research will solve everything. Hence the inevitable blind trust. Is there a problem with fossil fuel pollution? Electric power and hydrogen will be decisive: human intelligence will come to our aid. For every problem, 'mother' technology has a possible strategy. LS does not hesitate to warn us:

> The idea of promoting a different cultural paradigm and employing technology as a mere instrument is inconceivable because today the technological paradigm has become so dominant that it would be difficult to do without its resources and even more difficult to utilize them without being dominated by their internal logic. It has become countercultural to choose a lifestyle whose goals are even partly independent of technology, of its costs and its power to globalize and make us all the same. Technology tends to absorb everything into its ironclad logic, and those who are surrounded with technology "know full well that it moves forward in the final analysis neither for profit nor for the well-being of the human race", but "in the most radical sense of the term power is its motive –a lordship over all"[4].

The final quote from Romano Guardini in *The End of the Modern World* makes the moral logic underlying the technological paradigm understood. This does not mean that technology should not be valued. Indeed, the problem occurs when it is absolutized forgetting that technological conversion is not enough. Relational quality must be part of the game: humankind does not bend reality, but understands its meaning. It is the understanding of things that determines the value of technology and not technology that deludes us to think that that reality can be dominated and bent according to human needs. Ecological conversion is on a higher level. It needs the value of scientific research, but does not expect salvation from it.

A sixth temptation is 'individual conversion'. It's the one we met just a few years ago. There have been really virtuous, if not prophetic, people. They fully embraced the ecological cause through targeted choices. From those who have rethought city mobility, designing an efficient public service, to those who have perceived the goodness of abandoning fossil fuels through the electric car. From those who have carefully avoided any waste of water to those

[4] Pope Francis, *Laudato Si'*, n. 108.

who have understood the value of energy saving. From those who have taken responsibility for keeping a park, flower bed, garden or neighborhood clean to those who have chosen organic food. From those who prefer a Sunday walk in the countryside to a crowded shopping center to those who choose to invest their savings in green energy. The examples multiply. These are all laudable and generous attitudes: worthy of applause. And yet, they were not enough and are no longer enough. In time of integral ecology, where social and environmental issues intertwine, a deeper perspective is needed. LS describes the necessary change:

> Social problems must be addressed by community networks and not simply by the sum of individual good deeds. This task "will make such tremendous demands of man that he could never achieve it by individual initiative or even by the united effort of men bred in an individualistic way. The work of dominating the world calls for a union of skills and a unity of achievement that can only grow from quite a different attitude". The ecological conversion needed to bring about lasting change is also a community conversion[5].

So, we need community networks. Here also Guardini's words enable us to understand that in these days either conversion becomes community or it simply will not be. It is good to have places for meetings and discussion where community discernment can be made and shared choices arrived at. They concern families, condominiums, neighborhoods, cities, parishes, associations, workers and employers, communities, civil administrations: everyone must ask themselves which are the most responsible choices for a sustainable use of resources. It should be borne in mind that:

> Human choices are never peaceful. They are marked by the torment of imperfection. They suffer the ever present gap between the real and the ideal. It is precisely this gap that provides an opportunity to bring to fruition and offer increasingly adequate answers[6].

Community discernment would have the merit of challenging the two opposing attitudes. There is, on the one hand, the "that's the way it's always been done" attitude and on the other hand the "novelty at any cost" approach which may, in some cases, not be immediately feasible if not at the cost of a

---

[5] Pope Francis, *Laudato Si'*, n. 219.

[6] Bruno Bignami, *Terra, aria, acqua e fuoco. Riscrivere l'etica ecologica*, Bologna: EDB, 2012, p. 199.

'social butchery' in terms of jobs or economic accessability for all. Human decision making in the past has always been subject to the approximation of an *ongoing* research. The laborious and valuable task is to create community networks, places of sharing and open discussion. Urgency must not make us forget the difficulty: individual conversion is much easier. However, it fails to achieve its target: there is a need to renew at the same time both personal and environmental relationships.

Finally, the last temptation is to fall into an 'idealism' of nature. Sometimes one risks taking refuge in a divinization of creation, which is oblivious to its dangers. A virus invisible to the eye, Covid-19, has reminded everyone how this mythologizing is unrealistic. The sense of reality and fragility of creation cannot be lost. Just as we are aware of the threat when atmospheric or terrestrial events happen that destroy the human *habitat*: an earthquake, a cyclone, a storm, a hailstorm, a volcanic eruption. All this gives us the experience of an inhospitable world. And it leads humankind to equip itself accordingly: no houses are built at the foot of an active volcano, one doesn't climb the Tibetan mountains wearing flip-flops, one doesn't go off to the wide open ocean in a rowing boat, we follow anti-seismic architectural criteria in territories at risk. LS calls us to a reality check:

> If we acknowledge the value and the fragility of nature and, at the same time, our God-given abilities, we can finally leave behind the modern myth of unlimited material progress. A fragile world, entrusted by God to human care, challenges us to devise intelligent ways of directing, developing and limiting our power[7].

It is a question of preserving the sense of human limits and approaching nature with the fear of one who knows that not everything is under control. The human illusion of dominating nature remains the same. Nature comes in its multiple forms: beautiful and uncontaminated, ugly and raped by humankind, mysterious and terrible because it challenges our ability to enter into a relationship with the world. It becomes a matter of power. This is why Guardini in "*The Power*" invites us to learn that 'dominion of the world presupposes our dominion over ourselves'[8]. The challenge is in the human capacity for resilience, the ability to grasp the meaning of things, to study problems in depth, to direct our intervention. Power, in fact, can slide towards arrogance.

---

[7] Pope Francis, *Laudato Si'*, n. 78.

[8] Romano Guardini, *La fine dell'epoca moderna. Il potere.* Brescia: Morcelliana, 2015[12], 214.

We are experts on this because it is precisely despotic anthropocentrism that destroyed the care relationship between humankind and creation.

Every temptation focuses on one aspect and absolutizes it. It misunderstands the entirety. LS 138 itself shows that "the fragmentation of knowledge and the isolation of bits of information can actually become a form of ignorance, unless they are integrated into a broader vision of reality". Ecological conversion asks us to go further in assuming what Pope Francis calls the *desborde*, that is, the overflow. A spring, in fact, overflows and refreshes everyone in an irrepressible overabundance. The phrase, used during the Amazon Synod on October 2019 to get out of the quicksand of opposing arguments, can help us not to get bogged down in sectoral or partisan views. Not by chance, the post-Synodal exhortation *Querida Amazonia* turns conversions into dreams: dreams that are social, ecological, cultural and synodal. Already this focus could be precious: holding ecological conversion in tension with God the Creator's dream for humanity and history. We need a fruitful theology that can combine the protological and escatological. And we need a different point of view, which recognises new openings: a vision of the peripheries as real and true "laboratories of innovation".

2. The conversion of the Social Magisterium

The concept of conversion within the Social Thought of the Church (STC) is fairly recent. It emerges only in *Octogesima Adveniens* of Paul VI where in n.48 he speaks of personal conversion. However, a truly meaningful passage is found in *Sollicitudo Rei Socialis* (SRS 38) of John Paul II. Conversion is contrasted with the experience of sin and is described as 'change of behaviour or mentality or way of being'. It is characterised by the capacity to hold together the personal level and the social. Conversion as a renewed relationship with God demands the transformation of the heart in order to fulfill the prophecy of Ezechiel 36:26. The new Spirit is able to remove the heart of stone and replace it with a heart of flesh. So, a renewed relationship with God is united with a fraternal one, the inner dimension with the social. The new heart is the fruit of the action of God in humankind. John Paul II used this biblical category to express the need to overcome the moral obstacles for authentic development. A sign of conversion is awareness of the interdependence between people and nations. Injustices that take place in distant countries challenge the conscience of each person, above all within a social and economic system which is interconnected. He writes: "It is above all a question of interdependence, sensed as a system determining relationships in the contemporary world, in its economic, cultural, political and religious elements, and accepted as a moral category" (SRS 38). The virtue of solidarity springs from this, seen not as vague compassion or a shallow distress at the injustices suffered by people, but it as a "firm and persevering determination to commit oneself to the common good; that is to say to the good of all and of each individual, because we are all really responsible for all".

The encyclical goes on further to analyse the causes that put in jeopardy the development of people: they are the craving for profit and the thirst for power. These attitudes when they are shared by consciences crystallise in "structures of sin". To overcome and defeat them it is necessary to take on the opposite approach: "a commitment to the good of one's neighbour with the readiness, in the gospel sense, to "lose oneself" for the sake of the other instead of exploiting him, and to "serve him" instead of oppressing him for one's own advantage".

The merit of SRS was that of proposing an articulated analysis of conversion starting from the experience of social sin. Every time humankind seeks its own exclusive individual interest and lets itself be drawn by thirst for power, social relationships are excluded. Degraded humanity does not pervade just the single individual, but involves and structures social vision. Cities come out of it the worst, with economic and social consequences. The analysis is exquisitely moral. Sin, however, can only be conquered in a conversion offered by solidarity as a commitment to the good of the other. The biblical categories that express this conversion are those of losing one's life and free service: disinterest in oneself for the sake of the other as a person. Note the structural link from the point of view of moral theology not only between sin and possible conversion, but more in depth between the personal level and the social. Just as personal sin builds models of social life so personal conversion constructs strong relationships and strengthens the common good[9].

The very concept of "interdependence", as taken up in the analysis of John Paul II, is one that is closer to the "connections" proposed by Pope Francis in *Laudato Si'* for taking hold of the transformations in progress in our time and for indicating the way to possible solutions. If 'everything is connected', it is only by inhabiting relational connections can the road out of social in-

---

[9] Sergio Bastianel, *Moralità personale nella storia. Temi di morale sociale*. Trapani: Il Pozzo di Giacobbe, 2011, pp. 177-193.

justices and and environmental degradation be seen. Certainly, in the recent teaching of Francis, only with the Amazonian Synod (October 2019) is the concept of conversion associated explicitly with that of ecological sin, reconstructing the moral theological framework already defined in SRS. However, a close reading of LS does not leave room for misunderstandings of any kind. Francis, in fact, defines in the document different destructive and problematic approaches (instrumental reason, the technical paradigm and relational relativism ...) which display the same outcome: people and creatures are reduced to mere objects. The instrumentalisation of the creation of God, the reduction of people to human waste, consumerism, the destruction of habitats, the commodification of common goods, social exclusion are dramatic outcomes of a lack of fraternity. The reference is to the famous analysis of Paul VI in *Populorum Progressio* 66, taken up again by Benedict XVI and by Francis: "Human society is sorely ill. The cause is not so much the depletion of natural resources, nor their monopolistic control by a privileged few; it is rather the weakening of brotherly ties between individuals and nations".

LS 106-112 proposes a detailed analysis of the negative consequences of the technocratic paradigm. It produces a distorted relationship with reality. Humankind deludes itself that creatures are simply there at its disposal, promoting an attitude of manipulation. Human intervention in nature is not under discussion, but the relational model "presupposes the lie about the infinite availability of the goods of the planet" (LS 106). This paradigm structures the constitutive relationships of humanity in an optic of dominion. It tends to make things such that "nothing remains outside of its iron-clad logic" (LS 108): power becomes the way in which one conducts oneself in dealings with nations, people, creatures ... it is seen in the field of the economy and politics. "Finance suffocates the real economy" (LS 108): the workers become costs not richness, hunger continues to kill, inequalities increase and social imbalances take root ever more deeply in society. Furthermore, the capacity to have a collective view in the solution of the problems decreases: the sense of the totality is lost along with the existing relationships between things and the wider horizon. All this brings us to acquire partial vision.

Faced with such complex questions, the sixth chapter of LS on spirituality and ecological education dedicates a number of passages to the theme of ecological conversion (216-221). Christian spirituality allows us to gain a mysticism capable of animating and motivating the commitment of our times. There is need for a "way of thinking, of feeling, and of living", even before ideas and innovative proposals. Without this dimension which warms the heart, conversion appears in terms of inconclusive moralism. Spirituality nourishes a passion for care of the world: it moves interiorly, it encourages, it gives meaning to personal and community commitment. From it spring up different levels of the one ecological conversion.

The first is the interior one, which expresses itself in the transformation of the heart. Repentance is needed regarding sinful relationships so as to bring about change from within. This means knowing how to examine our lives so as to understand the contribution each one is making, negatively, to the ecological crisis. No-one can consider themselves excluded, but a truly innovative practice is born out of this profound interior meeting between humankind and God. A healthy relationship with the creator promotes an "integral conversion of the person" (LS 218). This level however is not enough. The involvement of the community is necessary because "social problems must be addressed by community networks and not simply by the sum of individual good deeds" (LS 219). Community conversion exceeds the initiatives of individuals in tightening different relational models which lead to cooperation, alliances, laws, sharing of intentions, joining of forces. The points of no return at the social, economic and political level spring from shared discernment. A process of enduring change is begun, a dynamism of growth in awareness and a better quality of social experience. The community networks correspond to a further maturing of conversion because they can affect the bonds that constitute human life.

In this way, ecological conversion leads to the development of creativity and enthusiasm, to recognising the relationships with everything that God has created and implies free gift and tenderness. Upstream there is an anthropology and a way of seeing things: awareness increases that "every creature reflects something of God and has a message to communicate" (LS 221). From the Christian point of view the light and strength of the Holy Spirit extends to relationships with all creatures. Saint Francis of Assisi reached the peak of this spirituality when, at the moment of his existential decline, he calls upon creatures as brothers and sisters. Just at the time when his strength began to weaken, his physical condition became fragile and his blindness advanced inexorably, the poor man of Assisi wrote the Canticle of the Creatures to the glory of God and in praise of the beauty of creation. It was the year 1224.

The encyclical LS represents a hymn to the greatness of God but it is also an act of trust in human capacity. One can always hope. One can still heal things.

Some different relational models are already present. They are recognised and supported. LS 112 tells us:

> Liberation from the dominant technocratic paradigm does in fact happen sometimes, for example, when cooperatives of small producers adopt less polluting means of production, and opt for a non-consumerist model of life, recreation and community. Or when technology is directed primarily to resolving people's concrete problems, truly helping them live with more dignity and less suffering. Or indeed when the desire to create and contemplate beauty manages to overcome reductionism through a kind of salvation which occurs in beauty and in those who behold it. An authentic humanity, calling for a new synthesis, seems to dwell in the midst of our technological culture, almost unnoticed, like a mist seeping gently beneath a closed door. Will the promise last, in spite of everything, with all that is authentic rising up in stubborn resistance?

There are good practices which enable us to hold on to hope. Conversion is already taking place. And it is as if we were asked: 'Are we aware of this?'

The encyclical *Fratelli Tutti* does not add any new elements either relating to ecological conversion or to the theme of conversion in general. There is an underlining which deserves to be considered: it is found in n.68, in the context of the commentary on the parable of the Good Samaritan (chapter 2). Francis wants our anger when someone is pushed to the margins of life. The personal dignity of each of us is at stake. Anger has to dig deep, "so as to make us descend from our serenity to get involved in human suffering" (FT 68). A close relationship is signaled between anger and conversion: human suffering, if truly encountered, never leaves things as they are. It leads us to act in order to change the situation and to raise up the downtrodden to dignity. Rather than speak of conversion, FT notes forms of risen life to inspire us in this our age and shows conversions experienced in the Church itself. In n.54, in fact, we are invited to recognise that God continues to sow goodness in the furrows of humanity marked by the pandemic. Here, conversion appears as a renewed look at human life: the hero with a high-sounding name is not to be found among the great and famous of the earth. Rather, between the folds of the ordinary there have been people who have been revealed as capable of being cured: 'doctors, nurses, pharmacists, supermarket and transport workers, men and women who work to provide essential services, volunteers, priests, religious .... they have understood that no one is saved alone'. The message is clear: salvation is not the same as security. Humanity capable of being cured contributes to the former, the latter represents a distraction for those who don't want to find themselves committed.

The involvement of the Church in the history of humanity shows how the path of Christian ethics is neither pacific nor linear. Without a critical mindset and without evangelical discernment the danger is always round the corner of ending up engulfed by worldly logic. An example is the slowness, recognised by the same Pontiff, with which the Church "condemned slavery and different forms of violence with any strength" (FT 86), as if Christian theology and spirituality were not able to act as a catalyst in arguments of cultural tendencies contrary to fraternity. The encyclical shows a successful ecclesial conversion around two themes: the just war and the death penalty. On both these questions the up to date message on war and justice is brought to our attention: on the one hand we can "no longer think of war as a solution" (FT 258) given the certain disproportionality that occurs with the use of ever more destructive armaments; on the other hand, the death penalty is "inadequate from a moral standpoint and no longer necessary from that of penal justice" (FT 263). Even while not dealing with ecological conversion, FT shows that conversion in general concerns above all the whole thought and moral teaching of the Church. Not a static ethic, defined once and for all for everyone, but, subject to evangelical discernment in the course of history, it can respond to the plan of God. It is not only the conversion of others that is asked for but one recognises it already present on one's path. Moreover, an important awareness for ecological conversion is also noted: within the scope of integral ecology the destructiveness of war and the recognition of the dignity of every person are key points not to lose sight of.

3. The Phenomenology of Conversion

Conversion is not born from nothing. It requires a way of thinking and a certain perspective. As Mauro Ceruti suggests, "thought in crisis is impotent faced with a world in crisis"[10]. A vision is needed which often is born from a radical disorientation confronted with the situation in which one finds oneself[11]. The courage of real change is only possible if the categories of thought and action, which have been used up to this point, are set aside. For Italo Mancini this is the favourable condition for the birth of utopia. If every crucial transformation has

[10] Mauro Ceruti, *Sulla stessa barca*. Magnano: Qiqajon, 2020, p. 14.

[11] Italo Mancini, *Teologia, ideologia, utopia*. Brescia: Queriniana, 1974, pp. 467-472.

its starting point in thought, it is similarly true that a thought that is not organised, not structured and which does not "bite" into reality ends up being sterile. Ideas alone are ineffective for changing the world and humanity: it is necessary to bring thought to the level of action and decisions.

Conversion puts its own credibility at stake at this level. It connects theory and praxis. One doesn't stop with just a good idea, but also organises good practice. Orthodoxy calls into question orthopraxis. In this way the close bond between the subject that acts and the world is highlighted. The world comes out transformed by human actions but the opposite is also true, that is, that the subject comes out converted by the fact that changes are possible. Thought does not carry the rejection of reality, but sees in it a static test of its possibilities of realisation. Perspectives of scope and meaning offer depth to history. Conversion is grafted onto the ability to join thought with commitment, ideas with militance, plans with decisive action. Thus one becomes capable of expressing something new that regenerates the world. This happens thanks to the power of a thought incarnate in praxis. For the believer all this is illuminated by the Word of the Gospel. Conversion assumes the characteristics of a gesture of hope, because thought is structured not as a simple instrument of dominion or knowledge as an end in itself, but it is the result of a restless heart that allows itself to be transformed and listens to reality. We see an example in Pope Francis, who in the book-interview *Terrafutura* with Carlo Petrini, founder of Slow Food, tells how his ecological sensitivity was born, and the origin of the encyclical *Laudato Si'*. Attention to environmental themes has matured in the meantime. He remembers:

> When, in 2007, I went as Bishop of Buenos Aires to the 5th Episcopal Conference of Latin America and the Caribean at Aparecida in Brazil, I remember the forcefulness with which the Brasilian Bishops spoke of the great problems of Amazonia. With every pious breath they expounded the argument dedicating streams of words to the environmental and social implications of the issues in question. I remember well having been irritated by this approach and to have also commented: "These Brazilians are driving us mad with their speeches!" Indeed I didn't understand why our meeting of bishops had to be given over to the theme of Amazonia. For me the health of the green lung of the world was not a preoccupation, or at least I didn't understand how it came into my role as a bishop. With the passage of time however, to the team redacting the final document there continued to arrive concerns on this front, also from the columbians and equadorians. I insisted on leaving them aside,

> telling myself that I didn't understand this urgency and insistence. A lot of time has passed since 2007 and I have completely changed my view of the environmental problem. Then I didn't understand, and seven years later I wrote the encyclical[12].

The conversion of Bergoglio happened thanks to the cry of the Brazilian bishops, interpreters of the sufferings of their people, and to his inner personal restlessness which knew how to react in the face of external provocation. Between irritation and acceptance there stands the capacity to listen. That very restlessness allows a person to be drawn towards new horizons, opening up to the other and their cry. One could speak with the philosopher Silvano Petrosino about the "intimately driving, *humanly fruitful* character of restlessness itself"[13]. Restlessness can go further than need, which stops instead at recognising what is not there, a limit, a lack which does not manage to grasp a huge appeal coming from reality. Human desire, when it is left to dwell in restlessness prepares for exodus and exile. There is an alternative. There is another who calls. There is a voice that makes an appeal and wants an answer. Inside this exodus there is personal conversion. We let ourseles be transformed by otherness, the presence of which seeks hospitality in its own sphere of living.

We come across this paradigm of conversion in many biographies. For example, Bishop Oscar Arnulfo Romero was converted to the cause of the Salvadorian people after having seen the testimony of the Jesuit priest Rutilio Grande, assassinated on 12th March 1977. Conversion made him forget his many fears so as to take up with courage the fate of this oppressed people. And how many other experiences could we cite? Michel de Certeau speaks of "personal ecstasy", of experience of brokenness that breaks in. A breach is opened, a gash occurs through the presence of the other which refers to the presence of He who is "greater": "there occurs something that surprises and that creates a starting point"[14]. It is the privileged moment from which a pathway begins that defines in a decisive way its own personal history. A journey begins which is called vocation, a second call beyond that initial one of life itself.

It is a good moment to reveal the same dynamic which underlies conversion as an opening to the other and the command given by God "to cultivate and

[12] Carlo Petrini, *Terrafutura. Dialoghi con Papa Francesco sull'ecologia integrale*. Firenze-Milano: Giunti-Slow Food Editore, 2020, p. 24.

[13] Silvano Petrosino, *Dove abita l'infinito. Trascendenza, potere e giustizia*. Milano: Vita e Pensiero, 2020, p. 84.

[14] Michel de Certeau, *Mai senza l'altro. Viaggio nella differenza*. Magnano: Qiqajon, 2007, p. 25.

care for the earth" (Gen 2:15). Living on the earth is not the same as occupying it, but to put oneself at the service of its possibilities for life. 'Humankind exists and lives as humankind not because it is in existence and lives its life, but because it takes "care" of existence and life, that is it "lives" on it, and it "lives" on it in so far as it cultivates-and-cares for whether it be existence or life'[15]. What happens surprisingly, thanks to personal action, is that in cultivating and caring, humankind discovers how to build something new. We transform. We invent. We plan. There is no human action which can be configured as creation from nothing, because it is born within something that precedes it, which stimulates and represents a beginning for those who take hold of it. Conversion is made possible inside the experience that humankind is not enough in itself. There is something else around us that moves us and moves us on. There is more that calls us to action. It motivates us in making decisions. It calls us into question as a moral conscience. It gives us a glimpse of new beginnings, because we are able to open up different horizons and views previously overlooked and forgotten[16].

Sometimes it is anger in the face of injustice that brings about a reaction. Other times it is reference to possible traumas or tragedies of humanity that pushes the human conscience into action. Hans Jonas has already illustrated that the heuristic of fear can favour moral responsibility. Its moral imperative leads us to look at the consequences of action: Act in such a way that the consequences of your action are compatible with the lasting presence of authentic human life on the earth"[17]. Beyond the philosophical debate about the wisdom of founding responsibility on fear, it remains true that there is a role for moral emotions in human choices. It is Elena Pulcini who deserves credit for investigating the link between fear, fragility and global ethic[18]. Care is the response to the imperative to preserve life and humanity. And the reference to fragility calls into question the reflection of Paul Ricoeur, according to whom that which is fragile is by its nature threatened by the violence of history and

---

[15] Silvano Petrosino, *Lo spirito della casa. Ospitalità, intimità e giustizia*. Genova: Il melangolo, 2019, p. 39.

[16] On this subject: Chiara Giaccardi and Mauro Magatti, *Nella fine è l'inizio. In che mondo vivremo*. Bologna: Il Mulino, 2020.

[17] Hans Jonas, *Il Principio responsabilità. Un'etica per la civiltà tecnologica*. Torino: Einaudi, 1990, p. 16.

[18] Elena Pulcini, *La cura del mondo. Paura e responsabilità nell'età globale*. Torino: Bollati Boringhieri, 2009, pp. 220-269.

has the power to reawaken the heart, to cause pain, to provoke anger faced with something that "we denounce as deplorable, unsustainable, inadmissible, unjustifiable"[19]. Care corresponds to carrying the burden of the fragility of the other, vulnerability in our age is global such that responsibility for it being experienced and concrete demands that death of the individual is decreed as the single criteria for action and is master of the world. The necessary grief at the death of the "I" is the condition for safeguarding humanity at the global level. For this, Francis encourages us "to think of ourselves more and more as a single family dwelling in a common home" (FT 17).

These reflections, which can appear abstract, in reality find confirmation in certain positions held by the Magisterium of the Church itself. It is well established, in fact, that the encyclical *Pacem in Terris* (1963) by John XXIII responds to the need for a calming down in the face of the Cuban crisis right in the middle of the cold war. The arms race, in the eyes of the pope from Bergamo, puts at risk the common good of peace, especially in the light of what happened at the end of the Second World War, when the use of nuclear weapons opened up a threat to humanity. There was the concrete possibility that humankind could annihilate itself and destroy entire territories of the planet. For this reason the Pope recognises that "human beings live under the nightmare of a hurricane that could be unleashed at any moment with unimaginable consequences" (PT 60.) The danger of the cold war moved the Pontiff to make an intervention which then was deepened in the Pastoral Constitution *Gaudium et Spes* of the Second Vatican Council. Number 80 warns of the risks of total war, seeking a rethinking of the Catholic position on war. The theme according to the conciliar fathers had to be tackled with a "completely new mentality": an expression that leads directly to conversion[20].

Even more significant is the historical importance of the actual ecological question. The risk of the self annihilation of humankind is present in the com-

---

[19] Paul Ricoeur, "Le sfide e le speranze del nostro comune futuro", in *Persona, comunità e istituzioni. Dialettica tra giustizia e amore*, edited by Danese, Attilio. Fiesole: ECP, 1994, p. 108.

[20] The Council states in *Gaudium et Spes* 80: "The horror and perversity of war is immensely magnified by the addition of scientific weapons. For acts of war involving these weapons can inflict massive and indiscriminate destruction, thus going far beyond the bounds of legitimate defense. Indeed, if the kind of instruments which can now be found in the armories of the great nations were to be employed to their fullest, an almost total and altogether reciprocal slaughter of each side by the other would follow, not to mention the widespread devastation that would take place in the world and the deadly after effects that would be spawned by the use of weapons of this kind. All these considerations compel us to undertake an evaluation of war with an entirely new attitude".

plex relationship between the human species and the environment. Global warming, the pollution of the land and the seas, the consumption of land, the plundering of resources, social inequalities .... put the balance that keeps human life on this earth at risk. LS 55 speaks of "behaviour that at times seems like suicide": sustainability is proposed to safeguard the ecosystem. This means taking into account "the capacity for regeneration of every ecosystem in its diverse sectors and aspects" (LS 140). Taking care of the environment is taking care of humankind and its own possibility of life. Benedict XVI in *Caritas in Veritate* 51 already invited us to "protect above all humankind against the destruction of itself". Faced with fateful dangers to the planet the Magisterium proposes a model of humanity that can be thought of as a community of destiny, as the human family. On the same lines the encyclical *Fratelli Tutti* finds a place, which was published right in the middle of the pandemic. In this case too the Magisterium responds to the request for hope following this health crisis and asks for the courage of conversion regarding certain themes which surfaced at the beginning of this millenium: migrants, the crisis of democracy, economic inequalities, new forms of social exclusion. It becomes clear that the Magisterium itself intervenes and makes a discernment on the situations and warns about attitudes that are destructive or devoid of hope.

It does not scandalise us to note that conversion is born by listening to history, if not in reaction to what is happening or to avoid irreversible adversity. However, this needs to be put in a generative context. The rejection of existing evil is transformed in positive choices[21]. The culture of waste becomes the culture of care. The competition of the market society gives way to solidarity. The search for maximum individual profit in the economy is replaced by cooperation and social benefits. Consumerism is supplanted by a sense of moderation. The logic of waste has the double downside of destroying both creation and the world of humanity, depriving people of their *habitat*. Care transforms time from a simple succession of moments *(chronos)* to unique and paradigmatic opportunity *(kairos*). All this helps us to understand how there is no normal time, but every season can become "Passover" (Easter) of Christ, a call to conversion. And this call always passes through a wound in human life or society. For example, this season of the pandemic reveals the inconsistency of the individualistic illusion. We have a future only if we think

---

[21] On this subject see the precious pages in Luigina Mortari, *Educazione ecologica*. Bari-Roma: Laterza, 2020, pp. 151-168.

of ourselves as within a common destiny[22]. The interconnections in which we find ourselves involved, make the horizon "of unity and sharing a common destiny among the nations of the earth, more and more palpable" (FT 96). History presents the extraordinary opportunity to think about the common vocation of humanity, which is that of knowing how to welcome one another through reciprocal care.

4. The ecological conversion

In the environmental field conversion relates to sin. Usually it's called 'ecocide', which means pollution of water, air and Earth's resources. Similarly, the significant loss of biodiversity both of flora and fauna leads to the destruction of an entire ecosystem. Ecological disasters go on to affect the right relation between humankind and nature. Many cases have already occurred, where humankind have been forced to flee from territories that have become inhospitable or where human impact is so destructive as to turn ecosystems into desert places. Ecological sin is not an invention of our time in order to update the sin list to 'current fashion', but it is fruit of a renewed awareness. Together with the possibility of denying the relationship between man and man, current humankind can degrade this God - given gift. So pollution affects the constitutive relationship between God the Creator and the human creature, between the human creature and different creatures. The tragedy arises in a disharmonious backdrop between creaturehood and creativity. The first one situates man within the vital relationship with the Creator. Creativity separated from creaturehood absolutises the work of *homo faber*, who deludes himself that he is all-powerful. Ecological conversion brings the rainbow back into the relationship between creaturehood and creativity: from the divorce caused by despotic anthropocentrism to a union lived in the balance of integral ecology. After all, ecocides makes one aware of the bond that socially unites all the members of a community and that holds all living beings together.

Anyway, ecological sin relates to conversion as an outcome full of hope. It can trigger paths of conversion that bring into play bad habits, damaging behavior and dangerous choices. Every conversion is an invitation to personal moral responsibility called to take part in the care of creation. However, it has to deal with reality, therefore it is committed to promoting the concretely possible

---

[22] Mauro Ceruti, *Sulla stessa barca*, pp. 25-38.

good. Single behavior, virtuous and shared, can radically turn an ecological sin into state of conversion. Small steps in a horizon of constant progression can widely expand and become prophetic good practices. Obviously not everyone can influence things in the same way, but it is always wise to remind ourselves that our actions express who we are. Different levels of decision-making will have a different social and political weight. However, no consciousness acting responsibly can consider itself irrelevant. As Romano Guardini reminds us: "It is a horrible illusion to think that action stays 'outside' of who is acting, since actually that action penetrates him too, indeed himself before the object of his acting. Actually he constantly 'becomes' what he 'does'"[23]. If this is true for everyone, it is also true in the social field. Every good behavior makes one a good person and becomes an invitation to goodness. Ecological conversion comes true within this sharing. It's not about spreading good practices, but it's about transforming hearts: action gradually transforms the relation between man and environment and modifies social experience. Gratuitousness of gestures of good has a saving effect on the persons who carry them out and on community life. It provokes reflection, calls for change and drives us to do better.

Conversion to ecological culture is depicted by Pope Francis in this way in LS 111:

> Ecological culture cannot be reduced to a series of urgent and partial responses to the immediate problems of pollution, environmental decay and the depletion of natural resources. There needs to be a distinctive way of looking at things, a way of thinking, policies, an educational programme, a lifestyle and a spirituality which together generate resistance to the assault of the technocratic paradigm. Otherwise, even the best ecological initiatives can find themselves caught up in the same globalized logic. To seek only a technical remedy to each environmental problem which comes up is to separate what is in reality interconnected and to mask the true and deepest problems of the global system.

The complaint approach to environmental issues is no longer enough. It generates a culture of opposition, incapable of stirring minds, of making a leap in quality to accompany new paths. A different view is needed, a thought, a politic, an educational discourse, a life style and a spirituality opening an ecological culture. Everything leads to a new development model, to a vision of progress "which is healthier, more human, more social, more integral" (LS 112). Man is at the center, with his relationships that constitute him and make

[23] Romano Guardini, *La fine dell'epoca moderna*, 178.

him grow. This is the most serious opposition to the technocratic paradigm.

Environmental sustainability is often supported among contradictions hard to reconcile. So you end up playing the 'Cloven Viscount' by Italo Calvino: somebody embraces sobriety when you talk about climate change and is the champion of consumerism when it's about places of work. There are those who complain about global overpopulation, especially in the poorest countries, and at the same time calls for local birth growth. There are those who would like to limit the productive expansion of China and India and at the same time keep talking about economic growth as a criterion for resolving the European crisis. There are those who sponsor the measurement of well-being as a new social approach and continues to use GDP for the assessment of economic performance. How can we deal with persistent schizophrenias?

The cultural approach has an infallible verification method. You are on the right track if you carry out the principle that has been repeatedly invoked by Pope Francis, according to which "time is greater than space". It is a matter of carrying out processes, of triggering paths more than inhabiting spaces and conquering areas of power. To do so you need to put conscience at the center. The manipulative style that often characterizes humankind doesn't facilitate a change of pace. Generating virtuous processes is possible only if consciences are formed, not just through laws or rules.

Today more than ever it is necessary to take paths that allow us to escape from the duality of the past: development or sustainability, environmental crisis or social crisis, global or local dimension, health or work, environment or economy. A new thought, starting from LS, will lead to developing methods that keep things together, weave networks, build connections. It is worth recalling the good practices already existing in many places: authorities, communities, businesses, families, people who are leaders in the field of sustainability, of circular economy, of *green* finance. Moreover, why not tell of the good practices of particularly strong and attentive communities? These situations support discernment in concrete reality increasingly required. It is a chance to make the right choice, capable recognising the value of people, of their work, of their human investment, of their talents: from the simple act of doing the shopping to taking political decisions, it is possible to build a new model of development and to give proper recognition to people's faces. When the territory expresses different vocations and does not fall back on a single reality, we can see a reflourishing of economic and social enterprises. To be honest, they are a re-flourishing of life!

5. The pangs of novelty

The author, Emanuele Severino says this:

> Phenomena, even of crowd stultification, have a reason. We are abandoning values of traditions. But we are not yet making use of chances given by the inevitable future in which technology will dominate. We are in this intermediate time, I'm using the metaphor of those trapeze artists who, being initially attached to the trapeze, they leave it to grab the other one, whether or not there is a net under them, but in the meantime they are suspended. We are in this time of suspense that is full of meaning[24].

Can we agree with Severino? Is that really so? In any case, the image of the trapeze artist suspended between one support and the other looks appealing. The Christian concept of conversion clearly expresses the change needed to inhabit the present and the future. After all, the call addressed by Jesus to the first disciples to change from being fishermen to being fishers of men is no joke. The 'repent, and believe in the Gospel' represents not only a change of pace. It is a new horizon. Christianity is born on these premises. For the men of Galilee, it was neither a mere change of time and way of fishing, nor moving to another lake… But it was a different relational context.

We are living interesting times. Pope Francis often proclaims a change of epoch, much deeper than a time of change. Compared to the sequence of events, their ability to guard the meaning of life, even in discontinuity, is much more important. Actually, the rethinking of forms and times of work is at stake, together with the economic model, the role of the market and of the public, the vision of the environment, a social paradigm, a way of relating among people.

Can novelty be better represented by the metaphor of turning the page or by the one of birth pangs generating a new world? The nocturnal dialogue between Jesus and Nicodemus relates to the second image (John 3): Christ teaches about being born from above, 'from water and Spirit' (v. 5). The misunderstanding between the two nocturnal dialoguers is played out around a materialistic vision of birth. To this rebirth seen as 'entering a second time into his mother's womb' (v. 4), Jesus contrasts birth from the Spirit. It may appear as an abstract discourse, but it is what usually occurs in people's life, when they let themselves be visited by grace and come to understand a new

[24] Monica Mondo, "Anche se stupidi, sono tempi interessanti. Interview with E. Severino", *Avvenire* 22 gennaio 2020, p. 18.

path of life. They start a family, undertake a vocational path, leave a business for another one that they sense as their life's 'vocation', understand that they can dedicate years to political service, give the go ahead to a company or a cooperative… That is what happens to disciples called by Jesus in a moment of their ordinary working activity to a new configuration of their life. They set out again, overcoming the temptation of resignation. They simply continue on their way.

THE NEW CREATION IN CHRIST AS THE FOUNDATION OF AN INTEGRAL ECOLOGY

Ligita Ryliškytė, SJE

Teach us … to recognize that we are profoundly united with every creature as we journey towards infinite light[1].

1. Introduction

The time we live in is marked by a growing awareness that everything in our world is interconnected. It is not just that we know ourselves to be made of the dust of the stars that exploded billions of years ago. The recent experience of the pandemic makes us painfully aware how much life forms as we know them depend on the subtle equilibrium of natural forces, within the human organism and without. As moderns, we all tend to think of this interdependence in terms of probabilities and odds, of chance and contingency. How we make sense of this, however, differs. For instance, while many marvel at the scantiness of the chance that the rate of the expansion of the universe should fall precisely within the narrow range that makes the existence of our world possible, not everyone discerns in this the divine Providence that has arranged things in such a way that the world process is truly contingent and still upwardly oriented. Secularist and materialist outlooks are further buttressed by evolutionary accounts of the universe that overlook the significance of human morality or culture. Grounded in empiricist epistemologies, these accounts

[1] From "A Prayer for our Earth", Pope Francis, Encyclical Letter *Laudato Si' on Care for Our Common Home*, 24 May 2015, (hereafter LS), n. 246.

DOI 10.32060/C.S.9.521-548

eclipse the fact that human intelligent, reasonable, and responsible freedom (or the lack thereof) affects the probabilities of the emergent world order. Likewise, empiricist sciences are not well equipped to account for human freedom as mediated in and through culture—and by culture we mean "the set of meanings and values that informs a way of life"[2]. This omission frequently results in the lack of recognition that humans are an integral part of the evolving universe; an overzealous biocentrism easily follows. However, upholding the dichotomies nature-nurture, chance-law, and biocentrism-anthropocentrism is not as sound at it seems. To the contrary, reflection on our life in the 'common home' (LS, 164), especially as affected by the ecological crisis and the recent pandemic, mandates the rejection of these dichotomies as intellectually and morally an unacceptable choice.

The better alternative, as the encyclical *Laudato Si'* by Pope Francis abundantly demonstrates (and the most recent encyclical *Fratelli Tutti* endorses), is to conceive of the world process as a multi-layered and complex dynamism where human values and meanings are of key importance for preventing the imminent ecological crisis. Human free rationality and love can shift the evolving world order, accelerating, slowing down, or preventing the disaster. To use the idiom of the major interlocutor of this paper, Bernard J. F. Lonergan, SJ, the encyclical *Laudato Si'* supports the assumption that the universe is subject to both 'classical' and 'statistical' laws, that is, it represents the interplay of regularities (such as the laws of physics) and contingent events (such as quantum indeterminacy and human freedom). The result is that the world comes to be according to probabilities. We cannot change the laws of nature but, in certain circumstances, by changing the conditions under which these laws operate, we can affect the probabilities of their realization. Thus, the ecological crisis is not unavoidable but—if nothing changes in the way we live—highly likely. As Pope Francis argues, to prevent it from happening, humanity needs to embrace an integral ecology—the ecology that rejects false dichotomies and embraces a simple life style that is attuned to the Spirit, the liberator of all creation (cf. *Rom* 8:21-22). The present global ecological and health crises show us that we can no longer think of nature as always capable of self-healing, no matter what we do, or of ourselves as insulated from the effects of our own unbridled consumerism, technocracy, and instrumental

---

[2] Bernard J. F. Lonergan, *Method in Theology* [1972], Toronto: University of Toronto Press, 2013, p. xi.

approach to creation. Neither can we think of ourselves as agents of progress unless we redefine what the progress is (LS, 194) by pursuing the path of human authenticity and ongoing integral conversion.

Pope Francis is not alone in promoting a complex, integral, and dynamic approach to the world order. There is a growing interdisciplinary dialogue between scientists, philosophers, and theologians who reject false dichotomies between nonhuman and human creation and show that it is possible to integrate the evolutionary and emergentist worldviews with biblically grounded *fides quaerens intellectum* (faith seeking understanding). Recent contributions to a non-reductionist understanding of the world's coming-to-be include the works by Sarah Coakley, Martin Nowak, Denis Edwards, etc., to be briefly discussed later. From the perspective of systematic theology, these contemporary efforts represent a modern shift away from still further—and properly theological—dichotomy between the doctrines of creation and redemption, the dichotomy commonly accompanied by the subordination of creation to redemption. In my view, no one has articulated this much needed shift better than Edward Schillebeeckx, OP, who discerned in the Christ event "concentrated creation"[3].

2. The Rationale and Objective

To clarify the *raison d'être* of this paper, *Laudato Si'* also reminds us that the doctrines of creation and redemption form a unified whole: nature and nurture, creating and healing, human and nonhuman worlds are intimately interconnected (e.g., LS 117, 138). Humanity and its 'common home' with all its creatures, as the encyclical suggests (and as this paper further develops), can be conceived as a complex and multilayered network. There are three fundamental relationships which, according to the biblical creation narratives, are constitutive of human life in this network: with God, with our neighbor, and with nature; all three are affected by sin (LS, 66). As inseparable from the divine act of creation, salvation involves the restoration and healing of all three relationships (LS, 119). In the New Testament, the continuity between creation and redemption is especially endorsed by the Pauline parallel between the first creation and the new creation in Christ (Col 1:15-20; Eph 1:9-10). But how can we understand this continuity in explanatory terms?

[3] Edward Schillebeeckx, *I Am a Happy Theologian: Conversations with Francesco Strazzari*, New York: Crossroad, 1994, p. 50.

And how does that understanding illuminate the foundations of an integral ecology?

By drawing on biblical scholarship and Lonergan's theories of history and generalized emergent probability, this paper argues that integral ecology must be founded on re-thinking redemption as the restoration of the dynamic world order. In contrast to understanding sin and salvation with respect to human beings alone, this approach allows us to reclaim the biblical notion of redemption as the new creation of all things in Christ. Christ's resurrection as representative human being, raises up the world as a dynamic system because human beings are an integral part of that system. In unity with Christ the head, all things are renewed and held together in and through the whole Christ, the relational and interpersonal order grounded in the Triune *agape* love. This '*agape*-network' is instrumental in shifting the probabilities of the world's coming-to-be. In resonance with Pope Francis's recent encyclical *Fratelli Tutti*, Lonergan explains that this body comes to be through the diffusion of divine friendship to everyone, even to one's enemies. Because of sin, it is fitting that this friendship is diffused according to what Lonergan calls 'the Law of the Cross', that is, by transforming evil into good.

We proceed in three steps. First, we examine the biblical support for the unity between creation and redemption and the christomorphic dimension of the world's coming-to-be. Second, we creatively appropriate Lonergan's theories of history and generalized emergent probability as an explanatory counterpart to the biblical doctrine. Finally, Lonergan's contribution is brought into direct conversation with the encyclical *Laudato Si'*.

3. Creation and Redemption in the Bible

Biblical scholarship here is engaged to support my explanatory account of redemption as the new creation in Christ that emerges according to the Law of the Cross. Thus, this section does not seek to provide a comprehensive report on the biblical scholarship, in and through its development[4]. Furthermore, as

[4] An interested reader is advised to consult the works by biblical scholars, such as Richard J. Clifford, SJ, to whom this brief discussion is highly indebted; see especially: *Creation Accounts in the Ancient Near East and the Bible*. Washington D.C.: Catholic Biblical Association, 1994; "The Hebrew Scriptures and Theology of Creation", in *Theological Studies* 3/46-1985, pp. 507–523; "The Bible and the Environment", in Kevin W. Irwin and Edmund D. Pellegrino (ed.), *Preserving the Creation: Environmental Theology and Ethics*. Washington, D.C.: Georgetown University Press, 1994, pp. 1–26.

a systematic theologian, I look for patterns rather than particular instances in Scripture. While biblical scholars are concerned with Paul as Paul, my two major concerns here are (1) how the Pauline image of the new creation in Christ is informed by the Old Testament narratives of creation and (2) how this illuminates our understanding of redemption.

3.1 Creation and Redemption in the Hebrew Scripture

Let us start by noting how intimately the doctrine of creation in the Hebrew Scriptures is bound to the doctrine of redemption. This can be demonstrated by first recalling some of the key characteristics of biblical cosmogonies. But before we proceed any further, a word of caution is in place. Since the heightened modern sensitivity to nature and ecological issues has fueled interest in the Bible's creation narratives, biblical scholars warn us about the risk of inscribing a modern spirit into the ancient texts[5]. In response to an unqualified anthropocentrism and technocracy, such modern views frequently overlook the fact that the Bible's accounts of creation are predominantly anthropocentric, not biocentric. Biblical anthropocentrism follows from the primary intent of the biblical authors. The cosmogonies of the Bible and the ancient Near East did not seek scientific understanding about the origins of the world but attempted to explain under what assumptions human life as social and historical makes sense. They were primarily interested in the emergence of an organized society and thus placed human concerns at the center of their attention.

However, biblical anthropocentrism is relative, not absolute. It is relational and historical and, furthermore, qualified by theocentrism: the world is not just "the world for the human being" but, even more fundamentally, "the world of God"[6]. Not the least, this is evident in the biblical *imago Dei* doctrine: since human beings are made in the image of God (Gen 1:27), human existence is to be their relationship to God. As the divine mandate to stewardship (Gen 1:26, 29-30) further elucidates, our correspondence to God, "is not static but dramatic; it is shown in action"[7]. The *imago Dei* is fully realized when the human being acts as the representative or image of God, 'ruling' over the world

---

[5] Richard J. Clifford, *Creation Accounts,* p. vii; "The Hebrew Scriptures and Theology of Creation", p. 509.

[6] None of the books in the Bible qualify biblical anthropocentrism as radically as the Book of Job (see Chaps. 38–42); cf. Edward Schillebeeckx, *Christ: The Experience of Jesus as Lord.* New York: Seabury, 1980, p. 530.

[7] Richard J. Clifford, "The Hebrew Scriptures and Theology of Creation", p. 522.

not as a sovereign but as an intelligent, responsible and loving steward who is ultimately accountable to God. Consequently, as Pope Francis reminds us in *Laudato Si'*, "our 'dominion' over the universe should be understood more properly in the sense of responsible stewardship", not domination as 'mere power' (LS, 116)[8].

Biblical anthropocentrism shows itself as relational and historical in yet another way. In contrast to the scientific accounts of 'creation', which conceive of the world's emergence by appealing to impersonal forces of nature, biblical cosmogonies are narrated in dramatic, personal, and psychological terms. Most importantly for our purpose, several key texts in the Hebrew Scripture represent the process of creation as a conflict of wills in which one part is victorious—a common feature of ancient Near East cosmogonies[9]. Respectively, the outcome of creation is not simply the physical world but also an organized human society. In other words, the biblical accounts of creation do not dichotomize 'nature' and 'history', as they witness that "the human race and its natural habitat are indissolubly linked"[10].

This link is demonstrated already in the Genesis creation narratives, where humanity is but a part of God's good creation. But it is especially forged in the Psalms whose phraseology clearly communicates the unity between creation and redemption. As exemplified by Psalms 104 and 136, creation includes both God's making the physical universe and God's leading God's people into their place in that universe[11]. The emergence of the world and the emergence of a people who are redeemed by YHWH is one process, as the language of 'creation' and 'redemption' is used interchangeably to denote the same divine act for which the psalmist praises God. In a similar manner, Second Isaiah interprets the redemptive action of YHWH in bringing Israel back from the Babylonian exile as a new cosmogony: the restoration of Israel is a renewal and a continuation of the original creative act by which Israel was brought into existence—the Exodus-Conquest where God defeated the Sea (Is

---

[8] See also Lonergan's distinction between 'mere power' and 'authentic power' in "Dialectic of Authority" [1974], in *A Third Collection*, ed. John D. Dadosky and Robert M. Doran, Collected Works 16. Toronto: University of Toronto Press, 2017, p. 6.

[9] Richard J. Clifford, "The Hebrew Scriptures and Theology of Creation", p. 511.

[10] Richard J. Clifford, "The Bible and the Environment", in *Preserving the Creation*, pp. 1-26, at p. 3.

[11] Richard J. Clifford, "The Hebrew Scriptures and Theology of Creation", p. 515.

43:16-21)[12]. As Richard J. Clifford, SJ, argues, the return from exile is the new creation because "people do not properly exist scattered in exile, apart from Yahweh, without land and temple, ritual and officials"[13].

Though in the past biblical scholars have not always recognized how integrally human beings fit within the nonhuman world and, following Gerhard von Rad's influential 1936 essay, tended to subordinate the biblical doctrine of creation to soteriology, the foregoing provides a glimpse of how this has changed. Later biblical scholarship vigorously criticized von Rad's stance, rejecting the presence of the dichotomies 'nature-history' and 'creation-redemption' in the Hebrew Scripture[14]. Though there is relatively little said on the matter directly, the New Testament retains this seamless view of creation-redemption but interprets it in the light of the Christ event. This leads us to my second point, namely, that the New Testament, in particular, the Pauline epistles, understand redemption as the new creation in Christ.

3.2 Redemption as the New Creation in Christ

By analogy with the Hebrew Scripture's interpretation of the Exodus-Conquest as a new creation that defeats chaos and darkness, the New Testament links creation and redemption by interpreting Christ's victory over death as the defeat of the primordial enemy of the human race. Once again, the process of the new creation has to do with the victory of the personal will (or love), and its outcome is the emergence of a new society which is ordered to God and God's eschatological reign. In the New Testament, this emerging society—frequently spoken of as 'being in Christ', the body of Christ, and the reign of God—is integrally linked to all of creation.

Though creation and redemption in Christ are analogous and continuous, they are, however, not identical. As the Pauline epistles and, especially, the Prologue to the Gospel of John imply by drawing on the Genesis creation story and the Wisdom tradition[15], the first creation is out of chaos or nothingness, whereas the 'new creation' refers to creation out of the privation proper to human history as sinful and marked by suffering[16]. The New Testament's

[12] See more in Richard J. Clifford, "The Hebrew Scriptures and Theology of Creation", p. 518.

[13] Richard J. Clifford, "The Hebrew Scriptures and Theology of Creation", p. 519.

[14] Richard J. Clifford, "The Hebrew Scriptures and Theology of Creation", p. 507.

[15] E.g., see the phraseology in Jn 1:1-3, Col 1:15, Gen 1:27-28, and Prov 3:19-20, 8:1-36.

[16] See Richard J. Clifford, "The Bible and the Environment", p. 24.

interpretation of redemption as the new creation can be further illuminated by drawing on some specific examples that speak of the new creation as reconciliation of all things in Christ.

That Christ's role in creation is analogous to his role in redemption is attested in multiple New Testament passages, most conspicuously in Paul (Col 1:15-20; Eph 1:3-14; 2 Cor 5:17; Gal 6:15) and John (John 1:1-5,10-11,14). The hymn in Colossians 1:15-20 deserves our special attention. Creation and reconciliation here are presented in parallel stanzas, where verses 15-17 (creation) and 18-20 (redemption) constitute a chiastic structure:

15 He is the image of the invisible God, the firstborn of all creation.	*18 He is also head of the body, the church; and He is the beginning, the firstborn from the dead, so that He Himself will come to have first place in everything.*
16 For by Him all things were created, both in the heavens and on earth, visible and invisible, whether thrones or dominions or rulers or authorities-all things have been created through Him and for Him.	*19 For it was the Father's good pleasure for all the fullness to dwell in Him,*
17 He is before all things, and in Him all things hold together.	*20 and through Him to reconcile all things to Himself, having made peace through the blood of His cross; through Him, I say, whether things on earth or things in heaven*[17].

The unity of the divine act in creating and redeeming the world in Christ is conveyed through formal parallels repeated in chiastic order: "he is" (*hos/hotos estin*) in vv. 15, 17, and 18; the "firstborn" (*prōtotokos*) in vv. 15 and 18; and the triple acclamation "in him" (*en auto*), "through him" (*di autou*) and

[17] "15 ὅς ἐστιν εἰκὼν τοῦ θεοῦ τοῦ ἀοράτου, πρωτότοκος πάσης κτίσεως, 16 ὅτι ἐν αὐτῷ ἐκτίσθη τὰ πάντα ἐν τοῖς οὐρανοῖς καὶ ἐπὶ τῆς γῆς, τὰ ὁρατὰ καὶ τὰ ἀόρατα, εἴτε θρόνοι εἴτε κυριότητες εἴτε ἀρχαὶ εἴτε ἐξουσίαι· τὰ πάντα δι᾽ αὐτοῦ καὶ εἰς αὐτὸν ἔκτισται· 17 καὶ αὐτός ἐστιν πρὸ πάντων καὶ τὰ πάντα ἐν αὐτῷ συνέστηκεν, 18 καὶ αὐτός ἐστιν ἡ κεφαλὴ τοῦ σώματος τῆς ἐκκλησίας· ὅς ἐστιν ἀρχή, πρωτότοκος ἐκ τῶν νεκρῶν, ἵνα γένηται ἐν πᾶσιν αὐτὸς πρωτεύων, 19 ὅτι ἐν αὐτῷ εὐδόκησεν πᾶν τὸ πλήρωμα κατοικῆσαι 20 καὶ δι᾽ αὐτοῦ ἀποκαταλλάξαι τὰ πάντα εἰς αὐτόν, εἰρηνοποιήσας διὰ τοῦ αἵματος τοῦ σταυροῦ αὐτοῦ, [δι᾽ αὐτοῦ] εἴτε τὰ ἐπὶ τῆς γῆς εἴτε τὰ ἐν τοῖς οὐρανοῖς" (UBS4/Nestle-Aland 27th Ed. Greek New Testament). The English version follows the NASB Bible.

"to/for him" (*eis auton*) in vv. 16, 17 and 16, 20[18]. As Clifford notes, "the virtual identity of verses 16d and 20a, *ta panta di autou kai eis auton ektistai* and *kai di autou apokallaxai ta panta eis auton*, makes creating (*ektistai*) parallel to reconciling (*apokallaxai*)"[19]. Notably, reconciliation is further explicated as 'making peace' through Christ's cross (v. 20)[20].

Besides supporting the view that creation and redemption in Christ are not separated, this passage is relevant to our further discussion in three other ways. First, not unlike the description of the cosmic redemption in Rom 8:18-30, it speaks of creation and reconciliation of the whole world, both human and non-human: in Col 1:15-20, "all things" (*ta panta*) is repeated six times, and the message is reinforced by referring twice to "heaven and earth" (vv. 16 and 20), Hebrew idiom for the world[21]. Second, the whole process is thought of in christomorphic terms: all things hold together in Christ (v. 17) and the outcome of redemption is not simply an organized society but the very body of Christ (v. 18), constituted by the community of those assimilated and conformed to Christ in his dying and rising, the idea that Paul develops more fully elsewhere[22].

This christomorphism, however, cannot be isolated from the context of Paul's trinitarian references, such as the triadic structure of praising the saving action of the God who 'sums up' all things in Christ in Eph 1:3-14. Moreover, the whole Christ who comes to be in the divine act of the new creation has an ecclesial dimension: Paul primarily conceives the body of Christ as the church. This should not be taken to mean, however, that the new creation is identical with the church. As Clifford elucidates, "though the whole universe is created anew, creation begins at a particular point—the chosen people"[23]. Third, as Col 1:22-24 suggests and 2 Cor 5:17-18 clarifies, the ecclesial body is charged with carrying on Christ's ministry of reconciliation: "So then, if anyone is in Christ, he is a new creation; what is old has passed away—look, what is new has come! And all these things are from God who reconciled us to

[18] See more in Richard J. Clifford, "The Bible and the Environment", pp. 19–20.

[19] Richard J. Clifford, "The Bible and the Environment", p. 20.

[20] "εἰρηνοποιήσας διὰ τοῦ αἵματος τοῦ σταυροῦ αὐτοῦ".

[21] See also "all things" in Eph 1:3-14 and John 1:3. Though the synoptic Gospels typically do not reflect on Christ as creator, Christ's miracles signify that salvation means the restoration to health and flourishing of human beings and their environment. See Richard J. Clifford, "The Bible and the Environment", p. 22.

[22] For instance, see Col 1:24; Eph 1:22-23, 4:14-16; 1 Cor 10:16-17, 12:12-31; Rom 12:4-5.

[23] Richard J. Clifford, "The Bible and the Environment", p. 20.

himself through Christ and who has given us the ministry of reconciliation"[24]. Furthermore, according to Paul and John, reconciliation is the restoration of the human friendship with God that results in building a community founded on love rather than on self-absorption and self-interest[25].

To conclude this brief excursion into the biblical world, creation and redemption in the Bible are not identical but intimately connected. The unity of the doctrines of creation and redemption is reflected in the New Testament's teaching on redemption as a new creation in Christ. Redemption concerns the reconciliation of all things in Christ, both human and nonhuman creation. Human beings, however, are endowed with special dignity and responsibility, conferred to them as the images and stewards of God. This dignity and responsibility are exercised in action, as they participate in Christ's ministry of reconciliation: the restoration of the right relational and interpersonal order, where everything is subordinated to God. Such a ministry is realized in history through the community which conceives of itself as the body of Christ that shares in his afflictions, in the hope of rising with him.

4. The Law of the Cross and the World's Coming-to-Be

At first glance, Lonergan's philosophical accounts of history as the compound of progress, decline, and recovery, and of generalized emergent probability as the immanent intelligibility of the world's coming-to-be, have little to do with the biblical doctrine explicated above. However, his until recently relatively unknown *Redemption* treatises, composed at the Gregorian University in the late fifties and early sixties, demonstrate that this is far from true[26]. In these treatises, Lonergan not only explicitly draws on the biblical scholarship of his day (an inspiration for my previous section!) but also re-contextualizes the mystery of the cross in relation to cultural, social, and historical processes.

Without a doubt, the pinnacle of Lonergan's *Redemption* treatises is the 'Law of the Cross'; according to this 'law', the deepest meaning of redemption—and its intrinsic intelligibility—is to bring good out of evil. In developing this 'law', Lo-

[24] "ὥστε εἴ τις ἐν Χριστῷ, καινὴ κτίσις· τὰ ἀρχαῖα παρῆλθεν, ἰδοὺ γέγονεν καινά τὰ δὲ πάντα ἐκ τοῦ θεοῦ τοῦ καταλλάξαντος ἡμᾶς ἑαυτῷ διὰ Χριστοῦ καὶ δόντος ἡμῖν τὴν διακονίαν τῆς καταλλαγῆς" (2 Cor 5:17-18).

[25] See Col 1:21; Rom 5:6-13; 1 Cor 8:1; John 15:12-17.

[26] After years in obscurity, these treatises are available in English. See Bernard J. F. Lonergan, *The Redemption*, trans. Michael G. Shields, ed. Robert. M. Doran, Daniel H. Monsour, Jermey D. Wilkins, Collected Works 9. Toronto: University of Toronto, 2018.

nergan brings together the biblical summons to discipleship (take one's cross and follow Christ), the injunction to love one's enemies, and the biblical notions of the reconciliation and friendship, and of the restoration of the cosmic order with, in, and through Christ. Lonergan's soteriology complements *Insight's* proposal that the heart of the divine solution to the human problem is the transformation of evil into good through faith, hope, and, above all, charity, and his suggestion that this transformation is subject to emergent probability[27]. With an eye on Lonergan's post-*Insight* and, especially, post-*Method* developments, this section re-contextualizes his contribution to help us think through the foundations of an integral ecology. We shall first discuss the Law of the Cross, then Lonergan's theory of history, and, finally, the immanent intelligibility of the order of the universe.

4.1 The 'Law of the Cross'

The 'just and mysterious Law of the Cross', as Lonergan styles it in Thesis 17 of his Christology textbook *De Verbo Incarnato*, concerns the meaning of redemption as the transformation of evil into good[28]. Lonergan calls it a law "both because of the fittingness grasped in each step and because of the generality it has in this present economy of salvation"[29]. Though this law governs the entire economy of salvation, it is not absolutely necessary but rather fitting[30]. The *lex crucis* can be grasped as the unity of three transitions within the all-encompassing order governed by divine wisdom, goodness, and justice: 'evil from evil', 'good from evil', and 'good from good'.

First, 'evil from evil': As a natural consequence, the evil of sin ('evil of fault')—an absolute privation in itself—always results in some relative and secondary privation of the natural goodness of things ('evil of pain'), be it moral corruption, social decline, or suffering[31]. This first step can be discerned

---

[27] Bernard J. F. Lonergan, *Insight: A Study of Human Understanding* [1957], ed. Frederick E. Crowe and R. M. Doran, Collected Works 3. Toronto: University of Toronto, 1992, ch. 20.

[28] See Bernard J. F. Lonergan, *Redemption,* pp. 196–264.

[29] Bernard J. F. Lonergan, *Redemption*, p. 203.

[30] Bernard J. F. Lonergan, *Redemption*, p. 235.

[31] Lonergan here draws on St. Thomas's distinction between *malum culpae* ('evil of fault') and *malum poenae* ('evil of pain'). Of itself, the evil of fault (in Lonergan's idiom, 'basic evil' and 'contraction of consciousness') does not have a good reason or intelligibility and thus represents an absolute privation: a sheer failure of rational freedom. The 'evil of pain', on the other hand, is evil in a secondary sense: insofar as even corrupted things have some being, *malum poenae* is only a relative privation. See more in St. Thomas Aquinas, *Summa Theologiae* I.48.5c; *De malo* 1.4–5; Bernard J. F. Lonergan, *Redemption,* p. 450.

in Christ's crucifixion: as a friend of both God and the human being, Christ accepts the consequence of human sin, so that the world might be reconciled to God. Lonergan develops the theme of friendship by drawing on the Bible and the Augustinian-Thomist notion of a friend as another self, *alter ego*[32]. Just as friends share with each other everything they are and have, and can even compensate for the wrongdoing of a friend, so also Christ makes our cross his own (for the cross properly belongs to us as sinners) to restore us to friendship with God[33].

Second, 'good from evil': By grace, in and through Christ's humble acceptance of the cross, God draws good out of the 'evil of pain'. Whenever love does not shrink away from suffering but instead resists evil by refusing to reflect and multiply it, the 'evil of pain' is transformed into a supreme good, the new life with, in, and through Christ. This second step is seen in the victory of Christ's will over the darkness of sin, as manifested in his resurrection from the dead. To explain this second step, Lonergan weaves together the biblical themes of the new creation in Christ, discussed in the first section, with the later theological developments, especially by St. Augustine and St. Thomas, and his own historically-minded philosophical approach to social and historical processes, which will be briefly discussed in a moment. Lonergan is especially indebted to Augustine's insistence that God preferred 'justice over power' in choosing how to redeem the world[34], and that, in the final score, divine justice is grounded in the mystery of divine wisdom and love. In Augustine's own words, "God judged it better to draw good out of evils than not to allow evils to exist".[35] On the other hand, Lonergan's understanding of a supreme good into which evils are transformed is deeply Thomist: for him, this supreme good can be conceived as the order of the universe, the body of Christ, and the kingdom of God[36].

---

32 See St. Augustine, *conf.* 4.6.11 (CCL 27: 45.22); St. Thomas Aquinas, *Summa Theologiae* I-II.28.1–2; Aristotle, *Nicomachean Ethics*, 9.4 and 9.9; Bernard J. F. Lonergan, *Redemption*, pp. 241 and 606.

33 Bernard J. F. Lonergan, *Redemption*, pp. 239–41, 489.

34 St. Augustine, *trin.* 13.17 (CCL 50A: 404.1–2).

35 St. Augustine, *ench.* 8.27 (CCSL 46: 64.53–54). Cited in Bernard J. F. Lonergan, *Redemption*, pp. 135, 521, 525.

36 See Bernard J. F. Lonergan, *Redemption*, pp. 197, 199; Lonergan here echoes the claim of his early student essay *Pantōn Anakephalaiōsis* (1935), in which he speaks of the recapitulation of all history in Christ, the new Adam. See Bernard J. F. Lonergan, "*Pantōn Anakephalaiōsis*: A Theory of Human Solidarity", in *Archival Materials: Early Papers on History*, ed. Frederick E. Crowe,

Third, 'good from good': The third 'step' of the *lex crucis* signifies the transition that Lonergan frugally describes as the "blessing of the Father". [37] In light of his discussion of the historical order of justice and his analogy of friendship, this transition can be understood as the multiplication of the good that originates though the victory of the cross. Insofar as it is marked by eschatological proviso, the good of the Father's blessing is still in the process of coming to be: it is already present in the historical order and yet awaits the fulfillment at the *eschaton,* where all tears will be wiped away and where there is no longer suffering. As an earthly reality, this good can be thought of as the fruit of cooperative grace, by which the divine friendship is diffused in history in and through the whole Christ, head and members. As we shall see at the end of this section, the blessing also means the shift in the probabilities of the world order.

4.2 The Tripolar Dialectic of History

Though the 'Law of the Cross' is not explicitly mentioned in *Insight*, Lonergan's account of the tripolar dialectic of history clearly anticipates it. Lonergan explains that human history is the compound of progress, decline, and recovery[38]. For our purposes, three important aspects of this theory need to be clarified.

First, according to Lonergan, history is never simply a sheer progress or decline or recovery, but a system on the move whose actual direction is determined by the dynamic interplay of all three processes. Second, each of the determinants have their own principle. In a nutshell, the principle of progress is human authenticity, the principle of decline is human inauthenticity, while the principle of recovery (or, in theological idiom, redemption) is Christian authenticity. Inauthenticity, human and Christian, yields inauthentic traditions and, in turn, is conditioned by them. Third, according to Lonergan, because

Robert M. Doran, John D. Dadosky, Collected Works 25. Toronto: University of Toronto, 2019, pp. 38–64, esp. pp. 41, 54–64.

[37] This follows from the correspondence that Lonergan assigns between the different moral orders and the three 'steps' of the 'Law of the Cross': 'evil to evil' corresponds with the role of Christ's enemies, 'good from evil' with the role of Christ, and 'good from good' with the role of the Father. See Bernard J. F. Lonergan, *Redemption*, p. 519.

[38] See Bernard J. F. Lonergan, *Insight,* Chs. 7, 18, 20; Bernard J. F. Lonergan, *Method*, pp. 52, 55, 101, 117–18, 242, 288, 290–91; "The Transition from a Classicist Worldview to Historical-Mindedness" [1966], in *A Second Collection*, ed. Frederick E. Crowe and Robert M. Doran, Collected Works 13. Toronto: University of Toronto, 1996, pp. 3–10; "*Insight* Revisited" [1973], Collected Works 13, pp. 228–29.

the causes of social decline in history are complex—that is, besides the intelligible, they include the irrationality and unreasonableness of sin—and because humanity's existence is driven by an unquenchable desire for infinite intelligibility and love, merely human means (that is, human natural intelligence and creativity) are insufficient for reversing the decline. The solution to the human problem cannot be merely human but, in a sense, needs to be divine. To be fully human entails transcending—but not rejecting—the merely human: "to be just a man is what man cannot be"[39]. Lonergan's *Method in Theology* (1972) summarizes the tripolar dynamism as follows:

> As human authenticity promotes progress, and human unauthenticity generates decline, so Christian authenticity—which is a love of others that does not shrink from self-sacrifice and suffering—is the sovereign means for overcoming evil. Christians bring about the kingdom of God in the world not only by doing good but also by overcoming evil with good (Rom 12:21)[40].

The reader familiar with Lonergan will recognize that here he refers to a triple—intellectual, moral, and religious—self-transcendence. For him, commitment to self-transcendence is the key to human and Christian authenticity. Self-transcendence is achieved by following what he calls 'transcendental precepts': "Be attentive, be intelligent, be reasonable, be responsible, be in love"[41]. Furthermore, in the world as it exists *de facto*, being in love entails 'overcoming evil with good': surrendering to the 'Law of the Cross' is the criterion of Christian authenticity.

In other words, the principle of progress in history is human liberty as intelligent, reasonable, and responsible. The principle of decline is bias—the distortion of reasonable and responsible intelligence and loving, which truncates human self-transcendence. Lonergan discerns four kinds of bias (dramatic, individual, group, and general) that put the pursuit of the common good at risk[42]. Dramatic bias largely operates on the subconscious level, as a kind of psychological scotosis which represses insights that lead to personal growth. The other three biases pertain to conscious intentional acts. Individual bias truncates person's self-transcendence by ignoring any further questions and affections that upset the pur-

[39] Bernard J. F. Lonergan, *Insight,* p. 750.

[40] Bernard J. F. Lonergan, *Method*, p. 291.

[41] See Bernard J. F. Lonergan, *Method*, p. 268.

[42] For more on the fourfold bias, see Bernard J. F. Lonergan, *Insight,* pp. 204–31 and 244–63; for a summary, see Bernard J. F. Lonergan, *Method*, p. 231.

suit of egoistic interest. Group bias desists critical insights that would prevent the dominant group from seeking its own good, over against the common good. This yields group antagonism, as well as a plethora of social, economic, and ecological injustices. Finally, there is general (or 'common sense') bias. This bias is 'general' because it systematically rejects the intelligent, reasonable, and responsible solutions to the problems that human societies face in history, such as the ecological crisis. Because of its scope, this is the most dangerous kind of bias. In the grip of this bias, political leaders resent theoretical knowledge and extol short-sighted practicality at the expense of plunging their societies into the long cycle of decline. The experiences of ecological, political, or global health crises (such as the recent pandemic) lend many examples of such obtuseness and short-sightedness.

If the fourfold bias is so all-pervasive, how on earth are we not constantly sliding down the hill of decline? According to Lonergan, the answer lies with both human intelligence and receptivity to grace. While, ultimately, the reversal of decline hinges on the saving action of God in Christ by the power of the Holy Spirit, as the images of God, human beings are also endowed with free rationality and creativity that promote progress, not the least, by enabling humankind to learn from its mistakes. In his later essays, Lonergan speaks of this dynamism as the convergence of two vectors—the creative vector 'from below' and the healing vector 'from above'—thereby implicitly transposing the classical notion of grace as elevating and healing nature into the dynamic context of the emerging world order.

Though in *Insight* Lonergan avoids using special theological categories, even there he makes clear that the transformation of evil into good through love stands at the center of the divinely originated and humanly appropriated remedy from decline:

> If human historical process is … a compound of progress and decline, then its redemption would be effected by faith, hope, and charity. For the evils of the situation and the enmities they engender would only be perpetuated by an even-handed justice: charity alone can wipe the slate clean. The determinism and pressures of every kind, resulting from the cumulative surd of unintelligent policies and actions, can be withstood only through a hope that is transcendent and so does not depend on any human prop. Finally, only within the context of higher truths accepted on faith can human intelligence and reasonableness be liberated from the charge of irrelevance to the realities produced by human waywardness[43].

---

[43] Bernard J. F. Lonergan, "Transition from a Classicist Worldview", p. 9. Lonergan here refers to Ch. 20 of *Insight* (see ibid., 9n10).

In an important post-*Insight* essay on historicity, Lonergan makes fully explicit this connection between the 'Law of the Cross' and the tripolar dynamism of history, as he contends that the *lex crucis* provides a strictly theological complement to his "general analysis of the dynamic structure of human history"[44].

To conclude, decline in history follows upon the truncation of human self-transcendence without which there is no authentic human progress. Since authenticity is a precarious achievement, which in the *de facto* human condition cannot be sustained by human means alone, decline calls for a divine solution to the problem of evil. In the world as we know it, embracing this solution means living according to the 'Law of the Cross', which in turn requires an antecedent willingness to suffer for the sake of God's kingdom.

4.3. The New Creation in Christ as the Shift in the World's Coming-to-Be

This brings us to my third point concerning emergent probability as explanatory of a supreme good into which human evils are transformed through Christ's self-giving, the whole Christ. To clarify this, we need first to recall Lonergan's appropriation of the Augustinian-Thomist analogy with friendship for understanding reconciliation through the cross. Then we will outline his theory of generalized emergent probability as the immanent intelligibility of the universe[45].

In his soteriology, Lonergan explains that the kingdom of God is embraced as human beings accept and cultivate the gift of divine friendship, received through Christ by the power of the Holy Spirit. According to Lonergan, the analogy with friendship answers the 'why' of the cross: the Son of God became one with us in our humanity, to the point of making our cross his own, "*for the orderly communication of God's friendship to his enemies*"[46]. This friendship is communicated in an orderly manner for it follows the free plan conceived by God's loving wisdom. This plan, however, unfolds contingently and in time, in accordance to three principles: "the principle of cause, the principle of continuity, and the principle of the diffusion of friendship"[47]. These principles imply that God usually acts through secondary causes, in accordance with what is fitting to a created nature, and through God's friends. Notably, God's friend loves all potential friends of God (no exclusions!), because a friend loves her friend's

[44] See Bernard J. F. Lonergan, "Transition from a Classicist Worldview", p. 8.

[45] See Bernard J. F. Lonergan, *Insight,* pp. 270–294; pp. 148–49, etc.

[46] Bernard J. F. Lonergan, *Redemption*, p. 631. Emphasis original.

[47] Bernard J. F. Lonergan, *Redemption*, p. 633.

friends, even her personal enemies, for the sake of her friend. Lonergan applies these principles to the 'Law of the Cross', contending: "unless human psychological and social laws are suspended … people cannot be turned from evil to good except according to the rule [*ratio*] of the cross"[48].

As argued elsewhere, by applying the three principles of cause, continuity, and the diffusion of friendship to redemption, Lonergan tacitly invites us to think of his theory of generalized emergent probability—which is an intrinsic intelligibility of the universe as the world-coming-to-be—as an explanatory framework for understanding the historical agency of Christ and his followers[49]. This theory explains the dynamism of the universe in terms of the secondary causes, that is, genetic, dialectical, classical, and statistical laws (note that though somewhat confusing, 'genetic' here refers not so much to 'genes' as to 'genesis'). Insofar as these secondary causes work according to created natures, intelligent creatures (with their obediential potency for God) are not excluded from this account. With an eye to contemporary contributions, but without going into technical details, permit me now to briefly clarify what this means, so that we can proceed to illuminating our key question of integral ecology.

In line with the contemporary "re-emergence of emergence", as expounded by Philip Clayton, Niels Henrik Gregersen, Arthur Peacocke, and others, Lonergan conceives of the world order as a dynamic system on the move, an emergent continuity-in-discontinuity that is open to genuine novelty without necessarily excluding God's providence[50]. This genuine novelty emerges cumulatively from the dynamically linked sequence of 'lower' levels—in Lonergan's idiom, coincidental manifolds—that both limit and invite 'higher' transformations[51]. Everything is interconnected as the more complex and

---

[48] Bernard J. F. Lonergan, *Redemption*, p. 639.

[49] The argument, which unfortunately cannot be pursued here, is supported by a number of clues that Lonergan gives in his *Insight* and *Redemption* treatises. For more, see Ligita Ryliskyte, Doctoral Dissertation "*Cur Deus Cruciatus?* Lonergan's Law of the Cross and the Transpositions of 'Justice Over Power'", Boston College, 2020, http://hdl.handle.net/2345/bc-ir:108819.

[50] For a comprehensive discussion of the idea of emergence, see the essays collected in Philip Clayton and Paul Davies (ed.), *The Re-Emergence of Emergence: The Emergentist Hypothesis from Science to Religion*. Oxford: Oxford University Press, 2006; Nancey Murphy and William R. Stoeger (ed.), *Evolution and Emergence: Systems, Organisms, Persons*. Oxford: Oxford University Press 2007.

[51] Obviously, "lower" and "higher" levels here are metaphors pointing toward the increasing structural, functional, or explanatory complexity of the distinct sequential constellations in the emergent world order.

developed ('higher') levels sublate—that is, enrich and transform without negating—'lower' levels. Though 'higher' integrations depend on what the 'bricks' presented by coincidental manifolds are, these integrations also modify what happens on the 'lower' levels (e.g., think of how urbanization affects the natural environment). Thus, physical, chemical, biological, neurological, psychic, cultural, and spiritual realities constitute a multilayered, complex, and constantly developing system of hierarchically ordered and yet in many ways interdependent processes that account for the emergent world order. The development (genetic sequence) is dialectical inasmuch as there are blind alleys, failures, and conflict, too. Furthermore, emergence is subject to the interplay between 'classical' laws, that is, the laws governing regularities (such as the movement of the planets or the law of gravity) and 'statistical' laws, which describe frequencies and probabilities, thus accounting for the randomness and freedom involved in nonhuman and historical processes.

In terms of Lonergan's critical metaphysics, generalized emergent probability is the immanent intelligibility of the world's coming-to-be (in scholastic terms, something akin to formal causality). It is 'generalized' in the sense that it includes all events and all things, not the least, human conscious intentionality. It is '*emergent* probability' insofar as the probabilities of what is in fact actualized are constantly changing, as the world process is 'indeterminate' (includes genuine contingency) and yet 'upwardly directed' (tends to higher integrations). Given sufficiently long periods of time and the vastness of space, the emergence of the more complex systems can be reasonably expected. This is so because the world is constituted not merely by unrelated events and things, but also—and even more fundamentally so—by a conditioned and conditioning series of circular processes[52]. Some examples of such circular processes (in Lonergan's idiom, 'schemes of recurrence') include the Krebs cycle in the cell, circular planetary movements, circulation of water in nature, wildfires, and recurrent human actions, such as production and consumption cycles or human habits. In statistical terms, the schemes of recurrence account for a jump in probabilities, which undergirds the upward direction of the world process—an emergence of higher integrations. The main reason is that the probability of the interdependent events (such as events coiling in a circle) is the sum, not the product, of the probabilities of each event. Whenever the scheme of recurrence is established, the probabilities 'jump', making higher

[52] See Bernard J. F. Lonergan, *Insight*, pp. 2,94; 148–49.

integrations concretely possible and probable. These higher integrations, in turn, affect the way lower integrations function (think of how anxiety affects one's digestion). To bring this closer to home, the emergence of an organized society or community, such as Israel in Hebrew Scripture, or the ecclesial body, can be conceived in terms of the higher 'scheme of recurrence'. As both conditioned and conditioning, such a 'scheme' affects what happens on all other levels, changing the probabilities of the emerging world order.

Remarkably, Lonergan's theory of emergent probability includes both nonhuman and human creation. Though, in "a universe of being [humankind] is but an item"[53], it is in women and men that emergent probability becomes self-conscious and thus 'intelligent' and capable of foreseeing and even changing the odds of emergence[54]. In distinction from nonhuman nature, human beings do not have to wait for their environment to make them. Instead, they become the executors of the emergent probability for others, as they transform their natural environment, as well as social, economic, and political conditions of their historical existence. However, human beings have only a relative control over emergent probability: their insights and decisions "remain probable realizations of concrete possibilities" and "earlier insights and decisions determine later possibilities and probabilities of insight and decision"[55].

Since, according to Lonergan, the reversal of decline in history fittingly comes through surrendering to the 'Law of the Cross' emergent probability as 'intelligent' favors human progress insofar as humanity commits to self-transcendence in knowing and loving, to the point of self-sacrifice. This integral notion of emergent universe stands in complementarity to contemporary interdisciplinary contributions. For instance, a collaboration led by Sarah Coakley and Martin A. Nowak on evolution and altruism, proposes that the emergence of the human species on our planet hinges on the development of the capacity for (costly) cooperation[56]. Conceived as altruism, such a cooperation

[53] Bernard J. F. Lonergan, *Insight*, p. 498.

[54] Bernard J. F. Lonergan, *Insight*, p. 235.

[55] Bernard J. F. Lonergan, *Insight*, p. 235.

[56] The authors note that this development includes (but is not limited to) the transition from "direct" to "indirect" reciprocity, as illuminated by game theories. The biological condition for a possibility of such a development is that our brains are hard-wired for cooperation through the functioning of the "mirror neurons". Correspondingly, against reductionist "selfish gene" theories, cooperation is posited as the third principle of evolution, equally indispensable as mutation and selection. See more in Martin A. Nowak and Sarah Coakley (ed.), *Evolution, Games, and God: The Principle of Cooperation*, Cambridge: Harvard University Press 2013.

involves the willingness to forego personal advantage and is "motivated by good will or love for another (or others)"[57]. In light of the foregoing, we can add that Lonergan's 'Law of the Cross' corroborates these insights and expands them to include the supernatural dimension. The *lex crucis* foregrounds that the source of the willingness to forego immediate satisfactions and to accept hardships for the sake of terminal value—God's reign and, ultimately, God *in se*—is the gift of divine friendship, received in and through Christ.

Having re-contextualized the 'Law of the Cross' as constitutive of the world's coming-to-be, let us now consider Christ's historical agency and human responsibility in and for history. In continuity with Lonergan's thought, it seems right to propose that Christ's Paschal mystery shifts the probabilities of the world process by restoring the order of friendship. The world's ultimate destiny is decisively inaugurated in Christ's victory over death. Concretely, a new creation in Christ emerges according to probabilities, upwardly but indeterminately directed by divine Providence in accordance to the 'Law of the Cross'. Moreover, the new probabilities are realized in and through the whole Christ, head and members. It is in and through his body that Christ's personal offer of friendship in the calamity of his death on the cross and in his vindication sets up reverberations in history. The shift in probabilities is concretely actualized whenever the diffusion of friendship takes place, whenever human biases caused by egoistic and group self-interest or short-sighted practicality are uprooted through self-transcendent love, whenever the enemies-made-friends of God transform evil into good. This emergent order of friendship—the order that yields intelligent, reasonable, and responsible cooperation—is grounded in charity. As such, this order is a relational *agape* network, an order that imitates and participates in the interpersonal relations of the Trinity itself. Christ's rising lifts up the whole of creation, as assimilated and conformed to him, men and women turn to God as their friend and extend this friendship to the rest of creation, both human and nonhuman, to each in its own way.

In sum, the new finality, introduced by Christ's dying and rising, is realized in history through the dialectical unification of all things in Christ. Through the mystery of the cross, in Christ as the alpha and omega, the firstborn of all creation and the eschatological fulfilment of the world, all three fundamental human relationships—with God, with neighbor, and with nature—are renewed and healed. Insofar as humankind is the executor of emergent prob-

---

[57] Sarah Coakley and Martin A. Nowak, "Introduction" to *Evolution, Games, and God*, p. 4.

ability for others, the new creation in Christ is also a transformation of evil into good on the cosmic scale. Paraphrasing St. Paul, through the victory of his self-giving love, the cosmic Christ—that is, the whole Christ constituted by the ecclesial network of *agape* and by all friends of God—holds together *ta panta*, all that is and comes to be.

5. *Laudato Si'* and Lonergan's Contribution

The foregoing illuminates that the foundation of integral ecology is the new creation in Christ. By employing Lonergan's contribution, we were able to demonstrate that this claim, though articulated in figurative biblical language, has a strong explanatory grounding, provided by the integration of the 'Law of the Cross' with the theory of history as compound of progress, decline, and redemption, and with the theory of generalized emergent probability. Such a grounding, furthermore, reclaims the christomorphic or—in terms more agreeable with interreligious dialogue—agapic and cruciform aspects of human progress in history, often denied by the secularized and consumerist culture which finds it difficult to assign some value to suffering.

To show how the foregoing systematic reflection converges with Pope Francis's reflection on the ecological crisis, we will now directly bring Lonergan's contribution into dialogue with the encyclical *Laudato Si'*. Though the language of the encyclical frequently deliberately leaves out in-depth Christological considerations—which can be explained by the fact that Pope Francis addresses "every person living on this planet", not just the followers of Christ (LS, 3)—the encyclical does not leave out Christ. For instance, Pope Francis profoundly reflects on the biblical theme of a new creation in Christ in the following passage:

> The ultimate destiny of the universe is in the fullness of God, which has already been attained by the risen Christ, the measure of the maturity of all things. Here we can add yet another argument for rejecting every tyrannical and irresponsible domination of human beings over other creatures. The ultimate purpose of other creatures is not to be found in us. Rather, all creatures are moving forward with us and through us towards a common point of arrival, which is God, in that transcendent fullness where the risen Christ embraces and illumines all things. Human beings, endowed with intelligence and love, and drawn by the fullness of Christ, are called to lead all creatures back to their Creator[58].

---

[58] LS, 83. Internal citations omitted.

In complementarity to the notion of redemption as a shift in the world's finality according to probabilities in and through the whole Christ, Pope Francis warns against underestimating the importance of interpersonal relations (LS, 119) and against ignoring "an order and a dynamism" that God has written into the world in creation (LS, 221). He further argues that the ecological crisis is just a part of an ethical, cultural, and spiritual crisis. As such, it calls for a fundamental renewal, grounded in integral ecology—ecology which "clearly respects its human and social dimensions" and is "capable of taking into account every aspect of the global crisis" (LS, 137). The scope of the challenge demands a profound change of heart: a conversion at once personal and communal (LS, 217-19). In resonance with Lonergan's proposal that a supreme good into which human evils are transformed is the whole Christ, Pope Francis notes that a Christian recognizes that such a conversion is possible because "Christ has taken unto himself this material world and now, risen, is intimately present to each being, surrounding it with his affection and penetrating it with his light" (LS, 221).

The foregoing provides just a glimpse into a number of key points that intersect in Lonergan's contribution and the encyclical *Laudato Si'*—and the intersection is not simply coincidental but reveals common theological sources and a common commitment to Christ. For my limited goal, it is sufficient to note two such points that further focus our attention on the foundations of an integral ecology: the re-definition of human progress and the call for integral ecological conversion. While Lonergan provides a biblically grounded heuristic framework for understanding these themes in relation to the new creation in Christ, *Laudato Si'* offers concrete determination of the 'known unknowns' as regards our response to the ecological crisis.

5.1. The Ecological Crisis and the Re-definition of the Notion of Progress

The first point of intersection concerns the encyclical's call to redefine progress. As Pope Francis insists, when it comes to the ecological crisis, we cannot be happy with halfway measures. In such a deep crisis, we need to go to the roots, rethinking the causes of decline and what defines progress. Speaking of the new models of global development and the economic goals that would contribute to the long-term solution to the ecological crisis, Pope Francis writes:

> It is not enough to balance, in the medium term, the protection of nature with financial gain, or the preservation of the environment with progress. Halfway measures simply delay the inevitable disaster. Put simply, it is a matter of

> redefining our notion of progress. A technological and economic development which does not leave in its wake a better world and an integrally higher quality of life cannot be considered progress[59].

It is striking that in his own time and context, Lonergan similarly urged his contemporaries "to refuse half-measures and insist on complete solutions even though it has to wait"[60]. It is even more remarkable that he offered a theory of history that today is still relevant because it illuminates that the adequate notion of progress needs to be built on an adequate account of human anthropology that focuses on self-transcendence in knowing and loving.

The way *Laudato Si'* rises to meet the challenge of redefining human progress deeply resonates with Lonergan's proposal. Lonergan's insight that the principles of progress and recovery are, respectively, human and Christian authenticity, are of particular relevance. Pope Francis warns that to reduce progress simply to economic growth is to overlook the diminished quality of life that frequently occurs in the midst of economic growth. Not the least, this diminishment is caused "by the deterioration of the environment, the low quality of food or the depletion of resources" (LS, 194). To achieve a really sustainable growth that includes social and environmental responsibility for the most vulnerable, progress needs to be redefined in relation to the meanings and values that are truly worthwhile, as they point to the "ultimate meaning of our earthly sojourn" (LS, 160). *Laudato Si'* reminds us that to leave a livable world to those who come after us, the ecological issue cannot be approached piecemeal. The very notion of progress needs to be redefined in terms of self-transcendence for the "leap towards the transcendent ... gives ecological ethics its deepest meaning" (LS, 210). Such a leap opens us up to the other and, ultimately, to the mystery of God:

> A correct relationship with the created world demands that we not weaken this social dimension of openness to others, much less the transcendent dimension of our openness to the "Thou" of God. Our relationship with the environment can never be isolated from our relationship with others and with God. Otherwise, it would be nothing more than romantic individualism dressed up in ecological garb, locking us into a stifling immanence[61].

---

[59] LS, 194.

[60] Lonergan here speaks of the renewal of theology; see "Dimensions of Meaning" [1965], in *Collection*, ed. Frederick E. Crowe and Robert M. Doran, Collected Works 4. Toronto: University of Toronto, 1988, pp. 232–45, at p. 245.

[61] LS, 119.

Lonergan's theory of history helps to clarify what this and many other passages in *Laudato Si'* implicitly suggest: that the principle of true progress is a human self-transcendence that yields human authenticity, the foundation of genuine human flourishing.

As Pope Francis repeatedly shows in the encyclical, only authentic human living can prevent the ecological crisis. To recall, in Lonergan's terms, human authenticity means being attentive, being intelligent, being reasonable, being responsible, and being in love. Similarly, Pope Francis explains that human persons are "capable of rising above themselves" when they are able "to take an honest look at [themselves], to acknowledge [their] deep dissatisfaction, and to embark on new paths to authentic freedom" (LS, 205). This path of authentic freedom involves a cruciform letting go of instant gratification, commodities, and satisfactions that rob the poor of the basic necessities for living and are detrimental for the environment (LS 30, 189, 193, etc.). It teaches us about the peace that comes with appropriating the rule "less is more" (LS, 222). It invites us into contemplating nature (LS, 225) and into cultivating a cultural ecology, which involves protecting the cultural achievements of humanity (LS, 143). The path of authentic freedom, furthermore, calls for a human collaboration that resists the culture of rampant individualism, practical relativism, and compulsive consumerism, and is guided by critical, responsible, and loving intelligence. Such a collaboration, however, is not easy for it demands integral conversion, the theme to which we now turn.

5.2 The Call to Integral Ecological Conversion

Since, in the long run, only an ecology that is integral really matters, the call to ecological conversion, as *Laudato Si'* suggests, is also the call to integral conversion. As Pope Francis contends, "it is no longer enough to speak only of the integrity of ecosystems. We have to dare to speak of the integrity of human life" (LS, 224). Such an integrity demands an ongoing conversion which transforms all dimensions of human life and, in humility and sobriety, is ultimately ordered to God.

Lonergan's notion of the threefold—intellectual, moral, and religious—conversion, once again, deeply resonates with the encyclical's call for integral conversion and provides a heuristic structure for understanding what 'integral' means. According to Lonergan, religious conversion is an unconditional and unrestricted falling in love with God[62]. Religious conversion grounds and

[62] Bernard J. F. Lonergan, *Method*, p. 107.

sublates, without replacing, intellectual and moral conversion: "Just as an unrestricted questioning is our capacity for self-transcendence, so being in love in an unrestricted fashion is the proper fulfillment of that capacity"[63]. Intellectual conversion is a withdrawal from the self-deception that we can know without honest and critical search for correct understanding. Moral conversion consists in turning away from false criteria for decisions, such as instant satisfactions or gratifications, to true criteria for decisions: values formed by a good conscience[64]. Though the three modes of conversion are distinct, they all manifest the cruciform convergence of two 'vectors': the ordinary unfolding of human self-transcendence through intelligent, critical, and responsible inquiry (the vector 'from below') and the transformative and healing gift of divine love (the vector 'from above'). Essentially, then, conversion is a withdrawal from inauthenticity to authenticity according to the 'Law of the Cross'.

The encyclical *Laudato Si'* provides many occasions which specify what such an integral conversion entails in the face of the ecological crisis. As intellectual, such a conversion demands critically examining existing policies, economic models and arrangements, as well as social and cultural norms. As moral, ecological conversion entails turning away from greed and the search for instant gratifications to values that are refined as we develop good habits (LS, 209). *Laudato Si'* especially highlights the need for developing the habits that help us to conserve energy, and cultivate peace and care for the least privileged. As religious, conversion both instills humility and fills hearts with the joy of choosing a simple life style, just as Christ did. Humility, in turn, leads to acknowledging that we are not autonomous and cannot replace God with our own ego (LS, 224). The fruit of such an integral ecological conversion, and of the genuine repentance of the ways we are complicit in the disfigurement and destruction of creation, then, is solidarity with the poor and care for the common good (LS, 156 58). Citing the Ecumenical Patriarch Bartholomew, Pope Francis poetically sums up the summons for an integral conversion by bringing to light its cosmic significance:

> Bartholomew … asks us to replace consumption with sacrifice, greed with generosity, wastefulness with a spirit of sharing, an asceticism which "entails learning to give, and not simply to give up. It is a way of loving, of moving gradually away from what I want to what God's world needs. It is liberation

[63] See Bernard J. F. Lonergan, *Method*, p.106.
[64] See Bernard J. F. Lonergan, *Method*, pp. 238–240.

> from fear, greed and compulsion". As Christians, we are also called "to accept the world as a sacrament of communion, as a way of sharing with God and our neighbours on a global scale. It is our humble conviction that the divine and the human meet in the slightest detail in the seamless garment of God's creation, in the last speck of dust of our planet"[65].

This cosmic significance of ecological conversion as integral reminds us both of the Pauline vision of the cosmic Christ and Lonergan's explanatory notion of the convergence of the creative and healing vectors 'from below' and 'from above' operative in the emergent world order.

Because it regards an extremely complex relational network, integral conversion, to reiterate, is not just a personal affair but a "community conversion" (LS, 219). Neither individual good deeds nor the atomistic instances of conversion are capable of bringing about the change that the world in crisis needs. Lonergan's contribution, once again, might help us to understand this better. To recall his proposal, human beings are intelligent, social, and historical agents: agents capable of discerning the dynamism of the emergent world order by which they are conditioned; agents that partake in and create the series of the flexible circles of the human 'schemes of recurrence' (the major 'bricks' in the emergent world order); finally, agents that are capable of changing not just their environment but also cultural, social, and historical conditions. This agency and responsibility in and for history is always exercised in the context of traditions. Traditions are developed by communities—groups of people constituted by common experiences, meanings, values, and commitments. Hence, it is not only personal but also communal meanings and values that can get distorted. Just as there is human authenticity and inauthenticity, so there is also authentic and inauthentic traditions. An authentic human person can easily be caught in an inauthentic tradition—such as the consumerist or individualistic culture—which interferes with the achievement of genuine human authenticity. Even with the best of intentions, one might end up simply 'authentically' multiplying an inauthentic tradition. In terms of ecological concern, "isolated individuals can lose their ability and freedom to escape the utilitarian mindset, and end up prey to an unethical consumerism bereft of social or ecological awareness" (LS, 219). Effectively resisting inauthentic traditions that lead to global crises takes communal conversion.

To summarize, important key thematic links make Lonergan and Pope Francis interesting conversation partners in reflecting on the foundations of an

---

[65] LS, 9, internal citations omitted.

integral ecology. At the heart of these connections lies their mutual interest in the complexity and dynamism of the universe as the relational order that is in need of healing, restoration, and redemption. This brief attempt to bring together some of the resonating or complementary themes in Pope Francis's *Laudato Si'* and Lonergan's explanatory and biblically grounded accounts of the whole Christ, history, and the dynamic world order is not exhaustive or even comprehensive. Rather, it is exploratory and illustrative. Hence, if the reader feels that Lonergan has helped him or her to understand somewhat better the explanatory underpinnings of integral ecology and, in turn, if Lonergan's theory was somewhat illuminated by *Laudato Si'*, the limited goal of this section is achieved.

6. Conclusion

The foregoing has brought to our attention that integral ecology calls for human collaboration, networking, and communion. Reflecting on the need for a 'community conversion', Pope Francis reminds us that "social problems must be addressed by community networks and not simply by the sum of individual good deeds" (LS, 219). Not unlike Lonergan, he further elaborates on the meaning of this conversion in addressing its social, cultural, economic, civic, political, and religious dimensions, and by calling for interdisciplinary, national, and international cooperation. He also speaks to what Lonergan conceived as the key to the correct understanding that leads to right action—the necessity to ask and answer all relevant questions, such as: "What will [a proposed business venture or project] accomplish? Why? Where? When? How? For whom? What are the risks? What are the costs? Who will pay those costs and how?" (LS, 185). *Laudato Si'* thus illustrates what it takes to be truly intelligent, reasonable, and responsible social and historical agents in the face of the ecological crisis. In Lonergan's terms, the probabilities of long-lasting and sustainable solutions emerge as the result of a complex interaction between the series of conditioned and conditioning recurrences, constituted by the processes in the nonhuman nature and human habits, relationships, cooperation, and institutions.

But, as both Lonergan and Pope Francis would agree, the most important shift in emergent probabilities is brought about by love as friendship—a fraternal love that is grounded in the interpersonal Triune love itself (LS, 240). This is so because ecological concerns cannot be answered without human growth in solidarity, responsibility, compassionate care, and the capacity for communion (LS, 228). As Pope Francis reminds us, this fraternal love is always *agapic,* that is,

> [it] can only be gratuitous; it can never be a means of repaying others for what they have done or will do for us. That is why it is possible to love our enemies. This same gratuitousness inspires us to love and accept the wind, the sun and the clouds, even though we cannot control them. In this sense, we can speak of a "universal fraternity"[66].

In parallel to Lonergan's 'Law of the Cross' and the principle of the diffusion of friendship, the encyclical *Laudato Si'* highlights that such a love "sows peace and friendship", breaks "the logic of violence, exploitation and selfishness", and, in and through the Church, promotes a "civilization of love" (LS, 230-31). Hence, love for the society and commitment to the common good, according to Pope Francis, are the expressions of charity—the supernatural love that in Lonergan's Thomist terms is understood by the analogy with friendship. This love, as my recontextualization of the 'Law of the Cross' in relation to history has illuminated, transforms human reality on all levels, as it "affects not only relationships between individuals but also 'macro-relationships, social, economic and political ones'" (LS, 231). By way of shifting the emergent probabilities of the dynamic world order, this love reverberates throughout the entire universe as the new creation in Christ, the ultimate foundation of an integral ecology.

[66] LS, 228.

HINDUISM AND ECO-HUMAN HARMONY: IN DIALOGUE WITH *LAUDATO SI'*

Bryan Lobo, S.J.

1. Introduction

Hinduism, in its immense potentialty to be eco-friendly, can be linked to the fundamental vision of Pope Francis' Encyclical, *Laudato Si'*, that invites the readers to care for nature (Creation). Hinduism, as a "nature-based" religion is far more complex and diverse in its theories and practices if compared to other religions especially the semitic ones. Polytheism and monotheism coexist in Hinduism[1] which for many is a contradiction but not for the all encompassing Hindu scriptures and theologians. It is precisely because nature is so intrisincally involved in the conception of the Divine, that Hinduism holds in healthy tension the seemingly contradictory views on God and most importantly lends itself towards eco-respect.

In this article, aspects of Hinduism will be presented having the proposals of *Laudato Si'* as a backdrop. Since Pope Francis insists on entering "into dialogue with all people about our common home" (3)[2], dialoguing with Hindus through their scriptures, religious beliefs and practices, and their environmental activism, would be a further step towards united efforts for ecological health because according to the Pope "other religions… have expressed deep concern and offered valuable reflections on issues which all of us find disturb-

[1] There is even henotheism in Hinduism according to M. Müller which is the belief in one all-powerful God without denying the existence of lower level deities or gods.

[2] Pope Francis, Encyclical Letter *Laudato Si' on Care for Our Common Home*, 24 May 2015, http://www.vatican.va/content/francesco/it/encyclicals/documents/papa-francesco_20150524_enciclica-laudato-si.html, (accessed 28/12/2020). Henceforth, all quotations will be taken from this online text of the Vatican site and only the numbers will be mentioned after the quotations.

DOI 10.32060/C.S.9.549-572

ing" (7). It is our contention that Hinduism has a lot to offer in the ongoing reflection and activity for ecological concerns.

Before we proceed, it must be stated that this article favours a positive reading of Hinduism towards ecology against which arguments will be touched upon wherever necessary. It is true that the question remains whether there is an actual ecological concern, as shared by modern environmentalists, in the Hindu scriptures, beliefs and practices. Even though in their religious perspectives and praxis, Hindus are more prone to live in harmony with nature, healthy ecology must therefore not necessarily be the presumed intention of Hindu religiosity. There are a number of researchers who are critical of a positive stand on Hinduism as an ecological religion[3]. At the same time, there are others who are supportive of such a stand[4]. There is however a general acknowledgement among researchers regarding Hinduism's pro-environmental potential.

2. The Vedic view on creation and nature deities for eco-respect

Chapter two of *Laudato Si'* which is entitled "The Gospel of Creation", begins by upholding the proposition of the "rich contribution which religions can make towards an integral ecology and the full development of humanity"

[3] To name a few, Lance E. Nelson, "The Dualism of Non-dualism", in Lance E. Nelson (ed.), *Purifying the Earthly Body of God: Religion and Ecology in Hindu India*. Albany: State University of New York Press, 1998, pp. 61-88; Id., "Reading the Bhagavadgita from an Ecological Perspective", in Christopher K. Chapple and Mary Evelyn Tucker (eds.), *Hinduism and Ecology: The Intersection of Earth, Sky, and Water*. Cambridge MA: Harvard University Press, 2000, pp. 127-164; Emma Tomalin, "Bio-Divinity and Biodiversity: Perspectives on Religion and Environmental Conservation in India", in *Numen* vol. 51/3-2017, pp. 265-295; Gerald Larson, "'Conceptual Resources' in South Asia for 'Environmental Ethics'", in Baird J. Callicott and Roger T. Ames (eds.), *Nature in Asian Traditions of Thought: Essays in Environmental Philosophy*. Albany NY: State University of New York Press, 1989, pp. 267-277.

[4] To name a few, David Kinsley, *Ecology and Religion: Ecological Spirituality in Cross-Cultural Perspective*. New Jersey: Prentice Hall, 1995, pp. 54-67; Harold Coward, "The Ecological Implications of Karma Theory", in Lance E. Nelson (ed.), *Purifying the Earthly Body of God: Religion and Ecology in Hindu India*, pp. 39-49; T.S. Rukmani, "Literary Foundations for an Ecological Ayurveda, the Arts, and Abhijnanasakuntalam", in Christopher K. Chapple and Mary Evelyn Tucker (eds), *Hinduism and Ecology: The Intersection of Earth, Sky, Water*, pp. 101-126; Philip Lutgendorf, "City, Forest, and Cosmos: Ecological Perspectives from the Sanskrit Epics" in Christopher K. Chapple and Mary Evelyn Tucker (eds), *Hinduism and Ecology: The Intersection of Earth, Sky, Water*, pp. 245-268; O.P. Dwivedi, "Human Responsibility and the Environment: A Hindu Perspective", in *Journal of Hindu-Christian Studies*, 6-1993, pp. 19-26; Banwari, *Pancavati: Indian Approach to the Environment*. New Delhi: Shri Vinayaka Publications, 1992; Vandana Shiva, *Staying Alive: Women, Ecology and Survival in India*. New Delhi: Kali for Women, 1988.

(LS 62). Furthermore, given the following affirmation of the Pope, reflecting on the Vedic view of creation and nature deities would be in keeping with the spirit of this section of the Encyclical:

> Given the complexity of the ecological crisis and its multiple causes, we need to realize that the solutions will not emerge from just one way of interpreting and transforming reality. Respect must also be shown for the various cultural riches of different peoples, their art and poetry, their interior life and spirituality. If we are truly concerned to develop an ecology capable of remedying the damage we have done, no branch of the sciences and no form of wisdom can be left out, and that includes religion and the language particular to it[5].

The various hymns found on the topic of creation in the Vedas are in the form of philosophical, theologico-spiritual and deistic speculations that present a variety of perspectives that come to play in the diverse understandings of the act and the fact of creation. We cannot enter into all the perspectives here, but we shall choose from them those, that according to us, are intrinsically connected to the theme of this paper[6]. The choice of the hymns or verses will be made on the basis of one primary theme and that is divinity in creation. Connected to this theme, we will present the verses of the hymns of the nature deities which form an important part of the Ṛg Veda for ecological reflections.

For the purpose of our paper, a detailed exegesis of the following texts is not needed. We shall first present these hymns and verses, then offer reflections that would require a brief hermeneutical reading of the texts wherever necessary.

Puruṣa Sūkta – Hymn to the Primordial Person (Ṛg Veda X, 90)[7]

In this hymn the primordial person or "God" goes through a self-sacrifice for the sake of creation.

---

[5] Pope Francis, *Laudato Si'*, n. 63.

[6] Among the variety of hymns and myths present in the Vedas on this topic some could be mentioned here: *Viśwakarma* (The architect of the universe) Ṛg Veda X, 82; *Puruṣa Sūkta* (hymn to the primordial person) Ṛg Veda X, 90; *Hiraṇyagarbha* (The Golden Germ) Ṛg Veda X, 121; *Nāsadīya Sūkta* (The hymn of the origins) Ṛg Veda X, 129; *Tapas* (Creative fervor) Ṛg Veda X, 190; *Skambha* (The cosmic pillar) Atharva Veda X, 7; In the following verses of the Brāhmaṇas we find the self-sacrifice of the Lord of creatures (*Prajāpati*) for the sake of creation: Śatapatha Brāhmaṇa II, 2.4.1; III, 9.1.1; X, 4.2.2; VI, 1.2.12-13; Taittirīya Brāhmaṇa II, 3.6.1.

[7] The translations of this and the following Ṛgvedic hymns are from Ralph Griffith, *The Hymns of the Ṛgveda*. Delhi: Motilal Banarasidass, 2004[6].

> ...This Purusa is all that yet hath been and all that is to be;.... three-fourths Purusa went up: one-fourth of him again was here... From that great general sacrifice the dripping fat was gathered up. He formed the creatures of the air, and animals both wild and tame... From it were horses born, from it all cattle with two rows of teeth. From it were generated kine, from it the goats and sheep were born... The Moon was gendered from his mind, and from his eye the Sun had birth. Indra and Agni from his mouth were born, and Vāyu from his breath[8] ... from his navel came mid-air; the sky was fashioned from his head; Earth from his feet and from his ear the regions. Thus they formed the worlds[9].

Hiraṇyagarbha[10] – Birth of the Golden Womb (Rg Veda X, 121)

In this hymn the Supreme Lord and his creation is glorified.

> In the beginning rose Hiranyagarbha, born Only Lord of all created beings, he fixed and holdeth up this earth and heaven... Giver of vital breath, of power and vigour, he whose commandments all the God's acknowledge. The Lord of death, whose shade is life immortal... Who by his grandeur hath become Sole Ruler of all the moving world that breathes and slumbers, he who is Lord of men and Lord of cattle... through his might, are these snow-covered mountains and men call sea... he in his might surveyed the floods containing productive force and generating Worship. He is the God of gods, and none beside him.

Among the number of nature deities (devas – the shining ones) addressed and invoked in the Ṛg Veda, we shall consider Sūrya (Sun), Vāyu (wind), and Agni (Fire).

Sūrya (Ṛg Veda I, 50)

> Surya that all may look on him. The constellations pass away, like thieves, together with their beams, before the all-beholding Sun... Swift and all beautiful art thou, O Surya, make of the light, illuming all the radiant realm... O thou farseeing One, God, Surya, with the radiant hair...remove my heart's disease, take from me this yellow hue.

---

[8] Indra, Agni and Vāyu are nature deities (Indra – Thunder/Rain; Agni – Fire; Vāyu – Wind).

[9] Among the various verses not inserted here, mention can be made of the verse that is considered the scriptural foundation of the caste system in India: "The Brāhman (priest) was his mouth, of both his arms was the Rājanya (warrior) made. His thighs became the Vaiśya (merchant), from his feet the Śūdra (laborer) was produced" (verse 12).

[10] This word is also interpreted as "cosmic egg" and "God". Therefore, Panikkar does not hesitate to use the expression, "Birth of God" in his interpretation of this word in the light of the hymn. See Raimundo Panikkar, *The Vedic Experience - Mantramanjari: An Anthology of the Vedas for Modern Man and Contemporary Celebration*. London: Darton, Longman & Todd, 1979, 67-68.

Vāyu (Ṛg Veda I, 2)

Beautiful Vayu, come, for thee these Soma drops[11] have been prepared. Drink of them, hearken to our call.

Agni (Ṛg Veda I, 75)

How may the mind draw nigh to please thee, Agni? What hymn of praise shall bring us greatest blessing? Or who hath gained they power by sacrfices? Or with what mind shall we bring thee oblations?...

In the Atharva Veda a whole hymn of 63 verses is dedicated to the Earth (Bhūmi Sūkta – Atharva Veda XII, 1). We quote the 12th verse: "Impart to us those vitalizing forces that come, O Earth, from deep within your body, your central point, your navel; purify us wholly. The Earth is mother; I am son of Earth"[12].

The creation hymns of the Vedic age offer the readers a cosmogonical contemplation of Hinduism where the Vedic authors in a spiritual and intellectual state construct in deep amazement and wonder images of the beginnings of the universe that involve God, man and the cosmos. It is no wonder that Panikkar calls the Puruṣa Sūkta a cosmotheandric hymn[13]. But it is this very cosmotheandrism, according to Panikkar, that has led some commentators to brand this hymn as pantheistic. Pantheism would be an unjust and hasty conclusion on the part of the commentators here because this hymn maintains the immanence and the transcendence of God in the phrase: "Three-fourths Purusa went up: one-fourth of him again was here" where the major part of God does not become the universe. Even Pope Francis is against pantheism when he accredits the Judeo-Christian thought to have demytholized the perception of "nature as divine" (78). Panentheism would be a proper judgement in this regard. However, it is important to bear in mind for our discussion here that a sacred or holy perception of anything, in this case the environment, will lead

---

[11] *Soma* was a juice used as a ritual drink extracted from a plant. Researchers are divided over the specific type of plant used for the juice.

[12] This translation is taken from Panikkar, *Mantramanjari*, 124.

[13] Panikkar, *Mantramanjari*, p. 73. Panikkar states: "It (the hymn) reveals to us the character of the creation-sacrifice: its all-embracing function in which the entire universe is involved. It is neither a merely divine affair, not a purely human endeavor, nor a blind cosmic process; it is human, divine and cosmic all in one. That is, it is cosmotheandric. God, Man and universe are correlates. God without Man is nothing, literally "no-thing". Man without God is exclusively a "thing", not a person, not a real human being, while the world, the cosmos, without Man and God is "any-thing", without consistency and being; it is sheer unexisting chaos. The three is constitutively connected". pp. 72-73 (brackets mine).

to respect of that thing or environment. Furthermore, within the context of the above hymn, Anand sees nature as human: "If humans and the rest of creation are so akin, it is because all creation comes from the Primordial Man. The Puruṣa-sūkta is an etiological protology: the cosmos is so human, so similar to man, because it comes from Anthropos"[14]. If God creates from nothing, as stated by the Christian dogma on creation, it maintains the total transcendence of God but reduces the value of creation since creation is coming forth from nothing. Creation by emanation where the world is seen as emanating from God just like the spider web emanating from the spider (Muṇḍaka Upaniṣad 1: 1,7) is not a pantheistic image because the web is not the spider and therefore the world emanating from God is not God. Such images in Hindu Scripture according to Eliot Deutsch leads to reverence for all living beings[15].

We need to maintain a transcendence-immanence perspective of God in the act of creation to elevate the dignity of creation. The Bible tells us that God looked upon his creation and it was very good (Gen 1:31). But this is not enough. A sacred reverence for creation, as we find in Hinduism, will go a long way in upholding the values propogated by *Laudato Si'* towards Creation. Quoting John Paul II, Francis affirms the following: "We can say that 'alongside revelation properly so-called, contained in sacred Scripture, there is a divine manifestation in the blaze of the sun and the fall of the night" (85). Later in agreement with the bishops of Brazil, Francis points out "that nature as a whole not only manifests God but is also a locus of his presence" (88). We could also add the famous phrase of Gerald Manley Hopkins here: "The world is charged with the grandeur of God". For the Hindus, instructed by the wonderful creation hymns of the Vedas, such expressions come naturally. It leads to a profound respect for creation because the whole of creation and not only humans are made in the image and likeness of God[16]. "Discovering the presence (of God in creation) leads us to cultivate the 'ecological virtues'" (88 brackets mine).

Furthermore, the interrelationship of creatures stressed by Francis (85,86) is intrinsic to the Puruṣa-sūkta. Commenting on this hymn Subhash Anand writes:

---

[14] Subhash Anand, *Hindu Inspiration for Christian Reflection: Towards a Hindu-Christian Theology*. Subhash Anand: Gujarat Sahitya Prakash, 2004, p. 8.

[15] Cf. Eliot Deutsch, "Vedanta and Ecology", in elliyavaram *Mahadevan* Ponnambalam *Mahadevan* (ed.), *Indian Philosophical Annual*. Madras: The Centre for Advanced Study in Philosophy, 1970, pp. 80-83; Lance Nelson, "Ecology" in *Brill's Encyclopedia of Hinduism*, vol. IV, K. Jacobsen et al (ed.). Boston: Brill Leiden, 2012, p. 664.

[16] Cf. Subhash Anand, *Hindu Inspiration for Christian Reflection*, p. 21.

> Our earliest ancestors lived in very close dependence on nature. It provided them food (fruits, leaves, roots), shelter (caves) and the light they needed (the sun and the moon)... This dependence led to an experience of communion and they felt that they were in many ways like the nature that surrounded them... This experience explains why in the Puruṣa-sukta different constituents of nature and the various groups within the society are related to the various parts of the human body, and thereby of the Puruṣa[17].

Can this experience become part of our own everyday life? It is through such an experience that one realizes an interconnectedness with the whole cosmos. Even though Pope Francis feels that "the different aspects of the planet – physical, chemical and biological – are interrelated, so too living species are part of a network which we will never fully explore and understand" (138), the Vedic seers in the Puruṣa-sūkta seemed to have understood this interrelatedness.

The nature deities, to whom the hymns of the Vedas are dedicated, are an expression of that spiritual experience. A certain reverence created by that experience is what led the Vedic people to personify and "worship" the natural elements as gods. They were not gods in the sense of the Christian God or the Ultimate Reality. The Ultimate Reality, as the Vedic seers progressed in their spiritual experience, was called Brahman and these gods were not Brahman. They were given "divine" respect and hailed with terminologies similar to "lord" or any other equivalent term just as a King would be given royal respect and addressed as "Lord" or the Judges in the court of law addressed as "my Lord". Furthermore, the "worship" of these gods must not be perceived in the same way as the worship of the highest God. For example genuflection before the Blessed Sacrament and genuflection before a King would not be considered as a similar act of adoration or worship. In this regard, the personification of nature, which is evident in the Vedic hymns and may appear problematic to Western Christian researchers, should not be understood as if the Vedic people saw the natural elements as living beings and "worshipped" them. The hailing of natural elements could be perceived not so much as an expression of personification but as an expression of reverential elevation to fulfill ones inherent longings for the divine. For instance, in the Atharva Veda fearlessness is addressed in the following way – "Breathe on us fearlessness..." (Atharva Veda VI.40.1). This did not mean that "fearlessness" was regarded as an animate being or a god. Such examples in Hinduism

[17] Subhash Anand, *Hindu Inspiration for Christian Reflection*, 3-4.

need not prevent us from interpreting the veneration of nature deities as the Vedic people's deep respect for nature. The nature deities of the Vedic era or of Hinduism in general could, therefore, be acknowledged as a great potential for environmental concerns.

Iti Chattopadhyay, concludes her article "Environmental Concern in the Vedās" in the following words:

> Thus, we find that the attitude of the Vedās is holistic. Not only human beings, they attached great importance to the air, fire, water, land, flora and fauna in order to protect, preserve and nourish the environment. Man is considered as a member of this vast family. In fact, here man is taught to live in harmony with nature and recognize that divinity prevails in all elements including plants and animals that help to combat social as well as environmental problems. Thus, the Vedic perspective and belief system about both the living and non-living elements can act as a safeguard for environmental protection[18].

The notion of the presence of God in the universe and in the smallest of all creatures with which Pope Francis begins his "A prayer for our earth" (246), is a notion that must be religiously utilized for enviornmental concerns. It should make us perceive "the world as sacrament"[19] (*Sacramentum Mundi*). Such a perception would doubtlessly lead us to protect the environment when it is exploited, abused, and eventually destroyed.

3. The Post-Vedic ethico-moral view for eco-respect

The post-Vedic smṛti literature of Hinduism, the Dharmaśāstras and the Purāṇas, contain exhortations and regulations on the practise of Dharma (ethico-moral and social duties)[20] which from the perspective of our paper could be understood as helpful in some instances for environmental protection and eco-respect. The practise of Dharma is geared towards improving the quality of life. The Dharma upheld by the Dharmaśāstras and the Purāṇas is radically against the "decline in the quality of human life and the breakdown of soci-

---

[18] Iti Chattopadhyay, "Environmental Concern in the Vedās", in *Journal Of Humanities And Social Science* (IOSR-JHSS), Vol. 22/6 (2017), 21-23

[19] This is the title of the book by Francis D'Sa, Isaac Padinjarekuttu and Jacob Parappally (ed.), *The World as Sacrament: Interdisciplinary Bridge-building of the Sacred and the Secular*. Pune: Jnana-Deepa Vidyapeeth Theology Series, 1998.

[20] *Dharmaśāstras* are those theological texts that contain rules and regulations for personal and social behaviour, moral and ethicl norms, civil and criminal law. The word *purāṇa* means "old" or "ancient". These sacred writings belong to the genre of mythology and folklore.

ety" which is the title of point four of chapter one of *Laudato Si'* where the Pope states: "Human beings too are creatures of this world, enjoying a right to life and happiness, and endowed with unique dignity. So we cannot fail to consider the effects on people's lives of environmental deterioration, current models of development and the throwaway culture" (43). An integral ecology would be a favourable theme for the post-Vedic literature especially in the concept of the "welfare of all beings" of the Bhagavad Gītā which will be treated briefly later in this section.

A healthy environment would be a sign of prosperity and a consequence of following Dharma. Therefore, "the Dharmaśāstras include laws designed to protect – in various ways – animals, trees, rivers, land and so on. These prescriptions have been translated into homilies and edifying stories in the Purāṇas, ..."[21] Vasudha Narayanan, quotes a relevant passage from the Matsya Purāṇa[22].

> The goddess Parvati planted a sapling of the Asoka tree and took good care of it. The divine beings and sages came and told her: "O [Goddess] ... almost everyone wants children ... What do you achieve by creating and rearing trees like sons?" Parvati replied: "One who digs a well where there is little water lives in heaven for as many years as there are drops of water in it. One large reservoir of water is worth ten wells. One son is like ten reservoirs and one tree is equal to ten sons[23].

Narayanan also discovers that the Dharmaśāstras do not only recommend and celebrate the planting of trees but condemns their cutting as well: "The Varaha Purāṇa[24] says that one who plants five mango trees does not go to hell and the Vishnu Dharmottara (3.297.13) claims that one who plants a tree will never fall into hell"[25]. Lance Nelson, mentions the Matsya Purāṇa that "describes a tree festival (vṛkṣotsava) in which existing trees are gaily decorated and saplings are planted. The text proclaims, 'A person who plants even one (ekam api) tree will assuredly dwell in heaven (svarga) for 30,000 years of [the god] Indra' (MatP. 59.18)"[26]. It is as though the "secondary" scrip-

[21] Lance Nelson, "Ecology", 666.

[22] Matsya Purāṇa is the story of the Fish Avatāra of Lord Viṣṇu.

[23] Vasudha Narayanan, "One Tree is Equal to Ten Sons", in Harold Coward and Daniel Maguire (eds.), *Visions of a New Earth: Relgious Perspectives on Population, Consumption, and Ecology*. Albany NY: State University of New York Press, , 2000, p. 114.

[24] The Varāha Purāṇa is the story of the Boar avatāra of the Lord Viṣṇu.

[25] Vasudha Narayanan, "One Tree is Equal to Ten Sons", p. 114.

[26] Lance Nelson, "Ecology", p. 666.

tures of Hinduism are making eco-sustainence mandatory for salvation. The King, according to the titles used for him in the Dharmaśāstras, was the earth protector (bhūpa) or earth maintainer (bhūbharaṇa)[27]. Protection of the earth would inevitably lead to benefit the Kingdom. The glory of the King's reign depended in a certain sense on the protection and good management of the earthly resources. "It is clear from many classical Indian texts on the duties of kings that a king had a responsibility to reap the benefits of the nature's resources for the sake of his kingdom and its citizens; yet it is also clear that the successful and ideal king had to carefully manage these resources and respect the earth"[28]. In the Varāh Purāṇa (172) the encounter of the merchant Gokarṇa with a group of distressed tree nymps is narrated where the nymphs give Gokarna the following instructions: "He who plants a peepal, a margosa, a banyan, ten flowering plants, two pomegranates, two oranges, and five mangoes never goes to hell… The learned people declare a tree to be like a son because it provides materials for domestic life, provides shelter for small animals, twice in a year birds breed in it and it provides fruits to the originator throughout the year"[29].

The epic Mahābhārata and The Bhagavad Gītā which is a part of the Mahābhārata are texts that could also be considered from an ecological perspective for our paper. It is true that the fundamental vision of the Bhagavad Gītā could be interpreted as world denying subbordinating or relativizing the emphasis on the "this-worldly" dharma, by "the traditions's final mokṣa (liberation) orientation"[30]. We could agree with Nelson when he states: "It goes without saying that the Gītā, like other ancient religious texts, does not address contemporary ecological issues directly, for it was composed long before the crisis emerged. In reading the Gītā from an ecological perspective, as when with similar concerns we read other classical religious texts, we reframe its teachings in light of problems it was not intended to address"[31]. However,

---

[27] Mary McGee, "State Responsibility for Environmental Management: Perspectives from Hindu Texts on Polity", in Christopher K. Chapple and Mary Evelyn Tucker (ed.), *Hinduism and Ecology: The Intersection of Earth, Sky, and Water*, p. 63. In Lance Nelson, "Ecology", p. 666.

[28] Mary McGee, "State Responsibility for Environmental Management: Perspectives from Hindu Texts on Polity", 73. In Lance Nelson "Ecology", 666.

[29] Varāha Purāṇa 172. 39. 43-44. In Lance Nelson, "Ecology", p. 667.

[30] Lance Nelson, "Reading the Bhagavadgītā from an Ecological Perspective", in Christopher K. Chapple and Mary Evelyn Tucker (ed.), *Hinduism and Ecology: The Intersection of Earth, Sky, and Water*, p. 144. Brackets are mine.

[31] Lance Nelson, "Reading the Bhagavadgītā from an Ecological Perspective", p. 131.

we would like to focus on the potential of the Bhagavad Gītā for an ecologically positive outlook for the sake of our paper which Nelson himself does not totally reject when he recognizes that,

> There is no doubt that a deeply held belief in the superordinate value of a nonmaterial spiritual reality has the potential – if rendered sufficiently compelling – to do much to undercut the ecologically damaging ethos of the dominant consumerist, growth-oriented "religion of the market". It is equally clear, as I will be neither the first not the last to point out, that the Gītā provides such a spiritual vision[32].

The all-inclusiveness of Hinduism is its biggest strength due to which contrary propositions need not be taken as mutually exclusive. They must be viewed within a positive dialectical tension utilizing any proposition wherever needed.

Ahiṃsā - Non-Injury (Non Violence), is an important concept found in the Mahābhārata. Firstly, it must be noted that the concept of ahiṃsā entered Hinduism through the renunciant Indic tradition[33]. Secondly, "in many Hindu texts and traditions, violence is not condemned, and many texts appear to have remained untouched by the doctrine of ahiṃsā"[34]. However, it does not mean that ahiṃsā is not appreciated and proclaimed worthy of practise. In the Mahābhārata ahiṃsā is called the highest virtue, the highest self-control, the highest gift, the highest power, the highest happiness and truth[35]. Ahiṃsā, therefore, has had a significant influence on Hinduism. It can be interpreted in many ways. To some it is the mere abstinence from eating meat and practising vegetarianism. To some, such as Mahatma Gandhi, it can be applied to a wider social and political context where ahiṃsā supported by satyāgraha (way of truth) were the two pillars of Gandhi's political agitation against colonial injustice and exploitation where one fights the enemy through non-violent resistance. To some it is non-injury to fellow human beings and to nature. If ahiṃsā is understood as non-injury to nature, then it carries with it great scope to be utilized in the biological context. It is true that for Hinduism and also for Gandhi, ahiṃsā was a means to be embraced to attain the highest truth – God.

---

[32] Lance Nelson, "Reading the Bhagavadgītā from an Ecological Perspective", p. 131.

[33] Cf. Bryan Lobo, "L'induismo: tra violenza e non violenza", Roberto Cipriani et al., *Religioni e Conflitti: Conoscere la divisione pre progrettare l'incontro in un mondo in guerra nel nome di Dio*. Roma: Aracne, 2016, p. 283.

[34] Knut Jacobsen, "Ahimsa", in Knut Jacobsen et al (ed.), *Brill's Encyclopedia of Hinduism*, vol. 2, Brill, Leiden-Boston 2010, p. 675

[35] Cf. Knut Jacobsen, "Ahimsa", p. 675.

To some environmentalists then, non-violence towards nature is only a means to "salvation" and therefore practised not for the sake of nature but for salvation. Nature is instrumentalized for another goal and consequently does not enjoy intrinsic value. In this kind of an intrepretation the following statement of Lal is well founded: "The Hindu recommendation to cultivate a particular kind of attitude (namely, *ahiṃsā*) toward animals is based *not* on considerations about the *animal* as such but on considerations about how the development of this attitude is part of the purificatory steps that bring men to the path of *mokṣa*"[36]. To support the intrinsic value of nature, Framarin deals with the problem of the instrumentalist interpretation using a logical method[37]. After arguing how the instrumentalist interpretation is circular and arbitrary[38], the author concludes that since plants, animals and human beings have intrinsic value ahiṃsā practised towards them brings merit which leads us closer to the highest good - mokṣa[39]. As is evident from the intention of the argument, ahiṃsā as a Hindu virtue that helps to attain salvation necessarily leads to enviornmental protection and must therefore be freely propagated by ecologists.

Connected to the concept of non-violence is the concept of the *Welfare of all beings* present in the Bhagavad Gītā 3.20, which has become quite popular among the Hindus: "…if again you consider the welfare of the world, then you should work [and act]". The idea of working for the "welfare of the world" (loka-saṁgraham) exists for the Hindus as a scriptural norm. The welfare of the earth and its people resonates with the integral ecology and sustainable development proposed by *Laudato Si'*: "An integral ecology is inseparable from the notion of the common good, a central and unifying principle of social ethics. The common good is 'the sum of those conditions of social life which allow social groups and their individual members relatively thorough and ready access to their own fulfilment'" (156). What is important to note here is that to work for the welfare of the world, in Hinduism, is present in the Bhagavad Gītā and therefore is applied to all the devotees as an ethical injunction from God Kṛṣṇa

---

[36] Basant K. Lal, "Hindu Perspectives on the Use of Animals in Science", in Tom Regan (ed.), *Animals and Sacrifices: Religious Prespectives on the Use of Animals in Science*. Philadelphia: Temple University Press, 1986, p. 200. In Christopher Framarin, "The Value of Nature in Indian (Hindu) Traditions", in *Religious Studies* 47/3-2011, p. 286.

[37] Cf. Christopher Framarin, "The Value of Nature in Indian (Hindu) Traditions", pp. 285-300.

[38] For this argument see Christopher Framarin, "The Value of Nature in Indian (Hindu) Traditions", pp. 285-294.

[39] Cf. Christopher Framarin, "The Value of Nature in Indian (Hindu) Traditions", p. 299.

himself. The impact of scriptural teaching is much greater than the teaching of a religious head. However, it does not mean that every Hindu, especially the Vaiṣṇavites and in this case the followers of Kṛṣṇa, are bound under the pain of sin to work for the welfare of the world. Rather, what needs to be realized is that the scriptural injunction creates a consciousness of an integral ecology. Even though the understanding of integral ecology may not be identical to the one proposed by *Laudato Si'*, it however shares in many ways the ultimate perspective of *Laudato Si'*. To live in harmony with God, human beings and nature is central to the Hindu way of being. "Cosmotheandric" harmony or Ṛta, which is in many ways at the heart of the awareness of *Laudato Si'* as well, is the result of a healthy ethico-moral life. If eco-harmony necessarily entails harmony with God and human beings or vice-versa, then Hinduism, in comparison to other religions is in a much better position to offer the building blocks of such harmony to further the ideals of *Laudato Si'*.

The discussion on Hinduism and environmental ethic wrestles, from various angles, with the main question whether Hindu texts and traditions confer intrinsic value to the enviornment[40]. Two main premises gathered from our reflections which bestow direct moral standing or intrinsic value to nature are unity with the Divine and interconnectedness with humans. It is argued therefore that harming nature would mean harming God and humans directly. Once again, Framarin, after presenting pro and con arguments of various scholars on this topic, favours the argument because of its initial conviction that a plausible enviornmental ethic presumed by some scholars as entailed in the Hindu texts and traditions, attributes direct moral standing to nature.

Taking off from Framarin's argument, if nature enjoys intrinsic moral value then eco-preservation has a lot to benefit because it creates a reverence for nature that prevents its exploitation. It is true, that such reverence need not lead to scruplosity in utilizing nature for our purposes. But, it could at least lead to scruplosity regarding the abuse of nature which is the main concern for all enviornmentalists and even the document *Laudato Si'*. Hinduism, in comparison with other religions, and from what we have explained in the previous sections, holds by far the greatest potential to uphold the intrinsic and ethical value of the enviornment.

---

[40] For a detailed discussion of this topic see, Christopher Framarin, "Hinduism and Environmental Ethics: An Analysis and Defense of a Basic Assumption", in *Asian Philosophy* 22/1-2012), pp. 75–91.

4. The sacrality of nature and Hindu Religious activities

The sacredness of nature due to God's presence theologically developed in Hinduism as briefly reflected upon in the first section, is at the same time a profound spiritual experience and therefore requires a separate but brief presentation in this paper.

On the one hand, there is a world-denying spirituality in classical Advaita – non-dualism (of the Vedānta school) where Śaṅkara, one of the foremost proponents of this school sees the world as māyā (illusion, unreal) and a block to final liberation or mokṣa because it seduces the soul into passions, desires and cravings that bind the soul with harmful karma leading to cycles of rebirths and thereby suffering instead of liberation. To be liberated would therefore mean liberation from this world which causes suffering. In this concept, the human body is considered as the prison of the soul. As a spiritual exercise, the body must be necessarily dominated through ascetical practises to achieve mokṣa. Other activities if believed to render mokṣa to a person would be unhesitatingly carried out. For example, throwing the remains of the cremated bodies in the Ganges. Here, one has a classical case of how spirituality can lead to environmental degradation.

On the other hand, there is another form of Advaitic spirituality that belongs to the Tantric consciousness of Hinduism where the world is the creative energy (Śakti) of God distinct but at the same time one with God. Within this dialectical reality, corporeality is regarded as positive and also thereby as a means to mokṣa[41]. This kind of an inclusive non-dualism is popular among Hindus and helps to create a positive enviornmental consciousness where nature as intrinsically sacred or divine cannot be seen as a mere metaphor or a philosophical concept but as a "testable reality"[42]. It is within such a reality of oneness and distinction between the divine, human and cosmos, that the non-dual spiritual experience is functional and transformative and has led the Hindu saints to find God not only in oneself but in Creation as well. Such an experience is evident, for instance, in the following hymn of Kabir a North Indian saint of the 15th century:

---

[41] Cf. Virgilio Agostinelli, "Tantrism and Corporeality", in Bryan Lobo, et al (ed.), *Enlightenment and Tantra: Hindus and Christians in Dialogue*. Rome: Gregorian & Biblical Press, 2018, pp. 223-241.

[42] Maria Cristina Kaveri, "Warp and Weft: Role and Destiny of Human Beings", in Bryan Lobo, et al (eds.), *Enlightenment and Tantra: Hindus and Christians in Dialogue*, p. 46.

O God, your Godliness
Pervades every body
Like the red unseen
In the green henna leaf
... In flowers, in branches
In this whole world
Wherever I look – you only you!
In an elephant, you become an elephant
An ant is just a little you!
... In earth, in ocean, in every
Creature you alone reside[43].

In a somewhat similar way St. Francis of Assisi wanted his followers to contemplate God when they were in contact with Creation. Pope Francis reminds us that St. Francis "... asked that a part of the friary garden always be left untouched, so that wild flowers and herbs could grow there, and those who saw them could raise their minds to God, the Creator of such beauty" (LS 12).

Along with the above idea of God in Creation, central to the Hindu ideal is sacrifice, self-restraint and asceticism; aspects common to the major religions. However in Hinduism, Creation is realized by the primordial sacrifice, as explained above, which implies that it is sacrifice that is responsible for harmony (ṛta) and life in creation. Sacrifice, therefore, is intrinsically linked to self and eco-harmony. Consequently, harmony in self leads to harmony with nature. The basic element for self-harmony is non-attachment. Not to be attached to the fruits of one's labour (Niṣkāmakarma) is the central idea of the two verses found in the Bhagavad Gītā (Bh.G. XII. 11; II.47-48) which could be an invitation to live a life of non-attachment in relation to Creation. These were the verses that also inspired Mahatma Gandhi to live a life of non-attachment. To live in non-attachment is to live a life where greed is not allowed to dictate one's activities. When Gandhi teaches, "there is enough for everyone's need but not enough for everone's greed", he appeals to a simple way of life that is not attached to excess of goods and luxuries for which nature is exploited. Non-attachment is subsequently translated into an eco-spirituality that becomes the backbone of a balanced attitude to avoid amassing of goods and wealth for growing personal and social consumption. This eco-spirituality

[43] Linda Hess and Shabnam Virmani (trs.), *Ajab Shahar, Kabir in America: 12 Kabir Songs by Prahlad Singh Tipanya*. Bangalore: Publisher, 2004, pp. 43-44. In Lance Nelson, "Ecology", p. 671.

challenges Industrialization, consumerism, growing demands by increasing populations which has led to enviornmental disasters. A return to the eco-spirituality of the past is what is required today for which the formation that Rabindranath Tagore recommends shall be quoted to the full:

> Besides practising brahmacharya a boy should live in the midst of nature. Towns are not our natural abodes, and have been built to supply our material needs. That we should be born in towns and be brought up in the lap of stone and brick was never intended by Providence. Towns snatch us away from the bosom of nature and consume us in their hungry furnace…
> But nature's help is indispensable when we are still growing up, and still learning, and before we are drawn neck and crop in the whirlpool of affairs. Trees and rivers, and blue skies and beautiful views, are just as necessary as benches and blackboards, books and examinations.
> From ancient times the Indian mind has developed through close and unbroken contact with nature, and the desire to be one with animal and vegetable life has been inherent in the Indian spirit.
> The four elements of earth, water, air and fire, form a whole and are instinct with the universal soul – this knowledge cannot be gained at a school in town. A school in town is a factory which can only teach us to regard the world as a machine[44].

It is the above way of thinking that led Rabindranath Tagore, guided by the Hindu ideals of education in the past, to start his own school – Shantiniketan (the abode of peace). It is true that a majority of towns which grew due to the onslaught of industrialization were creating a nature-less ambience in which human beings grew oblivious of the wholesome effects of living in close proximity with nature. In recent years, however, people have realized this unhealthy fact and therefore have made efforts whether personally or through governmental organizations to open up parks and natural resorts and other green zones either within the cities and towns or at the periferies.

Various religious activities of the Hindus involve nature. Plants, trees and rivers are worshipped. Worshipping nature does not necessarily lead to eco-sensitivity as argued by Lance Nelson, mentioned above, nor would it lead to the prevention of pollution as noted in the case of the river Ganges. Vijaya Nagarajan finds a similar kind of a contradiction in the Kōlam ritual

---

[44] Rabindranath Tagore, *Towards Universal Man*. Bombay: Asia Publications, 1961, pp. 72-73. In Subhash Anand, *Hindu Inspiration for Christian Reflection*, p. 24

of Tamil Nadu where women make beautiful ritual designs near the threshold of the house[45] out of devotion for the earth who is considered in this case as a goddess Bhū Devī. Nagarajan finds a "contradiction between women's reverence for Bhū Devī and their seeming disrespect to her throughout the day, as they threw trash and garbage on the earth, the very same place that they considered to be sacred"[46]. Nonetheless, the women are aware of this contradiction in which the patience of the earth goddess is presupposed in the following opinion of a woman who shares it with Nagarajan: "We draw the Kōlam as our first ritual act in the morning to remember Bhū Devī. We walk on her. We spit on her; we poke here; we burden her. We expect her to bear us and all the activities we do on her with endless patience. So, we do the kōlam. (Jaya Mami, Thiruvaiyāru village, Thanjāvūr District"[47]. Nagarajan arrives at the concept of "intermittent sacrality" to define this kind of a behaviour inherent in those practising such rituals which seem eco-friendly. By intermittent sacrality, Nagarajan means a sacrality of an object that is temporarily created by the individuals or community and does not last forever. We could quote Nagarajan at length here in her interaction with the villagers on this point:

> For a long time I puzzled over the contradictions between religious ideology and religious practice. If the perception of natural objects were steeped in sacrality, why did that not lead naturally to a consummate sense of care and protection of that natural object? In conversations, when I pointed out what seemed to me to be a contradiction between belief and action, both women and men would laugh and repeatedly tell me, oh what we believe does not have to rule at all times and at all places. Just because one place is made sacred at one particular time does not necessarily mean that it remains special throughout the day. You have misunderstood. How could we live on this earth otherwise? We could not do anything, now, could we? There are so many rules and rituals that we have to follow. It is understood that we are human and that we need to live too. So, there is no expectation that we need to care for everything that is sacred all the time. We could not do it. We could not live. We have human needs and we must fulfill them[48].

[45] It is also made "at the feet of divine images in the main domestic shrine, and on temple grounds". For more on the Kolam ritual see Vjaya Nagarajan, "The Earth as Goddess Bhū Devī: Toward a Theory of 'Embedded Ecologies' in Folk Hinduism", in Lance Nelson (ed.), *Purifying the Earthly Body of God: Religion and Ecology in Hindu India*. Albany NY: State University of New York Press, 1998, pp. 270-271.

[46] Vjaya Nagarajan, "The Earth as Goddess Bhū Devī…", p. 275.

[47] Vjaya Nagarajan, "The Earth as Goddess Bhū Devī…", p. 270.

[48] Vjaya Nagarajan, "The Earth as Goddess Bhū Devī…", pp. 277-278.

In this case, sacrality and pollution can go together, but it therefore does not mean that it is a religious practise that has no ecological character to it. The pollution aspect could be controlled by other secular education, practices and conscientization, but the aspect of sacrality can go a long way in being in touch with nature enhancing a particular bond helpful for environmental concerns that otherwise would not have existed. Similar views could be considered in other ritual practices of the Hindus regarding the veneration of a plant called Tulsi[49], a plant that is normally found in Hindu homes. It is a medicinal herb used for anxiety, stress and fatigue. The veneration of this plant by the Hindus helps, as it is believed, to acquire favors and graces from God. Veneration is normally done by circumambulation of the plant or even of a tree considered sacred normally found within the precints of a temple. "Almost every temple in South India dedicated to the gods Shiva or Vishnu, or to a manifestation of the Goddess, has *sthala vriksha* ("the tree of a [sacred] place"), a particular tree that was sacred to that area. This is the "official" tree of the temple and is usually a grand old specimen that is surrounded by a beaten path used for circumambulation by pilgrims and devotees"[50].

Along with plants and trees, there are even groves that are regarded as sacred because they are believed to be associated with some deities and spirits. These groves cannot be disturbed or destroyed, again not for ecological reasons, but perhaps because of the fear of facing the wrath of the deities. In exchange the sacred groves have become "hotspots of biodiversity"[51]. "In addition to their actual or potential function as refuges for diverse flora and fauna, sacred groves have attracted considerable attention among conservationists as offering other environmental benefits, including the protection of water sources and the retention of ground water"[52]. Nugteren notes, "sacred groves have become something of a vogue among botanists and ecologists. They have become part of the romantic imagination about a pristine, bountiful, and respectfully treated natural environment in India"[53]. However, even

---

[49] "Ocimum tenuiflorum (synonym Ocimum sanctum), commonly known as holy basil or tulsi, is an aromatic perennial plant in the family Lamiaceae. It is native to the Indian subcontinent and widespread as a cultivated plant throughout the Southeast Asian tropics". https://en.wikipedia.org/wiki/Ocimum_tenuiflorum (accessed 04/01/2021).

[50] Vjaya Nagarajan, "One Tree is Equal to Ten Sons", p. 116.

[51] Emma Tomalin, "Biodivinity and Biodiversity: Perspectives on Religion and Environmental Conservation in India", in *Numen* 51/3-2004), p. 273.

[52] Lance Nelson, "Ecology", p. 669.

[53] Albertina Nugteren, *Belief, Bounty, and Beauty: Rituals around Sacred Trees in India*. Leiden: Brill, 2005, p. 421. In Lance Nelson, "Ecology", p. 669.

though religious fear helps the protection of these groves, exploitation and destruction of these groves are still threats that loom large in the present Indian commercial and political scenarios.

Moving on to the issue of rivers in India, it is a fact that among all the rivers of India, the Ganga is by far the most popular of them all for various reasons but most importantly for religious activities of ritual and moral cleansing. The following story highlights this point clearly:

> A King goes to sleep on the banks of the River Ganga. When he wakes up in the middle of the night, he sees some women covered in filth taking a dip in the holy river. They emerge from the river cleansed and then disappear. The king returns on several nights and see the same thing. Eventually he asks them who they are; they reply that they are the embodiments of the rivers of India. Everyday, they tell him, human beings bathe in the rivers and their sins are absolved by that act. The rivers – embodied as women – absorb the moral dirt and then come to the Ganga, the grand purifier, to purify themselves[54].

It is true that the Ganga today is physically polluted playing the role of a moral purifier[55]. Religion, in this case, has proved to be a cause of pollution because religiously speaking, if "Ganga is the supreme (divine) purifier, she cannot be polluted"[56]. Furthermore, industrial and other wastes are freely discarded into the river without qualms of conscience. Not only religion, therefore, but other factors as well are involved in the pollution of the river Ganga. Environmentalists and other individuals and groups[57] are making the Government and the people aware of this dangerous phenomenon that has led to some stray efforts at cleaning up the Ganga. To avoid further pollution as the first step to clean the Ganga, religious sanction, which would be the most effective at least in this case, could be used where the people need to be told to stop polluting their goddess (mother) Ganga or face her wrath in future. There are even religious texts to show how the people while maintaining physical cleanliness should avoid polluting the rivers[58].

---

[54] Vjaya Nagarajan, "One Tree is Equal to Ten Sons", p. 116.

[55] Cf. Vjaya Nagarajan, "One Tree is Equal to Ten Sons", p. 116.

[56] Lance Nelson, "Ecology", 668. Brackets mine.

[57] One such group is the CGC (Clean Ganga Campaign – Swatcha Ganga Abhiyan). Cf. Kelly D. Ally, "Idioms of Degeneracy: Assessing Ganga's Purity and Pollution", in Lance Nelson (ed.), *Purifying the Earthly Body of God: Religion and Ecology in Hindu India*, p. 299.

[58] Cf. Kelly D. Ally, "Idioms of Degeneracy…", p. 307.

5. Hinduism and Environmental Activism

In moments of great danger and impending destruction of nature, Hindus of various denominations have embarked upon activities to protect nature because of their religious beliefs.

A classic case of how a religious belief system influenced environmental protection is found among the Bişṇoi community, the most popular among the many that favor eco-protection, who "practise a religion of environmental conservation. They believe that cutting a tree or killing an animal is a blasphemy. Their religion, an offshoot of Hinduism, was founded by Guru Maharaj Jambaji, who was born in 1451 CE in the Marwar area"[59]. Jambaji, while suffering the consequences of famine and the felling of trees decided to lay down 29 injuctions and "principal among them was a ban on the cutting of any green tree and on the killing of any animal or bird"[60]. Furthermore, "a vegetarian diet is prescribed, along with the practice of compassion for all living beings. Jambhesvara (Jambaji) warned, in particular, that on their deathbeds, those who killed animals would be haunted by their painful cries"[61]. This practice eventually led to a rich, green and healthy environment with a wide range of flora and fauna. In 1730 the King of Jodhpur (near Rajasthan) who wanted wood to build a new palace sent his soldiers to cut the trees of the Bişṇois to which the Bişṇois protested. When the soldiers did not heed to their protests the Bişṇois under the leadership of Amṛta Devi, a Bişṇoi woman, hugged the trees. Around 363 of them were killed by the soldiers. When the King got this news he called off the operation and gave the Bişṇoi community protection for their beliefs. "Today, the Bishnois community continues to protect trees and animals with the same fervour. Their community is the best example of the true Hindu-based ritual defence of the environment in India, and their sacrifices became the inspiration for the Chipko movement of 1973"[62].

---

[59] Dwivedi, "Human Responsibility", p. 24.

[60] Dwivedi, "Human Responsibility", p. 24.

[61] Pankaji Jain, *Dharma and Ecology of Hindu Communities: Sustenance and Sutainability*. Surrey. Farnham, U.K: Ashgate, 2011, p. 59. In Lance Nelson, "Ecology", p. 672, bracket mine.

[62] Dwivedi, "Human Responsibility", 24.

63

The Chipko[64] movement began in the 1970's in the Garhwal region of Uttarakhand at the foothills of the Himalayas to protect local forests which were the means of livelihood for the people living there from being exploited. The event was sparked when the government gave an Allahabad-based company for sport-goods a licence to cut the trees for wood to make tennis rackets to be exported. Chandi Prasad Bhatt, the leader of the Dasholi Gram Swarajya Sangh (DGSS), a local Gandhian cooperative movement announced, "Let them know that we will not allow the felling of ash trees. When they aim their axes upon them, we will embrace the trees"[65]. The company was finally not allowed to cut the trees. This success spread like wildfire and similar movements began in other parts of India leading the then government to impose a 15 year moratorium on commercial tree felling in the Himalayan forests. Gandhian non-violence was practised by the DGSS in their struggle to save the trees and the forest from destruction. Sunderlal Bahuguna, another well-known Gandhian leader, with his slogan "ecology is the best economy", was also responsible for spearheading the chipko movement in the Himalayas. Slowly this movement has had an impact not only on the whole of India but

63 This picture is taken from the following site: https://en.wikipedia.org/wiki/Chipko_movement#/media/File:Big_chipko_movement_1522047126.jpg : (accessed: 16/01/21).

64 Chipko means to "cling" or to "stick" in Hindi.

65 Thomas Weber, *Hugging the Trees: The Story of the Chipko Movement*, New Delhi 1988, 40. In Lance Nelson, "Ecology", 673.

on the world as well. Among the many songs and poems of the Chipko poet, Ghanshyam Raturi Sailani, we could cite one:

To weave garments for barren earth
Plant new trees and make new forests.
To save the earth from Impending danger
Come one come all and get together.
For forests are our life and soul
They are the sources of rivers all.
Rains and springs are created in them
The soil, water and air are because of them.
Trees are blessed even by God
He came once to save them from the odds.
They are worshipped by mankind
Without considering their safety all planning is blind.
Apricot, apple, walnut are their treasure
We derive from them fodder wood and water.
Once Uttarkhand was Heaven on earth
Where in deep forests, Vedas and Purans took birth.
To drive away the ruthless cutters
Come on, come all and get together.
Now it has become our religion
To save the forests from destruction[66].

Inspired by such Gandhian enviormental activism, opposition to the dams that the government intended to construct to utilize the river waters for large cities with a preference for affluent societies became a great means not only to save the environment but also the poor from being exploited and destroyed. The Tehri Dam and the Narmada Dam oppositions are classic cases of such enviornmental activism.

Sunderlal Bahuguna, is one of the main campaigners against the construction of the Tehri Dam which is built on the Bhagirathi river considered part of the river Ganges. The complain of many Hindus is that the government has compromised the sanctity of the river Ganges for the sake of electricity. Bahuguna in his conversation with George James states, "It is for the sake of Delhi that this dam [the Tehri Dam] is being constructed. Water is being diverted to Delhi. I ask our rulers, 'What have you done to the Yamuna? You have pol-

---

[66] https://www.himalmag.com/chipko-poet-and-his-songs/: (accessed 09/01/21).

luted the Yamuna and now you want the Ganges water to wash off your sins in Delhi"[67]. In support, again, of the eco-system and the prevailing tribes of that Himalayan region, he complains, "…how long will this dam work? It is temporary. But if they plant the whole Himalaya with trees, the water renewing capacity of Himalaya will increase. And that will be permanent. It will make the hill people prosperous also. Today what has happened to the hills by felling the trees is that there are landslides and the soil, which is the flesh and blood of the mountains, is flowing down to the sea"[68].

Based on the ethics and principles of Satyāgraha, the Narmada Bachao Andolan – NBA (Save the Narmada movement) was spearheaded by a woman activist Medha Patkar against the various dam projects on the Narmada river especially the construction of the *Sardar Sarovar Dam*. Hinduism may not have played a defining role in this movement, however, it is regarded as "perhaps the longest living Gandhian non-violent movement in the history of the world"[69]. Besides, Gandhi's non-violent struggle, as is well-known, was inspired by the Hindu ideal of ahiṁsa. The NBA, therefore, could be considered as a movement activated in a remote way by Hindu principles. The fight against the Narmada projects was indeed a fight against the exploitation of the environment and the poor. The *Right Livelihood Foundation*, that awarded Medha Patkar the "Right Livelihood Award" for her continuous and committed work to save the Narmada region had this to say about the project:

> The Narmada projects are the epitome of unsustainable development. Two of the largest proposed dams, Sardar Sarovar and Narmada Sagar, have been under construction since 1961. According to Narmada Bachao Andolan, the dams force the displacement of about a million people and affect many more, largely poor peasants and tribals. They also cause immense ecological damage through the inundation of forests, including prime habitats of rare species. Resettlement and compensation have been totally inadequate and there is not the remotest prospect that the displaced people, the 'oustees', will be adequately resettled, nor that the ecological damage can be compensated for[70].

---

[67] https://www.india-seminar.com/2004/533/533%20in%20coversation.htm: (accessed 09/01/21).

68 https://www.india-seminar.com/2004/533/533%20in%20coversation.htm: (accessed 09/01/21).

[69] https://www.nationalheraldindia.com/opinion/the-longest-gandhian-satyagraha-over-narmada: (accessed 09/01/21).

[70] https://www.rightlivelihoodaward.org/laureates/medha-patkar-and-baba-amte-narmada-bachao-andolan/#:~:text=According%20to%20Narmada%20Bachao%20Andolan,prime%20habitats%20of%20rare%20species (accessed 09/01/21).

Environmental activism in India is remotely sustained by Hindu ideals. The activists are aware of such religious backing and are likely to to use it to good effect. Sometimes such activism in its liason with religion is lobbied by political parties for political gains. To give one example, the Tehri Dam posed a great danger for certain sacred sites to the Hindus. The Vishwa Hindu Parishad (World Hindu Council), went all out to support the activism against this Dam presenting it "as a danger to Hinduism and hindutva, the very "Hinduness" of India"[71]. Still, environmental activism continues in India with a zeal that is upheld by the Hindu religion.

6. Conclusion

In this paper, Hinduism, as mentioned in the introduction, is presented in a positive light since it has the foundations for ecological and overall health. It is already wellknown that Hinduism, through its Scriptures, traditions and religious activities is viewed by researchers as a religion that holds principles for the benefit of eco-human harmony. Those principles are and need to be further utilized for environmental activism whether in India or abroad. It is by those very foundations and principles that Hinduism holds the privilege to be in perfect sync with the ideals and vision of *Laudato Si'*.

---

[71] Lance Nelson, "Ecology", p. 674.

SHORT BIOGRAPHIES

Jacquineau Azetsop, SJ, was the Dean of the Faculty of Social Sciences at the Pontifical Gregorian University. He has authored or coauthored a number of academic articles, and his books include Structural violence, population health and health equity (2010) and HIV and AIDS in Africa (2016). He passed away in October 2021.

Bruno Bignami, priest of the diocese of Cremona, professor of moral theology, has been director of the National Office for Social Problems and Work of the Italian Episcopal Conference since 2018. From 2010 to 2020 he was president of the Fondazione don Primo Mazzolari in Bozzolo (MN). He has published books and articles on ethical issues, as well as on social and ecological issues. He is an invited professor at the Pontifical Gregorian University.

Louis Caruana, SJ, started his education with a first degree in mathematics and physics, and then proceeded with a master's degree in philosophy (London) and another one in theology (Paris). He was ordained a Jesuit priest in 1991 and then obtained his doctorate at the Department of History and Philosophy of Science, University of Cambridge. He is now Professor at the Philosophy Faculty of the Gregorian University in Rome and Adjunct Scholar at the Vatican Observatory. From 2005 to 2013, he was at Heythrop College, University of London, where he was appointed Reader in 2013. From 2014 to 2020, he served as Dean of the Faculty in Rome. He is on various advisory boards, including the one for the *International Philosophical Quarterly*. In 2020, he was the holder of the St Ignatius Loyola Chair at Fordham University, New York. His research deals mainly with the interaction between philosophy of science, philosophy of nature, metaphysics, and philosophy of religion.

Elisabetta Casadei, professor of Philosophical Anthropology and Political Philosophy at the Higher Institute of Religious Sciences "A. Marvelli" (Rimini) and at the Pontifical Gregorian University (Rome); she cooperates on the Master of Philosophical Consultancy and Existential Anthropology of the Pontifical University Regina Apostolorum (Roma). Director of the "S. Cecilia"Female Theological College and of the Studium "Evangelii Gaudium" of the Diocese of Rome, females colleges where cultures and knowledges meet. She joins her teaching and pastoral service as a Postulator for the Causes of Saints. Among her publications: Interior life, political discernment and civil commitment (Rubbettino, 2013).

Lorella Congiunti (PhD in Philosophy and Theology) since Academic Year 1999-2000 she has been teaching Philosophy of Nature (and since 2019-2020 also Metaphysics) at the Pontifical Urbaniana University in Vatican City. She is President of the International Society of St. Thomas Aquinas (SITA) and she was VicePresident (the first woman VicePresident) of Pontifical Urbaniana University (2013-2017). She is also a member of the Scientific Committee and the Teaching Staff of the Joint Diploma on Integral Ecology "Laudato Si'", the Joint Diploma on Thomistic Studies and the Joint Diploma on Women and Church. Her monographs include: *Soggettività e ontologia* (1996); *La ragione e il suo contrario* (2004), *Soggetto del sapere e scienze moderne* (2005), *Ateismo ateo* (2014), *Lineamenti di filosofia della natura* (2010, II edition 2016), *Estudo de temas tomistas. Da Razão à Fé em Tomás de Aquino* (2015), *Oulines of Philosophy of Nature* (2020), as well as a number of chapters in books and articles in scientific journals.

Paolo Conversi completed graduated degrees in Economics from 'La Sapienza' University of Rome and in Social Sciences from the Pontifical Gregorian University. He then obtained a Ph.D. in Social Sciences from the Pontifical Gregorian University and a Ph.D. in Economics from 'Roma Tre' University. In 1999, after taking part in development and humanitarian projects in Mexico, Colombia and Bosnia Herzegovina, he became an Official of the Section for Relations with States of the Secretariat of State of the Holy See. He has been a member of many Holy See delegations to various international meetings, dealing with sustainable development, climate change, disarmament, and humanitarian assistance. He is teaching 'Integral Ecology' and 'Economics' at the Faculty of Social Sciences of the Pontifical Gregorian

University, where he is the coordinator of the *Laudato Sì Observatory*. He is also teaching 'International Environmental Law' and 'History of Economic Thought' at the Pontifical Lateran University.

Giacomo Costa, SJ, is born in Genoa in 1967. He joins, in 1992, the Society of Jesus (Jesuits). After the philosophical and theological studies, he attends a postgraduate course (DEA) in Political and Moral Sociology with Luc Boltanski at the EHESS (Ecole d'Hautes Etudes en Sciences Sociales) in Paris, living and working at the same time in the banlieue (Saint-Denis). From 2004 to 2006 he is member of the Institute of Political Education Pedro Arrupe in Palermo, teaching in the Master on "Governance and Territorial Development". From 2010 to December 2021 he is Director of the monthly Aggiornamenti Sociali, a Jesuit magazine based in Milan that offers social, political, ethical and ecclesial analysis. Since 2008 he is also President of the Fondazione Culturale San Fedele in Milan, an institution committed since immediately after WWII in articulating faith and justice in dialogue with cultures and society. In 2013 he takes part in the establishment of the Fondazione Carlo Maria Martini, of which he is Vice President. In November 2017 he is appointed by Pope Francis Special Secretary of the 15th Ordinary General Assembly of the Synod of Bishops; in 2019 Secretary of the Commission for Information of the Synod for the Pan-Amazon Region and now he is part of the Steering Committee of the Synod 2021-2023.

Paolo Foglizzo, born in Turin in 1964, graduated in Economics from the local University and later obtained a Licentiate in Social Doctrine of the Church and Public Ethics from the Pontifical Gregorian University in Rome. After working for three years at the Social Justice Secretariat of the General Curia of the Society of Jesus in Rome, in 2001 he moved to Milan and joined the editorial staff of the Jesuit-run monthly journal *Aggiornamenti Sociali* <www.aggiornamentisociali.it>, where he deals in particular with economic issues and Catholic Social Teaching. He co-ordinated the activities of *Aggiornamenti Sociali* within "The Future of Work – Labour after *Laudato Si*'" Project and sat in the Editorial Board in charge of assembling the report *Care is Work, Work is Care*.

Francesco Gambino is professor of Private and Civil Law at the University of Macerata. He wrote three monographs, several papers and contributions on civil law, comparative law, sociology and general theory of law, some of which in English, French, Spanish, Portuguese and Chinese. Member of the

Italian Association of Civil Law Scholars; member of the Association Henri Capitant des amis de la culture juridique française; member of the Union of Private Law Lawyers; member of the ISLL - Italian Society for Law and Literature.

Jakub Gorczyca, SJ, is professor emeritus at Gregorian University. In addition to several articles on ethics and philosophical anthropology, he has published: *L'uomo e la risposta al valore,* Roma 1984; *Chrystus i ethos. Szkic o etyce filozoficznej w kondycji chrześcijańskiej*, Kraków 1999; *Edyta Stein*, Kraków 2008[2]; *Zarys etyki fundamentalnej*, Kraków 2014; *Essere per l'altro. Fondamenti di etica filosofica,* Roma 2016[2].

Adrien Lentiampa Shenge, SJ, a Congolese Jesuit, is a professor of the history of contemporary philosophy and of the philosophy of law in the Faculty of Philosophy, at the Gregorian University (Rome) since 2014. Between 2008 and 2014, he taught philosophy at several academic institutions in his home country, the Democratic Republic of Congo. Author of a thesis on the thought of Paul Ricoeur, defended in 2007 at the Gregorian University, he is a proven specialist of Paul Ricoeur and Emmanuel Levinas. He is currently working on the mutual contributions between African rationality and Western rationality. His publications include: *Paul Ricœur. La justice selon l'espérance, Bruxelles,* Ed. Lessius, 2009 ; « La culture de la vie et la promotion du pardon en Afrique » in Nouvelle Revue Théologique n° 133 (Juillet-Septembre 2011), 402-420 ; « La philosophie et la communication de la vérité aujourd'hui » in Revue Philosophique de Kimwenza n° 7 (Février 2012), 75-87 ; « De l'ère de l'individu à l'ère de la personne. Pour une économie à visage humain » in Gregorianum 94/4 (2013), 833-847; « La conception kantienne de la responsabilité » in Gregorianum 101/2 (2020), 295-315.

Bryan Lobo, SJ, is presently the Dean of the Faculty of Missiology at the Gregorian University Rome. His research involves theological and missiological themes on which he has published various articles. As Editor of the book series of the Faculty of Missiology, *Documenta Missionalia*, he has edited two books: *Maximum Illud. La missione tra storia e attualità*, G&B Press, Rome 2020; *Enlightenment and Tantra: Hindus and Christians in Dialogue*, G&B Press, Rome 2018.

Pierre Martinot-Lagarde, SJ, has been Special advisor for socio-religious affairs at the International Labour Organisation in Geneva since 2008. In this position, together with Rev. Robert J. Vitillo, Director of the International Catholic Migration Commission, he initiated and facilitated "The Future of Work – Labour after Laudato Si'" Project. Previously he directed CERAS (Centre de recherche et d'action sociales, the Jesuit centre for social ethics based in the Paris region) and was editor-in-chief of its journal Projet. He holds a PhD in Demography from the University of Pennsylvania (Philadelphia, USA) and a canonical Licentiate in Theology from Centre Sèvres (Paris). He joined the Society of Jesus in 1984.

René Micallef, SJ, is a Maltese priest and member of the Society of Jesus. He lives and works in Rome as an associated professor in Moral Theology and Social Ethics at the Faculty of Theology of the Pontifical Gregorian University. Among his research interests and teaching commitments are topics such as the sources and foundations of moral reasoning, the relevance and shortcomings of the various "natural law" traditions in contemporary ethical debates, migration policymaking, and the grounding of human rights. He holds degrees in Biology and Chemistry (University of Malta), Philosophy (University of London) and Theology (Facultés Jésuites de Paris; Universidad Pontificia Comillas - Madrid), and a doctorate in Sacred Theology from Boston College (2013).

Ciara Murphy is the Environmental Policy Advocate for the Jesuit Centre for Faith and Justice, Dublin. She works on influencing policy change at a national level as well as focusing on community-based local initiatives. She achieves this through commentary, public consultation submissions and collaborating with peer NGOs and community organisations. She holds a BSc. in Environmental Biology and a PhD. in Environmental Microbiology which have equipped her with a broad knowledge on environmental and biodiversity issues.

Ferenc Patsch, SJ, is a Jesuit priest and member of the Hungarian Province of the Society of Jesus. He holds a Licentiate degree in philosophy and a Doctorate in theology from the Pontifical Gregorian University in Rome. After 7 years of studies in Italy, he served in Hungary for 5 years as Associate Lecturer at the Sapientia College of Religious Orders and as Editor-in-Chief of the journal of the same College, and as Visiting Lecturer at the Pázmány

Péter Catholic University of Budapest. He also taught at Hekima College, the Jesuit's Theology Faculty in Nairobi (Kenya) and in the Instituto Superior Pedro Francisco Bonó in Santo Domingo (Dominican Republic). Since 2012, he has been teaching theology at the Pontifical Gregorian University in Rome (since 2018 Professor Extraordinarius) where he served also as Chair of the Department of Fundamental Theology (2018-2021). His fields of interest and research are: the theology of religions, philosophical hermeneutics, and fundamental theology.

Gaetano Piccolo, SJ, a Jesuit, is a Full Professor of the Faculty of Philosophy of the Pontifical Gregorian University. After his studies on St. Augustine (I processi di apprendimento in Agostino d'Ippona, Aracne 2014 2), he dedicated himself to Questions of Epistemology (Significato e interpretazione. Indagine sulla conoscenza, Carocci 2011) and of Philosopy of Language (Fatti di parole. Filosofia del linguaggio, GBPress 2019). He is currently working on the metaphysical foundations of decisional processes.

Ligita Ryliškytė, S.J.E., a native of Lithuania, is a Professor of Theology at Vytautas Magnus University (Kaunas, Lithuania) and a Visiting Assistant Professor of Systematic Theology at Boston College (Boston, U.S.A.). Her research focuses on the Augustinian-Thomist tradition (especially as transposed by Bernard J. F. Lonergan) and its contemporary developments in the areas of Christology and soteriology. She has authored a number of articles in theological journals and has recently contributed to a collected volume on critical realism. Currently, she is editing a collected volume on the theological method. Her first book, *Why the Cross? Divine Friendship and the Power of Justice*, is under contract with Cambridge University Press.

Paul Tang Abomo, SJ, holds a Master in Philosophy (St. Pierre Canisius, Jesuit School of Philosophy, Kinshasa, DR Congo), a Master in Theology (Weston Jesuit School of Philosophy/Harvard University, Cambridge, MA, USA), a Master and a PhD in Political Science and Government (Loyola University Chicago, USA). He has taught at Loyola University (Chicago, USA), Hekima Peace Institute (Nairobi, Kenya). He is now Associate Professor at the Gregorian University. He is the author of R2P and the U.S. Intervention in Libya. New York Palgrave Macmillan, 2018.

Gerard Whelan, SJ, is an Irish Jesuit and Director of the Department of Fundamental Theology at the Pontifical Gregorian University, Rome. He directs the "Lonergan Project" at the university and is author of, *A Discerning Church: Pope Francis, Lonergan, and a Theological Method for the Future* (Paulist Press, 2019.) Before coming to Rome, he lived in Africa for fourteen years. This work included being pastor of a poor parish in Nairobi and assisting the Holy See in its presence at the Governing Council of UN-HABITAT, the UN agency for human shelter. He has worked in Ignatian spiritual direction for both lay people and seminarians, and was the Ecclesiastical Assistant of the World Union of Catholic Women's Organizations for ten years. His current research interest is in integral ecology, as outlined in the encyclical *Laudato Si'*.

Prem Xalxo, SJ, is lecturer in Moral Theology at the Gregorian University in Rome. He had his Licentiate in Moral Theology from Accademia Alfonsiana, Rome and Doctorate in Moral Theology from the Gregorian University. His doctoral thesis, Complementarity of Human Life and Other Life Forms in Nature, was published in 2007 from Editrice Pontificia Università Gregoriana, Rome. A revised version of it, Current Ecological Crisis and Its Moral Dimension, was published in 2008 by Xavier Publications, Ranchi. His primary fields of interest, study and research are Ecological Ethics and Ethical-Theological Vision of Social Communications. After the promulgation of Laudato Si', he has been coordinating Joint Diploma in Integral Ecology, a joint venture of the Pontifical Universities of Rome to create an interuniversity and interdisciplinary collaboration for a common cause – Care for Our Common Home.

Finito di stampare nel mese di Settembre 2022
presso Printbee - Noventa Padovana (PD)